UNDERSTANDING PHYSICS

Robin Millar

Lecturer in Physics Education
University of York

Collins Educational

An imprint of HarperCollinsPublishers

Published by Collins Educational
An imprint of HarperCollins*Publishers*
77–85 Fulham Palace Road
London W6 8JB

First published in 1989 by Unwin
Hyman Ltd.
Published in 1993 by Collins Educational

Reprinted 1994

© Robin Millar 1989

All rights reserved. No part of this
publication may be reproduced, stored
in a retrieval system, or transmitted in
any form or by any means, electronic,
mechanical, photocopying or otherwise,
without the prior permission of Collins
Educational.

British Library Cataloguing in
Publication Data

Millar, Robin
    Understanding physics
    1. Physics – For schools
    I. Title
    530

    ISBN 0–00–322362–0

Diagrams by RDL Artset Ltd
Cartoons by Jane Cope
Typeset by August Filmsetting,
Haydock, St Helens
Printed and bound in Hong Kong

Contents

1. Energy, Matter, Fields: Three Big Ideas — 1
2. Measurement — 15
3. Forces — 22
4. More about Forces: Balance and Stability — 31
5. Describing Motion — 37
6. What Causes Motion? — 57
7. Laws of Motion — 61
8. Force, Mass and Acceleration — 68
9. Circles, Projectiles and Satellites — 76
10. Momentum — 80
11. Work and Energy — 92
12. Pressure — 110
13. Floating and Sinking — 122
14. Thermal Properties of Matter — 127
15. The Behaviour of Gases — 136
16. Heating and Cooling — 145
17. Heating and Internal Energy — 154
18. Light and Reflection — 165
19. Refraction — 174
20. Lenses and Optical Instruments — 184
21. Waves — 196
22. The Electromagnetic Spectrum — 207
23. Sound — 214
24. Electrostatics — 226
25. Current Electricity: Basic Ideas — 235
26. Ohm's Law and Resistance — 248
27. Electrical Power and Domestic Electricity — 264
28. Sources of Electric Current — 278
29. Magnetism and Electromagnetism — 281
30. Electromagnetic Induction — 297
31. Electron Beams and the CRO — 310
32. Electronics — 320
33. Radioactivity — 336
34. Energy and Energy Supply — 357

Appendices:

1. Mathematical skills — 371
2. Microcomputer Methods in Mechanics — 374

Answers to Questions — 375
Index — 376

Examination Questions

The author and publisher thank the following examination bodies for permission to use selected examination questions at the end of chapters.

- AEB   Associated Examining Board
- JMB   Joint Matriculation Board
- LEAG London and East Anglian Group
- MEG   Midland Examining Group (Oxford and Cambridge Schools Examination Board: Salters' Chemistry Project specimen questions)
- NEA   Northern Examining Association (Associated Lancashire School Examining Board, Joint Matriculation Board, North Regional Examinations Board, North West Regional Examinations Board, Yorkshire and Humberside Regional Examinations Board)
- NISEC Northern Ireland Schools Examinations Council
- O&C   Oxford and Cambridge Schools Examination Board
- SEB   Scottish Examination Board

Photo credits

The author and publisher thank the following for permission to reproduce photographs.

- 1 far left: Steve Ashton
- 2 middle: Express Lift Company Ltd; upper right: British Gas
- 3 seventh: Hitachi UK
- 5 Shell
- 6 top: Christopher Sykes
- 8 first: Ralph Wyckoff; second: G. K. L. Cranstoun/SPL; fourth: Science Museum; fifth: Tate & Lyle; sixth: R. H. MacDougal
- 15, 16, lower, 17 National Physical Laboratory
- 25 Ford Motor Company
- 26 upper right: Costain
- 27 Costain
- 31 Shepherd Building Group
- 35 left: Leyland Bus; right: Boots Company plc, Baby Business Centre
- 37 Allsport UK Ltd
- 38 lower: Metropolitan Police
- 44 Austin Rover Group
- 57 Science Museum
- 58 lower: NASA/USIS
- 59 top left: Goodyear Tyre & Rubber Co; lower: Sealink British Ferries Ltd
- 60 upper: Motor Industry Research Association
- 61 upper: Science Museum; middle: Shell; lower: NASA/BYRON
- 66, 81 Dunlop Fort
- 78 lower: Kodansha
- 79 lower: NASA
- 82 left: RoSPA Media Services; right: Taken from: Verkeer en Veiligheid (Rijksuniversiteit Utrecht PLON/BV Uitgeverij NIB Zeist, 1981)
- 83 lower: The Independent
- 86 left: Allsport UK Ltd; right: Metropolitan Police
- 101 left: MIRA; right: Steve Ashton
- 102 left: Lotus; right: Rover
- 104 left: Stanley Tools; right: Black & Decker
- 110 second left: John Shaw/NHPA; right: Allsport UK Ltd
- 111 upper right: Shell
- 115 Science Museum
- 117 right: Fisons plc/Griffin & George
- 119 upper: Smiths Industrial Defence Systems; middle: British Aerospace plc/Civil Aircraft Division; lower: St Bartholomew's Hospital
- 124 top: Illustrated London News; middle: Xinhua News Agency; lower: Ken Lambert/Barnaby's Picture Library
- 129 first: Potterton International; second: Fisons Scientific Equipment
- 131 lower: British Pipeline Agency
- 132 upper: British Rail
- 133 right: Osram GEC
- 147 The Royal Aeronautical Society
- 149 first: Steve Ashton; second: U Böcker/RSPB
- 150 Rockwool
- 156 upper: National Centre for Alternative Technology; lower: The Electricity Council
- 165 left: Syndication International
- 166 upper: Popperfoto; lower: Orville/Andrews/SPL
- 168 lower: David Scharf/SPL
- 170 upper left: Sandra Wegerif; lower right: BBC Hulton Picture Library
- 171 left: Leterrier/Odeillo; right: I. T. Power/Bernard McNelis
- 178 upper left: Telefocus; lower left: Adrian Meredith Photography/British Airways; right: Lennart Nilsson/Bonniér Fakta
- 184 upper: Science Museum
- 191 upper: Olympus Optical Company UK
- 192 top right: Gene Cox; lower: Science Museum
- 194 Royal Greenwich Observatory
- 196 Eric Thorburn/South of Scotland Electricity Board
- 208 upper: Telefocus
- 209 left: Geoff Williams & Howard Metcalf/SPL; upper right: Agema Infrared Systems/SPL; lower right: EOSAT
- 210 third & fourth: St Bartholomew's Hospital
- 215 centre & right: English Chamber Orchestra
- 219 Royal Albert Hall
- 221 St Bartholomew's Hospital
- 222 lower left: Forestry Commission, Edinburgh; lower right: Sandra Wegerif
- 223 London Mozart Players
- 232 left: Peugeot Talbot Motor Company; right: ICI Agrochemicals
- 233 Mullard, Southampton
- 235 Mary Evans Picture Library
- 266 upper: Electricity Council Appliance Testing Laboratories; lower: Philips
- 268 Schlumberger Industries Electricity Management
- 271 EMACO Ltd
- 279 Peter Fraenkel
- 284 lower left: Jon Prosser
- 288 first: Moorfields Eye Hospital; second: Barnaby's Picture Library; third: Birmingham Airports Authority; fourth: Haymarket Motoring Photo Library
- 294 Black & Decker
- 301 CEGB
- 305 upper: St Bartholomew's Hospital
- 306 first, second, fourth: CEGB
- 336 right: Mansell Collection, BBC Hulton
- 338 lower left: Patrick Blackett/SPL; lower right (2): C T R Wilson
- 339 CERN
- 342 lower: BBC Hulton
- 347 Larry Mulvehill/SPL
- 348 upper left: Sabre International Products; right: UKAEA; lower right: IAEA
- 349 top left: St Bartholomew's Hospital; centre left & right (2): UKAEA; bottom left: Black & Decker
- 350 The Royal Society
- 352 left: UKAEA; right: British Nuclear Fuels plc
- 353 JET/UKAEA
- 357 top: Adam Hart-Davis/SPL
- 358 top: Mary Evans Picture Library; lower left: Barnaby's Picture Library; lower right: Science Museum
- 365 top: Leterrier/Odeillo; centre: Texaco; bottom: N Scotland Hydroelectric Board
- 366 Michael Flood/National Centre for Alternative Technology
- 367 Building Research Establishment, Crown Copyright

Jim Jardine: **41** upper, **48**, **50**, **58** right, **78** upper, **79** centre & right, **198**
Project Physics: **199**, **200**, **202**

Robin Millar supplied the following photographs: **26** lower, **33**, **38** upper, **59** middle right, **149** lower right, **162**, **165** last, **180**, **190** lower, **229** lower left, **275**, **284** top four, **306** third & fifth

Particular thanks are due to John Prosser who took all the remaining photographs in the book

Front cover: Adam Hart-Davis
Back cover: Science Photo Library (spectra) Adam Hart-Davis (prism)

The quotation on page 6 is taken from *The Feynman Lectures on Physics*, by R. P. Feynman, R.B. Leighton and M. Sands (Addison Wesley, Reading, Mass. 1963).

Preface

This textbook is written as an introduction to physics. It covers all the physics required by most GCSE physics and balanced science courses. It may also be useful for someone who has completed GCSE science and is beginning to study physics at A level. I hope it will also be of interest to anyone who just happens, for whatever reason, to want to know more of what physics is about.

The book begins and ends with perhaps the 'biggest' idea in physics—*energy*. In between, it has groups of chapters on forces and motion, heating, light and waves, sound, electricity, electronics and radioactivity. The order of the chapters is just one possible route through the material; there are many ways of studying the ideas which the book covers.

I have tried to write the book *for* the learner, so that it can be used for self-study. The basic ideas about energy, matter, forces, light and electricity are explained in greater detail, with many diagrams to illustrate the important ideas. Every chapter makes links between the physics ideas and everyday life—showing how we can understand natural phenomena or technological applications using our physics knowledge.

Each chapter ends with some graded questions, including several from various Examination Boards. Many of these are designed as diagnostic questions, to test understanding of basic concepts. Answers to all the numerical questions are given at the back of the book.

Writing a book like this takes time! I am grateful to Liz, Neil and Ruth for their tolerance and patience whilst 'the book' was in preparation. I should also particularly like to thank Pat Winter and Jane Gregory at Unwin Hyman for their encouragement and their attention to detail in putting the book together.

I imagine that every author has some reader in mind as he or she writes. And I suppose that it is to that reader that the book should properly be dedicated. So this book is dedicated to the memory of my father, the first person to whom I tried to explain many of things which it contains.

Robin Millar
York, 1989

1: Energy, Matter, Fields: Three Big Ideas

ENERGY

Energy is one of the most important ideas in science. It is also an idea we meet in everyday life. ▷

Although the word **energy** crops up in everyday situations, it has a more exact meaning in physics. We will begin this book by taking a closer look at what physicists mean by **energy**.

1.1 Being energetic

It is hard to define **energy** – to say exactly what energy *is*. So let's begin by thinking about what we mean when we use the word **energy**. If we say that we have a lot of energy, we mean that we feel fit and lively and able to run around and do things. Someone who is always active and busy is said to be **energetic**. After a hard day's work, or playing a game, we say we have 'no energy left'.

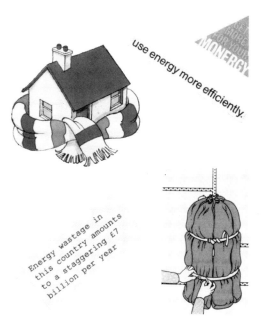

Where do we get our energy from? It comes from eating certain types of food. Carbohydrates and fatty foods are our energy supplies. Our bodies **digest** the food and this provides us with the energy to walk, run, jump, and so on.

1.2 Where our energy goes

What sorts of things are we able to do when we feel we have lots of energy?

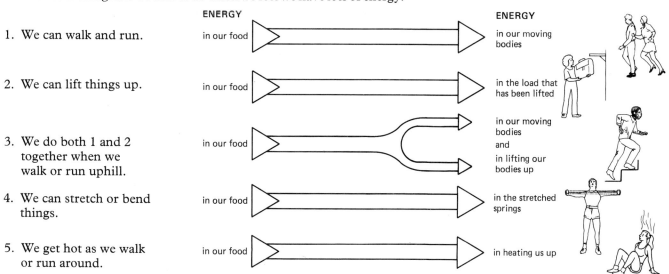

	ENERGY		ENERGY
1. We can walk and run.	in our food	⟹	in our moving bodies
2. We can lift things up.	in our food	⟹	in the load that has been lifted
3. We do both 1 and 2 together when we walk or run uphill.	in our food	⟹	in our moving bodies and in lifting our bodies up
4. We can stretch or bend things.	in our food	⟹	in the stretched springs
5. We get hot as we walk or run around.	in our food	⟹	in heating us up

1.3 Making things happen

When we have energy, we can make things happen. But there are other ways of making things happen, or of doing jobs that we want done.

Moving
This model buggy moves without being pushed. It has a small electric motor and runs from batteries.

A car has an engine which uses petrol. The fuel supplies the energy to make the car move.

Lifting
A lift uses mains electricity. This runs the motors which allow the lift to raise heavy loads from one floor to another.

Heating
This central heating boiler burns gas to provide hot water and heating for all the rooms in the house.

The torch runs on batteries. When it is switched on, the tiny wire filament inside the bulb heats up and gives out light.

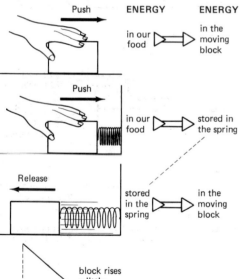

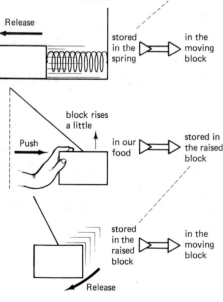

1.4 Storing energy

Fuels (such as petrol, gas, coal and oil) and batteries are energy stores. But there are still other ways we can store energy.

◁ Imagine pushing a wooden block along a smooth table. We are making the block *move*. We are transferring the energy in food we have eaten into energy in the moving block.

◁ Imagine now that the same block is attached by a spring to the wall. As we push, we compress the spring. This time we are transferring some of our energy into energy stored in the spring. We know there is energy stored in the compressed spring because of what happens if we let go of the spring! The block will move back without any need for us to do anything. The energy stored in the spring is being transferred into energy ◁ in the moving block.

Another way that the same block stores energy is when it is lifted up to a higher level. If the block is tied to a string, it will rise as it is pushed aside. We are transferring energy from our bodies into the stored energy of the block. This time it is harder to say exactly *where* the energy is stored. It is stored in the position of the block above the ground. When we release it, ◁ this stored energy is transferred back into energy in the moving block.

Storing and heating

Let's look at one other kind of 'stored energy'. Imagine we pushed the wooden block back and forth over the table several times. When we finish pushing, the block is stationary – there is no movement. So where has our energy gone this time? The block *has* changed in one way – if we feel it, we find that it has become slightly warmer. So has the surface of the table. The energy is now inside the block (and the top of the table), making it hotter. We need some way of explaining how this works – how it is possible for the energy to be 'inside' the block – **internal** energy. We will come back to this later in the chapter.

1.5 An energy 'way of looking'

Arrow diagrams help us to look at events from an energy point of view. We can describe what we actually see happening, and we can also think of it in energy terms. Let us look at some common devices and situations from an energy point of view.

1. Electric kettle
When the kettle is switched on, the water inside heats up. The kettle itself and the surrounding air also become hotter.

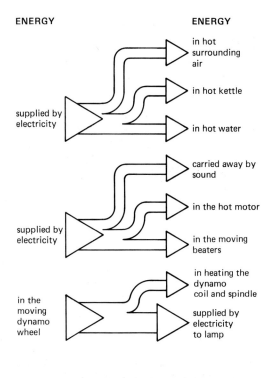

2. Food mixer
When the electricity is switched on, the beaters of a food mixer turn rapidly. Inside the mixer is a small electric motor which makes the beaters move. The motor also gets hot and produces some sound as it runs.

3. Cycle dynamo
This does the opposite of the motor. The dynamo wheel is turned by the cycle tyre and the dynamo produces electricity to run the cycle lamps.

4. Battery
A battery contains a special mixture of chemicals. When the battery is connected into an electric circuit, the chemicals react. The energy stored in the chemicals is transferred into electrical energy in the wires of the circuit.

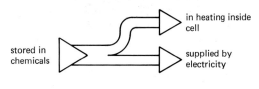

5. Camera light-meter
A photographer uses a light-meter to indicate the brightness of the scene she is to photograph. The energy carried by the light shining on the light-meter is transferred into electrical energy which makes the pointer of the light-meter move.

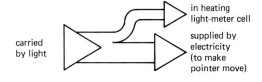

6. Guitar
To play a guitar, you pluck the strings. This makes them vibrate. The energy of the moving strings is carried away by the sound.

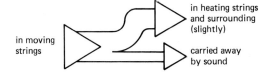

7. Transistor radio
This is a slightly more complicated situation. A radio picks up radio waves and converts these into sound. But it also needs batteries (or mains electricity) to work. The radio picks up the tiny amounts of energy carried by the radio waves and transfers these into tiny amounts of electrical energy in the circuits inside the radio. It then uses the energy stored in the batteries to make these electrical signals bigger (to amplify them). Energy is carried away by the sound from the radio's loudspeaker. The radio also heats up a little while it is switched on.

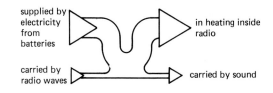

Where the energy is	A lot of energy	A little energy
Moving objects	heavy object moving fast	light object moving slowly
Object lifted up	heavy object lifted to a considerable height	light object raised just a little
Stretched spring	strong spring large stretch	weak spring small extension
Light	bright	dim
Sound	loud	quiet
Hot object	large object at high temperature	small object just a little above room temperature

The idea of *measuring* amounts of energy may seem a rather strange one. But we can easily imagine what large and small amounts of energy might mean. Many of the later chapters in the book will look at how to measure amounts of energy in different situations.

1.6 Energy patterns

In all the examples we have looked at, the energy viewpoint gives us a way of looking at things which happen in the world around us. But are there any general **patterns** in the energy arrow diagrams? Here are some things we notice:

Pattern 1: Nothing happens without an energy 'input'. There always has to be some energy to begin with.

Pattern 2: The energy never disappears completely; it always seems to turn up somewhere.

This leads us towards a very important and useful idea: perhaps the amount of energy at the end of an event (adding up the amounts in all the places it has got to) is the same as the amount we started with. This is called the **law of conservation of energy**.

There are two other important patterns in the energy arrow diagrams:

Pattern 3: The arrow diagrams generally branch as they go from left to right. The energy spreads out and ends up in a whole variety of places.

Pattern 4: Almost all energy transfers result (eventually) in something being heated up.

So there is a second energy law, the **law of spreading of energy**. When energy is released from concentrated energy stores (like fuels and food), it tends to spread out and end up in a number of different places. It is very difficult, and may even be impossible, to get it all back together again! ▽

Energy spreading

If we set a cup of hot tea on the table, it gradually cools down. As it does, the table below the cup and the air all around it become a little warmer. Later still, the table and the cup will have cooled down too by sharing their extra energy with the surroundings.

Eventually the energy will have spread so far that it is almost impossible to notice it any more – but everything *has* got just a little bit hotter.

Hot things always cool down and their surroundings warm up, so that everything ends up at the same temperature. The opposite never happens. What would you think if you were sitting in your bath when the water suddenly began to freeze at one end of the bath and to boil at the other? This wouldn't break the law of conservation of energy, but it *would* break the law of spreading. It never happens!

So energy spreading means that hot things will cool down by spreading their energy around to the things nearby. Hot things are valuable and it is hard to keep them hot. Fuels are especially valuable because they are *concentrated* stores of energy.

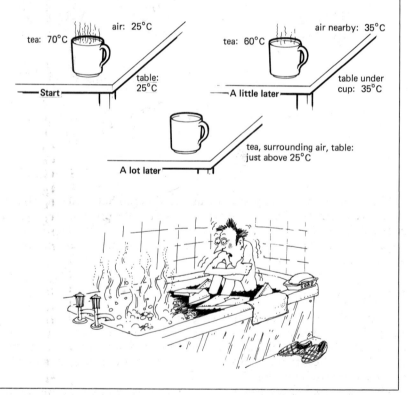

The two energy laws are so important that we will summarise them:

There is always the same amount of energy at the end of a process as there was at the beginning. (Conservation of energy)

In any process, energy tends to spread out into more and more places. (Spreading of energy)

1.7 Forms of energy

It is sometimes useful to talk of different **forms** of energy when we are thinking of the different places where energy can appear.

Kinetic energy: the energy in moving objects. Anything which is moving has kinetic energy.

Potential energy: stored energy which can be released. **Elastic potential energy** is stored in springs or any other springy material when it is stretched or compressed. **Gravitational potential energy** is stored when any object is raised up to a higher level.

Chemical energy: energy stored in mixtures of chemicals. It can be released by making the chemicals react. For example, a mixture of coal and air (oxygen) has stored chemical energy. When the coal and oxygen react (burning), energy is released. All fuels store chemical energy.

Electrical energy: in electrical circuits, energy is transferred along wires from place to place.

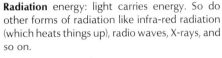

Radiation energy: light carries energy. So do other forms of radiation like infra-red radiation (which heats things up), radio waves, X-rays, and so on.

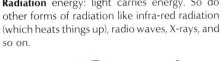

Internal energy: a hot object has more energy than a cold one. We can often see no difference between a hot and cold body from outside (unless it is very hot). The energy is 'stored' inside the hot body – it is **internal** energy. (This is sometimes called **heat**, but it is better to think of it as internal energy inside the body.)

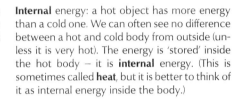

Sound energy: energy is carried by sound.

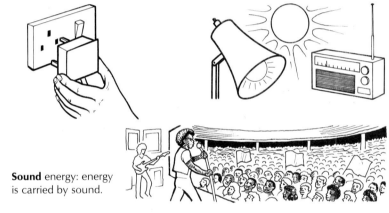

These **forms of energy** are just useful labels to help us think about energy in different situations. Energy doesn't really come in a range of different forms. In fact we discover that some forms of energy are really special cases of another form when we look at them more closely (see section 1.15).

1.8 Energy crisis?

Now and then we see posters advising us to 'Save Energy!' But the first energy law is that energy is always conserved: there is the same amount of it after an event as there was at the beginning. Why do we need to be careful about the way we use something which is conserved? If energy is really conserved, how could it run out?

The answer lies in the second energy law: when energy is transferred from one place to another, it always spreads out. There is the same amount at the end, but it has become dispersed. It is our **fuels** – our *concentrated* energy stores – that we need to use with care, looking for ways to cut down the rate at which we use them. Coal, oil and gas won't last forever. We face not an 'energy crisis' but a 'fuel crisis'.

In the final chapter of the book, we will come back to take a further look at this question of energy supply.

6 Energy, Matter, Fields: Three Big Ideas

Richard Feynman (1918–88), American Nobel prize-winning physicist: 'If all of scientific knowledge were to be destroyed, and only one sentence passed on to the next generation of creatures, what statement would contain the most information in the fewest words? I believe it is the atomic hypothesis... that all things are made of atoms – little particles that move around in perpetual motion, attracting each other when they are a little distance apart, but repelling upon being squeezed into one another. In that one sentence, there is an enormous amount of information about the world, if just a little imagination and thinking are applied.'

MATTER

We call the world around us 'the material world'. It is made out of **matter**: wood and stone, metals and plastics, water and air, living tissue. People have always been curious about what matter was really made of, and how the properties of all materials might be explained. Over the centuries, we have collected evidence and gradually built up a picture of what matter is.

1.9 Arguments about matter

Around 2500 years ago, the ancient Greek philosophers discussed and speculated on questions about nature. One philosopher, Democritus, argued that everything was made up of tiny particles, or **atoms**. There was no experimental evidence for or against, and many other philosophers disagreed with him. Even so, the idea persisted for over two thousand years, until eventually the evidence of experiments showed that this atomic 'picture' or 'model' of matter was a useful one.

A **particulate theory of matter** is one in which eventually we reach a point where it is impossible to divide a piece of matter into smaller pieces. Matter is made up of tiny particles. All scientists nowadays accept and use this sort of 'picture' or 'model' of matter in their work.

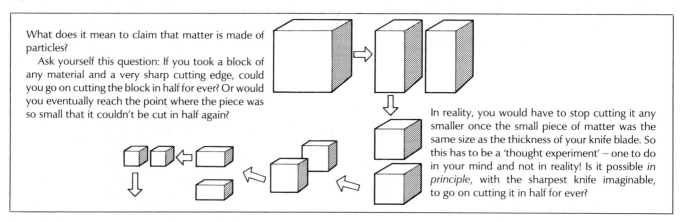

What does it mean to claim that matter is made of particles?

Ask yourself this question: If you took a block of any material and a very sharp cutting edge, could you go on cutting the block in half for ever? Or would you eventually reach the point where the piece was so small that it couldn't be cut in half again?

In reality, you would have to stop cutting it any smaller once the small piece of matter was the same size as the thickness of your knife blade. So this has to be a 'thought experiment' – one to do in your mind and not in reality! Is it possible *in principle*, with the sharpest knife imaginable, to go on cutting it in half for ever?

1. A block of wood or iron has a fixed shape.

2. It may seem that a pile of sugar can change its shape; but each individual solid grain of sugar has a fixed shape.

3. Liquids take the shape of the container they are poured into, though their volume stays the same.

4. If the gas tap is turned on briefly, the small volume of gas released will gradually spread all round the room – it can be smelt anywhere in the room.

What sorts of things should a useful theory of matter be able to explain?

● One thing we notice is that all examples of matter fall into one of three groups: solids, liquids and gases. These are the three **states** of matter and a theory of matter should be able to explain the differences between them.

● A second thing we know is that the same substance can exist in all three states. For example, water (liquid) changes into ice (solid) if cooled, and into steam (gas) if heated. The same substance (water) can be either solid, liquid or gas depending on its temperature. A theory of matter should be able to explain how solids can be changed into liquids and liquids into gases by heating them up. ▽

Ice, water, steam: all forms of the same substance

1.10 The molecular theory of matter

The molecular theory of matter says that all matter is made of very tiny particles called **molecules**. See the descriptions and diagrams. ▷

These explain the most obvious differences between solids, liquids and gases. Solid objects stay a constant shape because the molecules are held quite tightly together and must keep the same neighbours, so the solid keeps its shape. A liquid hasn't a fixed shape but can flow and take up the shape of the vessel which holds it. This is because the molecules are not so tightly held together and can move past each other, allowing the liquid to flow. However, the molecules cannot escape entirely from each other. When the liquid flows into a new container, it still has the same volume because its molecules are still closely packed together. A gas, on the other hand, is able to flow to fill completely any container it is put in. The gas molecules are completely free of each other and can flow in all directions (including upwards).

1.11 Molecules and atoms

Matter is made of tiny particles called **molecules** which are the basic building blocks. But molecules themselves are made of something even smaller. Molecules are groups of **atoms** very tightly bound together. The study of how atoms combine to form molecules is what **chemistry** is all about. For example, a molecule of water is made of two atoms of hydrogen and one of oxygen; a molecule of methane (the main gas in North Sea gas) contains four atoms of hydrogen and one of carbon.

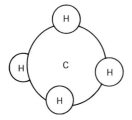

methane molecule (1 carbon atom + 4 hydrogen atoms)

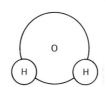

water molecule (2 hydrogen atoms + 1 oxygen atom)

Chemists have found that all the materials in nature are made from a quite small number of different atoms (about 100 altogether) combined in different ways. These 100 or so distinct substances are called **elements**: hydrogen, oxygen, carbon, iron, copper, uranium are some examples. Much of the important evidence for the atomic and molecular theory of matter comes from chemistry – studying the ways atoms combine to form molecules of different materials. We will not go into this in this book, but it is important to realise that there is a lot more evidence for the molecular theory in addition to what we will look at here.

Although molecules have a structure of their own (they are made of atoms bonded together), we will continue to picture a molecule as a small sphere, because that is a good enough picture for many purposes. Later in the book, we will even need to look at what the atom itself is made of, for it turns out to be built out of still smaller bits again!

Molecules in solids, liquids and gases

In a solid, the molecules are arranged in a regular pattern, with each molecule attached to its neighbours. Imagine the solid is made of a large number of spheres packed closely together, and connected together by springs. The spheres represent the molecules, and the springs are the forces holding the molecules together. The molecules are not stationary, but are constantly vibrating.

A liquid is similar to this except that the spheres are not arranged so neatly and regularly. Each molecule is still weakly joined to its neighbours, but the links are not nearly as strong as in the solid. Again the molecules are constantly vibrating. Because they are not tightly bound to their neighbours this means that they can move slowly around, jostling against each other as they do so.

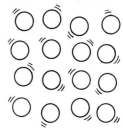

In a gas, the molecules are so far apart that they are completely free of each other. They move around randomly at high speed.

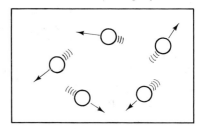

THE 'PARTICLES' IN MATTER

Talking of matter being made of **particles** may be misleading because of the way we use the word 'particles' in everyday speech. We often talk, for example, of 'particles of sand'. When sunlight streams into a room we see 'dust particles' floating in the air. The 'particles', or molecules, of which all matter is made are much much smaller than this – unimaginably tiny. A single grain of sand contains more than 1 million million of them!

Energy, Matter, Fields: Three Big Ideas

1.12 Evidence for the molecular theory

Some of the evidence for the molecular theory of matter is very direct, but most of it is in the form of clues which we have to interpret as best we can.

1. Very high magnification photographs

Until recently no one had ever seen the image of a molecule. Recently it has been possible to take photographs using special types of very high magnification microscopes which actually show the molecules.

Notice how the molecules are arranged in a neat orderly pattern. Photographs like this can only be taken of very large molecules. The picture shows the molecules looking like little spheres – we cannot see any individual atoms.

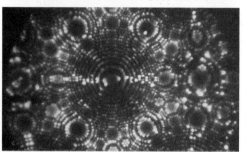

One type of microscope, called the field-ion microscope, enables even greater magnification to be achieved. On the left is a photograph, taken by this method, of the tip of a tungsten wire. Each of the small white dots is a single atom!

Evidence: High magnification photographs allow us to see the individual particles (molecules) in some solids.

2. Crystals

Rock crystals

Granite

Rock crystals are among the most beautiful natural objects. Under a low magnification microscope, we can see that many other rocks are also made up of small crystals. Substances like sugar and salt are also crystalline in form.

Crystals have smooth flat faces with sharp edges. Crystals of the same substance are all the same shape.

Sugar crystals   Salt crystals

Why are crystals such regular shapes? If we build a model out of polystyrene balls, packing them together in a regular pattern, we get a shape which is very like a crystal. A polystyrene ball model would break more easily along certain planes, leaving a flat surface like we find on a crystal.

Evidence: The regular shape of crystals suggests that they are made of closely packed spheres (molecules).

3. Brownian motion

Brownian motion is named after the Scottish botanist Robert Brown who first observed it in 1827. He was using a microscope to look at pollen grains in water, when he noticed that the grains kept moving about. At first he thought the pollen was alive, so he boiled it in water and tried again. They still moved! Brown was unable to explain it.

We can see Brownian motion by looking at smoke under a microscope. Smoke is really small pieces of solid matter floating in air. If some smoke is put into a smoke cell and lit from the side, the individual specks of smoke can be seen with a microscope.

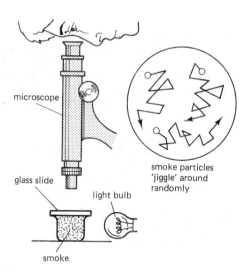

When you look down the microscope you see tiny white dots of light moving around randomly. The motion is constant and doesn't settle down after a time. Don't be misled – these are **not** molecules you are seeing! But the experiment **is** an exciting one for it is direct evidence for the molecular model of a gas. The white dots are specks of smoke; they are moving about because the molecules of air in the smoke cell are bumping into the tiny pieces of smoke.

We have to use tiny objects to see this effect – this is why smoke is used. With a larger object there would be so many collisions with the air molecules on all sides that their effects would cancel out and no motion would be seen at all.

Evidence: Brownian motion provides good evidence that a gas (air) consists of invisible molecules moving around randomly at high speed.

4. Diffusion experiments

In gases
The smell from an open perfume bottle will soon fill a room. Our nose detects the vapour (a gas) from the perfume. How does it spread from the bottle to other parts of the room? This process is called **diffusion**.

If we join a gas jar full of a coloured gas (such as nitrogen dioxide, which is brown) and another gas jar full of air, the gases slowly mix. The process occurs no matter what position the two connected gas jars are in.

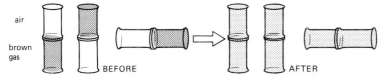

These experiments show that the molecules of gases are in motion. When both containers are full of gas, diffusion is slow. This is because the molecules of nitrogen dioxide collide with the molecules of air and this greatly slows down their rate of mixing. When the second container is empty, we see just how fast gas molecules move. In fact the speeds are several hundred metres per second!

It is a bit like the difference between trying to walk through a shop on the last Saturday before Christmas, and doing the same thing when the shop is empty. You don't make so much progress when there are 'lots of other people milling around in all directions.

In liquids
Diffusion takes place in liquids too. ▷

Liquid molecules are also moving about randomly all the time. Diffusion is much slower than with gases because the molecules in the liquid are much closer together and cannot move around so freely.

Evidence: Molecules in gases and liquids are not at rest but are constantly moving.

5. Compressing solids, liquids and gases

If we try to compress a solid, we find that it is difficult to change its volume. Hard materials, like steel or wood, don't seem to change shape at all if we squeeze them. Softer materials, like rubber and Plasticine, deform easily but their volume stays the same.

If a syringe is partly filled with water and the nozzle is sealed, it is impossible to compress the water into a smaller space. Liquids are difficult to compress. This suggests that the molecules in solids and liquids are already very close together and cannot be forced any closer. ▷

On the other hand, if some air is trapped in a syringe, the plunger can be pushed in fairly easily, compressing the air into a smaller space. When the plunger is released, the air expands again to its original volume. Gases are much easier to compress than solids or liquids.

Evidence: In gases, the molecules are far apart; they can be forced closer together. In both solids and liquids, the molecules are closely packed and there is little room to push them any closer together.

6. Change in volume during a change of state

If the molecules of solids and liquids are both closely packed, then we would expect little change in volume when a solid melts. This is exactly what we find. When a block of ice melts, the volume of water obtained is much the same.

However, the molecules in a gas are much further apart, according to the molecular theory, and so we would expect a big increase in volume when a liquid or a solid changes into a gas. ▷

Evidence: The molecules in a gas are much further apart than the molecules in a solid or a liquid.

If we use two flasks instead of gas jars, pump all the air out of one of them and then open the tap connecting them, the nitrogen dioxide spreads very rapidly into both.

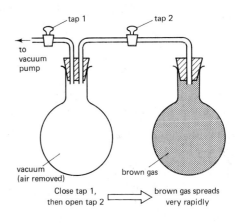

We can put some water into the bottom of a test-tube and then carefully introduce a layer of ink. (It is actually easier to add the bottom layer second using a dropper.) If the test-tube is left for several days, the two layers will mix.

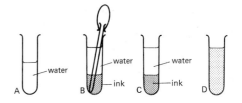

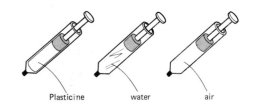

One way to show the increase in volume when a liquid or solid changes into a gas is to cut a 1 cm cube of solid carbon dioxide ('dry ice') from a block. The carbon dioxide gas obtained from it can be collected in an upturned measuring cylinder. Usually about 600 cm³ of gas is collected – 600 times the volume of the solid carbon dioxide.

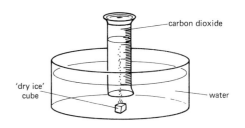

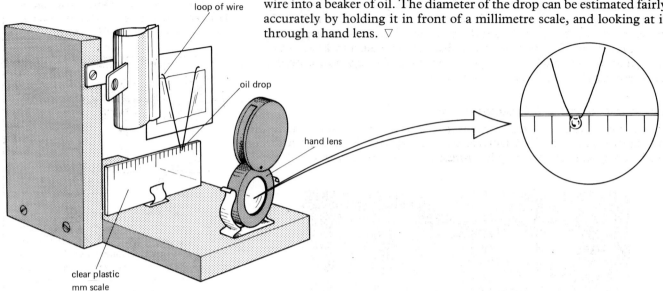

1.13 Estimating the size of a molecule

We can estimate how big a molecule is by taking a very small drop of oil and allowing it to spread on a water surface. Fill a tray with water to overbrimming. Then collect a small drop of olive oil by dipping a loop of wire into a beaker of oil. The diameter of the drop can be estimated fairly accurately by holding it in front of a millimetre scale, and looking at it through a hand lens. ▽

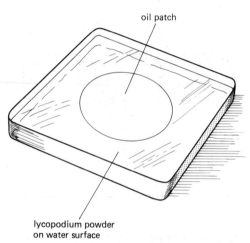

◁ Then spread some fine dust (chalk dust or lycopodium powder) all over the water surface and touch the loop of wire carrying the oil drop on to the centre of it. The oil spreads into a roughly circular patch on the surface of the water. As it spreads, it pushes the dust aside and so we can see clearly where the oil has got to.

Why does the oil spread a certain distance and then stop? The answer is that the drop spreads until it forms a patch one molecule thick. By measuring the diameter of the oil patch, we can estimate the diameter of an oil molecule.

Here are some results from a typical experiment. If you are not sure about volume or how to calculate the volume of a sphere or a cylinder, this is explained more fully in Chapter 2.

Diameter of spherical oil drop on the loop of wire = 0.5 mm
$(= 0.5 \times 10^{-3} \text{ m})$

Diameter of circular oil patch on the water surface = 0.3 m

◁ Volume of the original oil drop = $\frac{4}{3}\pi r^3$

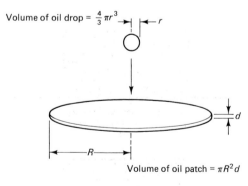

The oil patch on the water surface has the shape of a flat cylinder of radius R and thickness d.

The volume of this cylinder is $\pi R^2 d$

But, of course, the volume of the oil patch must be the same as the volume of the original drop. So:

$$\frac{4}{3}\pi r^3 = \pi R^2 d$$

Divide both sides by π: $\quad \frac{4}{3} r^3 = R^2 d$

Divide both sides by R^2: $\quad d = \frac{4r^3}{3R^2} = \frac{4(0.25 \times 10^{-3})^3}{3(0.15)^2} \text{ m}$

This leads to a value of approximately 10^{-9} metres for d.

10^{-9} m (or 1 nm) is 0.000 000 001 m.

If the oil spreads until the patch is one molecule thick, then 10^{-9} metres must be the diameter of one oil molecule.

Energy, Matter, Fields: Three Big Ideas

1.14 Changes of state

If we heat a block of ice, it melts and becomes a liquid. If we go on heating, the water boils and changes to steam. Many materials do this. By heating, we can change many solid materials into liquids, and liquids into gases.

The molecular model of matter can help us imagine what is happening inside the material during each of these changes of state. We need to add one extra idea: as the material gets hotter, something must be happening to the molecules. We cannot see the molecules so we must make a reasonable guess at what might be happening to them. Perhaps **as we heat a material up, its molecules move faster.** ▷

A practical illustration of changes of state

We can illustrate these ideas by putting some small polystyrene spheres into the upturned cone of a loudspeaker. The speaker is switched on to produce a quiet sound. At very low sound levels, the polystyrene balls lie

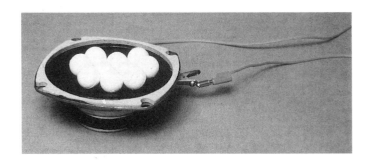

together in the speaker cone in a regular pattern. As the sound gets louder, the balls begin to vibrate slightly. Although they remain as a group, fairly closely packed together, they can move around a little. If you follow the motion of one marked ball, you will find that it wanders through the group. The balls are behaving like the molecules of a liquid. If the sound is made louder still, the balls will begin to jump around quite violently and their separation will increase greatly. They are like the molecules in a gas.

1.15 Another look at internal energy

In section 1.4, we talked of **internal** energy as energy stored inside a body when it is heated. Now we can picture much more clearly how this energy is stored. It is really kinetic energy – energy of the moving molecules. The hotter the body, the faster the molecules move. In a hot body, the molecules all have kinetic energy, but they are vibrating randomly in all directions, so the body itself does not move. ▷

1.16 Believing in the molecular theory

We have been able to look at just a small part of the evidence which leads scientists to trust in the molecular theory of matter and use it in their work. We have seen a few of the things which the molecular theory can help us to explain and understand. We will use the theory again throughout the rest of the book.

Of course, there are many things about materials which this simple molecular theory does not explain. For example, some solids are hard (like stone and steel), others are soft but rigid (like plastics), others still are soft and easily shaped (like plasticine or putty). Some hard materials are brittle (like glass), others are pliable (like metal sheet). To explain these differences we would need a more complicated theory – well beyond the level of this book. The simple molecular theory of matter is a beginning in understanding what matter is made from and why it behaves the way it does; it is not the whole story!

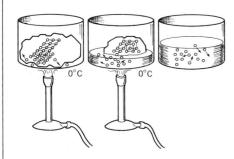

If a solid is heated, the molecules move faster and this makes the 'spring-like' links between them stretch more. At some point in the heating, the vibration of the molecules will become so great that the links can no longer hold and they begin to break. The solid has melted and become a liquid.

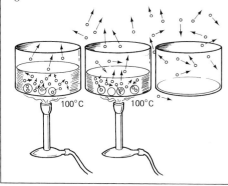

As the liquid is heated further, the vibration and motion of the molecules increases further. They bump around more and more rapidly, colliding with each other. Eventually the point is reached where the molecules are able to break away from each other and move freely around. The liquid has now become a gas.

The molecular theory of matter gives us a way of understanding internal energy and picturing what happens when an object is heated up.

Hot body: molecules moving randomly in all directions

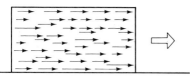

Moving body: molecules all moving together

FIELDS

What holds everything together? What keeps the stars together in galaxies? What holds the solar system together, with the Earth and the other planets going round the Sun? What keeps the Moon going round the Earth? And on a much smaller scale what holds the molecules together in a solid and, more weakly, in a liquid? We have used the 'picture' of springs holding neighbouring molecules together. Of course, we don't imagine that there really are springs between the molecules! So what keeps it all together?

1.17 Pushing and pulling without touching!

Usually if we want to move something, we have to touch it – we push it or pull it directly. But we can also have forces (pushes and pulls) between objects which are not touching. Let's look at some of these.

1. Static electricity

Static electricity

If you rub a plastic comb on your sleeve, you can use it to pick up light pieces of paper, or feathers. For an even more dramatic effect, rub a plastic pen and hold it close to a fine stream of water coming out of a tap. The water is bent to one side.

To investigate this further, take a polythene rod, rub one end of it and then balance it on a watch-glass. Rub another polythene rod and bring the two rubbed ends close together but not touching. The rods push each other apart. If we use a polythene rod and a perspex rod, we find that the rods attract each other. The force between them is very weak. They must be close together if we want to see any effect.

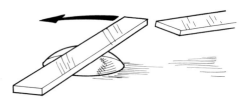

The attraction and repulsion between rubbed plastic rods are rather strange effects if we think about it – the rods push and pull each other without touching. It is sometimes called 'action at a distance'. But how do the rods exert forces on each other? How does each rod 'know' that the other is there?

Perhaps our first idea is that the influence is carried by the air (or some other material) between the rods. But, in fact, two rubbed rods in a vacuum would do just the same. It turns out that we cannot completely solve this puzzle, but we *can* find a more useful way of looking at it. There is a region around a rubbed rod where its influence can be felt. So we say that there is a **field** around the rod – an **electrostatic field**. If another rubbed rod comes into the field it will feel a push or a pull. But the field itself is completely invisible. The strength of the field gets gradually weaker as we move away from the rod.

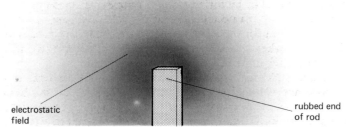

Molecules in solids and liquids are held together by electrostatic attractions. If a molecule moves too far from its neighbour, an electrostatic attraction force pulls it back again. On the other hand, if it comes too close, electrostatic repulsion pushes it back to its usual distance. This keeps the molecules firmly together and also means that the material is not easily compressed.

2. Forces between magnets

◁ Magnets influence each other without touching. This is another example of 'action at a distance'. In the region around a magnet there is a **magnetic field**. If another magnet comes into this field it will feel a force. The influence of one magnet on another is carried by the invisible magnetic field.

Forces between magnets

If you really want to 'feel' a field, then take two large magnets and hold them so that they are repelling each other.

As you try to push the magnets together it feels as if there really is something 'spongy' but totally invisible between them. With strong magnets it is impossible to push them together.

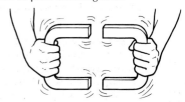

Magnetic forces can be investigated using bar magnets. Hang up a bar magnet. A second bar magnet will repel one end of this hanging magnet and attract the other end. The forces are much stronger than with rubbed rods.

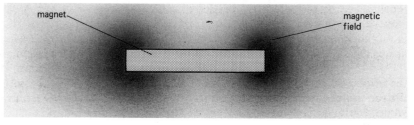

3. Gravity

Everything falls to the ground when it is released, because of the force of gravity. Gravity is another example of 'action at a distance'.

Around the Earth there is a **gravitational field**. Any object in the field will experience a force pulling it towards the Earth. The force is strongest near the Earth and gets gradually weaker as we go further away. One difference between gravity and both magnetism and electrostatics is that gravitational forces are always attractions. No one has ever discovered a gravitational repulsion force. The gravitational force is what holds the universe together. The galaxies, the solar system, the Earth and the Moon – all are held together by the force of gravity between objects.

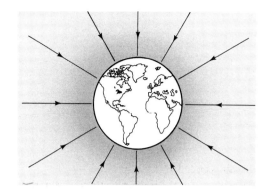

Gravitational forces

There is not just a gravitational field around the Earth, but also around any object. Of course the field is much weaker if the object is a lot smaller than the Earth. Isaac Newton was first to suggest that there is a force of gravity between *all* masses, but the first person to measure the size of the force was Henry Cavendish (1731–1810).

He took a beam with two large spheres hanging from its ends. The beam itself was then hung up so that it balanced. The suspension of the beam was made from a metal wire fastened very securely to the beam and to the ceiling of Cavendish's laboratory. Left to itself without any draughts or vibrations from outside, the beam would eventually settle in one particular position. Cavendish then brought another large sphere close to one of the hanging spheres. Because of the gravitational attraction between the two spheres, the beam twisted round slightly. The force is very small and so the amount of twisting is also very small, but Cavendish was able to measure it. As a result he was able to show experimentally that there really *is* a force between two masses.

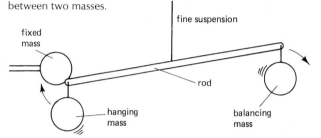

1.18 Similarities between the field forces

Electrostatics, magnetism and gravity are examples of 'action at a distance' – forces between bodies which are not touching each other. The **field** idea is useful in picturing what is happening. Perhaps these three different things are connected in some way? It turns out that the mathematical equations for the electrostatic and gravitational forces look very similar. This makes physicists think that all these forces might really be related, if only we understood them completely.

In Chapters 29 and 30, we will see that there is a connection between magnetism and electricity. But a theory to link up all the field forces (gravity, electricity, magnetism and two other types of force which only appear deep inside the atom) is one of the goals of modern physics. **Fields** are a very useful and important idea. But we don't yet know all there is to know about the field forces!

1.19 Fields and energy

Energy can be stored by lifting an object up to a higher level. This is called gravitational potential energy. But *where* exactly is the energy stored? The object itself hasn't changed. The energy is stored in the object-and-the-field together.

Electrically charged objects (like rubbed rods) also have some stored energy if they are in the field of other charges. Again the energy is stored in the position of the charge-in-the-field.

By lifting the object up to a higher place in the field, we have stored energy in the whole system. If we release the object, it falls down in the field and its stored energy is transferred into kinetic energy as it speeds up.

Energy can be stored in other fields too. If two repelling magnets are pushed together, energy is stored in the magnets-and-field. If we let go, the magnets spring apart and the stored energy is transferred into energy of the moving magnets.

1.20 Forces

In talking about fields, we have used the idea of a force – a push or a pull. **Force** is a very important idea in physics and we will have more to say about forces in Chapter 3.

Energy, Matter, Fields: Three Big Ideas

Questions

1. Draw energy arrow diagrams (like the ones on page 3) for the following situations:
   (a) cycling along a level road;
   (b) walking up a steep hill;
   (c) stretching an elastic band.

2. Draw energy arrow diagrams (like the ones on page 3) to show the energy changes in each of the following:
   (a) a gas cooker
   (b) a vacuum cleaner
   (c) a bow and arrow
   (d) a microphone
   (e) a bicycle – when you put on the brakes and stop
   (f) a candle
   (g) a cassette player
   (h) a solar-powered pocket calculator.

3. Look at the list of 'forms of energy' on page 5. (There are 7 altogether.)
   (a) Give an example of a process or a device which is **started** by each of these energy forms.
   (b) Give an example of a process or a device which results in each of these energy forms being **produced**.

4. If someone said to you that there was no need to switch off the light when you weren't in the room because 'energy is conserved anyhow, so you can't waste it', how would you answer them? Explain clearly why we need to 'Save It!'

5. Use the information on pages 8 to 10 to draw up a table with the headings shown below, summarising the evidence that all matter is made of molecules which are constantly moving around randomly in all directions:

Name of experiment	What we observe	What this tells us about matter

6. Make a poster to illustrate the **particulate model of matter** – the theory that all matter is made from tiny particles.

7. What is in the spaces between the molecules of a gas?

8. (a) Where does the sugar go when you stir your tea? How could you get the sugar back again? Would it be the same sugar as you started with?
   (b) If you put a glass of water on a window sill and leave it overnight, there will be less water in the glass in the morning. You may also notice some condensation (tiny drops of water) on the window pane just above the glass. Where has the 'missing' water from the glass gone? Is the condensation on the window the same water which has been lost from the glass?

9. A crystal of sugar is shaped roughly like a cube of side 0.5 mm. If sugar molecules are the same size as the oil molecules used in the experiment on page 10, estimate the number of sugar molecules in one sugar crystal. (Hint: Work out first how many sugar molecules in a row would fit along one side of the sugar crystal.)

10. What is it that pulls one magnet towards another? How does a charged rod influence another one nearby? What is it that pulls falling objects towards the ground?

    Explain how the idea of a **field** can help to account for all these observations.

> **INVESTIGATION**
> Can electric or magnetic fields pass through all materials, or is it possible to 'screen them off'?

Time

Measuring time is always based on something which 'beats' at a steady rate. The second used to be defined as a fraction (1/86 400) of a day. However, it was later discovered that days are not all the same length: the Earth does not rotate on its axis at a steady speed.

Laboratory clocks are based on the swings of a pendulum, or the vibrations of a tiny quartz crystal. However the very accurate clock which *defines* the second uses the frequency of some radiation emitted from caesium atoms, and doesn't look like a conventional clock at all! ▷

Shorter time intervals are measured in milliseconds, microseconds or nanoseconds.

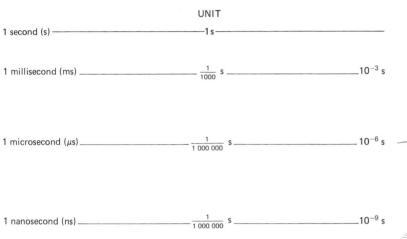

UNIT			APPROXIMATE LENGTH OF TIME
1 second (s)	1 s		time between human heartbeats
1 millisecond (ms)	$\frac{1}{1000}$ s	10^{-3} s	duration of a photographic flash
1 microsecond (μs)	$\frac{1}{1\,000\,000}$ s	10^{-6} s	duration of contact between golf club and ball
1 nanosecond (ns)	$\frac{1}{1\,000\,000}$ s	10^{-9} s	radio waves travel 0.3 m in 1 nanosecond

2.2 Derived units

Area

We need not define any new unit for measuring area. Area is a measure of the size of a surface. For example, if you want to know how much carpet you need to cover a floor, then it is the area of the floor which matters. Area is measured in **metre squared** (m^2). $1\,m^2$ is the area of a square with sides 1 m long.

For measuring smaller surfaces, the units centimetre squared (cm^2) or millimetre squared (mm^2) are more convenient.

```
     1 m   = 100 cm       = 1000 mm
So   1 m²  = 10 000 cm²   = 1 000 000 mm²
     1 m²  = 10⁴ cm²      = 10⁶ mm²
```

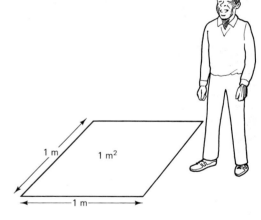

MEASURING AREAS

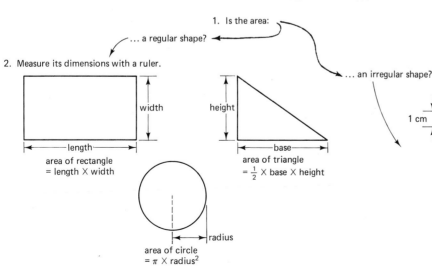

1. Is the area: ... a regular shape?
2. Measure its dimensions with a ruler.

area of rectangle = length × width

area of triangle = ½ × base × height

area of circle = π × radius²

... an irregular shape?

3. To measure the area of an irregular shape, place the shape on graph paper and draw round it.
4. Each little square is 1 cm². Count the squares lying inside the outline — if more than half of a square is in, count it; if less than half is in, don't count it.

Measurement

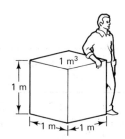

Volume

Volume is a measure of the amount of space a body takes up. It is measured in **metre cubed** (m³). 1 m³ is the volume of a cube with sides 1 m long – about the size of a large cupboard.

The metre cubed is too large a unit for most laboratory work and so we use either the centimetre cubed (cm³) or the litre (l). 1 cm³ is the volume of a cube with sides 1 cm long. 1 litre is 1000 cm³.

$$1\,m = 100\,cm = 10^2\,cm$$
So $1\,m^3 = 1\,000\,000\,cm^3 = 10^6\,cm^3$
$1\,litre = 1000\,cm^3$

Just to complicate matters further, volumes are sometimes measured in millilitres (ml). The prefix milli- always means $\frac{1}{1000}$th.

So $1\,ml = \frac{1}{1000}\,l$

But this means that 1 ml is exactly the same size as 1 cm³. You can use either millilitres or centimetres cubed to measure a volume – they are equivalent.

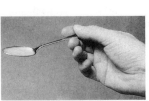

Spoonful: 5 cm³   Bottle: 600 cm³

Volumes of liquids

In the laboratory, volumes of liquids are usually measured using a measuring cylinder. Liquid is poured in, the cylinder is placed on a flat surface, and the level of the liquid surface on the scale gives the volume. Usually the liquid surface is not flat but curves upwards where it meets the glass sides (a 'meniscus'). The volume reading should be taken from the bottom of the meniscus.

Swimming pool: 200 000 l

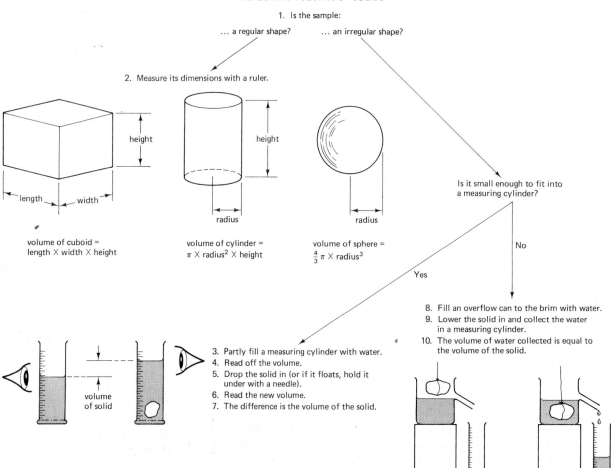

2.3 Density

We all know that foam polystyrene is light stuff and lead is heavy stuff. But a small lead washer is lighter than a large block of polystyrene! For a fair comparison, we must compare equal volumes of lead and polystyrene. The property we are really comparing is called **density**.

The density of a material is the mass of 1 metre cubed of it.

$$\text{Density} = \frac{\text{mass}}{\text{volume}}$$

The SI units of density are kilograms per metre cubed (kg/m^3).

In the laboratory it is often more convenient to measure masses in grams and volumes in centimetres cubed, and then to calculate densities in grams per centimetre cubed (g/cm^3). It is easy enough to convert from one unit to the other:

$$1\,g/cm^3 = 1000\,kg/m^3$$

Table 2.1 shows the densities of a number of common substances in both g/cm^3 and kg/m^3. It is not a coincidence that the density of water happens to be $1\,g/cm^3$. The standard kilogram was originally designed to have the same mass as $1000\,cm^3$ of pure water. However, it was not entirely accurate and so the standard kilogram is now known to be not exactly the same as this.

Table 2.1

Substance	Density kg/m^3	Density g/cm^3
air	1.3	0.0013
foam polystyrene	100	0.1
cork	250	0.25
wood	700	0.7
ethanol (alcohol)	800	0.8
ice	920	0.92
water	1000	1.00
glass	2500	2.5
aluminium	2700	2.7
steel	7800	7.8
lead	11 400	11.4
mercury	13 600	13.6
gold	19 300	19.3

EXAMPLE 2.1

A window pane is 120 cm wide by 80 cm high. It is made of glass 0.3 cm thick. The density of glass is $2.5\,g/cm^3$. What is the mass of the window pane?

First we calculate the **volume** of glass in the pane:

Volume of pane of glass = width × height × thickness
= 120 cm × 80 cm × 0.3 cm
= $2880\,cm^3$

To find its mass, we now use: $\text{density} = \dfrac{\text{mass}}{\text{volume}}$

$$2.5\,g/cm^3 = \frac{\text{mass}}{2880\,cm^3}$$

Multiply both sides by $2880\,cm^3$: $2.5\,g/cm^3 \times 2880\,cm^3 = \text{mass}$

$$\text{mass} = 7200\,g = 7.2\,kg$$

The mass of the pane of glass is 7.2 kg.

EXAMPLE 2.2

An iron bar has a volume of $120\,cm^3$. Its mass is 900 g. What is the density of iron?

By definition, $\text{density} = \dfrac{\text{mass}}{\text{volume}}$

$$= \frac{900\,g}{120\,cm^3}$$

$$= 7.5\,g/cm^3$$

The density of iron is $7.5\,g/cm^3$, or $7500\,kg/m^3$

Measuring density

To measure the density of a material, we have to measure the mass and the volume of a sample of it (see right and page 20). The easiest way to measure mass is by 'weighing'.

Density of air

Table 2.1 includes a value for the density of air. Gases also have density, and this can be measured. The steps are the same; again we have to find the mass and the volume of a sample of air. (See page 20.)

The densities of gases are much lower than those of solids and liquids. However, if you use the value for the density of air given in Table 2.1 to calculate the mass of air in the room you are now sitting in, you may well be surprised at the result!

Measuring the density of a solid

1. Find the mass of the sample by weighing it.

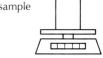

2. Measure the volume of the solid by one of the methods shown on page 18.

3. Calculate the density of the solid using the equation: density = mass/volume.

Measuring the density of a liquid

1. Find the mass of an empty measuring cylinder by weighing it.
2. Pour a sample of liquid into the cylinder. Weigh again to find the mass of cylinder plus liquid.
3. Subtract these two results to find the mass of the liquid.
4. Read off the volume of the liquid using the scale on the measuring cylinder.
5. Calculate the density of the liquid using the equation: density = mass/volume.

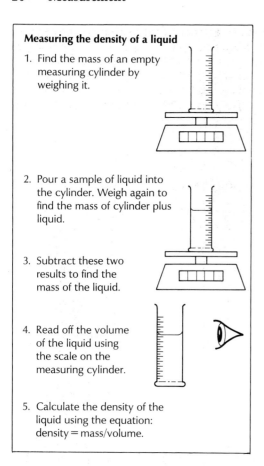

Measuring the density of air

1. Fit a strong round-bottomed flask with a rubber bung, glass tube and sealing clip. Then find the mass of the flask full of air by weighing it on an accurate balance.
2. Now pump the air out of the flask using a vacuum pump. (The flask must be strong enough to withstand this.) Weigh it again. Its mass is now slightly less.
3. The difference is the mass of the air inside. Subtract to find this.
4. To measure the volume of this air, fill the flask with water and then measure the volume of the water by pouring it into a measuring cylinder.
5. Calculate the density of air using the equation: density = mass/volume.

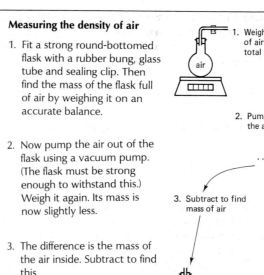

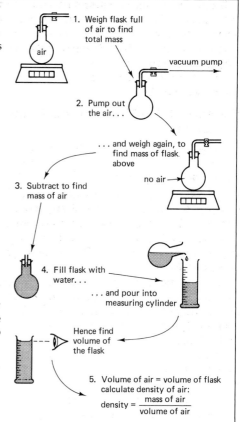

Questions

1. State which length-measuring instrument would be most suitable for measuring each of the following:

   (a) the length and breadth of your bedroom (to work out the cost of a carpet);
   (b) the distance a trolley has moved along the laboratory bench;
   (c) the inside and outside diameter of a copper water pipe;
   (d) the circumference of a 2p coin;
   (e) the thickness of a wire.

2. Estimate the following (in SI units) and then check each one by measurement.

	Estimate	Measured value
width of a page in your jotter		
your handspan		
circumference of a bicycle wheel		
mass of a pencil		
mass of your shoe		
time to count aloud to 50		
area of the base of a baked beans tin		
area of the sole of your foot		
capacity of a tea cup		
volume of a key		

3. One way to measure how far away a building is from you (if it is too far to use a tape measure) is by making a simple range finder.

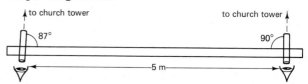

Using the range finder in the diagram, the left-hand straw has to be set at 87° to sight the church tower. By drawing a scale diagram, or by calculation, work out how far away the tower is.

How accurate is this measurement? What difference would it make to your answer if the angle really should have been 86° instead of the 87° you measured?

4. An even simpler rangefinder can be used to measure the height of a building or a tree. Standing 25 m from the tree, the straw has to be pointed upwards at 30° to sight the topmost branch. Draw a scale diagram (or do a calculation) to work out how tall the tree is.

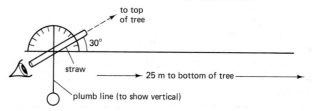

If the masses are then removed from the spring one at a time, it returns in stages to its original unstretched length. There is no permanent stretch, because we have used only fairly small loads and not applied too much stretching force. We will see later what happens to springs if we use larger stretching forces.

The stretching of materials was investigated over 300 years ago by Robert Hooke. He discovered:

the extension of a material is directly proportional to the stretching force.

This result is known as **Hooke's Law**.

Making a force meter

Force is measured in units called **newtons**, or N for short. The unit is named after the famous physicist, Isaac Newton (1642–1727) who first stated the laws of force and motion (see Chapters 6 and 7). 1 newton is roughly equal to the force of gravity on an average sized apple (i.e. the apple weighs about 1 N). The force of gravity on a 1 kg mass is almost exactly 10 N.

A spring makes a convenient force-meter. But first it must be calibrated – a scale must be marked on it. We can do this by applying some known forces and noting the extension each time. Then when an unknown force is applied, we can calculate its size or read it off from a graph or from a scale. The results on page 22 are a calibration graph for the spring. Imagine that we now hang a large potato on the spring and it causes an extension of 50 mm. From the graph we can read off the force exerted by the potato (its **weight**) – 2.5 N.

Spring balances are just springs which have been calibrated and provided with a marked scale. As the extension is proportional to the stretching force, the intervals on the scale are equal distances apart. This makes the balance easy to use and to read.

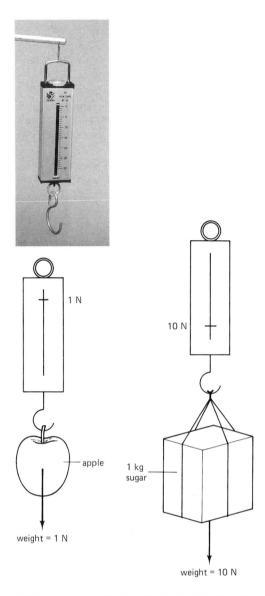

EXAMPLE 3.1

Two girls want to weigh a melon they have bought. They don't have kitchen scales, but they do have a spring and a 1 kg bag of sugar. The bag of sugar stretches the spring by 60 mm. The melon produces an extension of 84 mm. How heavy is it?

The bag of sugar has a mass of 1 kg. So its weight (the force of gravity on it) is 10 N.

This stretches the spring by 60 mm, so we can sketch an extension/stretching force graph for the spring.

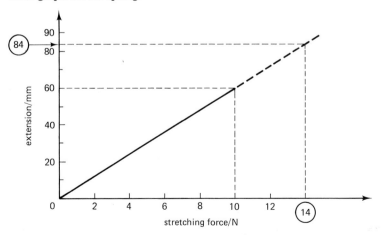

From this graph, the stretching force when the melon is hung on the spring can be read off. It is 14 N. The weight of the melon is 14 N. (This means that its mass is 1.4 kg.)

Spring constant

Different springs extend by different amounts if the same stretching force is applied. Strong springs have a small extension, weak springs have a large extension. For any spring we can define a **spring constant** in the following way:

$$\text{spring constant} = \frac{\text{extension}}{\text{stretching force}}$$

This quantity is constant up to the elastic limit because the extension is directly proportional to the stretching force. You can check this by calculating extension divided by stretching force for each of the results in the table on page 22. The result is the same each time. The spring constant is 0.05 N/mm, or 5 N/m.

If you know the spring constant for a spring, then it is easy to calculate the extension which any stretching force will produce:

extension = spring constant × stretching force

3.3 Elastic limit

In the spring experiment on page 22, the spring goes back to its original unstretched length as the load is removed. If we repeat the experiment, but go up to larger stretching forces this time (Table 3.2), we notice two changes:

1. The graph of extension against stretching force stops being a straight line after a certain point. This point is called the **elastic limit** of the spring.

2. As the spring is unloaded again, the extensions measured will be larger than they were previously for the same force. When all the masses have been removed, the spring will be longer than it was to begin with. It will have a permanent stretch.

Table 3.2

Stretching force/N	2	4	6	8	10	8	6	4	2	0
Extension/mm	40	80	120	170	225	185	145	105	65	25

An **elastic material** is one which returns to its original length when a stretching force is removed. The spring is elastic for small stretching forces – up to its elastic limit. Hooke's Law – the extension of a material is directly proportional to the stretching force – holds when the material is not stretched beyond its elastic limit.

3.4 Elastic materials

Do other elastic materials behave in the same way as springs? To find out, we can do stretching experiments on other materials.

Stretching elastic and polythene

The load on the elastic band or polythene strip is increased in stages. We find that the results follow the same pattern. For small loads (small stretching forces), the extension increases steadily with the stretching force – Hooke's Law. If the load is removed, the sample comes back to its original length. Eventually a point is reached where the material begins to stretch more as each extra load is added. It has reached its elastic limit and is now permanently deformed.

Elastic plastic!

The **definition** of an elastic material is one which comes back to its original shape when the deforming force is removed. A **plastic** material is one which is permanently deformed. The everyday meaning of **plastic** is a type of man-made material which is used for making all sorts of objects. An example of a common plastic is polythene. But experiment shows that, for small stretching forces, polythene is **elastic** – it comes back to its original shape!

In fact, it is probably true that **all** materials are elastic if the force is small enough. The materials we label 'plastic' are just the ones which have a low elastic limit.

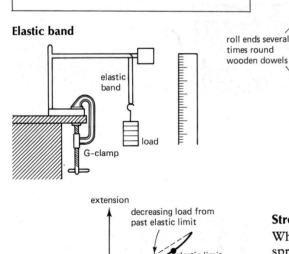

Elastic band

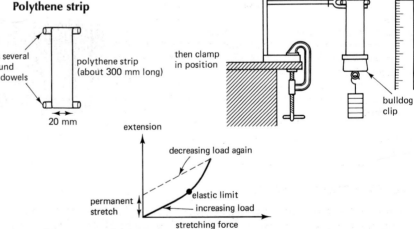

Polythene strip

Stretching a metal wire

When a spring is stretched, the extension is clearly due to the coils of the spring being pulled slightly apart. But does the metal of the spring actually get any longer? The stretching of a single metal wire can be investigated in much the same way as other materials. Much larger forces are needed and the extension produced is much smaller. There is also a risk of injury if the wire suddenly snaps, so it is wise to wear protective glasses if you are doing this.

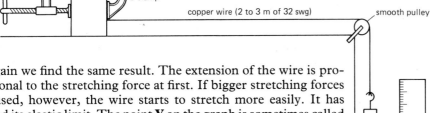

Again we find the same result. The extension of the wire is proportional to the stretching force at first. If bigger stretching forces are used, however, the wire starts to stretch more easily. It has passed its elastic limit. The point **Y** on the graph is sometimes called the yield point. Small increases in stretching force now cause quite large extensions. If the stretching force is increased still further, the wire will snap.

3.5 Explaining stretching

When a solid material is stretched, its molecules move slightly further apart. Our model of a solid is an array of molecules linked by forces (like tiny springs) to their neighbours.

The extension of these links is proportional to the applied force (for small extensions) and so the extension of the whole material is proportional to the stretching force. This model of a solid also explains why the material returns to its original length when the force is removed.

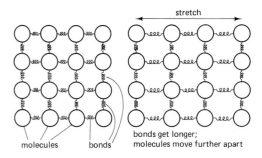

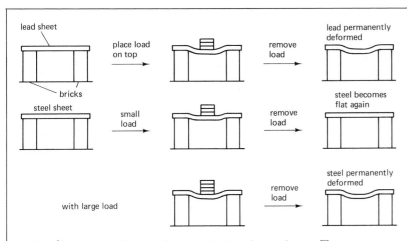

Mild steel sheets used for making car bodywork panels are ▽ shaped in this sort of way. A flat sheet is forced into a shaped mould by applying pressure. The forces are greater than the elastic limit of the steel and it is permanently deformed. One of the reasons why steel is used for car bodies is because it is a very easy material to shape permanently in this way.

Another similar technique is used to shape plastics. These materials are often more pliable when warm. In the **vacuum forming** process, a sheet of plastic is shaped by pumping the air out of the mould so that air pressure forces the sheet into shape. If you look closely at some plastic ▷ objects, you may be able to spot the point where the outlet from the mould to the vacuum pump was.

Shaping materials

In many manufacturing processes, we want to change the shape of a piece of material. Sometimes this is done by melting the material and pouring it into a cast to set. But often it is done by simply deforming the material – stretching it beyond its elastic limit.

Imagine a thin lead sheet placed across two bricks. If a small load is placed on top, the lead will bend a little. When the load is removed, the lead remains bent. Lead is a very easy metal to deform – it has a very low elastic limit.

We can try the same thing with a thin sheet of springy steel. Again it bends when a small load is placed on top. But this time when the load is removed, the plate becomes flat again. If we increase the load, however, we can bend the plate permanently.

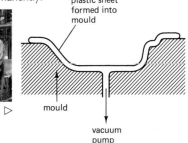

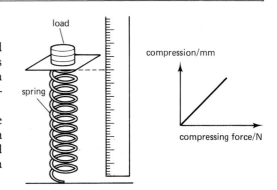

3.6 Compressing solids

Solids can be stretched by applying forces. They can also be compressed by forces which squeeze them. If a spring is squeezed by a pair of forces applied at each end, it will become shorter. The graph of compression (reduction in length) against compressing force applied is a straight line – just like the graphs for stretching forces.

When a solid block (e.g. a block of rubber) is compressed, the molecules are forced closer together. The spring-like links between them are now compressed. If the solid is released, it will return to its original shape and size. The molecular model of solids predicts that compression will follow the same pattern as stretching.

26 Forces

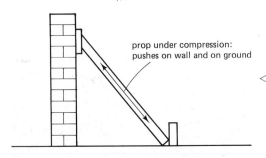

guy ropes under tension: pull on tent and on the ground

prop under compression: pushes on wall and on ground

BUILDING STRUCTURES

Tension and compression

In any building or structure, some parts will experience stretching forces whilst others have to withstand compressing forces. The guy ropes of a tent are taut; they are being slightly stretched. We say that they are in **tension**. The tension force pulls on the rope in both directions, tending to stretch it. As a result, the rope itself also pulls back on the tent and on the ground.

A prop used to support a wall during building operations is being compressed slightly by the weight of the wall. We say it is under **compression**. The compression force pushes on the prop in both directions, tending to compress it. So the prop pushes back on both the wall and the ground.

Strengths of materials

Ropes and cables can only be used under tension. You can pull with a rope but you can't push with it! Metal, stone or concrete beams, however, can be either in tension or compression.

Steel beams are strong in both tension and compression. Very large forces are needed to stretch or to compress them. Stone and concrete are very strong in compression but are weaker in tension. This is because they may develop small cracks and flaws which weaken the material.

◁ Steel is strong under both tension and compression; concrete is strong under compression.

Building bridges

Engineers use their knowledge of the properties of materials when designing bridges.

The simplest bridge is a straight beam across a gap. Many bridges are of this type.

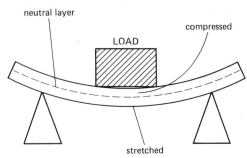

◁ When a load is placed on the middle of a beam bridge, the beam will sag a little. (Even motorway bridges bend when a car goes across, though the bending is too small to notice.) This means that the material on the underside of the beam is stretched and the material on the upper side is compressed. The material in the middle layer of the beam is not under any stress at all!

If the beam is made of concrete, the place where it is most likely to break is on the underside (because concrete is weaker under tension). To prevent this, bridge beams are made from reinforced concrete. It is made by allowing the concrete to set around some steel rods. The concrete sticks very firmly to the steel. Steel is strong when under tension, and it strengthens the beam.

A reinforcing rod in the lower half of the concrete beam will be more effective because this region is under tension and needs strengthening.

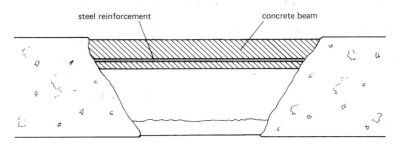

Pre-stressed concrete

A further improvement is to stretch the steel reinforcing rod before casting it in concrete. Once the concrete is set, the stretching force on the steel rod is released. The steel rod then holds the concrete in compression all the time.

Designing stronger beams

The middle section of a bent beam is neither stretched nor compressed. In a sense, the material in this layer is having no effect and might as well not be there! An **I**-shaped beam is just as strong as a solid one but is much lighter. Some of the material has been removed from the neutral layer in the middle.

Look around at any buildings where you can see the metal framework. You will see a lot of **I**-shaped metal girders in use.

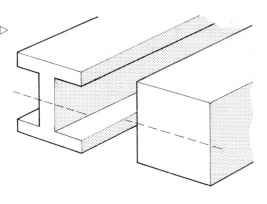

Strengthening a bridge

Another way to strengthen a bridge and prevent it bending is to add extra supports. A pillar at each end of the bridge can support struts or cables fixed to the centre of the span. These will be in tension. The two tension forces pulling upwards at an angle balance the downwards force of the load. The two tension forces in the cables *add together* to provide an upwards force.

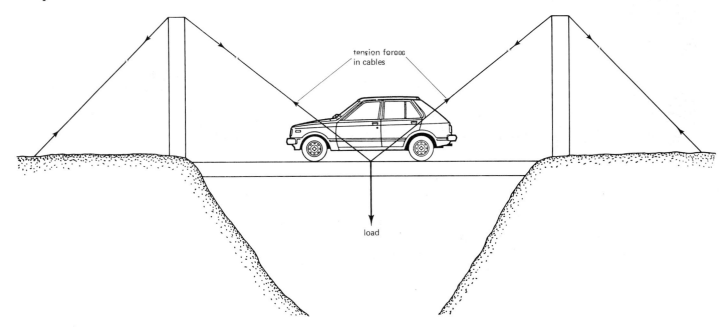

3.7 Adding forces

The last example involved the idea of adding forces. Let us look in more detail at how we can add forces together. We will begin with a simple situation.

Imagine an object being pulled by two forces, one of 3 N and one of 4 N. What is the total force pulling the body? The answer is that we cannot tell from that information alone. We need to know more about the **directions** of the forces.

If the forces are parallel and in the same direction, the total is 7 N. If the two forces are in exactly opposite directions, the total is 1 N (in the direction of the larger force). But what happens if the forces are not in line? It seems obvious that in the third drawing, the body will move off at an angle somewhere between the 3 N and 4 N forces, slightly closer to the larger force. The two forces are equivalent to a single force in this direction.

The single force which is equivalent to two separate forces is called the **resultant** of the two forces.

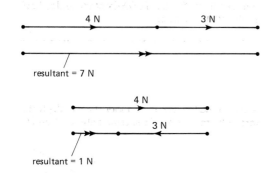

3.8 Vector addition

A force has a direction as well as a size (or magnitude).

Quantities like this are known as **vectors**. Others, like distance or volume or mass, have no direction and are called **scalars**. Forces have to be added by a method which takes their direction into account – vector addition.

If two forces are in the same line, vector addition is easy. We simply add the forces if both pull or push together; subtract them if one is in the opposite direction. This is just common sense. If they are at an angle, the easiest method is to draw a scale diagram. The procedure is as follows:

1. Draw a line parallel to one of the forces. Its length should be in proportion to the size of the force. Mark the direction of the force by an arrow.
2. Starting from the tip of this arrow, draw a second line parallel to the second force. Again make its length proportional to the size of the force. Mark its direction by another arrow.
3. Join the tail of the first arrow to the tip of the second. This line represents the resultant in both size and direction. These can then be measured off the diagram.

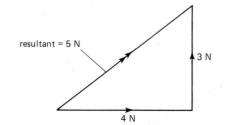

The easiest way to see how this works is by an example.

EXAMPLE 3.2

A box is pulled by two forces. What is the total force (the resultant force) on the box?

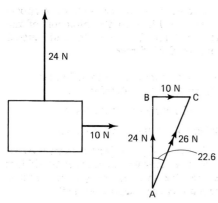

First we draw a line AB, 2.4 cm long, to represent the 24 N force. Then we draw a second line BC, 1.0 cm long, starting at B and in the direction of the 10 N force. The resultant force is represented by the line AC.

As the diagram is drawn to scale, we can measure AC to find how big the resultant force is. It turns out to be 2.6 cm long, so the force is 26 N. If we measure its direction, we find it is at approximately 23° to the 24 N force.

We could get the same results from a calculation, instead of doing a scale drawing:

By Pythagoras' theorem,
$$AC^2 = AB^2 + BC^2$$
$$= 24^2 + 10^2$$
$$= 576 + 100$$
$$= 676$$
$$AC = 26 \text{ units}$$

So, the resultant force is 26 newtons.

We can also calculate the size of the angle BAC:

$$\tan BAC = \frac{BC}{AB} = \frac{10}{24} = 0.42$$

Therefore angle BAC = 22.6°

The resultant force is 26 N at an angle of 22.6° to the 24 N force.

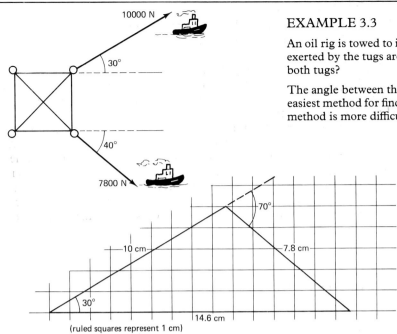

EXAMPLE 3.3

An oil rig is towed to its position in the North Sea by two tugs. The forces exerted by the tugs are shown in the drawing. What is the total force from both tugs?

The angle between the directions of the two forces is 70° in this case. The easiest method for finding the resultant is by scale drawing; the calculation method is more difficult.

Steps:

1. Draw a line of length 10 cm to represent the 10 000 N force. Mark its direction by an arrow.
2. Draw a second line of length 7.8 cm to represent the 7800 N force. This must start from the end of the first line.
3. Join the starting point of the first line to the end of the second. This represents the resultant force.
4. Measure its length and direction.

The resultant force is 14 600 N at an angle of 30° to the 10 000 N force.

3.9 Components of a force

Often (as in Examples 3.2 and 3.3) we want to add two forces to produce a single resultant force. Sometimes, on the other hand, we want to replace a single force by two forces which are equivalent. The two forces are said to be **components** of the original force. Often a pair of components at right angles turns out to be useful in tackling a problem.

Rule: To find the size of a component P, multiply the force F by the cosine of the angle between P and F.

The force F is equivalent to the two component forces X and Y. In other words, X and Y add to give F. X is equal to $F \cos \theta$ and Y is equal to $F \sin \theta$.

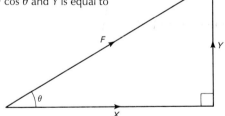

$X + Y = F$

$\dfrac{X}{F} = \cos \theta$

$\therefore X = F \cos \theta$

$\dfrac{Y}{F} = \sin \theta$

$\therefore Y = F \sin \theta$

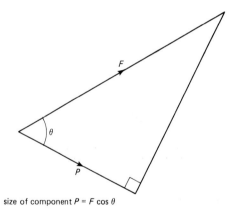

size of component $P = F \cos \theta$

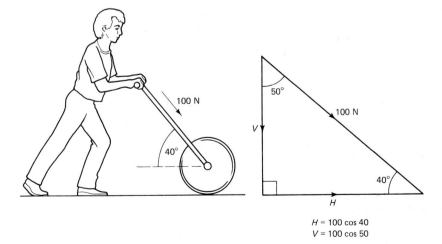

$H = 100 \cos 40$
$V = 100 \cos 50$

EXAMPLE 3.4

A garden roller is pushed across a lawn by applying a force of 100 N in the direction of the handle. What are the horizontal and vertical components of this force?

The horizontal component of the force is:
$H = 100 \cos 40° = 100 \times 0.766 = 76.6$ N

The vertical component of the force is:
$V = 100 \cos 50° = 100 \times 0.643 = 64.3$ N

So, the force actually pushing the roller forward is 76.6 N. The vertical component of 64.3 N just pushes the roller down into the ground. As far as the motion is concerned, it is wasted!

Questions

1 Look at these drawings. For each, write a sentence or two to say where forces are involved. Explain how you know there is a force acting by stating what you can see the force **doing**.

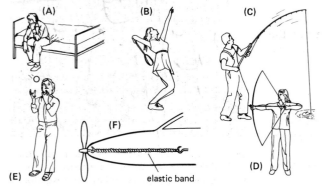

2 A group of students hang up a spring and attach various loads to the bottom end. Each time, they record the length of the spring in a table.

Reading	Load/N	Length/cm	Extension/cm
1	0	10.0	0.0
2	2	12.4	
3	4	14.8	
4	6	17.2	
5	4	14.8	
6	0	10.0	
7	6	17.2	
8	8	20.4	
9	10	26.0	
10	4	19.0	

(a) Complete the final column of the table, headed 'Extension/cm'.
(b) What would have been the extension if the third reading had used a 3 N load? What would the length of the spring have been?
(c) The first time there was a load of 4 N (reading 3), the extension was 4.8 cm. The second time (reading 5), it was the same. Why was it different the third time (reading 10)?
(d) Draw a graph showing how the extension changes with different loads. Use readings 1, 2, 3, 4, 8 and 9 for the graph. Estimate the elastic limit for this spring.

3 Look back at the graph shown on page 22. A student uses his shoe as a load and finds that it causes an extension of 70 mm. What is the stretching force caused by the shoe?

4 The springs in the diagram are identical and the same load is used each time. How does the extension in diagrams (B) and (C) compare with the extension in (A)?

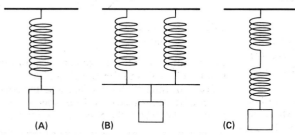

5 Experiments were carried out on the materials bone, tendon and skin to investigate the way they stretch when loads are added to them.
The graph below shows the relationship between load and stretch obtained for each.
(a) For which material is the amount of stretch directly proportional to the load, over the range shown on the graph?

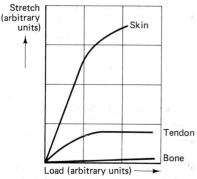

(b) Use the graph to describe the behaviour of *each* of the following materials when it is stretched, and explain how this is related to its particular function in an animal: (i) bone, (ii) tendon, (iii) skin. (AEB: SCISP)

6 Look at the four diagrams of bridges. For each of the bridge parts labelled, state whether you think this would be compressed (in compression) or stretched (in tension) when a heavy vehicle crosses the bridge. Explain your answer each time.
For which of the labelled parts could a steel rope be used instead of a steel or concrete beam?

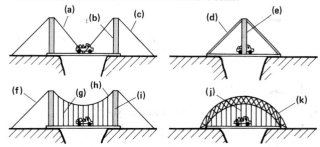

7 What is the total force pulling the box in each of the three situations below?
In which direction will the box begin to move?

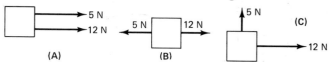

8 To close a sliding door, a child pulls with a force of 30 N in the direction shown. What is the size of the **component** of this force which is actually pulling along the direction of the rail?

From above: [diagram showing rail, 30°, door, 30 N]

INVESTIGATION

How strong is hair? Is hair elastic? Are long hairs stronger than short ones? Is dark hair stronger than fair hair?

4: More About Forces: Balance and Stability

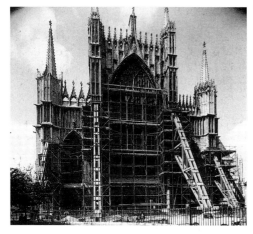

Most of the objects around us are stationary. They are kept at rest by the forces which act on them. In this chapter we will look at how forces act together to keep things from moving – to hold them stationary.

4.1 Keeping stationary

Think of a light hanging from the ceiling of a room. What holds it stationary? The weight of the light fitting and shade is a force pulling it downwards (the force of gravity). This is balanced by the tension force in the flex, pulling it upwards. The two forces balance each other – they add to zero.

Another similar situation is a television sitting on a table. The weight of the television pulls it downwards. If the table wasn't there, it would fall. But the table holds it up. The table exerts an upwards force on the television, exactly balancing its weight. This upwards force is called the **reaction** of the table on the television. Of course, if the table is flimsy and the television is heavy, the table may not be able to exert a large enough reaction force and will break!

During major repair work on the foundations, the walls of York Minster are held stationary by the forces exerted by these enormous buttresses (or props).

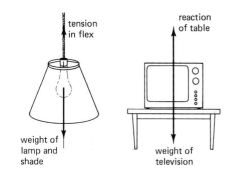

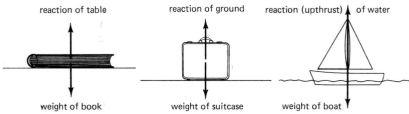

If something is stationary – at rest – then the forces on it must balance. They must add to zero.

4.2 Equilibrium

If two people want to hold and carry a heavy bucket of water, they might both hold the handle and pull upwards. Each could exert a force equal to half of the weight. Together the two forces would balance the weight.

$F_1 + F_2$ = weight of bucket and water

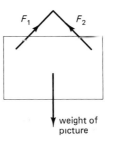

$F_1 + F_2$ (added by vector addition) must still be equal to the weight of bucket and water

$F_1 + F_2$ (added by vector addition) = weight

It is more likely that they will pull upwards at an angle. The two forces are no longer parallel. But if we add them by vector addition, they must still be equal to the downward force of the bucket's weight. There are many situations where the forces acting on a body to hold it steady are not all parallel.

When a body – like the oil tanker in Example 4.1 – is at rest, we say that it is **in equilibrium**. This term is also used to describe the system of forces acting on the body – the forces are also said to be **in equilibrium**.

EXAMPLE 4.1

The forces exerted on an oil tanker by the hawsers to buoys **2** and **3** are 50 000 N each. What is the force exerted by the hawser to buoy **1** if the tanker is stationary?

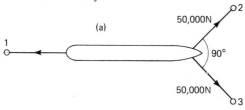

We begin by adding the two forces exerted by the hawsers to buoys **2** and **3**. This can be done by a scale drawing:

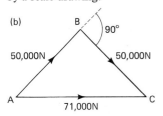

The resultant force is represented by the line AC. It is 71 000 N in the direction of AC.

So, if the total of all three forces on the tanker is zero, the force exerted by the hawser to buoy **1** must be 71 000 N in the opposite direction.

31

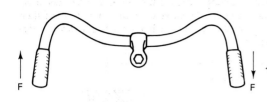

4.3 Turning forces

To keep a body in equilibrium, all the forces acting on it must add to zero. But this is not enough. For instance, think of what happens when you apply a force to one handlegrip of your bicycle, and an equal and opposite force to the other handlegrip.

These forces add to zero. But the handlebars don't stay at rest. They turn. A system of forces which add to zero may still make an object rotate. So equilibrium depends on more than just having the forces add to zero. Let's consider the turning effects of forces in a little more detail.

The turning effect of forces

If you want to undo a nut, you don't usually try to do the job with your fingers. Using a spanner it becomes much easier. In fact, the longer the spanner, the easier it is. The same force has more turning effect if it can be applied further away from the turning point (or **pivot**).

The turning effect of a force is called the **moment** of the force. It is defined in this way:

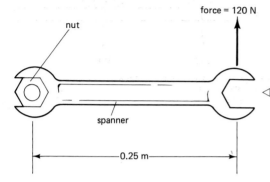

moment of a force = force × distance from the turning point to the line of the force

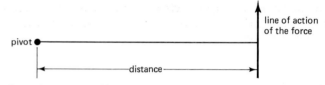

Since force is measured in newtons and distance in metres, the moment of a force is measured in newton-metres (Nm). The moment of the force applied to the nut is 30 Nm (120 N × 0.25 m). In this example, the force is turning the nut anticlockwise (as we look at it from this side). So this force has an anticlockwise moment of 30 Nm.

4.4 Principle of moments

If we take a bar and balance it at its mid-point, we can apply turning forces to it by placing masses on it at different points. The moments of these forces will depend on the distance from the pivot to the masses. We discover that the bar balances with two masses on it, when the anticlockwise moment is equal to the clockwise moment.

This diagram shows one possible arrangement which results in balance. The table shows a series of results from an experiment of this kind.

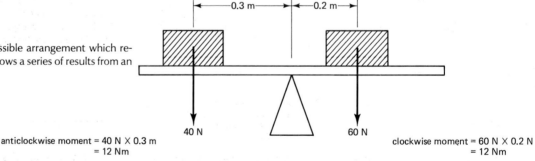

anticlockwise moment = 40 N × 0.3 m
= 12 Nm

clockwise moment = 60 N × 0.2 N
= 12 Nm

Table 4.1

Anticlockwise			Clockwise		
Force/N	Distance/m	Moment/Nm	Force/N	Distance/m	Moment/Nm
40	0.3	12	60	0.2	12
30	0.5	15	50	0.3	15
60	0.1	6	20	0.3	6
25	0.6	15	30	0.5	15
15	0.4	6	20	0.3	6

When there are more than two masses, we still find that the total clockwise moment must be equal to the total anticlockwise moment to make the bar balance.

Notice that we must calculate the moment of each force separately and then add the moments. We cannot add the forces first and then try to calculate moments, because each force acts at a different distance from the pivot.

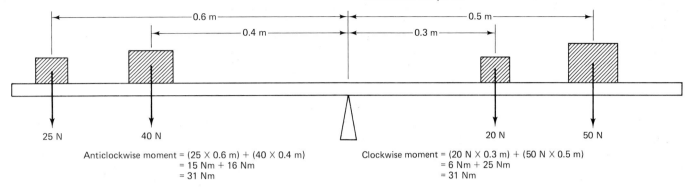

Anticlockwise moment = (25 × 0.6 m) + (40 × 0.4 m)
= 15 Nm + 16 Nm
= 31 Nm

Clockwise moment = (20 N × 0.3 m) + (50 N × 0.5 m)
= 6 Nm + 25 Nm
= 31 Nm

This result is called the **principle of moments**. It states:

When a body is in equilibrium, the sum of anticlockwise moments about the balance point is equal to the sum of clockwise moments.

4.5 Couples

Two equal but opposite parallel forces make a **couple**. The two forces applied to the bicycle handlebars in the drawing on p. 32 form a couple. To work out the turning effect of a couple, take the moments of the two forces about any point, and add them together. In fact, it doesn't matter which point you choose, the answer is always the same. ▷

EXAMPLE 4.2

What is the turning effect of a couple consisting of two forces of 20 N at a distance of 2 m apart?

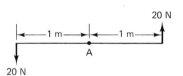

Total anticlockwise moment
about B = (20 N × 0.5 m) + (20 N × 1.5 m)
= 10 Nm + 30 Nm
= 40 Nm

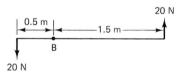

Total anticlockwise moment
about A = (20 N × 1 m) + (20 N × 1 m)
= 20 Nm + 20 Nm
= 40 Nm

The turning effect of the couple is 40 Nm. In fact, the simplest way to calculate the turning effect of a couple is to multiply one force by the perpendicular distance between the forces.

If you have ever tried to undo a nut using a 'tommy-bar' spanner, you will know that the couple is the same size about any point. Sliding the bar through the spanner head doesn't make any difference to the force needed to turn the nut.

Conditions for equilibrium: Summary

For a body to stay in equilibrium:

the sum of all the forces acting on it must be zero

and **the sum of all the clockwise moments about any point on the body must be equal to the sum of all the anticlockwise moments about that point.**

The moment of a couple is sometimes called a **torque**. This is a term which is used particularly by engineers. Torque is measured in newton-metres. Some typical values of torque are:

to turn a radio volume control	0.2 Nm
to turn a door handle	1 Nm
produced by a food mixer	2 Nm
produced by an electric drill	6 Nm
to undo an axle nut on a bicycle	30 Nm

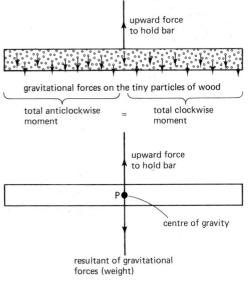

4.6 Centre of gravity

In the **principle of moments** experiment on p. 32, we did not include the weight of the bar as one of the forces. We started off with a balanced bar with no masses attached to it. Let us now consider what difference the weight of the bar would make.

If we are using a wooden bar – for example, a metre rule – we can think of the bar as being made up of a lot of tiny particles of wood. Each of these particles will have a force of gravity pulling it downwards (its weight). The bar balances at one particular point **P** where the sum of all the clockwise moments of these little gravitational forces on the wood particles is equal to the sum of all the anticlockwise moments.

The effect is exactly the same as if we had a single force acting downwards at **P**. So we can think of the gravitational force on the bar as a single force acting downwards at **P**. This resultant force must act exactly at **P**, otherwise it would have a moment about **P** and the bar would turn. The size of this single force is the sum of all the little forces; that is, it is the weight of the bar itself.

So the weight of a bar can be thought of as a single force acting downwards through the point **P**. This point **P** is called the **centre of gravity** of the bar.

4.7 Finding the centre of gravity of a flat shape

It is fairly obvious that the centre of gravity of a metre ruler will be at the centre. But for less regular shapes, it is not so obvious where the centre of gravity is. However, it can be found by a simple experiment.

If a shape is pinned up so that it hangs, it will come to rest with the centre of gravity directly below the pin. This must happen for the following reason: when the card is released, the weight of the card behaves like a single force through the centre of gravity. This is balanced by an equal and opposite force – the reaction of the pin. If the centre of gravity is not directly beneath the pin, these two forces will form a couple which will make the card swing downwards, and it will finally settle with its centre of gravity vertically below the pin. If we now hang a plumb line in front of the card we can draw a vertical line downwards through the pinhole. The centre of gravity lies somewhere on this line AB. We repeat this with the pin in a different position and obtain a second line CD. Since the centre of gravity must lie on both lines, it is at the point where the lines cross.

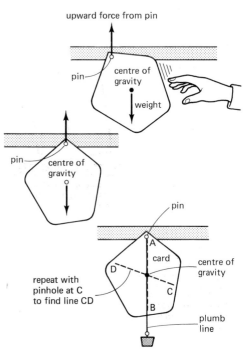

EXAMPLE 4.3

A trapdoor into the loft of a house hinges downwards from one edge. The trapdoor is a flat uniform piece of wood, 0.8 m square, of mass 4 kg.

What force is needed to push the door closed, if the force is applied at the edge furthest from the hinge?

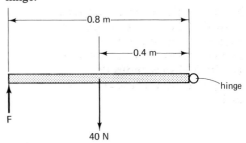

The mass of the door is 4 kg. Remembering that the weight of any 1 kg object is 10 N, this means that its weight is 40 N.

The door is made of a uniform sheet of wood. 'Uniform' here means that its mass is evenly distributed – in other words, the centre of gravity will be at the centre of the door, 0.4 m from the hinge. So the weight of the door can be represented by a single force of 40 N acting downwards at this point.

If the door is closed by a force at the outer edge, the moment of this force must balance the moment of the gravity force.

$$\text{clockwise moment} = \text{anticlockwise moment}$$
$$F\,\text{N} \times 0.8\,\text{m} = 40\,\text{N} \times 0.4\,\text{m}$$
$$= 16\,\text{Nm}$$

Divide both sides by 0.8 m:
$$F\,\text{N} = \frac{16\,\text{Nm}}{0.8\,\text{m}}$$
$$= 20\,\text{N}$$

The force needed to close the trapdoor is 20 N.

4.8 Stability

The position of the centre of gravity of a body is important in determining how **stable** the body will be. We say that a body is stable if it is difficult to topple over; it is unstable if it topples easily.

Imagine a box being tilted. The reaction force of the ground acts upwards at the corner; the weight acts downwards through the centre of gravity. If the tilt is small, this couple will tend to restore the box to its original position. But if the box is tilted so far that the centre of gravity moves beyond the pivoting edge, then the couple will make the box tilt further, and fall over.

There are two ways to make a body more stable. One is to design it so that the centre of gravity is as low as possible; the other is to give the body as wide a base as possible. The next diagram shows how both of these features allow the object to be tilted further before the centre of gravity is directly above the pivot point. So they can be tilted further before they reach the critical point where they will topple over.

The bus is designed so that most of its mass is low down. As it has a low centre of gravity, it doesn't easily topple over.

The legs of a baby's high chair are set wide apart so that the chair is very stable.

4.9 Types of equilibrium

A drawing pin can be in equilibrium in three different positions.

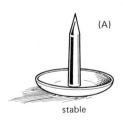

stable

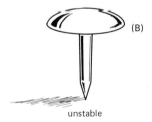

unstable

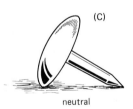

neutral

Pin **A** is said to be in **stable equilibrium**. If it is pushed slightly to the side, it will return to its original position.

Pin **B** is said to be in **unstable equilibrium**. If it is pushed slightly to the side, it will topple over completely. Indeed it would be almost impossible to balance it in position **B**, but theoretically this **is** an equilibrium position.

Pin **C** is said to be in **neutral equilibrium**. If it is pushed slightly to the side, it will stay in its new position. The centre of gravity is always vertically above the point of contact with the table.

An object is in stable equilibrium if a small push to the side makes its centre of gravity *rise*. It is in unstable equilibrium if a small push to the side makes its centre of gravity *fall*.

36 More about Forces: Balance and Stability

Questions

1 (a) An inn sign hangs from its support. What is the upward force exerted by the support?

(b) Two workmen carry a ladder. What is the **sum** of the upwards forces exerted by the workmen?

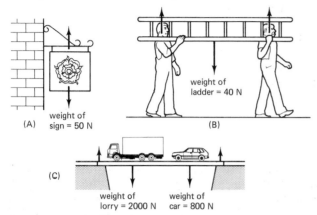

(c) A simple beam bridge is supported by the piers at either end. A lorry and a car are crossing the bridge. What is the **sum** of the two upwards forces exerted by the bridge piers?

2 A mail-order company advertises a spanner with a telescopic handle to make it easier to remove the wheel nuts if you need to change a car wheel. Explain why the extending handle makes the job easier.

3 The table records some observations made by placing coins on either side of a balanced ruler. For each experiment, state which side of the ruler will go down.

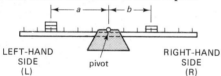

Expt.	LEFT-HAND SIDE		RIGHT-HAND SIDE	
	No. of coins	Distance a/units	No. of coins	Distance b/units
1	3	4	3	3
2	2	5	3	5
3	3	5	4	4
4	2	4	3	3
5	7	3	5	4

4 Diabetics have to be very careful about the amount they eat of certain kinds of food. Some use a simple balance at the table to check the amounts they are about to eat.

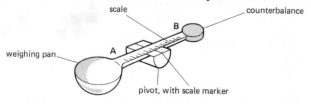

In one such balance, the pan is connected to a counterbalance by a scale which is calibrated (marked with a scale) to show the weights directly. To use it, you slide the scale through the pivot until it balances, then read the weight off the scale.

(a) Explain how the balance works.
(b) At which end of the scale (**A** or **B**) are the larger weight readings?
(c) Harder: The scale markings on a balance like this are not equally spaced. Can you explain why this is so?

5 The diagram shows a tower crane on a building site. You can think of the jib as a lever pivoted at the tower.

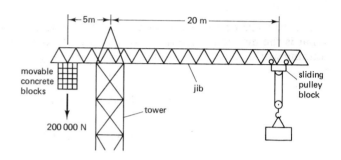

(a) What load could be safely lifted when the sliding pulley block is 20 m from the tower, with the movable concrete blocks in the position shown?
(b) If the crane had to lift a bigger load than this, suggest two courses of action which the crane driver could take.
(c) The safety regulations state that the maximum load which the crane is allowed to lift with the sliding pulley block 20 m from the tower is less than that calculated in (a). Suggest one reason why this is so.

6 Your aunt has a table lamp made from a bottle, into which a lamp holder has been fitted. Unfortunately it is rather unstable and falls over easily. Write a short note to her explaining why the lamp is so wobbly. Tell her how she can make it much more stable by part-filling the bottle with sand, and explain to her in simple terms how this works.

7 These five objects have all been cut from the same wood and all have the same height and maximum width. Which of them is the most stable? Which is the least stable? Explain your answers.

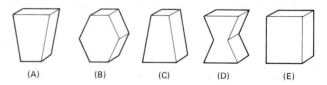

INVESTIGATION

Where is the centre of gravity of Britain?

5: Describing Motion

We all notice things that move. A baby's eyes will follow a moving object. Motion attracts our attention. Questions about motion are among the oldest scientific questions. Physicists have always been interested in how and why things move. Can we find a very small set of rules (or laws) which can explain all the different examples of motion we observe, and will allow us to predict what will happen in new situations? This is what the branch of physics called **mechanics** is all about.

Before we can attempt to **explain** motion, we need to be able to **describe** motion as clearly and accurately as we can. This is where we will make a start – by looking at how we can describe and measure the motion of moving objects.

5.1 Speed

Let us start with one question about motion: How fast can a human move? What is the maximum **speed** at which a human can run? The table of running world records gives some information which can help to answer the question.

MEN	Record	Name and country	Date
100 m	9.83 s	Ben Johnson (Canada)	Aug 1987
200 m	19.72 s	Pietro Mennea (Italy)	Sept 1979
400 m	43.29 s	Butch Reynolds (USA)	Aug 1988
800 m	1 min 41.73 s	Sebastian Newbold Coe (GB)	June 1981
1000 m	2 min 12.18 s	Sebastian Newbold Coe (GB)	July 1981
1500 m	3 min 29.46 s	Said Aouita (Morocco)	Aug 1985
WOMEN			
100 m	10.49 s	Florence Griffith-Joyner	July 1988
200 m	21.34 s	Florence Griffith-Joyner	Sept 1988
400 m	47.60 s	Marita Koch (GDR)	Oct 1985
800 m	1 min 53.28 s	Jamila Kratochvilava (Czech.)	July 1983
1000 m	2 min 30.60 s	Tatyana Providokhina (USSR)	Aug 1978
1500 m	3 min 52.47 s	Tatyana Kazankina (USSR)	Aug 1980

Two quantities are involved: – distance moved,
– time taken.

The speed of a moving object is the distance it travels in each unit of time.

Average speed

Let's work out the speed of an athlete who can run 100 m in 12.5 s. On average, she has covered 8 metres each second ($8 \times 12.5 = 100$).

So her average speed is $\dfrac{100 \text{ m}}{12.5 \text{ s}}$ or 8 m/s.

$$\textbf{Average speed} = \frac{\textbf{distance moved}}{\textbf{time taken}}$$

Why have we called this **average speed** rather than simply **speed**? If we think about the athlete's 100 m sprint in more detail, the reason becomes clear. She starts from rest (0 m/s) and builds up to a maximum speed as she gets into her stride. From then on she will try to maintain a steady speed for the rest of the race. Two things are apparent:

– her initial speed is less than 8 m/s
– so her maximum speed must be greater than 8 m/s

The value 8 m/s represents her average speed.

EXAMPLE 5.1

Peter cycles to school. The distance from his house to the school is 3 km. If the journey takes him 15 minutes, what is his average speed?

Distance travelled = 3 km
Time taken = 15 minutes
= 0.25 hours

$$\text{Average speed} = \frac{\text{distance}}{\text{time}}$$
$$= \frac{3 \text{ km}}{0.25 \text{ h}}$$
$$= 12 \text{ km/h}$$

We could also work out Peter's average speed in metres per second:

Distance = 3 km = 3000 m
Time = 15 minutes = 15×60 s = 900 s

$$\text{Average speed} = \frac{\text{distance}}{\text{time}}$$
$$= \frac{3000 \text{ m}}{900 \text{ s}} = 3.33 \text{ m/s}$$

38 Describing Motion

Instantaneous speed

The speedometer of a car doesn't tell you the car's average speed. The pointer indicates the speed of the car at that **instant** – its **instantaneous speed**. Often we are much more interested in instantaneous speed than average speed.

For example, a driver may drive through a town with an average speed of 30 miles per hour, yet still break the speed limit. His journey may have gone like this:

- travel at 30 m.p.h. along a straight road
- stop at traffic lights
- start again and build up speed to 30 m.p.h.
- overtake another vehicle by building up speed to 40 m.p.h.
- get stopped by the police for speeding!

It won't do the driver any good to plead that his average speed was 30 m.p.h. for the whole journey; it is his instantaneous speed that matters.

5.2 Distance–time graphs

Let's think a little more about the athlete who runs 100 m in 12.5 s. To find her average speed, we needed just two measurements: the length of the track (100 m) and the total time of her run (12.5 s).

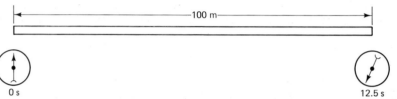

To examine her run in more detail, we could place a camera beside the track to take photographs at 1-second intervals. These photographs show us exactly where the athlete was throughout the run. This information can be summarised in various ways:

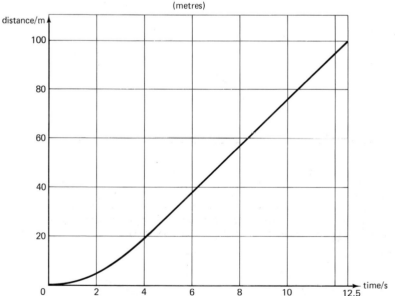

time/s	distance/m
0	0
1	1
2	5
3	11
4	19
5	29
6	38
7	48
8	58
9	67
10	77
11	86
12	95
12.5	100

(to nearest metre)

The distance–time graph presents the information very clearly. It shows how the distance increases slowly at first, then more quickly. From 4 seconds onwards, the distance increases steadily with time (the graph is a straight line). This is what we would expect: the runner gradually speeds up until she reaches her top speed and, if she is fit enough, she will maintain this speed right to the tape.

Can we find the instantaneous speed?

We can find the average speed of a moving object by making one measurement of distance and one of time. But how could we measure its instantaneous speed? The best we can do is to measure the average speed over a very short distance (and also, therefore, over a very short time). For instance, if we wanted to know the athlete's instantaneous speed as she passes the 10 metre mark, we could time her over a short distance at this point: from a mark at 9 metres to another at 11 metres.

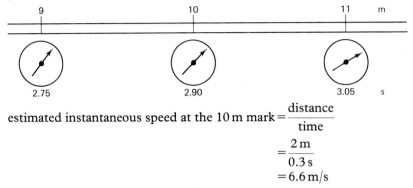

$$\text{estimated instantaneous speed at the 10 m mark} = \frac{\text{distance}}{\text{time}}$$
$$= \frac{2\,\text{m}}{0.3\,\text{s}}$$
$$= 6.6\,\text{m/s}$$

The shorter we make the distance, the better the estimate of instantaneous speed is likely to be. If we could measure the athlete's times at the 9.9 and 10.1 metre marks, the estimate of instantaneous speed would be more accurate; from 9.99 to 10.01 metres would be better still.

Distance–time graph: slope = instantaneous speed

The slope of the distance–time graph tells us the instantaneous speed. (If you are not sure what is meant by the slope of a graph, it is explained more fully in Appendix 1 at the back of the book.) The slope is small at first, gradually increases and then settles down to a steady value after 4 seconds. In the same way, the athlete's speed is small at first, gradually increases and then settles down to a steady value after 4 seconds.

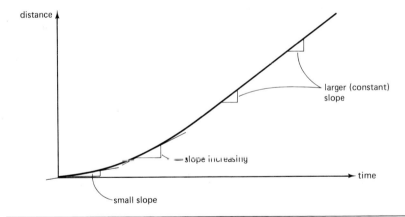

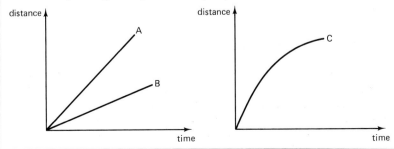

Interpreting distance–time graphs. Graph **A** is for an object moving at a steady (or constant) speed. The speed of object **B** is also constant but it is slower than **A**. Object **C** starts off moving quickly, but gradually slows down as time goes on. The slope is large to begin with and gets gradually smaller.

5.3 Measuring speed in the laboratory

1. Ruler and stop-watch

This is the simplest method:

◁ We time the object over a measured distance using a stop clock.

Then calculate: average speed $=\dfrac{\text{distance}}{\text{time}}$

The method is inaccurate for short time intervals because it is impossible to start and stop the watch at exactly the right moment. However, for intervals of 5 s or more, it can be quite satisfactory.

2. Ticker timer

Ticker timers give us a permanent record of the motion which we can then study at leisure afterwards.

◁ A ticker-timer makes a series of dots on a piece of tape as it is pulled through the timer. The time interval between one dot and the next is $\frac{1}{50}$ s (or 0.02 s). From the marks on the tape we can work out the speed.

First, find the part of the tape which corresponds to the bit of the motion you are interested in. Then, count a certain number (in this case 10) of spaces-between-dots. Mark this off. The time taken for this motion is 0.02 s × number of spaces. Measure the total length of the marked piece of tape with a ruler to find the distance moved. Then divide distance (length) by time to find the average speed.

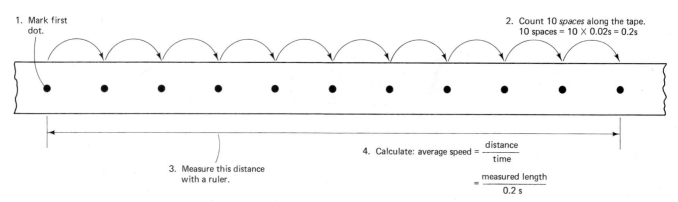

1. Mark first dot.
2. Count 10 *spaces* along the tape. 10 spaces = 10 × 0.02s = 0.2s
3. Measure this distance with a ruler.
4. Calculate: average speed $=\dfrac{\text{distance}}{\text{time}}$ $=\dfrac{\text{measured length}}{0.2 \text{ s}}$

Note: We could use 5 spaces, or 2 spaces, or just 1 space — rather than 10.

We know:	
1 space	0.02 s
2 spaces	0.04 s
5 spaces	0.1 s
10 spaces	0.2 s
etc.	

Strictly speaking, it is the average speed which we measure from the ticker-tape record. However, the average speed of a moving object during a $\frac{1}{50}$ s time interval is a good estimate of its instantaneous speed.

Typical ticker tapes

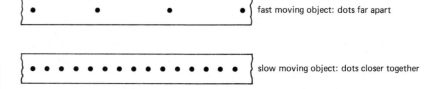

fast moving object: dots far apart

slow moving object: dots closer together

3. Strobe photographs

◁ A stroboscope shows us a series of glimpses of a moving object at regular intervals. We don't see it all the time. A xenon flashing-light stroboscope illuminates the moving object with a series of brief flashes of light. Each flash lasts a very short time (as little as one ten-thousandth of a second); the time interval between the flashes can be varied by turning a knob on the stroboscope.

The other stroboscope shown in the photograph is a rotating disc with a number of narrow slits cut round its rim. As we look through the slits, we see the moving object each time a slit is in front of our eye.

A strobe photograph can be taken using either form of stroboscope. With the flashing-light type, the experiment is carried out in a darkened room. The camera is mounted on a tripod, the moving object is lit by the stroboscope only, and the camera shutter is kept open. The photograph shows where the moving body is each time the light flashes. If we include a scale in the photograph, we can work out the speed of the moving object.

Again, in this method we are really measuring average speed, but since the time interval can be made quite small, it provides a good estimate of instantaneous speed.

4. Electronic timer

One of the problems of using a stop-watch to measure a short time interval is switching the stop-watch on and off at exactly the right moment. There is a **reaction time** between seeing an event and being able to respond to it. For accurate timing, we need an automatic method of starting and stopping the stop-watch.

Many stop-watches and electronic timers can be started by making a connection between two START terminals and stopped again by connecting two STOP terminals.

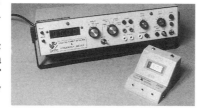

A **photodetector** (a photodiode or a phototransistor or a light-dependent resistor) can be used to switch a timer on and off. When light shines on the photodetector, the timer is OFF; when the light beam is interrupted, the timer is ON.

We can use this to measure the speed of a laboratory trolley (see above right). A card is mounted on top of the trolley. As it moves past the photodetector, the card interrupts the beam, switching the timer ON and OFF again.

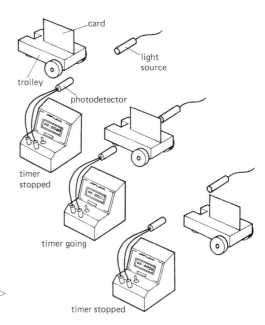

5. Using a microcomputer

Most microcomputers have an internal clock, and so they can be used for timing experiments. The timer must be switched on and off automatically and a photodetector switch is a good way of doing this. The photodetector switch is connected to the User Port of the microcomputer. (Details of a suitable circuit are given in Appendix 2 at the end of the book, together with a programme for measuring speeds in this way using a BBC microcomputer.) When light shines on the photodetector, the computer timer is OFF; when the beam is interrupted, the timer is ON.

The computer has one advantage over the automatic stop-watch method. If you tell it how long the card on top of the trolley is, then the calculation of speed (dividing distance by time) can be done by the computer and the speed appears directly on the screen.

Although the calculation is done by the computer automatically, this doesn't make the result any more accurate. In fact, it will be no more accurate than the automatic timer method.

A similar method can be used to measure times and speeds with VELA.

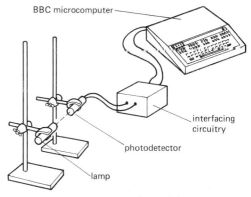

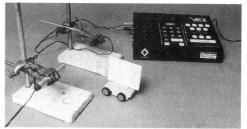

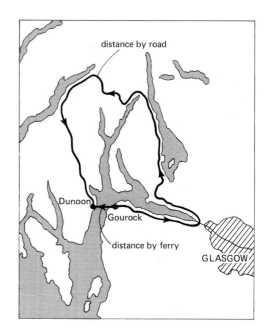

5.4 Distance and displacement

Measuring speed involves measuring **distance**. But what exactly do we mean by **distance**?

Take a look at the map. How far is it from the town of Gourock to Dunoon in Scotland? There are two possible answers to this question. If you go by car, then you will have to follow the road and the distance will turn out to be 98 miles. But if you have a boat, or if you choose to take the ferry, you will go by the direct straight-line route – a distance of 4 miles.

So there are two different meanings to the term 'distance from Gourock to Dunoon'! In physics we have two different words for these two distinct meanings.

If an object moves from **P** to **Q** along the dotted path, the **distance** it has moved is defined as the length of the actual path it has followed. Its **displacement**, on the other hand, is represented by the straight line from **P** to **Q**. Unlike the distance, the displacement has a direction. To describe a displacement fully, this direction should always be given. So, in the diagram, the displacement is 12 m at 60° east of north; the distance is simply 20 m.

Quantities which have a direction as well as a size (or magnitude) are called **vector quantities** (or **vectors**). Force is a vector we have already met. **Displacement** is another vector. Quantities which have a magnitude only are called **scalar quantities** (or **scalars**). **Distance** is a scalar.

5.5 Speed and velocity

We have defined average speed by the equation:

$$\text{average speed} = \frac{\text{distance moved}}{\text{time taken}}$$

But now that we have made a distinction between distance and displacement, there is a second, rather similar, quantity which we could also define:

$$\frac{\text{displacement}}{\text{time taken}}$$

This is given a different name; it is called **average velocity**.

$$\textbf{average velocity} = \frac{\textbf{displacement}}{\textbf{time taken}}$$

Like speed, velocity is measured in metres per second.
Average velocity is a vector, in the same direction as the displacement.

EXAMPLE 5.2

Unlike the 100 m race we considered earlier, the 200 m is usually run on a curved section of track. An athlete runs a 200 m race in 25 s on the track shown in the diagram.

What is:
(a) his average speed,
(b) his average velocity?

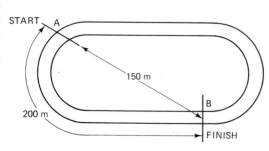

The athlete runs from A to B in 25 s. This is a distance of 200 m.

So, his average speed is $\frac{1200 \text{ m}}{25 \text{ s}}$

$= 8 \text{ m/s}$

But the total displacement of the athlete when he reaches **B** is not 200 m. It is the vector $\overrightarrow{AB}$: 150 m in the direction $\overrightarrow{AB}$.
So his total displacement is
$\qquad$ 150 m in direction $\overrightarrow{AB}$
Therefore his average velocity

$= \frac{150 \text{ m}}{25 \text{ s}}$ in direction $\overrightarrow{AB}$

$= 6 \text{ m/s}$ in direction $\overrightarrow{AB}$

As in Example 5.2, the average velocity is often quite different in size from the average speed. It also differs in being a vector, not a scalar.

However, the instantaneous speed and instantaneous velocity of a moving object are always equal in size. The instantaneous velocity differs simply in having a direction as well as a magnitude.

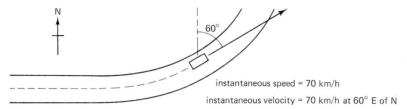

Velocity gives us more information than speed because it also tells us about direction.

5.6 Changing speed and changing velocity

When a moving object changes its speed, we say that it **accelerates**. When traffic lights change to green, a car will accelerate away. Its speed is increasing.

If the car later has to brake and slow down, this is also an acceleration – a negative acceleration this time. For a car moving along a straight road, a positive acceleration means that it is speeding up; a negative acceleration means that it is slowing down.

However, a moving object can change its motion in another way. Imagine a stone tied to a string being whirled round in a circle at a steady speed.

The stone's speed is the same at **P** and **Q**. But its velocity at **Q** is **not** the same as its velocity at **P** because the *direction* of motion has changed. Speed is constant but velocity is changing.

In this situation we also say that the stone is accelerating. An object is accelerating if: its speed is changing
      OR its direction of motion is changing
      OR both.

In other words, **an object is accelerating if its velocity is changing**. There is a good reason for defining acceleration in this way. As we shall see later (in chapters 6, 7 and 8), a force produces an acceleration. If there is a force, there will be an acceleration. If there is an acceleration, there must be a force causing it. But we cannot say the same thing about a change-of-speed. It is true that to cause a change-of-speed, there must be a force. But when there is a force, there may or may not be a change-of-speed. A force (the tension in the string) is needed to keep the stone flying round in a circle at a steady speed. Without this tension force, it would simply fly off (think what happens if you let go!). We can summarise this in the following way:

If there is a $\begin{cases} \text{change of speed,} \\ \text{change of velocity,} \end{cases}$ there MUST be a force causing it.

If there is a FORCE, $\begin{cases} \text{there MUST be a change of velocity.} \\ \text{there might (or might not) be a change of speed.} \end{cases}$

Force and change of velocity always go together. This is why velocity is a useful idea, and why we say there is an acceleration when **velocity** changes.

5.7 Acceleration

Apart from price, what is the difference between these two cars?

Both are capable of cruising at 70 m.p.h. which is the maximum permitted speed on our roads. One advantage of the Turbo version is its superior acceleration. The acceleration of a car is a measure of how quickly it can speed up. Performance figures from the manufacturers tell us how many seconds their new model needs to reach a speed of 60 m.p.h. from rest. The car on the right takes 17.5 s, while the Turbo version on the left accelerates from 0 to 60 m.p.h. in only 9.9 s.

If we want to measure acceleration in a laboratory expcriment, we need to know the change of speed in a certain time. That statement is almost correct. More exactly, we have an acceleration when the **velocity** changes. We define acceleration as the rate of change of velocity (and NOT as the rate of change of speed).

$$\text{average acceleration} = \frac{\text{change of velocity}}{\text{time taken for the change}}$$

What units would we therefore use to measure acceleration? In the car advertisement the average acceleration would be measured in miles per hour per second! That is to say, the average acceleration is the number of miles per hour by which the car speeds up each second. In SI units acceleration is measured in metres per second per second (or m/s^2 for short).

An object has uniform acceleration if it has equal changes of velocity in equal intervals of time.

Speedometer readings – at equal time intervals – as a car speeds up. The car's acceleration is **uniform**.

EXAMPLE 5.3

A lorry starts from rest along a straight road and reaches a speed of 4 m/s after 5 s.
(a) What is its average acceleration?
(b) If its acceleration is uniform, what is its velocity after 1 s, 2 s, 3 s, 4 s?

Since the journey is along a straight road, the speed and velocity are the same size. Velocity has a direction – 'the direction the lorry is travelling' – but to avoid repetition we don't write this down each time.

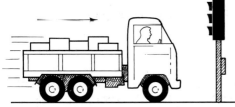

(a) Average acceleration $= \dfrac{\text{change of velocity}}{\text{time taken}}$

$= \dfrac{4 \, m/s}{5 \, s}$

$= 0.8 \, m/s^2$

(b) The lorry has uniform acceleration of $0.8 \, m/s^2$. That means that its velocity increases by 0.8 m/s each second.
So, after 1 s its velocity is 0.8 m/s
after 2 s its velocity is 1.6 m/s
after 3 s its velocity is 2.4 m/s
after 4 s its velocity is 3.2 m/s

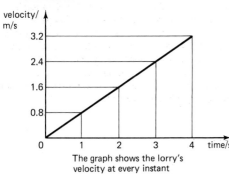

The graph shows the lorry's velocity at every instant

5.8 Velocity–time graphs

Slope and acceleration

We have seen on p. 38 how a distance–time graph is a useful method of summarising the motion of an object. A **velocity–time** graph is another way of recording the same information. It shows the velocity of the moving object at every instant during its motion.

EXAMPLE 5.4

Here is the velocity–time graph for a car accelerating away from traffic lights.

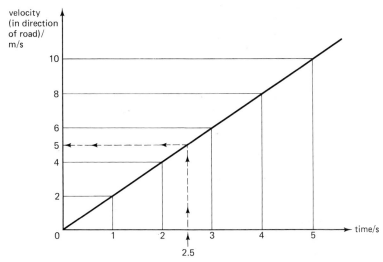

(a) What is its velocity after 2.5 seconds?
(b) What is the average acceleration of the car?

(a) We can read the velocity after 2.5 s straight off the graph: it is 5 m/s in the direction of motion.

(b) During the first 5 seconds, the car accelerates from 0 m/s to 10 m/s.
  Change of velocity $= 10 - 0 = 10$ m/s
  $$\text{Average acceleration} = \frac{\text{change of velocity}}{\text{time}}$$
  $$= \frac{10 \text{ m/s}}{5 \text{ s}}$$
  $$= 2 \text{ m/s}^2$$

The acceleration is uniform (the graph is a straight line). So we can simply say that the car's acceleration is 2 m/s^2.

The slope of a velocity–time graph is a measure of the acceleration.

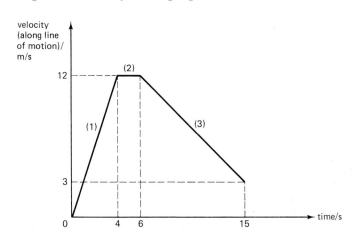

Interpreting a velocity–time graph: During stage (1) of the motion, the object accelerates uniformly from 0 m/s to 12 m/s in 4 s. So its acceleration is 3 m/s^2. During stage (2) there is no change of velocity. The acceleration is zero. During stage (3), the velocity drops from 12 m/s to 3 m/s in 9 s. The acceleration is -1 m/s^2. A negative value for acceleration means a deceleration – a slowing down. It is an acceleration in the opposite direction to the object's initial velocity.

The area under a velocity–time graph is always equal to the displacement, no matter how the velocity is changing. We can see this if we imagine dividing the motion up into a series of short stages, so short that we can regard the velocity as constant during each stage.

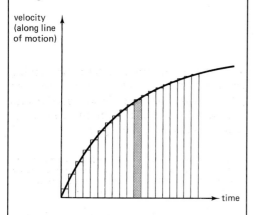

During each little time interval, the velocity is constant. The displacement in this brief time interval is equal to the area of the shaded rectangle. If we add up all these little rectangles we will have a measure of the total displacement. But, of course, the sum of the areas of the rectangles is just the total area under the graph. The area under the velocity–time graph is a measure of the total displacement in this case also.

Area under the graph

The area under a velocity–time graph is equal to the displacement of the moving object. It is easy to see this when the body has constant velocity.

For example, imagine a cyclist travelling at a constant velocity of 5 m/s for 30 s. She would travel 150 m (5 m/s × 30 s). The total displacement is 150 m along the direction of motion.

The area under velocity–time graph (A) is a rectangle whose sides are 5 units and 30 units. The area of this rectangle is 5 × 30 = 150 units. (The units should be m/s and s, which multiply to give m.) This is equal to the displacement.

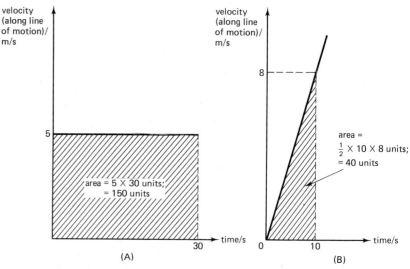

Does this still work if the velocity is changing? Velocity–time graph (B) is for a bus accelerating away from a stop. After 10 s it has reached a velocity of 8 m/s. Since it has speeded up steadily from 0 m/s to 8 m/s, we can say that its average velocity during this interval was 4 m/s. So in 10 s, it will travel 40 m. The area under the velocity–time graph is a triangle with base 10 units (s) and height 8 units (m/s). Its area is $\frac{1}{2} \times 10 \times 8$ units, or 40 units (s × m/s, or m). Again this is equal to the displacement.

The area under a velocity–time graph is equal to the displacement of the moving object.

EXAMPLE 5.5

While travelling along the road at 25 m/s, a car driver suddenly sees the road ahead blocked. He applies the brakes as fast as he can. His reaction time is 0.2 s. Once the brakes are applied, the car decreases uniformly in speed until it stops after a further 2.0 s.
(a) How far does the car travel during the driver's 'thinking time'?
(b) How far does the car travel during the 'braking period'?
(c) What is the total distance travelled after the driver sees the obstruction?

The 'reaction time' is the interval between noticing an event and actually reacting to it. During this 'reaction time' the car will continue travelling at 25 m/s. The velocity–time graph for the car will look like this:

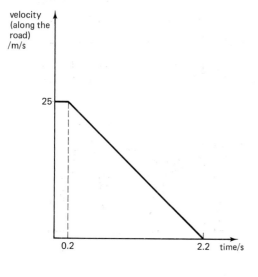

(a) Distance travelled during the driver's 'reaction time'

= area under graph between 0 s and 0.2 s
= 25 m/s × 0.2 s
= 5 m

(b) Distance travelled during 'braking period'

= area under graph between 0.2 s and 2.2 s
= $\frac{1}{2}$ × 25 m/s × 2.0 s
= 25 m

(c) Total distance travelled after seeing the obstruction: 5 + 25 m = 30 m

Velocity–time graphs in action: the tachograph

All lorries and buses in the EC are required by law to have a tachograph installed. This makes a complete record of the motion of the vehicle and can be used to check that the driver keeps to the rules about maximum speed and maximum period at the wheel without a break. The information is recorded on a special tachograph disc. This turns very slowly, one complete revolution in 24 hours. As it turns, a pen automatically records the vehicle's speed on the disc. What we are left with is a speed–time graph for the vehicle. It is circular, but we can easily imagine it straightened out to give the normal sort of speed– (or velocity–) time graph.

If we inspect the trace under a magnifying glass, we can tell quite a lot about how the vehicle was driven. Gentle acceleration and braking is much more economical and safer than sudden changes of speed.

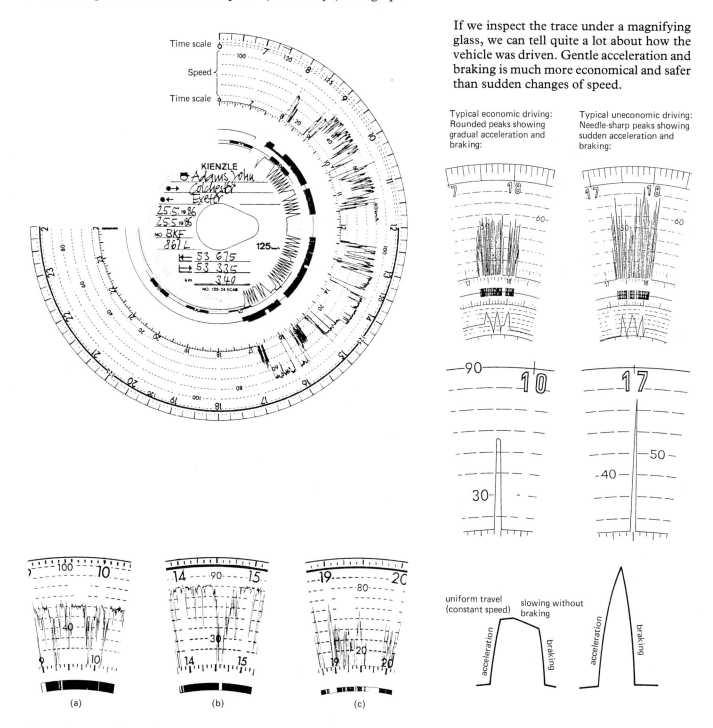

These three tachograph traces were recorded: on country roads, on a motorway, around town. Can you tell which is which by looking at the speed–time graphs? (See p. 56 for answers.)

48 Describing Motion

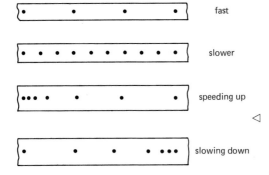

fast

slower

speeding up

slowing down

5.9 Measuring acceleration

If an object travels in a straight line in one direction, its speed and velocity are the same. We do not need to bother about the distinction. Some of the methods we used for measuring speed can also be used for measuring acceleration.

1. Ticker-timer

From a ticker-tape record it is easy to see if the object which produced the tape was accelerating. If the dots get gradually further apart, the object is moving faster; if they become gradually closer together, it is slowing down.

To measure the acceleration, we need to work out the velocity at two points and the time interval between them. ▽

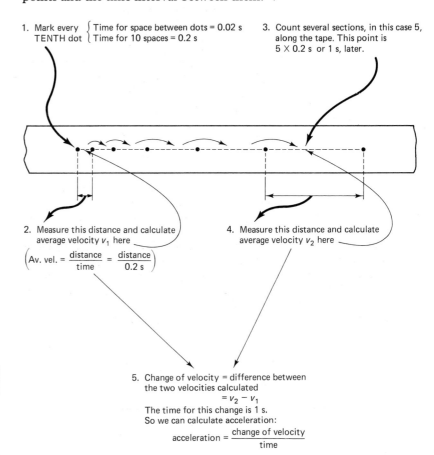

1. Mark every TENTH dot { Time for space between dots = 0.02 s / Time for 10 spaces = 0.2 s

3. Count several sections, in this case 5, along the tape. This point is 5 × 0.2 s or 1 s, later.

2. Measure this distance and calculate average velocity v_1 here

$$\left(\text{Av. vel.} = \frac{\text{distance}}{\text{time}} = \frac{\text{distance}}{0.2 \text{ s}}\right)$$

4. Measure this distance and calculate average velocity v_2 here

5. Change of velocity = difference between the two velocities calculated
 = $v_2 - v_1$
The time for this change is 1 s.
So we can calculate acceleration:

$$\text{acceleration} = \frac{\text{change of velocity}}{\text{time}}$$

2. Strobe photography

The method of calculating acceleration from a strobe photograph is almost exactly the same as that used for ticker-tape. We need to know the number of strobe flashes every second. The quick method is to calculate two velocities and the time interval between them. A more complete method is to work out the average velocity between each pair of images and calculate the acceleration from this.

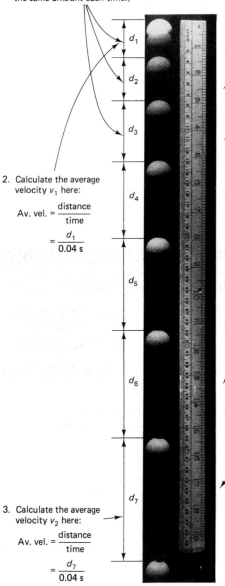

Strobe: 25 flashes per second
Time between flashes = 0.04 s

1. Use the scale in the photograph to work out these distances. (Useful check: they should increase by the same amount each time.)

2. Calculate the average velocity v_1 here:

$$\text{Av. vel.} = \frac{\text{distance}}{\text{time}}$$

$$= \frac{d_1}{0.04 \text{ s}}$$

3. Calculate the average velocity v_2 here:

$$\text{Av. vel.} = \frac{\text{distance}}{\text{time}}$$

$$= \frac{d_7}{0.04 \text{ s}}$$

4. The time for the change of velocity from v_1 and v_2 is 6 intervals = 6 × 0.04 s = 0.24 s

5. Work out the acceleration:

$$\text{acceleration} = \frac{\text{change of velocity}}{\text{time taken}} = \frac{v_2 - v_1}{0.24 \text{ s}}$$

3. Using a microcomputer

With a microcomputer, we can measure acceleration directly. The computer timer is used to measure times, and from these the velocities and accelerations can be calculated.

In a typical experiment, the computer timer is switched on and off by a photodetector switch. The timer is OFF while light shines on the photodetector and ON when the light beam is interrupted. To measure acceleration we need to find two velocities and the time interval between them. The card attached to the moving object has to cut the light beam twice:

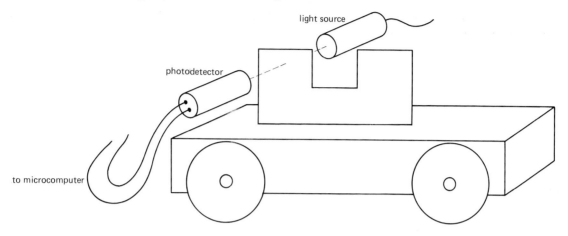

The photodetector switch is OFF at the beginning since the light beam is shining on it. As the card passes, it is switched ON, then OFF, then ON, and finally OFF again. The computer measures the two periods when the switch is ON and the time between these.

If the computer is given the length of the two pieces of card, it can work out two velocities for the moving vehicle. These will be different if the vehicle is accelerating. The change in velocity is divided by the time taken, to find the acceleration. The value for acceleration appears on the screen. (A suitable programme for the BBC microcomputer is listed in Appendix 2. VELA, if fitted with the physics EPROM, can be used in the same way).

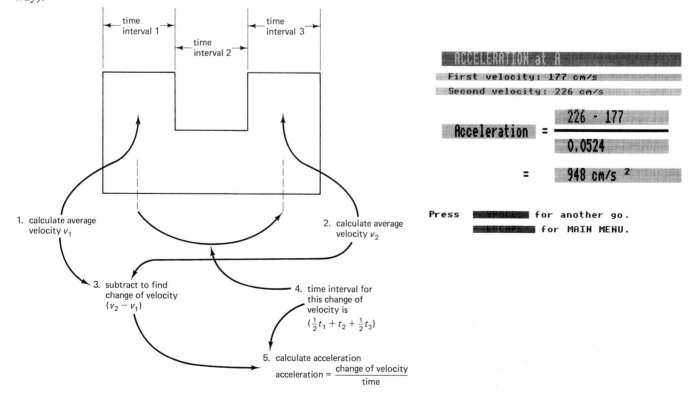

50 Describing Motion

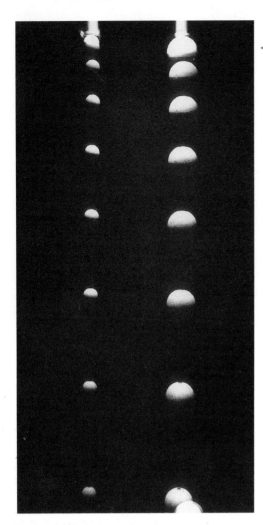

5.10 Free fall

◁ What is the motion of a freely falling object? The strobe photograph gives us two useful pieces of information:

- A falling object accelerates – the gaps between the ball's positions get bigger as it falls.
- Different objects fall with the same acceleration – the two balls stay side by side throughout their fall. (This is true for all compact objects. Large light objects, like feathers, falling through air are an exception. The reasons for this will be discussed later in Chapter 7.)

The acceleration of a freely falling object is called the **acceleration due to gravity**. It is usually given the symbol g. It can be measured in the laboratory by any of the methods for measuring acceleration. A falling object can pull a tape through a ticker-timer. A value for g can be calculated from a strobe photograph. To measure g by a microcomputer method, a metal (or heavy card) shape is dropped between a photodetector switch and a light.

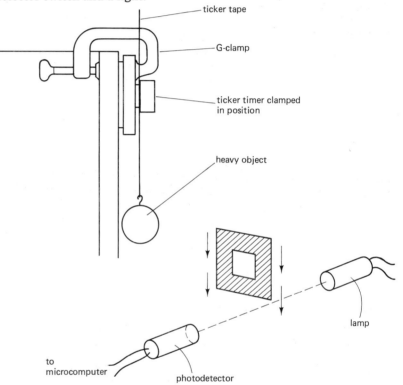

If the measurement is carried out very accurately, the value of g is found to be $9.81 \, m/s^2$. For most purposes, the value $10 \, m/s^2$ is accurate enough.

Any freely falling object increases its speed by $10 \, m/s^2$ every second; any object thrown vertically upwards will slow down by $10 \, m/s^2$ each second until it reaches the top of its flight. This is what a downward
◁ acceleration of $10 \, m/s^2$ means.

5.11 Equations of motion

Graphs are very useful for summarising information about the motion of an object and for doing calculations. Sometimes, however, we can get results more quickly by using equations. We will show how some equations of motion can be derived, though it is usually enough to be able to use the equations correctly. These equations apply to any object moving with uniform acceleration along a straight-line path.

If a moving object changes direction but continues to move in the same straight line (e.g. a ball thrown up and coming back down again), we can still use the equations and take account of this by calling all displacements and velocities in one direction positive (+) and in the other direction negative (−). We usually choose the positive direction as the direction in which the object begins to move.

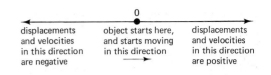

Deriving the equations

Let us consider an object which starts at velocity u and accelerates to velocity v after t seconds. Its velocity–time graph would look like this: From this graph we can work out two other quantities:

the acceleration — it is equal to the slope of the graph;
the displacement in time t — it is equal to the area under the graph.

Acceleration = slope of graph $\quad a = \dfrac{BC}{AC} = \dfrac{v-u}{t}$

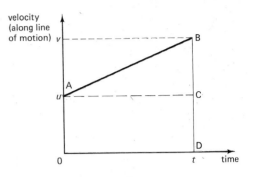

Note. This result is really just a way of writing, in symbols, the definition of acceleration:

$$\text{acceleration} = \frac{\text{change of velocity}}{\text{time taken}}$$

Multiply both sides by t: $\quad at = v - u \quad$ (1)
Add u to both sides: $\quad u + at = v \quad$ (2)
Rearrange: $\quad \boxed{v = u + at} \quad$ (3)

Displacement = area under graph
s = area OACD + area ABC
$= ut + \tfrac{1}{2}t(v-u) \quad$ (4)

But, in equation (1) above, $(v-u)$ is equal to at. So we substitute this into equation (4) to get:

$s = ut + \tfrac{1}{2}t(at)$
$\boxed{s = ut + \tfrac{1}{2}at^2} \quad$ (5)

From the graph we can see what the two parts ut and $\tfrac{1}{2}at^2$ in this formula 'mean'. ut is the distance the body would have travelled if it had simply kept moving at velocity u for the whole time interval t. $\tfrac{1}{2}at^2$ is the extra distance it has travelled as a result of its acceleration during the interval t.

We can obtain a third useful equation from (1) and (5) by eliminating t:

Divide both sides of equation (1) by a: $\quad t = \dfrac{v-u}{a}$

Substitute this into equation (5):

$$s = u\left(\frac{v-u}{a}\right) + \tfrac{1}{2}a\left(\frac{v-u}{a}\right)^2 = \frac{uv - u^2}{a} + \frac{\tfrac{1}{2}(v^2 - 2uv + u^2)}{a}$$

Multiply both sides by $2a$: $\quad 2as = 2uv - 2u^2 + v^2 - 2uv + u^2$
$\quad 2as = v^2 - u^2$
Add u^2 to both sides: $\quad \boxed{v^2 = u^2 + 2as} \quad$ (6)

Finally, there is one other simple equation which we can derive from the velocity–time graph.

The area under graph (**A**) is the same as that under the corresponding graph for a body whose velocity is constantly $\left(\dfrac{u+v}{2}\right)$ throughout the interval: graph (**B**). We can see this simply by looking at the two graphs. This means that if a body accelerates uniformly from velocity u to velocity v in time t, its average velocity during this interval is $\left(\dfrac{u+v}{2}\right)$.

$$\boxed{\text{average velocity} = \frac{u+v}{2}}$$

The precise moment at which it is travelling at an instantaneous speed of $\dfrac{u+v}{2}$ is at $\dfrac{t}{2}$ seconds, i.e. at the mid-point of the time interval.

Summary: equations of motion

The following equations apply to any object moving with uniform acceleration in a straight line. The object has initial velocity u and its acceleration is a. After a time t, its displacement is s and its velocity is now v.

Initial velocity = u $\qquad v = u + at$
Final velocity = v $\qquad s = ut + \tfrac{1}{2}at^2$
Time interval = t $\qquad v^2 = u^2 + 2as$
Displacement = s $\qquad$ Average velocity
Acceleration = a $\qquad = \dfrac{u+v}{2}$

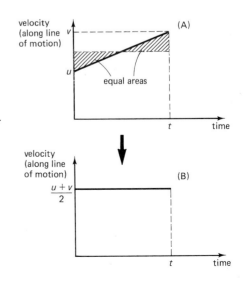

Let us now see a few examples of the equations of motion in action.

EXAMPLE 5.6

A motorist is travelling along a motorway with a steady velocity of 26 m/s. She pulls out and accelerates at 2 m/s² to pass a lorry. Overtaking takes 6 s.

(a) What is her speed after this time?
(b) How far has she travelled during overtaking?

From the information we are given: $u = 26$ m/s
$a = 2$ m/s²
$t = 6$ s

(a) $v = u + at$
$= 26 \times (2 \times 6)$ m/s
$= 26 + 12$ m/s
$= 38$ m/s

So, after 6 s, her velocity is 38 m/s.

(b) $s = ut + \tfrac{1}{2}at^2$
$= (26 \times 6) + (\tfrac{1}{2} \times 2 \times 36)$ m
$= 156 + 36$ m
$= 192$ m

The distance travelled is 192 m.

EXAMPLE 5.7

A cyclist accelerates uniformly from rest at 2 m/s² along a straight road.
(a) Sketch a velocity–time graph for the first 5 seconds of his motion.
(b) Calculate his displacement after 1, 2, 3, 4 and 5 seconds, and hence plot a displacement–time graph of the motion.

(a) See Graph (A) below.
(b) To calculate displacement s after 1 s: $u = 0$
$t = 1$ s
$a = 2$ m/s²

$s = ut + \tfrac{1}{2}at^2$
$= 0 + (\tfrac{1}{2} \times 2 \times 1)$ m
$= 1$ m

If we repeat this calculation for t = 2, 3, 4 and 5 s, we obtain these results:

Time t (s)	1	2	3	4	5
Displacement s (m)	1	4	9	16	25

So the displacement–time graph (B) is:

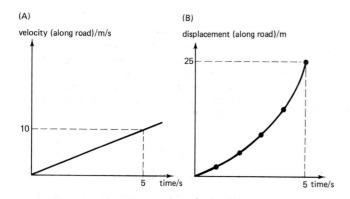

Notice that the shape of the displacement–time graph for something travelling with steady acceleration is a curve – a parabola.

EXAMPLE 5.8

An InterCity 125 train is cruising at 30 m/s. On reaching a long straight section of track, the driver begins to accelerate gently at 0.1 m/s². How far will the train travel before it reaches 40 m/s?

From the information we are given: $u = 30$ m/s
$a = 0.1$ m/s²
$v = 40$ m/s

$v^2 = u^2 + 2as$
$40^2 = 30^2 + (2 \times 0.1 \times s)$
$1600 = 900 + (0.2 \times s)$

Subtract 900 from both sides:
$1600 - 900 = 0.2 \times s$

Divide both sides by 0.2:
$s = \dfrac{1600 - 900}{0.2}$ m $= \dfrac{700}{0.2}$ m
$= 3500$ m

The train will travel a distance of 3500 m (3.5 km) before reaching a speed of 40 m/s.

EXAMPLE 5.9

A laboratory method of measuring the acceleration due to gravity g involves timing the free fall of a steel ball-bearing using an electronic timer. The ball drops when an electromagnet is switched off; this also starts the timer. When the ball strikes the hinged plate at the bottom, the timer stops.

In a particular experiment, the ball is allowed to drop a distance of 0.8 m. The timer records 0.42 s. What value does this give for g?

From the information we are given:

$u = 0$
$s = 0.8$ m
$t = 0.42$ s
$a = ?$

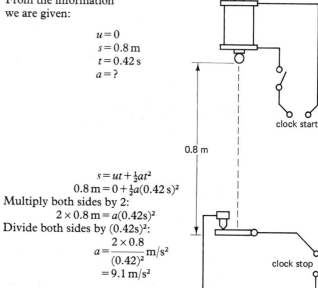

$s = ut + \tfrac{1}{2}at^2$
0.8 m $= 0 + \tfrac{1}{2}a(0.42\text{ s})^2$

Multiply both sides by 2:
2×0.8 m $= a(0.42\text{s})^2$

Divide both sides by $(0.42\text{s})^2$:
$a = \dfrac{2 \times 0.8}{(0.42)^2}$ m/s²
$= 9.1$ m/s²

This experiment gives a value of 9.1 m/s² for the acceleration due to gravity.

(**Note.** The most likely reason why this result differs from the more accurate value of 9.8 m/s² is that the electromagnet may not lose its magnetism immediately it is switched off. So the ball doesn't fall at once. This delay means that the time measured is slightly longer than it should be.)

5.12 Adding displacements

In section 5.4, we made a distinction between distance and displacement:

Distance is the length of the actual path a moving object follows.
Displacement is the length (and direction) of the straight line from its starting point to its finishing point.

If a journey is in several sections, how do we add distances and displacements?

Think of a group of hill-walkers setting off from their starting point at **A** and walking northwards along a path for 4 km to **B**. Then they turn eastwards and walk another 3 km to **C**, where they stop for a rest. How far have they gone?

The total **distance** they have travelled is 7 kilometres. Their total **displacement** is represented by the straight line from **A** to **C**. It is 5 kilometres in a direction 37 degrees east of north. We have added the displacements by vector addition.

In general, if we want to add two displacements, we do it by drawing vector diagrams. The sum of the displacement $\vec{PQ}$ and the displacement $\vec{QR}$ is the displacement $\vec{PR}$. By drawing the diagram to scale, we can work out what this sum is. The method is the same as the one we used in section 3.8 for adding forces.

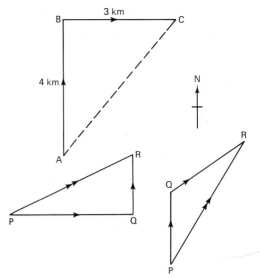

5.13 Adding velocities

Velocity is also a vector. Let us see how vector addition can be used to add two velocities.

Imagine that the load on a large building-site crane is being lowered steadily at 0.5 m/s, whilst the trolley which is carrying it is moving outwards along the beam of the crane at 1.2 m/s. The load has two separate velocities at the same time.

To calculate its total velocity, we draw a vector diagram, taking care that the two velocity vectors are drawn tip-to-tail. The total velocity is the vector sum. In this case it is 1.3 m/s in a direction at 67.4° to the vertical.

We could also do this by adding displacements. In 1 second, the load will move 0.5 m down and 1.2 m across. The sum of these displacements is 1.3 m at 67.4° to the vertical. If the load moves 1.3 m in one second, then its velocity is 1.3 m/s in this direction.

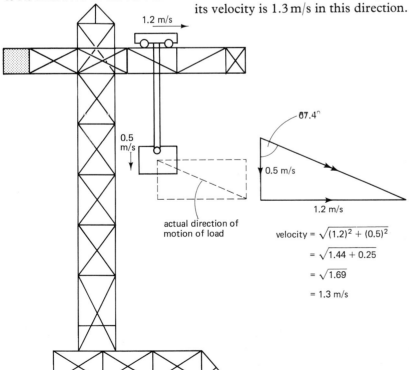

Resolving into components

To find the velocity of a crane's load, we added the vertical and horizontal velocities by vector addition. However, from the point of view of the crane driver the problem is almost exactly the opposite one! He knows where he has to move the load to, but he can only achieve this by moving it horizontally and vertically. He must break the motion up into a horizontal part and a vertical part.

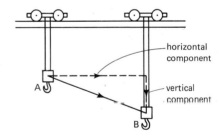

These are the **components** of the motion. The total displacement has a vertical component and a horizontal component. The total velocity of the load can also be split into a horizontal and a vertical component. When we do this, we are **resolving** the vector into components.

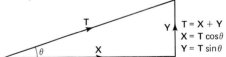

$T = X + Y$
$X = T \cos\theta$
$Y = T \sin\theta$

The vector **T** is equivalent to the two vectors **X** and **Y**. **X** and **Y** are the horizontal and vertical components of **T**. **X** is equal to $T \cos\theta$, and **Y** is equal to $T \sin\theta$.

5.14 Relative velocity

Strictly speaking, when we say that an object has a certain velocity, we should always say what this velocity is **relative to**. The reason for this is that it is impossible to tell whether any object is **really** moving; we can only tell whether it is moving **relative to** something else! Have you ever been sitting in a train which is stopped alongside another train at a station, and not been sure which train has started to move? All the time, all we can be sure of is that the trains are moving relative to each other.

Imagine a train passing through a station at a steady speed of 5 m/s. A passenger is walking along the corridor of the train at 2 m/s in the same direction as the train is moving.

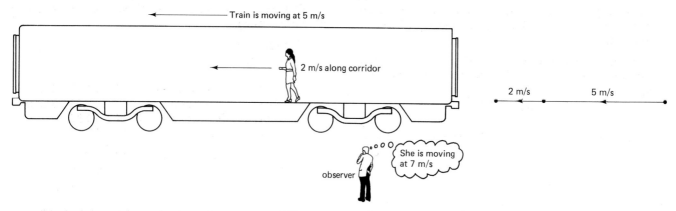

The passenger has velocity 2 m/s relative to the train. The train has velocity 5 m/s relative to the observer on the platform. The velocity of the passenger relative to the observer on the platform is 7 m/s in the direction of motion of the train. What we are really doing here is adding vectors.

If the passenger now turns and walks back in the opposite direction, again at 2 m/s, her velocity relative to the man on the platform is now 3 m/s. Again, this follows from vector addition.

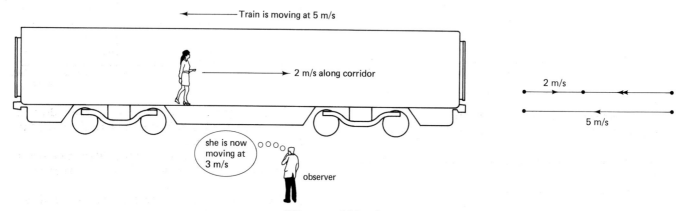

What would be the velocity of the passenger relative to the observer if she walked at 2 m/s across the carriage? This would once again be calculated by vector addition. Her velocity relative to the observer is 5.2 m/s in the direction $\vec{AC}$.

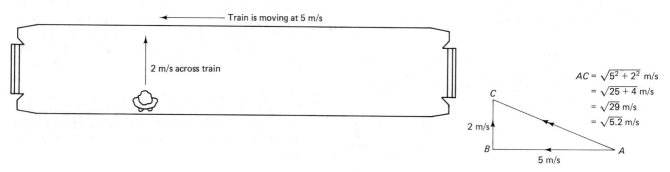

Questions

1. In a race, a swimmer covers 100 m in 1 minute 40 seconds. What is her average speed (in metres per second)?

2. How far could you cycle in 10 minutes at an average speed of 8 m/s?

3. The average speed of a bus on one particular route is 15 m/s. If the route is 3000 m (3 km) long, what time will the bus take to complete it?

4. Use the information in the table on p. 37 to calculate the average speed of each of the world athletics record-holders listed. (Note: a calculator will help in doing this.)

5. A girl measured the average speed of her bicycle. She found that she cycled 15.0 m in 16.1 s. She used the equation:
$$\text{speed} = \frac{\text{distance}}{\text{time}}$$
to calculate her average speed using her calculator. She wrote down that her average speed was 0.931677. Her arithmetic was correct but there are **two** things wrong with this answer. What are they? (Sp.NEA)

6. A man runs a race against a dog. Below is a graph showing how they moved during the race.

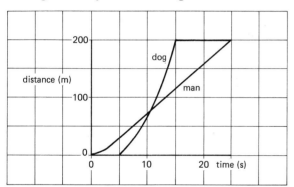

(a) What was the distance for the race?
(b) After how many seconds did the dog overtake the man?
(c) How far from the start did the dog overtake the man?
(d) What was the dog's time for the race?
(e) Use the equation: $\text{speed} = \frac{\text{distance}}{\text{time}}$
to calculate the average speed of the man.
(f) After 8 seconds is the speed of the man increasing, decreasing or staying the same?
(g) What is the speed of the dog after 18 seconds? (Sp.NEA)

7. Here are three sections of ticker tape. In each case the ticker timer was vibrating at 50 Hz. Calculate the average speed of the object attached to the tape between points **A** and **B**.

  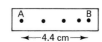

8. The plan below shows part of the railway system between London (King's Cross) and York.

For this question you can assume that the railway line consists of one track in each direction; that is, trains cannot overtake each other.

Here is part of the day's railway timetable showing three trains travelling along this route; the first goes to Doncaster, the second to York and the third to Peterborough.

Distance from
London/km

0	King's Cross	9.50	10.00	10.15
125	Peterborough	↓	11.15	11.30
250	Doncaster	11.30	12.30	
300	York		13.00	

(The 9.50 does not stop at Peterborough.)
Below is a distance–time graph for the 9.50 to Doncaster.

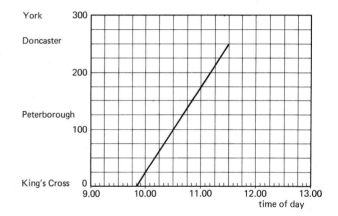

(a) Copy the graph and, on the same axes, draw a distance–time graph for: (i) the 10.00 to York; (ii) the 10.15 to Peterborough.
(b) What is the average speed of the 9.50 train between King's Cross and Doncaster? Give your answer in km/h.
(c) British Rail proposes to speed up the 10.15 train so that it travels to Peterborough at the same average speed as the 9.50 to Doncaster. Draw a new line on your graph to show the journey of this faster train. Label this line 'faster train'.
(d) What does your graph show you to be the disadvantage of trying to speed up the 10.15 to Peterborough in this way? Explain your answer.
(e) Suggest a possible way in which British Rail could overcome this disadvantage. (Sp.NEA)

9. To measure the speed of a model car, a piece of card 10 cm long is fixed to the car roof and it is allowed to cut through the light beam to a photodetector. The timer records a time of 0.125 s. What is the speed of the car?

10 A toy train runs round an oval track at a steady speed of 0.5 m/s. Is its velocity constant? What is its instantaneous velocity as it passes **P**?

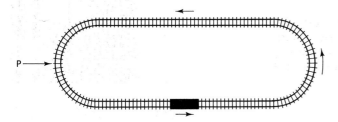

11 A cyclist starting from traffic lights accelerates at 1.5 m/s² for 5 s. What is her final speed?

12 A car, travelling at 25 m/s, accelerates to 30 m/s in 10 s to overtake a lorry. What is the car's acceleration?

13 An advertisement claims that a car can accelerate from 0 to 60 m.p.h. in 10.0 s.
(a) What speed in m/s is equivalent to 60 m.p.h.? (Take 1 mile to be 1500 m.)
(b) What is the car's acceleration in m/s²?

14 Describe the motion of the objects whose velocity–time graphs are as follows:

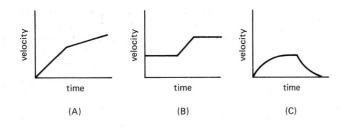

15 The sketch graph represents a journey in a lift in a department store.

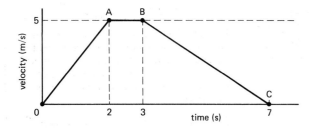

(a) Briefly describe the motion represented by (i) OA, (ii) AB, (iii) BC.
(b) Use the graph to calculate (i) the initial acceleration of the lift, (ii) the total distance travelled by the lift, (iii) the average speed of the lift for the whole journey. (Sp.NEA)

16 You are pedalling along on a bicycle at 6 m/s. You stop pedalling and freewheel for exactly 1 minute (60 s) before you stop. Assuming that you slow down steadily, what distance do you travel before you stop? (Hint: draw a velocity–time graph of your motion.)

17 A 100 m race is organised between an Olympic athlete and a car. The athlete finishes the race in 10 s. The car starts from rest and accelerates constantly at 2 m/s². Who wins the race?
(Hint: work out how far the car travels in 10 s.)

18 A ball is dropped from a bridge and takes 4 s to reach the river below. What is its speed as it hits the water? (The acceleration due to gravity is 10 m/s².) How high is the bridge above the water?

19 The runway on an airfield is 250 m long. A light aircraft has a minimum take-off speed of 30 m/s. What is the minimum acceleration it must have if it is to be able to take off?

20 The following information is adapted from the Highway Code for car drivers.

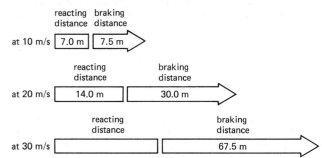

The distances shown are the shortest for an emergency stop from the speeds stated.
For simplicity, it is assumed that, while the driver is reacting, the car continues at a uniform speed but, while the brakes are applied, the car slows down at a steady rate.
(i) What is the 'reacting distance' for a speed of 30 m/s?
(ii) What reaction time is expected of a driver?
(iii) Draw the speed–time graph which corresponds to the information given in the diagram for a speed of 20 m/s. Assume that the brakes are on for three seconds.
(iv) Use this graph to estimate the speed at which a car, travelling at 20 m/s, would strike an obstacle three seconds after the driver first notices it.
(SEB)

INVESTIGATION

How does the stopping distance of your bicycle depend on the speed you are riding at? You will need a method of applying the brakes equally hard in each experiment, and a method for measuring your speed just before you brake.

The answers to the questions on p. 47 are (a) country roads, (b) motorway, (c) around town.

6: What Causes Motion?

Galileo Galilei
(1564–1642)

We can describe the motion of objects precisely in terms of their distance travelled, displacement, instantaneous speed, average speed, velocity and acceleration. But **describing** motion is not the same thing as **explaining** motion. In this chapter, we will begin to look at how we might explain and predict the motion of objects. What makes bodies move?

6.1 Making things move

If you want to move a heavy box across the floor, what do you do? The obvious answer is that you either push it or pull it! From our everyday experience, it seems that to move an object, a force – a push or a pull – is needed. ▷

But as we shall see, this 'common sense' idea is not quite correct. Even so, it is a good starting point. Indeed, it was what people believed for almost 2000 years from the time of Aristotle (384–322 B.C.) until Galileo, who lived from 1564 till 1642. Aristotle wrote that 'a moving body comes to a standstill when the force which pushes it along no longer acts'. And, indeed, we all know that a book will move across the table as long as we push it, but it stops when we stop pushing.

6.2 Force

The idea of 'force' seems to be important in explaining motion. The usual way to exert a force is by direct contact. To make a supermarket trolley move along, you push it with your hand. While your hand is in contact with the handle, you are exerting a force on the trolley. Similarly, if a golfer putts a golf ball, she exerts a force on the golf ball for the short period while the ball is in contact with the club. Once the ball has left the club, however, the golfer cannot influence it any further – she no longer exerts a force on it: see (B), right.

(A) force exerted by club   (B) force no longer exerted by club

But this is a problem for Aristotle's idea of motion. The golf ball doesn't stop moving once it leaves contact with the club – it keeps rolling for some distance, even though there is no longer any forward force on it.

So Aristotle's ideas about motion run into difficulties. We need a better way of explaining motion – better because it works in more situations. The man who took the first steps to discovering this better way of thinking about motion was an Italian called Galileo Galilei.

Galileo's study of motion began with two quite simple experiments. In the first one, he set up a pendulum using a metal ball tied to a light thread. He clamped a pin below the point of support of the pendulum, pulled the pendulum aside, and released it. He observed that the ball rose to its original height again after striking the pin.

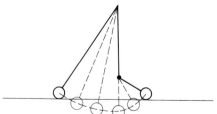

This led him on to his second experiment which was a 'thought experiment'. This means that he didn't actually do the experiment with real apparatus; he 'did' it in his imagination! Galileo's thought experiment is shown in the box on the right.

This leads to a conclusion which is rather surprising at first sight:

A body does not need a force to keep it moving. It is just as 'natural' for a body to keep moving steadily as to stay at rest.

Galileo's thought experiment

Imagine that we take a completely smooth rail with a groove in it which a ball can run along. (Of course, completely smooth rails do not exist, which was why this had to be a 'thought experiment'.) We bend the rail into shape **1**, hold a ball at the point **A** and let it go. Where will it rise to on the other side? From his knowledge of the pin and pendulum experiment, Galileo reckoned that the ball would rise to the point **B** which is at the same height as **A**, if the rail was completely smooth.

He then asked himself: what would happen if I bent the rail into shape **2** and repeated the experiment? Again, he argued that the ball would rise to the original height. To do this it must now travel a longer distance – it must travel to **C**.

But what will happen, Galileo wondered, if we bend the rail into shape **3**? The ball will keep on rolling until it reaches the original starting height. But of course it can never reach this height. So Galileo argued that it would keep on going – for ever. Once the ball had started moving, it would continue.

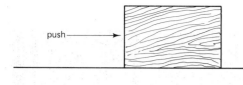
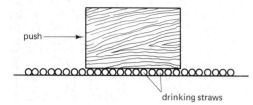
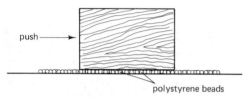

6.3 Rough and smooth surfaces

Galileo's thought experiment led him to this important conclusion about motion. But physicists are happier if they can carry out some 'real' experiments to confirm the results of a thought experiment. Indeed, one of the really useful things about thought experiments is that they give us ideas for real experiments which we can then try out.

◁ If we take a block of wood and push it gently across the bench, it stops moving as soon as we stop pushing. But if we place the same block on some rollers (some drinking straws for example) and give it the same push, it will continue moving for a short distance after we stop pushing. If we then place the same block on a flat tray covered with a layer of tiny round polystyrene beads and push it again, it moves much more freely and keeps on moving for quite a distance after we stop pushing.

What makes the difference in these three situations? The block and the push are the same each time, so the difference must be due to the surface on which the block is moving. If the surface is rough, as it is when the block is sliding on the bench, then the block stops quickly. But if we prevent rubbing between the surfaces by putting the block on rollers or making it roll on small beads, then it continues moving after the force stops.

We can reduce the rubbing between surfaces even further by keeping
◁ them completely apart. A Linear Air Track is a straight hollow metal (or plastic) tube which is shaped so that gliders can sit astride it. In the upper side of the tube there are two rows of holes. When air is blown into one end of the tube (usually from a domestic vacuum cleaner running in reverse), it comes out of the rows of holes and makes the glider float on a cushion of air. As a result the glider moves very smoothly along the track if you start it with a push.

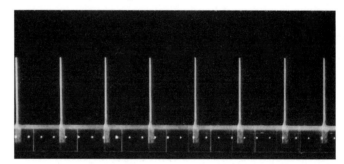

The strobe photograph shows a glider with a white straw attached to it. The images are equally spaced, showing that the glider is moving at steady speed along the track. Of course it does not continue moving for ever, but its motion is steady enough to make us believe that Galileo's idea is correct. If we could make the surface completely smooth, the moving object would continue moving at a steady speed in a straight line after the force stopped acting on it.

Until very recently it was impossible to do any experiment to confirm this, because a 'completely smooth' environment couldn't be produced. Nowadays, however, space laboratories give us the chance to do a real test experiment of this kind. The evidence agrees with Galileo's original idea.
◁ A moving body carries on moving at a steady speed after the force which made it move has stopped acting.

To summarise this very important idea about motion:

A force IS needed: to start a body moving
to stop a body moving
to make a body move faster
to make a body move slower
to make a body change its direction of motion.

But a force is NOT needed: to keep a body moving in a straight
line at a steady speed.

6.4 Friction

If the 'natural' thing for bodies to do is to keep moving at a steady speed in a straight line once they have been started, we then need to ask a different question about our everyday experiences:

Why do objects not continue moving at a steady speed?
Why do things slow down and stop after we finish pushing them?

The answer is that there is a **force** acting to slow them down. This force is called **friction**. Friction is always present when two surfaces rub over one another. It always tries to stop one body moving past the other. Friction tries to prevent the motion of one body relative to another.

We live in a world where we experience the effects of friction all the time. Friction between the soles of our shoes and the ground enables us to walk. If there is ice on the ground, friction is much less (though it doesn't disappear entirely) and we have a lot more difficulty moving around! The same is true of car tyres, which should give a good grip in all sorts of road conditions. Tyre treads are carefully designed to give plenty of friction between the tyre and the road. Although friction is a force which opposes motion, relative motion would be almost impossible without it!

The material used to make motorcycle brake-pads is chosen because it provides a lot of friction with a metal disc on the motorcycle wheel.

Of course, in some situations, we do not want friction. Where two surfaces are meant to slide smoothly over each other, too much friction is a nuisance. For example, a bicycle wheel should turn smoothly on its axle. A hinge is another example of a place where we want smooth motion. In these situations, we want to reduce friction as much as possible.

Two situations where we want a lot of friction

Two situations where we want to reduce friction

What causes friction?

Friction is caused by the roughness of surfaces. All surfaces are really quite rough if we look at them under a microscope, even those which we think are very flat and smooth.

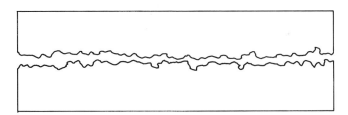

If two 'flat' surfaces are in contact, these humps and hollows will fit together. So when one surface tries to slide over the other they tend to stick. This is what causes friction.

We can reduce friction by preventing surfaces interlocking. The simplest solution is to make the two surfaces roll over each other, rather than sliding. It is easier to move something heavy, such as a cooker, by putting it on wheels, instead of dragging it along.

If we must have sliding contact, as in a hinge for example, oil is used to lubricate the movement. The oil fills the hollows on the two surfaces, and prevents them fitting together like a jigsaw. This makes it easier for them to slide over one another.

Friction and heating

You have probably noticed that friction between rubbing surfaces makes them heat up. Try feeling the brake blocks of your bicycle after you have used the brakes. They should feel warm – and the rim of the bicycle will also be warm. How does friction cause this heating?

When two surfaces slide over each other, the bumps and irregularities on one surface hit against the bumps on the other. The molecules on these parts of the two surfaces get knocked around by this and it sets them vibrating more vigorously. This, of course, means that the temperature of the surface layer has increased. (We know that when we heat something up, its molecules vibrate more strongly.) So friction causes heating by setting the molecules in the two surface layers vibrating more strongly, just as a result of their collisions with each other.

One form of transport uses the air track idea to reduce friction. The hovercraft floats on a cushion of air. This keeps it completely clear of the ground surface and friction is much less.

60 What causes motion?

6.5 Air resistance

If a body moves through a fluid (like water or air), it experiences a friction force. This force always acts to slow the body down, or to make it more difficult for it to move through the fluid. For a body moving through air, this friction force is usually called **air resistance**. Unlike friction between sliding surfaces, the size of the air resistance force depends on how fast the body is moving. When a car is travelling along a motorway, air resistance is the largest friction force acting on it. A lot of effort is used nowadays to design car and lorry bodies so that their air resistance is as low as possible. This makes the friction force of the air small, and gives the vehicle a higher top speed, and better fuel consumption.

Questions

1 Your uncle, who last studied science many years ago, believes that you need a force to keep something moving. As he says, 'If you stop pushing something, it stops moving.' Read sections 6.1 to 6.3 again and then write a letter to your uncle explaining to him why his idea is wrong. Try to include plenty of examples to convince him.

2 An astronaut while 'walking in space' pushes a spanner away from himself. Describe the motion of the spanner after it leaves his hand.

3 Write a short paragraph about each of these four diagrams explaining how forces are involved in these situations.

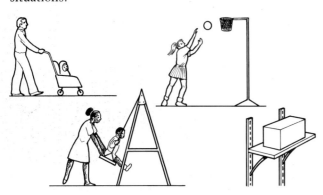

4 Imagine what it would be like to wake up one morning and discover that all friction had disappeared. Write a story about what you think it would be like living in a friction-free world.

5 Make a table with two columns, and list some situations where we try to reduce friction and others where we make use of friction. Try to think of at least **six** of each.

INVESTIGATION

Compare the friction 'grip' of different shoe soles on different floor surfaces. How much difference does it make if the shoe sole is wet?

---

6 (a) A pupil performs an experiment to discover the effect of different forces on the length of a thin piece of elastic. Table 1 shows the results of his experiment.

Table 1

Length of elastic/cm	15.3	20.2	24.7	29.4	34.1
Force pulling elastic/N	0.00	0.10	0.20	0.30	0.40

(i) Describe briefly how the pupil could have obtained the results in Table 1.
(ii) Draw a graph of 'length of elastic' against 'force pulling elastic'.

(b) In a second experiment the pupil uses the same piece of elastic to measure the force which an air stream exerts on an aluminium sail.

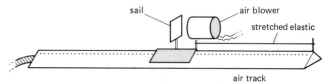

An air track vehicle with an aluminium sail sits on a horizontal frictionless air track. The vehicle is attached to one end of the track by the piece of elastic and an air blower directs a steady stream of air onto the sail. The pupil measures the length of the stretched elastic.

Table 2 shows the experimental results for sails of different area.

Table 2

Area of sail/cm²	20	50	80	100
Length of elastic/cm	18.6	23.6	28.5	31.8
Force on sail/N				

(i) Copy and complete Table 2 by writing down the force exerted by the air stream on each sail.
(ii) Use the results from Table 2 to show that the force on a sail varies directly as its area.

(c) In a final experiment the 100 cm² sail is bent backwards into a V-shape. Is the length of the elastic again 31.8 cm? Explain your answer.

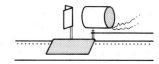

(SEB)

7: Laws of Motion

Galileo's work led to many important new ideas about motion. But Galileo never succeeded in putting all his ideas together into a simple law of motion. That was finally achieved by Isaac Newton, perhaps the greatest physicist who ever lived. ▷

7.1 Newton's First Law of Motion

The law of motion which we now call Newton's First Law of Motion can be stated in the following way:

If the total force acting on a body is zero, the body will stay at rest (if it is stationary), or will continue to move at a steady speed in a straight line (if it is already moving).

This means that unless a force acts, a body will continue with whatever motion it has. If it is stationary (at rest), it will stay stationary. If it is moving, it will continue moving at a steady speed in a straight line. This property of all bodies, their tendency to go on doing whatever they are doing, is called **inertia**. ▷

7.2 Inertia and mass

Inertia is a basic property of all matter. We can ask: Why do moving bodies tend to keep moving in a straight line at steady speed when no force acts? The answer – it is because of the body's inertia – isn't really an answer at all. It is just another way of saying the same thing! In fact, physics can provide no answer to the question. Bodies behave this way, because that is how they behave. We cannot explain inertia in terms of anything simpler.

The **mass** of a body is a measure of its inertia. If a body has a small mass, it will have a small inertia; if its mass is large, then so is its inertia. ▷ For example, an empty shopping trolley is easy to push across a floor; it is easy to start it moving and to stop it again. It hasn't got much inertia. But if the trolley is full of groceries, it is much harder to start and stop. The inertia is much larger. The trolley full of groceries has a large mass; the empty trolley has a small mass. ▽

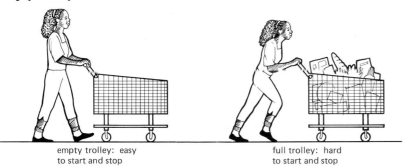

empty trolley: easy to start and stop

full trolley: hard to start and stop

In response to this, you may say that what we are really talking about is the trolley's 'weight'. But it may surprise you to know that the same result would apply in a 'weightless' environment. If an astronaut in space wants to move a full box, he finds that it is harder to start it moving and to stop it moving than an empty box – just the same as on Earth, even though the box is 'weightless'. Weight has nothing to do with it! The important property is **mass**.

Mass is difficult – almost impossible – to define. It is sometimes defined as 'the amount of material' in the body, the total number of atoms of stuff which are there! This may be useful, as it points out that the mass of a body is fixed – it does not change as the body is moved about. Even if it is taken into space, its mass will stay the same as on Earth.

Isaac Newton

Newton was born in Lincolnshire in 1642, the year of Galileo's death, and lived until 1727. As well as his work on the laws of motion, he is also famous for his work on the gravitational force and on light and optics. In his writings, Newton acknowledged that all his work had depended on the work of earlier scientists. He wrote: 'If I have seen further than other men, it is because I have stood on the shoulders of giants.' One of these 'giants' was Galileo.

This tanker has so much inertia that it has to stop its engines 3 miles from port so that it can slow down in time.

62 Laws of motion

7.3 Comparing masses (or inertias)

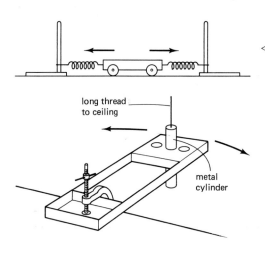

Although mass is hard to define, we can compare masses in the laboratory. The drawing shows two ways of doing this.

If the moving body has large inertia, it will resist constant changes of motion, and its vibrations will be slower. If the inertia is small, the vibrations will be faster. Using the inertial balance, we can confirm that the weight of the metal cylinder is not important, by suspending it from a long thread. This doesn't change the time of vibration. When the mass of the moving object is increased by adding a second identical trolley, or a second metal cylinder, the period of vibration becomes longer.

7.4 Motion caused by a steady force

When there is no applied force, a body's motion remains unchanged. But what happens when a force **is** applied? This can be investigated by an experiment. We can use a trolley as the moving body, apply a constant force to it, and obtain a record of its motion on ticker-tape.

Applying a constant force

The simplest method of applying a constant force to a trolley is to use an elastic string. If we keep the stretch constant, the force (the pull) is constant. So we pull the trolley with an elastic string stretched by a fixed amount. We attach one end of the elastic to a post at the back of the trolley, and stretch it until the other end is exactly between the two posts at the front of the trolley.

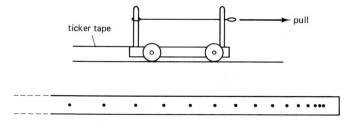

If we can pull the trolley along like this, keeping the stretch of the elastic constant, we are applying a constant force to it.

Force and acceleration

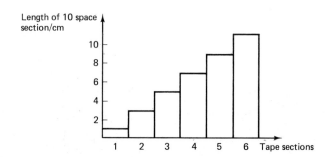

EXAMPLE 7.1

In an inertial balance experiment, the time for 20 oscillations of the balance is measured for loads of 1, 2 and 3 units of mass (metal cylinders). The results are shown in Table 7.1.

Table 7.1

Mass	Time for 20 osc./s
1	6
2	7
3	8

An object whose mass is not known is then placed in the balance. The time for 20 oscillations with this mass is found to be 6.6 s. What is the inertia of the 'unknown' mass? (How many metal cylinders is it equivalent to?)

The first step is to draw a graph of the results:

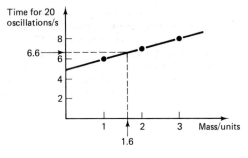

From the graph, we can see that a mass of 1.6 units would correspond to a time of 6.6 s for 20 oscillations. The 'unknown' mass is equivalent to 1.6 of the standard cylindrical masses.

A typical ticker-tape from this experiment has dots which get further and further apart, showing that the trolley is **accelerating**. If we measure the distances between dots, we find that the average speed increases steadily by the same amount from one interval to the next. The acceleration is **uniform**.

When a steady force is applied, a body has a uniform acceleration in the direction of the force.

Newton's Second Law of Motion is based on this result. It states how the force and acceleration are related. This is the subject of Chapter 8.

7.5 Balanced and unbalanced forces

The simplest idea of a force is simply a push or a pull. Using Newton's First Law, we can be more precise about what a force is. A force is 'something which changes the motion of a body'. This is useful because it gives us a way of knowing when a force is acting. If the motion of a body is changing, then there is a force acting on the body. If the body has constant motion (in a straight line at steady speed) or is at rest, then there is no total force acting on it.

Why have we used the words 'total force' rather than just 'force' in this last sentence (and in the statement of Newton's First Law earlier)? The reason is that in most practical situations, there will be more than one force acting on a body. To predict how it will move, it is the **total force** that matters – the sum of all the individual forces acting on it. Let us see how this works.

Balanced forces

The bag is not moving. It is at rest. So the total force acting on it must be zero. The force of gravity (its weight) pulls it downwards. So there must be a second force pushing it upwards which balances this. This second force is called the *reaction* of the table. Solid objects like tables exert an upwards reaction to balance the weight of any object placed on them.

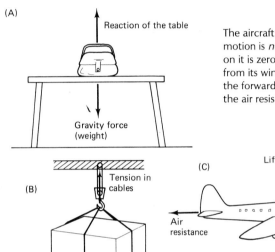

The aircraft is flying at steady speed. Its motion is *not* changing. So the total force on it is zero. This means that the lift force from its wings is equal to its weight, and the forward thrust of its engines is equal to the air resistance.

The load is hanging at rest. The downward force of gravity is balanced by the upward force of the tension in the cable.

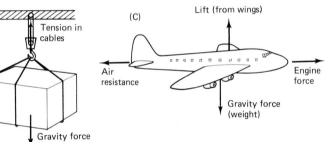

Balanced and unbalanced forces

The cyclist in (A) is riding along the road at a constant speed. There is no change of motion. So the forces must be balanced. The forward force exerted by the cyclist is equal to the frictional forces (rubbing friction at the axles and tyres and air resistance).

To speed up, the cyclist must exert a larger forward force (B). This is now bigger than the frictional forces. Part of the forward force is not balanced by the frictional forces. The total force is forwards – the cyclist accelerates.

To slow down or stop, the cyclist can either put on the brakes (larger frictional forces) or pedal less hard (smaller forward force) (C). The frictional forces are now bigger than the forward force. The total force is in the backwards direction – opposite to the motion. The cyclist slows down.

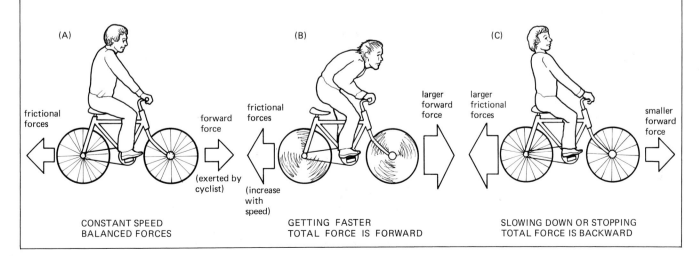

CONSTANT SPEED
BALANCED FORCES

GETTING FASTER
TOTAL FORCE IS FORWARD

SLOWING DOWN OR STOPPING
TOTAL FORCE IS BACKWARD

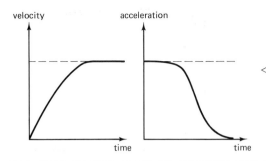

7.6 Terminal velocity

We often have the situation where an object starts from rest and speeds up until it reaches a steady speed. When we start off on a bicycle, we begin from rest and pick up speed as we pedal away. When we eventually reach a steady speed, we still have to pedal to maintain our speed. Driving a car is much the same. The car accelerates from rest. Eventually it reaches a cruising speed. The engine still has to run to keep the car going at this steady speed. Let us think about the forces involved in this sort of situation.

Reaching a terminal velocity

At the beginning, the bicycle is at rest. The forward force exerted by the cyclist is the only force, so the bike accelerates forward.

As soon as the bike starts to move, frictional forces begin to act. There is rubbing friction at axles and wheels, which is more or less constant at all speeds. The air resistance force is small because the cyclist is moving slowly. The total force is forward, so the bicycle keeps accelerating forward – building up speed.

As the bicycle gets faster, the air resistance force grows larger. Air resistance depends on the speed at which an object is moving through the air. The forward force stays the same but the total force is now smaller. The forward acceleration gets less.

Eventually the cyclist reaches a speed where the air resistance force is large enough to balance the forward force. The total force is zero. The cyclist now continues at a steady speed.

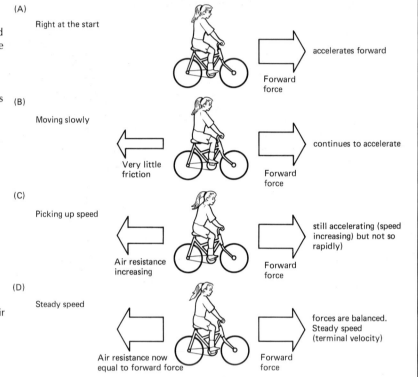

The steady speed which the cyclist reaches in the fourth drawing is called her **terminal velocity**. The harder she pedals, the faster she will move before the air resistance force balances her forward force. If there is a strong headwind, the air resistance force will be larger and the same forward force on the pedals will not produce so large a terminal velocity.

Falling objects and terminal velocity

Any object falling through the air also reaches a terminal velocity – and then falls at a steady speed. The diagrams on the left explain how this happens.

The air resistance force on a moving object depends on two things:
– how fast it is moving;
– how large a surface it presents to the air.

A stone or a ball presents a very small surface, compared with its weight. A feather on the other hand has a very large surface area. So the air resistance force on a feather is quite large even at low speeds. It will balance the feather's weight at a very low speed. The air resistance force will only balance the weight of a ball if the ball is moving very fast.

At the moment it is released, the only force on the ball is its weight. So it accelerates downwards. Initially this acceleration is 10 m/s² (see section 5.10).

As it speeds up, the air resistance force increases. The total downward force gets less. The ball still accelerates downwards – but its acceleration is now less than 10 m/s².

Eventually it reaches a speed where it is moving so fast that the air resistance force is equal to its weight. The total force is zero. The ball has reached its terminal velocity. From here on it will fall at a steady speed.

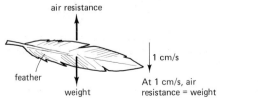

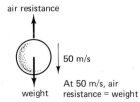

Guinea and feather experiment

If our explanation involving air resistance is correct, then a feather should fall just like a stone if there was no air resistance. It does! We put a coin and a feather into a long glass tube, sealed at both ends with rubber bungs. When we turn the tube upside down, the coin falls quickly to the bottom whilst the feather drifts down slowly – just as we would expect. Now we pump all the air out of the tube and try again. This time both coin and feather fall together. The feather accelerates downwards at $10 \, m/s^2$, just like the coin. ▷

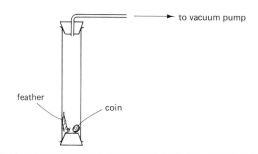

Sky-diving: a special case of terminal velocity

(A) At the moment when a sky-diver jumps out of a plane, the only force acting on him is gravity – his weight. So he accelerates downwards at $10 \, m/s^2$.

(B) As he starts to fall, the air resistance force opposing his motion begins to build up. He still accelerates, but no longer at $10 \, m/s^2$.

(C) Air resistance increases with speed. Eventually he is falling at a speed where the air resistance force is equal to his weight. The forces are balanced. He continues to fall at a steady speed – about 200 km/h (about twice the maximum permitted driving speed on a motorway!). This is his terminal velocity.

(D) The skydiver now opens his parachute. This has a much larger surface area than his body alone. As a result, the air resistance force at any particular speed is much larger. At 200 km/h, the air resistance force on the parachute is much bigger than the skydiver's weight. The total force is now upwards. The skydiver decelerates sharply – he slows down.

(E) He continues to slow down until he reaches a new steady speed – a new terminal velocity – when the air resistance force on his parachute is again equal to his weight. This happens at a speed of around 10 m/s. So the skydiver now drifts down to the ground at a steady speed of 10 m/s. Landing at this speed is about the same as jumping from a 5 m high wall.

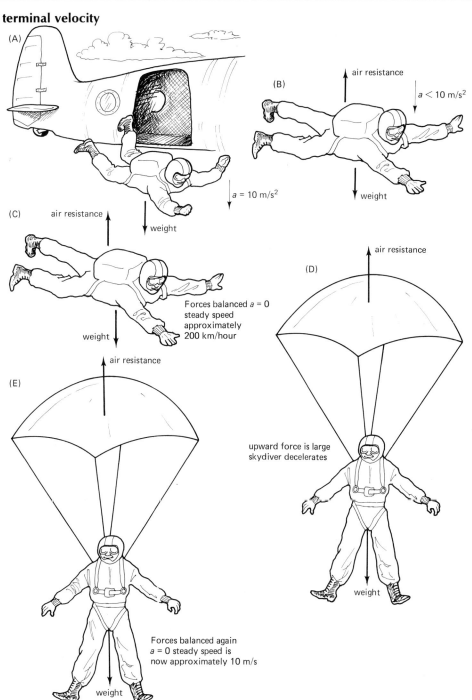

When the skydiver is travelling at his terminal velocity (either with his parachute closed, or open), the total force acting on him is zero. There is only an unbalanced vertical force when his motion is changing – when he is speeding up at the start of the jump, or immediately after he opens his parachute.

66 Laws of motion

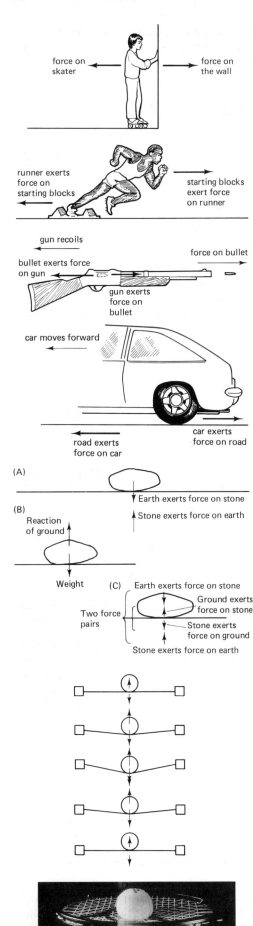

7.7 Forces in pairs

Forces always involve two bodies. If you push something, you feel it pushing back on you. For instance, imagine that you are wearing roller skates and standing beside a wall. If you push the wall, you will begin to move backwards. The wall has pushed you.

Notice too that the push on you is in the opposite direction to your push on the wall.

Another way of showing this is to use a trolley with a spring-loaded plunger. If the trolley is placed on a level bench and the plunger is released, the trolley does not move at all. It has nothing to push against. If a second trolley is placed beside the first, and the plunger is then released, **both** trolleys will move away. By pushing on trolley B, trolley A has been pushed in the opposite direction. Indeed, if the trolleys are identical, they move apart at the same speed. The forces on the two trolleys are the same size, but opposite in direction.

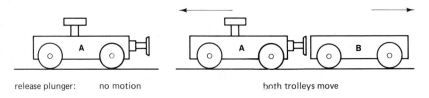

release plunger: no motion   both trolleys move

Forces **always** occur in pairs. Newton summarised this in his Third Law of Motion:

The force which body A exerts on body B is always equal and opposite to the force which body B exerts on body A.

◁ The two forces in the pair always act on different bodies.

7.8 Reaction

Sometimes Newton's Third Law is stated: 'To every action, there is an equal and opposite reaction.' This is not a very clear way to state the law. We have already used the word 'reaction' to describe the force which a table or the ground exerts upwards on an object. When 'reaction' is used in this book, this is the meaning which it always has.

To clarify our ideas, it may be worth looking at this situation in a little more detail. Imagine a stone lying on the ground. There is a force of gravity pulling the stone downwards. This force is exerted **by** the Earth **on** the stone. The other force in this force pair is the force exerted **on** the Earth **by** the stone. As we saw in Chapter 1, gravitational forces occur between any two masses. The stone really does pull the Earth up. Of course, the Earth is so large that we cannot see this force having any effect.

But we can also look at the situation from the point of view of the stone on its own. The stone is at rest. So the total force on it must be zero (according to Newton's First Law). The reaction of the ground balances the weight of the stone.

What we really have here are two sets of force pairs.

7.9 Force pairs are always equal

Newton's Third Law really means that when two bodies exert forces on each other, these forces are equal and opposite **at every moment**. So, for example, when a tennis player hits a ball, the force exerted by the racquet on the ball is equal to the force exerted by the ball on the racquet **at every moment** of their contact. High speed photography shows us clearly that the ball and racquet are quite distorted in shape during the time of contact.

The force on the ball probably builds up to a maximum and then gets smaller again, all within a very short time interval (around 5 milliseconds). According to the Third Law, the force on the racquet is the same size at every moment.

Questions

1 A demonstration experiment consists of a pen attached to a flat platform. This can move freely in any direction but has a 20 kg mass sitting on top of it. Explain why it is very difficult to sign your name with this pen.

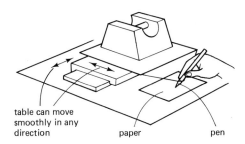

2 Here is a trick to try! Place a pile of 2p coins on a smooth table top. Then flick another 2p coin quickly along the table towards the pile. If you do it right, it should knock out the bottom coin and take its place – leaving the pile standing! Explain how this works.

3 Explain the following observations, using the idea of inertia:
(a) If you are standing in a bus or a train, you tend to fall backwards when it starts off and to fall forwards when it slows down again.
(b) When a car is involved in a collision, the passengers are thrown forward against the windscreen (unless they are wearing seat belts).
(c) If you have to brake very suddenly on your bike, you almost fly over the handlebars.
(d) When a car goes round a corner quickly, you feel as if you are being pushed outwards, away from the centre of the curve.

4 Some cars are fitted with padded headrests. These can prevent serious injury to the neck and back of the driver and passenger if another vehicle runs into the car from behind. Explain why the headrest is necessary.

5 Describe the motion of the box in each of the following situations.
(Note: If the box moves, the force of friction between the ground and the box is 40 N.)

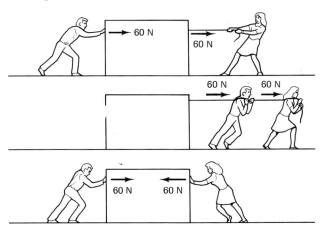

6 Copy the following diagrams and mark on them **all** the forces which are acting on the objects. Show clearly any pairs of forces which are equal.

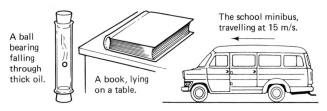

7 For each of the following questions, the answer is either:
    A. equal to its weight
    B. greater than its weight
    C. less than its weight
    D. zero

Choose the answer you think is correct to complete the following sentences.
(a) A cat is sitting on a soft comfortable armchair. The upwards force exerted by the chair cushion is ...
(b) A soap bubble floats motionless in the air. The upwards buoyancy force of the air is ...
(c) A parachute is dropped from a plane. As it drifts down at steady speed, the air resistance and frictional forces on it are ...
(d) After blast-off, a rocket rises at a steady speed from the surface of the Moon. The upwards force exerted on the rocket by its motors is ...

8 A man pushes a trolley with a force of 60 N and it moves at a constant speed – drawing (A). He then lets go – drawing (B).
What is the total force on the trolley in (A)?
What is the total force on the trolley in (B)?

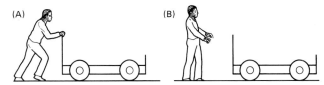

9 A boy throws a tennis ball straight up into the air and catches it again. In diagram (A) the ball is on the way up; in diagram (B) it is exactly at the top of its path; in diagram (C) it is on the way down again. Copy the diagrams and mark with an arrow the direction of the **total** force on the ball each time.

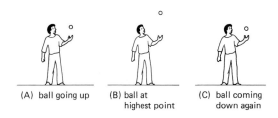

INVESTIGATION

Does the terminal velocity of a 200 g mass descending by (home-made) parachute depend on the area of the parachute?

8: Force, Mass and Acceleration

In the last chapter, we looked at how a body moves if there is no force acting on it, or if all the forces acting on it are balanced (and add to zero). In this chapter we will look at the motion of a body when a force acts.

Together, these two ideas enable us to explain almost all the motion we see, and to predict how things will move in many situations.

8.1 Force, mass and acceleration: an experiment

We saw in section 7.4 that a trolley has uniform acceleration if a constant force is applied to it. How big is this acceleration? It will depend on:

- the size of the force applied
- the mass of the moving body.

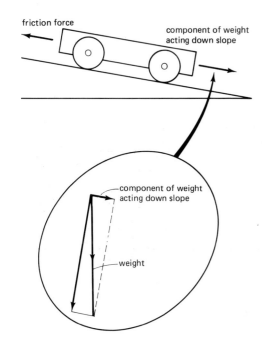

This can be investigated experimentally using trolleys and elastic strings. It is much better to investigate the effect of these two factors separately. Firstly, we investigate the effect of **different forces** on the **same mass**; then in a second experiment we investigate the effect of the **same force** applied to **different masses**. In each experiment, we are keeping one thing constant while we explore the effect of the other one.

In the experiments, we must make sure that the force applied by the elastic is the only force having any effect on the motion. If the trolley is on a level surface, there will also be a friction force opposing its motion. To eliminate the effect of this friction force, we put the trolley on a slightly sloping board. The slope should be just enough to ensure that the trolley runs down with a constant speed when it is started with a push. At this angle, the friction force is balanced by the component down the slope of the gravity force on the trolley. ◁

The total force along the board is now zero. The trolley moves at steady speed (if it is 'push started') as Newton's First Law says it should. This process is called **friction compensation**. We can check that the slope is correct by making a ticker-tape record of the trolley's motion; the dots should be equally spaced.

8.2 Force and acceleration

In the experiment (A), we pull a trolley along a friction-compensated board with one elastic string, stretched by a constant amount, and record its motion using a ticker-timer. From the ticker-tape we can calculate the acceleration.

We now want a means of applying a force which is twice as big. To do this we use two identical elastic strings side-by-side, both stretched by the same amount (B).

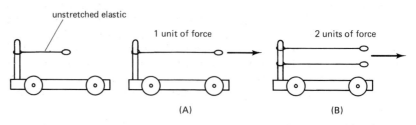

To get a force three times as big, we use three elastics, and so on. Each time, we obtain a ticker-tape from which we can measure the acceleration of the trolley (using the method explained in section 5.9). Table 8.1 shows a set of results from this experiment.

Table 8.1

Number of elastics	Acceleration/ mm/s²
1	34
2	62
3	100

The acceleration gets greater as the force gets larger. We can draw a graph of acceleration against force (number of elastics used). The points lie on, or close to, a straight line through the origin. Only experimental error prevents the graph from being a perfect straight line. ▷

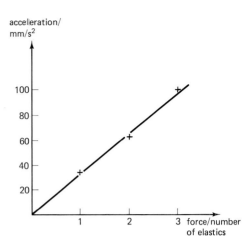

A straight line graph through the origin means that the two quantities plotted are **directly proportional** to one another. (If you are not quite sure about **proportion**, you might find it useful to read Appendix 1 at the back of the book, where it is explained more fully.) If the force is doubled, the acceleration is doubled; if the force is trebled, acceleration is also trebled, and so on.

So the experiment leads to the following result:

If different forces are applied to the same body, the acceleration of the body is directly proportional to the force applied.

acceleration $\propto$ force (if the same mass is used)

8.3 Acceleration and mass

Now we want to see what difference it makes if we use bodies of different masses. This time, we will keep the force the same throughout, by using a single elastic string all the time. First we pull a trolley down a friction-compensated runway using one elastic string, stretched by a constant amount, exactly as before. We measure its acceleration from a ticker-tape. Now we want to apply the same force to a body with twice the mass. We use two identical trolleys stacked together, which have twice the mass of one trolley on its own. Three trolleys have three times the mass.

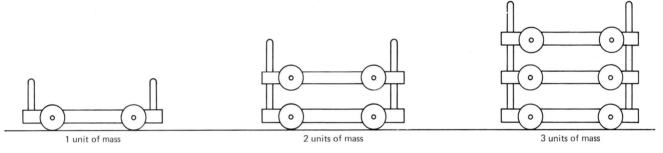

1 unit of mass   2 units of mass   3 units of mass

A sample set of results from the experiment is shown in Table 8.2 together with a graph of acceleration against mass. ▷

Table 8.2

Number of trolleys (n)	Acceleration (a)/ mm/s^2	$n \times a$
1	33	33
2	18	36
3	10	30

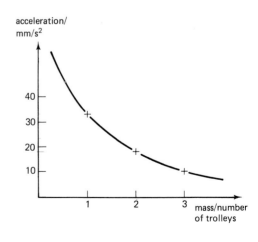

In this case, the acceleration gets smaller as the mass gets larger, just as we would expect. This is **not** direct proportion. But there **is** a connection between mass and acceleration. If we look at Table 8.2, we see that the acceleration is almost exactly halved when the mass is doubled. With three times the mass, the acceleration is about one-third of its original value. This is called **inverse proportion**. Notice that if we multiply the number in the mass column by the number in the acceleration column, the result is almost exactly the same each time (allowing for experimental error).

The results of the experiment can be stated as follows:

If the same force is applied to a number of different masses, the acceleration produced is inversely proportional to the mass.

mass $\times$ acceleration is a constant (if the same force is used)

Force, mass and acceleration

EXAMPLE 8.1

A sports car can accelerate at $2\,\text{m/s}^2$ away from traffic lights. The car has a mass of 800 kg. What is the total force needed to give it this acceleration?

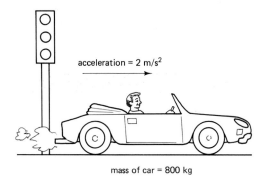

According to Newton's Second Law, the total force F is given by:

$$F = \text{mass} \times \text{acceleration}$$
$$= 800\,\text{kg} \times 2\,\text{m/s}^2$$
$$= 1600\,\text{N}$$

The total force on the car is 1600 N.

EXAMPLE 8.2

A force of 7 N pulls a trolley of mass 3 kg along. The trolley has an acceleration of $2\,\text{m/s}^2$. What size is the friction force opposing the trolley's motion?

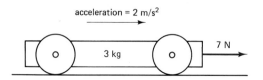

We can calculate the total force on the trolley needed to give this acceleration:

$$F = ma$$
$$= 3\,\text{kg} \times 2\,\text{m/s}^2$$
$$= 6\,\text{N}$$

But the force pulling the trolley is 7 N. So the friction force must be 1 N.

8.4 Unit of force

A force is something which causes a mass to accelerate. We can use this to define a unit of force.

In the SI system of units, the unit of force is called the **newton (N)**. 1 newton is the force needed to give a mass of 1 kilogram an acceleration of $1\,\text{m/s}^2$. From this definition, and the results of the two experiments above, we can derive a very important equation:

A force of 1 newton acting on a mass of 1 kg produces an acceleration of $1\,\text{m/s}^2$

A force of F newtons acting on a mass of 1 kg produces an acceleration of $F\,\text{m/s}^2$ (since acceleration = force)

A force of F newtons acting on a mass of m kg produces an acceleration of $\dfrac{F}{m}\,\text{m/s}^2$ (since mass × acceleration is constant)

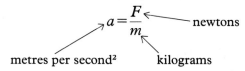

This result is Newton's Second Law. **The acceleration of a body is proportional to the unbalanced force acting on it, and inversely proportional to its mass.** It is often written in the form:

$$F = ma$$

force = mass × acceleration

8.5 Mass and weight

If we drop something it falls. To be more precise, it accelerates downwards. We have seen earlier (in section 5.10) that all compact bodies fall with the same acceleration if released in the lab. This acceleration – called the acceleration due to gravity – can be measured by an experiment. It is approximately $10\,\text{m/s}^2$.

An acceleration is always caused by an unbalanced force. The force which makes falling bodies accelerate is **gravity**. Gravity acts on all masses. The force of gravity on a body is usually called its **weight**.

Since weight is a force, it is measured in newtons. Mass on the other hand is measured in kilograms. This may appear confusing because we use these words differently in everyday life. If someone asks you your weight, you would probably give it in kilograms and not in newtons. What we usually call our 'weight' in everyday language is the quantity that physicists call 'mass'.

The two things **are** different:

Mass is a measure of a body's inertia (or reluctance to change its motion) and is measured in kilograms.

Weight is the force of gravity on a body and is measured in newtons.

But of course there is a connection between mass and weight. Imagine that we release a 1 kg mass and allow it to fall. It will accelerate downwards at $10\,\text{m/s}^2$.

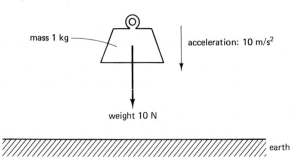

According to Newton's Second Law, the total force on the kilogram mass is given by:

$$F = ma$$
$$= 1 \text{ kg} \times 10 \text{ m/s}^2$$
$$= 10 \text{ N}$$

The force of gravity is 10 N. So a 1 kilogram mass has a weight of 10 N on Earth.

If we do a similar calculation for a 2 kilogram mass, we find that it has a weight of 20 N; a 3 kg mass has a weight of 30 N, and so on. ▷

The acceleration due to gravity is usually given the symbol g. It follows from $F = ma$ that the weight of a mass m kilograms is mg newtons:

$$\underset{\text{newtons}}{\text{weight}} = \underset{\text{kilograms}}{m} \underset{\text{metres per second}^2}{g}$$

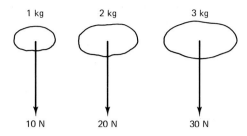
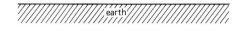
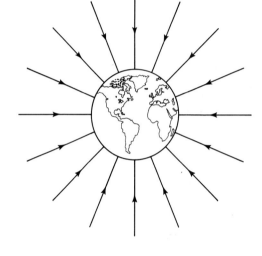

8.6 Gravitational field strength

We saw in Chapter 1 that we can think of the force of gravity near the Earth as caused by a **gravitational field** around the Earth. Any mass near the Earth is pulled towards the centre of the Earth. We can draw a field 'map' to show the direction of the gravitational field. ▷

The lines show the direction of the force which a mass would experience at every point. The force is always towards the Earth's centre, so the 'field lines' have arrows pointing inwards to indicate this. Usually we are interested in what happens near the Earth's surface. The curvature of the Earth can be ignored, and the gravitational field lines are almost exactly parallel.

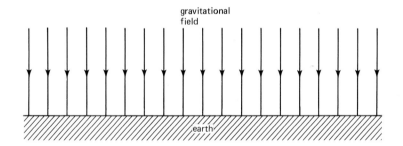

The weight of a body depends on how strong this gravitational field is. We have just seen that there is a gravitational force of 10 N acting on every kilogram of mass. The strength of the Earth's gravitational field is therefore equal to 10 newtons per kilogram (10 N/kg). ▷

In general:

$$\text{gravitational field strength} = \frac{\text{force of gravity (weight)}}{\text{mass}}$$

Since we already know that weight = mg, it follows that the value of gravitational field strength (in N/kg) must always be the same as the acceleration due to gravity (in m/s²).

So it follows directly from Newton's Second Law, that the following two statements are just different ways of saying the same thing:

Each kilogram of mass near the Earth's surface has a gravitational force of 10 N acting on it.

All freely falling bodies near the Earth accelerate downwards at 10 m/s².

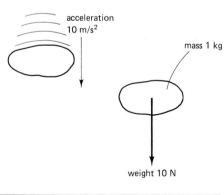

8.7 Mass is constant, weight can change

The weight of a body can change from place to place. The strength of the Earth's gravitational field gets less as we go further away from the Earth. The field lines get further apart as we go further away from the surface – the field gets weaker. So the weight of an object decreases as its distance from the Earth increases.

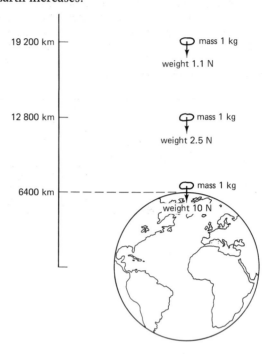

This point is worth noting quite carefully, for many people have the idea that things get heavier if they are lifted up to a greater height. In fact, the opposite happens. If we measure the acceleration due to gravity, g, at the top of a mountain, it is slightly less than at the bottom. Remember that g is also a measure of the gravitational field strength, so this too is smaller on the mountain top.

The weight of an object also varies very slightly from point to point on the Earth's surface. There are several reasons for this:

1. The Earth is not a perfect sphere, and these small irregularities affect the gravitational field pattern slightly.
2. In Chapter 1, we saw that there is a gravitational force between **any** two masses. If there are some particularly massive rocks below the Earth's surface, these can make the gravitational field strength slightly larger at the surface. By measuring g at the surface, geophysicists can find out about the rocks under the surface. This technique is used in prospecting for deposits of minerals and oil.

The Moon has a smaller mass than the Earth. Its mass is about one sixth of the Earth's mass. As a result, the gravitational field strength on the Moon's surface is about one sixth of the value on Earth, approximately 1.6 N/kg. A falling object on the Moon would have an acceleration downwards of 1.6 m/s². If an object could be taken into deep space, far away from all large masses, to a place where the gravitational field strength was zero, then it would have no weight at all.

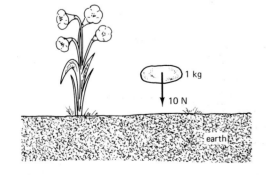

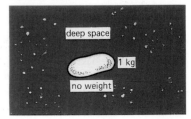

For a manned spacecraft, this is not the case. It is still much too close to Earth to be outside the gravitational field. The apparent 'weightlessness' of people and objects inside a spacecraft has an entirely different explanation, as we will see in Chapter 9.

Although the weight of a body changes as it is moved from place to place, the mass of the body does not change. The inertia of a body is constant, no matter where the body is. This means that an experiment which depends on a body's mass could be carried out anywhere and should give the same results. For instance, the experiment described in sections 8.1–8.3 to show how acceleration depends on force and mass would give the same result ($F = ma$) if we did it on the Moon, or even in a space laboratory where the bodies used were 'weightless'.

It is not strictly true that the mass of a body is constant. According to Einstein's theory of relativity, a very fast moving body has more inertia than the same body when it is moving slowly! However, the changes are too small to have any noticeable effect in any of the situations we will be considering.

EXAMPLE 8.3

If we pull a 4 kg mass with a force of 12 N, what acceleration will it have if we do the experiment:

(a) in a laboratory on Earth
(b) on the Moon
(c) in a space laboratory under 'weightless' conditions?

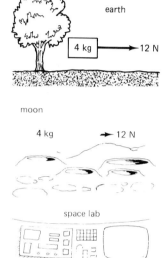

(a) According to Newton's Second Law,

$$F = ma$$
$$12\,\text{N} = 4\,\text{kg} \times a$$
$$a = \frac{12\,\text{N}}{4\,\text{kg}}$$
$$= 3\,\text{m/s}^2$$

On Earth, the mass has acceleration $3\,\text{m/s}^2$.

(b) On the Moon, the body still has mass 4 kg. The force applied is still 12 N. So the acceleration is the same as on Earth, $3\,\text{m/s}^2$.

(c) In a space laboratory, the body still has mass 4 kg. The force applied is still 12 N. So the acceleration will again be $3\,\text{m/s}^2$.

Is Newton's Second Law obvious?

It is always easy to be wise after the event! After someone has told you to apply different forces to a trolley and measure its acceleration each time, it almost seems obvious that acceleration will turn out to be proportional to the force. But the result really does tell us something very important, and not at all obvious, about the natural world. It tells us that there is a direct connection between the increase in distances travelled by a body between successive ticks of a clock (its acceleration), and the extension of an elastic (the force applied to the body). We could not have predicted any simple connection between these two quite separate things just by thinking about it; they are only shown to be connected by experiment.

The evidence we have for believing that Newton's Second Law is accurate is not just based on a trolley experiment. Physicists use this law because they have found that it works well for explaining and predicting motion in an enormous number of situations, most of them much more complex than our simple trolley experiment.

Questions

1 What force is needed to accelerate:
  (a) a 20 kg barrow at $0.5\,\text{m/s}^2$;
  (b) a 2 kg ball at $8\,\text{m/s}^2$;
  (c) a 800 kg car at $2\,\text{m/s}^2$?

2 What is the acceleration of:
  (a) an 8 kg mass when a 2 N force acts on it;
  (b) a 50 kg mass when a 400 N force acts on it?

3 A force of 1 N is applied to a brick sitting on a very smooth sheet of ice. (Assume that friction is so small it can be ignored.) The brick accelerates in the direction of the force.

Look at the situations on the right in turn, and match each diagram with one of the descriptions in the list.

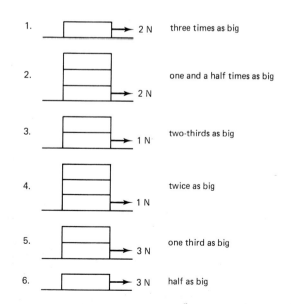

4 A total force of 0.4 N acts on a laboratory trolley. The trolley has an acceleration of 0.5 m/s². What is the mass of the trolley?

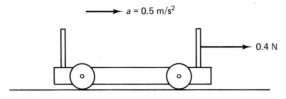

5 A car of mass 800 kg is travelling along a motorway at 12 m/s. To overtake a bus, the driver increases her speed to 20 m/s in 16 s.
(a) What is the car's acceleration?
(b) What total force is needed to produce this acceleration?
(c) If the friction and air resistance force on the car is 400 N, what is the driving force exerted on the car by the engine?

6 A cyclist and her bike have a total mass of 60 kg. Starting from rest, she can produce a forward force of 140 N. The friction force on the bike is 20 N.

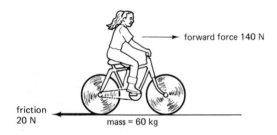

(a) What is the total force on the bike and rider?
(b) What is the cyclist's acceleration?
(c) If she keeps up this acceleration for 5 s, how fast will she then be travelling, and what distance will she have gone?

7 A small child steps out in front of a car which is moving at a speed of 57.6 km/h along a good road. The driver puts on his brakes and, without swerving, just manages to stop in time before hitting the child. The mass of the car is 1000 kg and the brakes operate for 3.2 s.
(a) Calculate:
  (i) The initial speed of the car in m/s;
  (ii) the acceleration of the car if the brakes operate for 3.2 s;
  (iii) the force needed to stop the car;
  (iv) the distance in which the car stops from the position where the brakes are first applied.
(b) Explain briefly why your answer to (a) (iv) represents a **minimum** stopping distance for the car.
(c) State **four** factors which might affect a driver's ability to stop under similar circumstances.
(d) Many motorways and major trunk roads are provided with crash barriers positioned between carriageways. Explain why such barriers are designed to slow a car down rather than to stop it almost immediately. (AEB: SCISP)

8 A supermarket trolley full of groceries has a total mass of 24 kg. To get it moving, a shopper pushes the trolley with a force of 50 N. The trolley's initial acceleration is 1.5 m/s². What size is the friction force on the trolley?
(Hint: you know the mass and the acceleration, so you can calculate the **total** force on the trolley.)

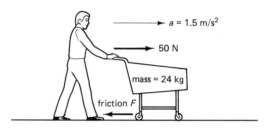

9 A car travelling at 12 m/s crashes into a wall and is stopped in 0.1 s. A front seat passenger anticipates the crash and tries to push against the dashboard to prevent himself being 'thrown' into the windscreen. The passenger's mass is 70 kg.
(a) With what speed is the passenger originally travelling?
(b) In what time interval must the passenger's body slow down to zero?
(c) What is the acceleration of the passenger's body during the crash?
(d) What force will he need to apply against the dashboard to stop himself?

How many times is this force greater than his weight? Do you think he could apply so large a force? What safety feature could have enabled his body to slow down rapidly without serious injury?

10 The acceleration of a vehicle on a sloping air track may be measured using the apparatus shown.

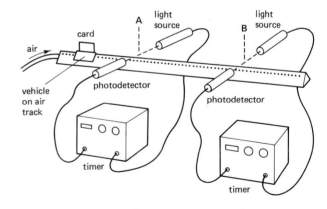

The vehicle starts from rest at the top of the track.
The time taken by the card on the vehicle to pass through a light beam is recorded by the timer.
The time taken by the vehicle to travel from **A** to **B** is measured to be 2.0 s.
Explain how you could find the acceleration of the vehicle. Your answer should include a list of the measurements you would take and a description of how you would calculate the acceleration. (SEB)

11 A girl on a sledge slides down a slope. The total mass of the girl and the sledge is 100 kg. The record of their journey from **A** to **D** is indicated on a combined stopwatch–speedometer attached to the sledge. The readings of this instrument at positions **A**, **B**, **C** and **D** are shown.

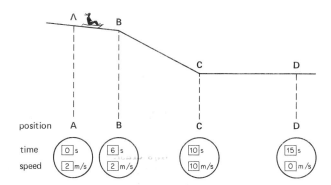

(a) Describe the motion of the sledge during each of these stages.
(b) Calculate the average retarding force acting on the sledge during stage **CD**.
(c) If the girl lies flat along the sledge after passing **A**, suggest possible new readings on the stopwatch–speedometer as the sledge passes **B**. (SEB)

12 An astronaut who is very fond of potatoes takes a 30 kg sack with him on a flight from Earth to a space station on the Moon and then on to the planet Jupiter. On the Moon, the gravitational field strength is about one-sixth of that on Earth; on Jupiter it is about 2.5 times greater than on Earth. Complete the table:

	Mass of potatoes	Weight of potatoes	Acceleration of potatoes when pushed with a force of 60 N
On Earth			
On the Moon			
On Jupiter			

Note: remember to include **units**.

13 (a) Unbalanced forces of 1, 2, 3, and 4 units were applied to a trolley, in turn, producing changes in its speed from rest. The following speed–time graphs were obtained.

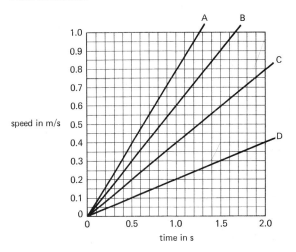

(i) Calculate the value of the acceleration for each speed–time graph.
(ii) Copy and complete the following table.

Unbalanced force/units	1	2	3	4
Acceleration/m/s²				

(iii) State the relationship between unbalanced force and acceleration which these results illustrate.
(b) A rider and his moped have a combined mass of 85 kg.
  (i) The rider finds that his average acceleration from rest is 0.6 m/s². Calculate the average unbalanced force acting on rider and moped.
  (ii) Eventually the moped reaches top speed so that, although the engine is still applying a force, the moped and rider do not accelerate. Draw a diagram showing the **horizontal** forces acting on the moped and rider at the top speed.
  (iii) Hence explain why they do not accelerate at this speed. (SEB)

INVESTIGATION
In a car accident, the car may be brought to a standstill in 0.1 s or less. Could you push hard enough on the dashboard to stop yourself in this time?

9: Circles, Projectiles and Satellites

This chapter looks at two important kinds of motion which Newton's Laws can help us understand:

– motion in a circle;
– projectile motion.

From this we can go on to see how it is possible for satellites to orbit the Earth and why people and objects inside them appear weightless.

9.1 Motion in a circle

In Chapter 5, we made a distinction between **speed** (a scalar quantity) and **velocity** (a vector). In section 5.6, we noticed that because velocity is a vector (with both magnitude **and** direction), a body can have **constant** speed but **changing** velocity. One example is a body moving along a curved path. As acceleration is defined as:

$$\frac{\text{change of velocity}}{\text{time taken}}$$

it is possible for a body to have constant speed but still be accelerating. Does this make sense, or is it just an example of physicists playing with words – making things seem more difficult than they really are? In fact, it does make sense: for a body moving in a circle, we can work out the direction of its acceleration and we will see that the result is entirely consistent with Newton's Laws of Motion.

The tension force in the string keeps the stone moving in a circle.

Imagine a stone being whirled round in a circle. We know from experience that the string is taut while the stone is moving round. The tension in the string pulls the stone inwards. The tension is a force, always pulling the stone towards the centre of the circle. If the string suddenly breaks, the stone will fly off in the direction it happens to be going at that moment.

This is exactly what Newton's First Law of motion predicts. When there is no force supplied by the string, the stone moves in a straight line at a steady speed.

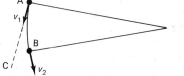

change of velocity

Let us think a little more about the motion while the string is taut. Consider the little section of the stone's motion shown in the diagram – as it moves from **A** to **B**. If the string broke when the stone was at **A**, the stone would fly off along the direction $\overrightarrow{AC}$. But because of the force supplied by the string, the stone is pulled inwards from this straight-line path, and follows the circular path. We can draw a vector diagram for the velocities during this short interval. The sum of the original velocity v_1 and the change of velocity must equal the final velocity v_2. The change of velocity is a vector pointing towards the centre of the circle. So the acceleration vector must also be in the same direction.

This means that the stone has an acceleration towards the centre of the circle **and** there is a force (the tension in the string) towards the centre keeping the stone in its circular path. This is consistent with Newton's Second Law ($F=ma$), which says that a force is required to produce an acceleration.

Examples like this show how velocity, acceleration and force are all linked up. This is why velocity is a more useful idea than speed, and why acceleration is defined as the rate of change of velocity (and not as the rate of change of speed).

If the string breaks, the stone will fly off in a straight line in the direction it is travelling at that moment.

9.2 Projectiles

The drawing summarises the effects on a body's motion of a force...

(A) ... in the same direction as the body's motion
(B) ... always at right angles to the body's motion
(C) ... constant in direction, but at right angles to the body's initial motion.

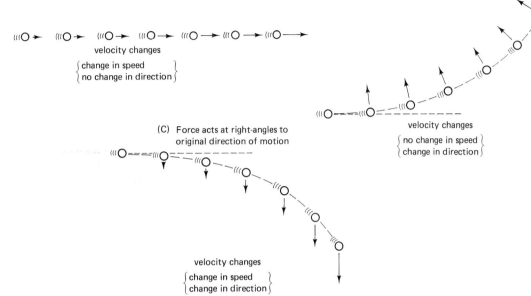

This last situation is the one we want to turn to now. A body that is thrown forwards is called a **projectile**; its path is sometimes referred to as its **trajectory**. From the moment it is released, only one force acts on a projectile – gravity, acting vertically downwards. The gravity force (weight) is constant, and gives the projectile a constant acceleration downwards (of $10 \, \text{m/s}^2$). There is no horizontal force (if we can neglect air resistance), so according to Newton's First Law, the projectile should have constant speed in the horizontal direction.

These deductions are correct only if the vertical motion and the horizontal motion are completely independent. Does the sideways motion of a projectile have any influence on its motion up-and-down? The only way to answer this is by experiment.

One simple experiment is illustrated in the diagram. You can try this for yourself at home. Take two identical coins and a ruler, and place them in the positions shown in the diagram. Then give the ruler a sudden sharp tap, pushing it sideways. The coin on top of the ruler will fall; the other coin is a projectile. If you listen carefully, you will hear them both land on the floor at the same instant. The projectile has fallen downwards as quickly as the falling body. The sideways motion has had no effect on its rate of fall.

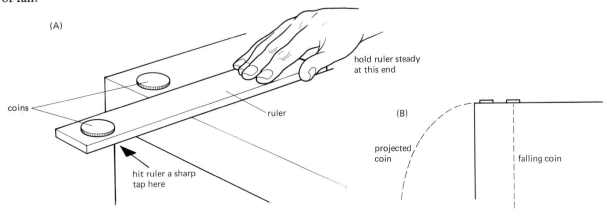

78 Circles, projectiles and satellites

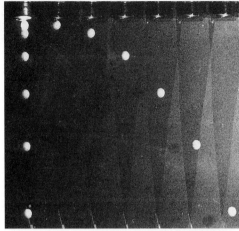

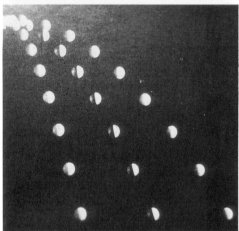

Strobe photography allows us to investigate this sort of situation more thoroughly.

The first picture shows a strobe photograph of a falling ball and a projected ball. Notice that their vertical positions are the same each time. This is confirmed by the second photograph which shows the motion of three projectiles with different horizontal speeds. The vertical motion is the same each time.

To summarise these results:

(a) The horizontal and vertical components of a projectile's motion are completely independent.
(b) The horizontal component of a projectile's velocity is constant.
(c) Vertically, a projectile has uniform downward acceleration.

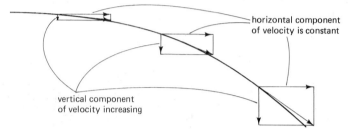

9.3 Newton's thought experiment

Today there are hundreds of satellites in orbit around the Earth. The Moon is a natural satellite of the Earth. In 1957, the Russian Sputnik 1 became the first man-made satellite. Satellites are used for a variety of purposes:

- communications (telephone and television links between different parts of the world;
- photographing the Earth's surface (for weather-forecasting, to study the growth of crops, and for espionage);
- for carrying out scientific experiments (a satellite is a mini-laboratory outside the Earth's atmosphere, and particularly useful for studying space).

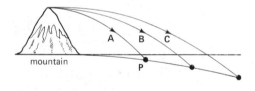

But why don't they fall? What keeps a satellite in position? In fact, the possibility of satellite motion was first proposed by Newton, who described the following 'thought experiment'. Imagine a very high mountain from which a shell can be fired horizontally. The shell is a projectile, so we would expect it to follow the parabolic path **A**. However, the Earth's surface is curved, not flat, and so the shell will travel further before landing – to the point **P**. Shells fired faster still would travel along path **B**, or path **C**. If the firing speed was faster still, Newton suggested that the shell would go right round the Earth and return to its starting point!

This is how a satellite orbits. It must be fired into orbit by a rocket so that it is travelling at exactly the right speed to follow a circular path. The satellite is high enough to be outside the Earth's atmosphere, so that there is no air resistance to slow it down. The satellite has **not** escaped from the Earth's gravity. In fact, it is falling all the time. But the curved surface of the Earth 'falls' away from the satellite at exactly the same rate as it falls towards the surface! So it stays at the same height!

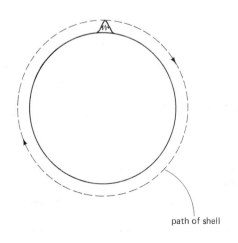

The satellite moves at constant speed in a circular path. As we have seen in section 9.1, this means that it is accelerating towards the centre of the circle (because although its speed is constant, its velocity is always changing – the **direction** of motion is changing all the time). The force needed to produce this acceleration is the force of gravity on the satellite. So a satellite stays in orbit round the Earth because of the pull of the Earth's gravity, and **not** because it has escaped from the pull of gravity.

The speed of a satellite in its orbit depends on how high it is. The pull of the Earth's gravity gets smaller, the further above the Earth you go. For a satellite in a high orbit, a slower speed is required to balance the effect of the gravity force.

Some satellites are placed in what is known as an 'Earth-stationary' orbit. They are in an orbit at just the correct height to ensure that their orbital speed matches the rotation speed of the Earth on its axis. As the satellite orbits, the Earth rotates under it at exactly the same rate. So from the point of view of an observer on the Earth, the satellite appears to be stationary. Its relative velocity is zero. This is very useful for communications satellites – they are always up there when you need them!

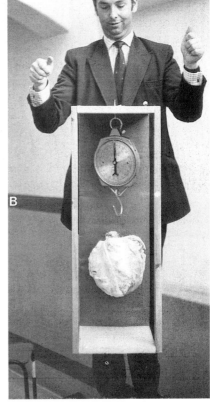

This satellite has a lower orbital speed, because it is further above the Earth.

Gravity provides the force needed to keep the satellite moving in a circle.

9.4 'Weightlessness'

An orbiting spacecraft is just a special kind of satellite. We know that astronauts in orbit experience 'weightlessness'. But we have just explained how satellites orbit because of gravity – not because they have escaped from gravity. So what is the explanation of weightlessness?

To answer this, we should look first at a simpler experiment carried out on Earth.

The spring balance in the box in photograph A is carrying a 1 kg mass. The balance reads 10 N. But when the box is dropped, the spring balance reads zero. As the box falls (B), there is still a force of 10 N acting downwards on the kilogram. But the spring balance and box are falling also. They are all accelerating downwards at $10 \, m/s^2$. Inside the box, the conditions really are the same as you would experience in deep space, so far away from the Earth and all other large masses that there really was no gravitational force.

Inside an orbiting spacecraft, both the spacecraft itself and the astronauts inside it are accelerating at exactly the same rate towards the centre of the Earth. They are falling, but because of the curve of the Earth's surface, the path they follow as they fall keeps them moving in a circular orbit. Normally we experience our weight, because of the reaction force of the floor. In an orbiting spacecraft, the floor is falling at the same speed as the astronauts, and so they experience 'weightlessness'.

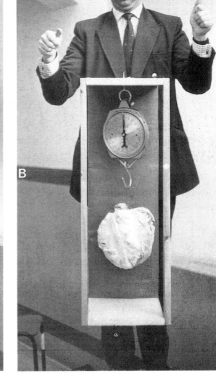

Questions

1 Your grandfather, who is a keen stamp collector, sends you a letter with this stamp on the envelope. He asks you to explain to him what the design on it is about. The stamp was issued to mark the 300th anniversary of the publication of Newton's most famous book, *Principia*. Write a letter back to your grandfather explaining the meaning of the stamp's design.

2 Copy the diagram of a satellite in orbit, and mark clearly all the forces acting on the satellite. Objects inside this satellite appear 'weightless'. Explain why.

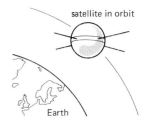

INVESTIGATION

Devise a cheap and easy method of plotting accurately the path of a projectile.

10: Momentum

mass of lorry = 10 000 kg
→v = 20 m/s
momentum = 200 000 kg m/s

mass of car = 800 kg
→v = 25 m/s
momentum = 20 000 kg m/s

mass of cyclist and bike = 100 kg
→v = 5 m/s
momentum = 500 kg m/s

mass of hockey ball = 0.5 kg
→v = 6 m/s
momentum = 3 kg m/s

mass of golf ball = 0.1 kg
→v = 0.5 m/s
momentum = 0.05 kg m/s

Momentum is a word we sometimes use when talking about moving objects. A heavy object travelling fast has a lot of **momentum**. We can measure momentum and we find that there is an important link between momentum and force. The scientific idea of **momentum** turns out to be very useful in understanding what happens when bodies interact – in collisions and explosions.

10.1 The idea of momentum

If you are cycling along slowly on your bicycle, you don't feel you have much momentum. But if you are freewheeling downhill, you have a lot of momentum. A heavy lorry has more momentum than a bicycle travelling at the same speed. Momentum depends on the **mass** of the moving body and how fast it is moving (its **velocity**).

The momentum of an object is defined as follows:

momentum = mass × velocity
momentum = *m v*

Mass is measured in kilograms and velocity in metres per second, so momentum is measured in kilogram metres per second (kg m/s). There is no shorthand name for this unit.

10.2 Newton's Second Law and momentum

When a force acts on a body, it accelerates – in the direction of the force. The size of the acceleration depends on the size of the force and the mass of the body. The equation which sums this up is: $F = ma$ (Newton's Second Law).

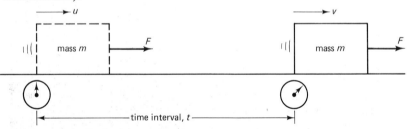

The constant force F makes the object speed up from an initial velocity u to a final velocity v in t seconds.

From this we can work out an important link between force and momentum.

By definition, $$\text{acceleration} = \frac{\text{change in velocity}}{\text{time taken}}$$

$$a = \frac{v - u}{t}$$

So, $F = ma$ can be written $$F = \frac{m(v - u)}{t}$$

or $$F = \frac{mv - mu}{t}$$

mu is the original momentum of the body; mv is its momentum after the force acts. $(mv - mu)$ is the change in momentum. So the equation above can be written in words:

$$\text{force} = \frac{\text{change of momentum}}{\text{time}}$$

or force = rate of change of momentum

Impulse

If:
$$\text{force} = \frac{\text{change of momentum}}{\text{time}}$$

then: force × time = change of momentum
$$Ft = mv - mu$$

The quantity 'force × time' is called the **impulse** of the force.

In the SI system, momentum has units kilogram metres per second (kg m/s). The units of impulse (force × time) are newton seconds (N s). But since force × time = change of momentum, these units must be the same:

$$1 \text{ kg m/s} = 1 \text{ N s}$$

10.3 Two versions of Newton's Second Law

There are two different ways of writing Newton's Second Law of motion.

The 'acceleration' version: Force = mass × acceleration
$$F = ma$$

The 'momentum' version:
$$\text{Force} = \frac{\text{change of momentum}}{\text{time}}$$
$$F = \frac{mv - mu}{t}$$

These are entirely equivalent.

10.4 Varying forces

The momentum version of the second law of motion really comes into its own when we are dealing with varying forces. In Example 10.1 the force is constant. In many real-life situations, the force is more likely to change during the interaction. For instance, when a tennis racquet strikes a tennis ball, the force will increase to a maximum as the strings stretch and the ball compresses, and then decrease again – all within a few milliseconds (thousandths of a second).

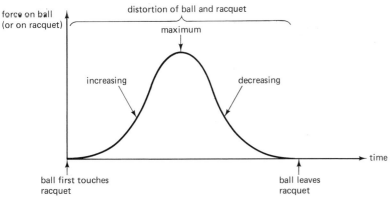

But if we know the total change of momentum of the ball and the time the collision lasted, this is all we need to work out the **average force** on the ball during the collision.

$$\text{average force} = \frac{\text{total change of momentum in collision}}{\text{total time collision lasts}}$$

The quantity 'force × time' is called the **impulse**. In this case, it is 5000 Ns. So the change of momentum in this time is also 5000 Ns (or 5000 kg m/s).

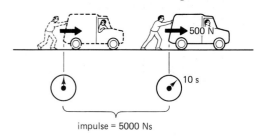

impulse = 5000 Ns

EXAMPLE 10.1

A spacecraft has mass 800 kg. Its rocket motors fire for 4 s and increase its velocity from 20 m/s to 25 m/s. What force do the motors exert?

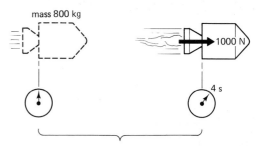

Before the rocket motors fire, the spacecraft's momentum is:

$$800 \text{ kg} \times 20 \text{ m/s} = 16\,000 \text{ kg m/s}$$

After the motors fire, its momentum is:

$$800 \text{ kg} \times 25 \text{ m/s} = 20\,000 \text{ kg m/s}$$

So, the change of momentum

$$= (20\,000 - 16\,000) \text{ kg m/s}$$
$$= 4000 \text{ kg m/s}$$

We can then use the equation:

$$\text{force} = \frac{\text{change of momentum}}{\text{time}}$$
$$\text{force} = \frac{4000 \text{ kg m/s}}{4 \text{ s}}$$
$$= 1000 \text{ kg m/s}^2 = 1000 \text{ N}$$

The rocket motors exert a force of 1000 N.

(**Note.** Another way to find the answer is to begin by working out the spacecraft's acceleration, then use $F = ma$ to find the force. Check that you get the same result if you do it this way.)

The Second Law in action

Seat belts

By law, front-seat passengers in cars must wear seat belts. Belts are becoming standard equipment in new cars for back-seat passengers also. But why is it safer to wear a belt? Imagine that you are in a car travelling at 70 km/h. Your body is also travelling along at 70 km/h. If the car is involved in a collision, it will suddenly be brought to a stop. But because of its inertia, your body will continue travelling at 70 km/h – until it hits an obstruction, usually the car's windscreen. Without a seat-belt, you can be seriously injured. The belt exerts a force on your body, bringing it to a stop without hitting the windscreen or the steering wheel.

Estimate how long it took for the dummy passenger to be stopped by the belt. If its mass is 80 kg and the car was travelling at 30 m/s before the crash, what average force did the belt have to exert?

(Answer: The dummy is stopped by the belt in photo 5. From photo 1 to photo 5 is four intervals, or 4×0.05 s = 0.2 s. The average force exerted by the belt is 12 000 N)

Testing a seat belt using a dummy passenger. The photos are at intervals of 0.05 s. Notice how the dummy moves forward but is prevented by the belt from hitting the steering wheel or the windscreen.

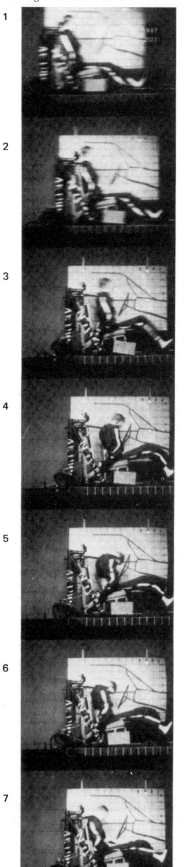
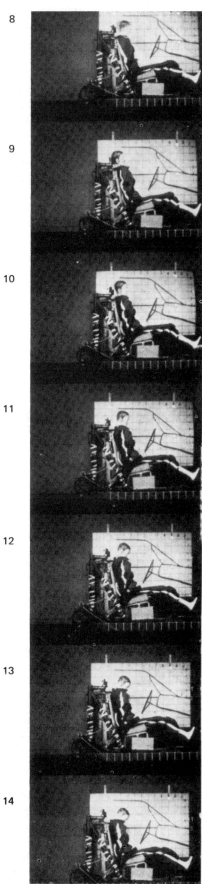

In an accident, it is important that the passenger's body is not held completely still by the belt. It moves a short way. By spreading the passenger's 'slowing down' over a longer period of time, the belt reduces the size of the force on the passenger. We can see how this comes about.

When a force F acts for time t, and causes a change of speed from u to v:

$$Ft = mv - mu$$

For the passenger in the car accident, the change of momentum $(mv - mu)$ in the collision depends only on her mass and the car's speed before the collision. If the time interval t in which she is brought to a stop is very small, then the force F will be large. On the other hand, if the time t can be made larger, the force F will be less.

This table summarises the position:

Duration of collision	Force on car and passengers
very short	very large
slightly longer	smaller

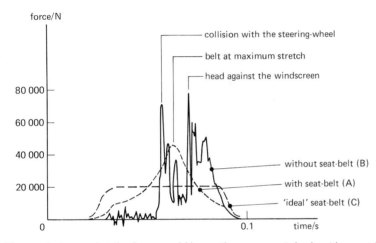

The graph shows what the force would be on the passenger's body with a seat belt (**A**) and without a seat belt (**B**). An 'ideal' seat belt would produce the graph (**C**), where the force is spread evenly over the whole time of the crash, and the force is smallest as a result.

Crumple zones

Another safety feature of modern cars which uses the same principle is the 'crumple zone' at the front and back. In a crash, the bonnet of the car is designed to crumple, making the collision last a slightly longer time. The force exerted on the car (and on its passengers) is then smaller.

Safety helmets

Safety helmets are compulsory for motor cyclists. There is a growing amount of evidence to suggest that many injuries to pedal cyclists would also be much less serious if they wore safety helmets too. The purpose of a safety helmet is to protect the wearer's head from large forces in an accident. It works on exactly the same principle as the seat belt and crumple zone.

Inside the helmet there is a layer of expanded foam or other similar padding. The cyclist's head is held in a webbing support. In a crash, if the cyclist's head hits another vehicle or the road, the webbing and the padding inside the helmet allow the head to move a short distance before stopping. The time allowed for the cyclist's moving head to slow down and stop is longer. So the average force on the head is smaller.

The best helmet is one which 'spreads' the slowing down force over the longest period of time – this way the force is smallest.

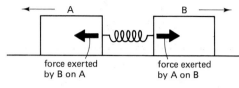

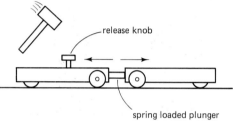

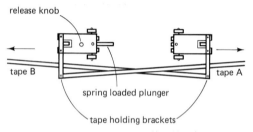

10.5 Interactions between bodies: forces in pairs

When a force acts on a body, it causes a change in the body's momentum. It is often easier to think of just one body and look at how its motion changes when a force acts. But we already know (section 7.7) that forces always occur **in pairs**. If I push a trolley, the trolley pushes back on me. If body **A** exerts a force on body **B**, then body **B** exerts a force back on body **A**. The two forces are equal in size but opposite in direction – Newton's Third Law of motion.

Instead of just considering one body, it is sometimes more useful to think of both bodies involved in an interaction.

10.6 Explosions

One simple interaction between two bodies is an explosion – where a single body at rest springs apart into two separate moving pieces. For example, if we place a spring-loaded trolley against another trolley and release the spring, the two trolleys move apart in opposite directions. This is a fairly gentle 'explosion' but it illustrates all the general principles.

We can measure the speeds of the trolleys by any of the usual methods (see section 5.3) such as a ticker-timer or an electronic timer with a photo-detector switch.

Because the trolleys move in opposite directions, it is impossible to do this experiment on a friction-compensated runway. The best we can do is to work on a level bench and measure both velocities shortly after the explosion, so that friction has had little time to cause much slowing down. With ticker-tape, we can take the first ten-space section from each tape and calculate the velocities from these. We can repeat the experiment using two and then three trolleys, with the same spring-loaded trolley. The diagram shows typical ticker-tapes from this experiment and the velocities measured from them.

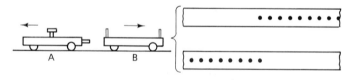

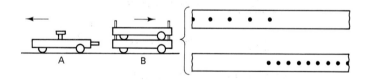

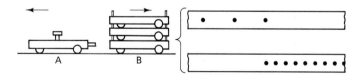

The easiest case is where the two trolleys have the same mass. The two trolleys move apart with exactly the same speeds. The velocity of trolley **A** is equal and opposite to the velocity of trolley **B**. It follows that the momentum of trolley **A** (mass × velocity) must be equal and opposite to that of trolley **B**.

In the second experiment, trolley **B** has twice the mass of trolley **A**. After the explosion, its velocity is half that of trolley **A**. So the momentum of trolley **B** is the same size as the momentum of trolley **A** – but in the opposite direction.

The same thing happens in the third experiment. This time trolley **B** has three times the mass of trolley **A**. Its velocity is one-third of trolley **A**'s. The momentum of **B** is the same size as the momentum of **A**, but opposite in direction.

This is an important result: when two bodies 'explode' apart, the momentum of one body is equal in size but opposite in direction to the momentum of the other body. But there is another way of putting this which is even more useful. Remember that velocity is a vector quantity. Momentum is mass × velocity, so it must also be a vector. After the explosion, the momentum of **B** is equal and opposite to the momentum of **A**. If we add the two momenta, the total is zero. Before the explosion, the bodies were stationary, so the momentum was zero. The total momentum is the same before and after the explosion.

Momentum is conserved in explosions.

After explosion:

'Exploding' trolley			Other trolleys			TOTAL
Mass	Velocity	mv	Mass	Velocity	mv	mv
1	−0.8	−0.8	1	0.8	0.8	0
1	−1.0	−1.0	2	0.5	1.0	0
1	−1.2	−1.2	3	0.4	1.2	0

EXAMPLE 10.2

A boy of mass 60 kg and a girl of mass 40 kg stand facing each other on a skating rink. The girl pushes the boy and he moves backwards at 2 m/s. What is the girl's velocity?

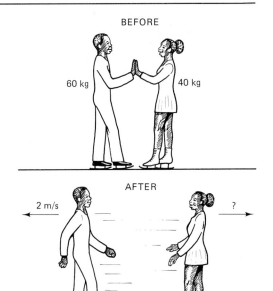

Before the push, they are both at rest. The total momentum is zero. After the push (an 'explosion'), the total momentum is also zero. So the girl's momentum is equal to the boy's, but in the opposite direction.

First we calculate the boy's momentum:
$$\text{momentum} = mv$$
$$= 60 \text{ kg} \times 2 \text{ m/s}$$
$$= 120 \text{ kg m/s}$$

So, the girl's momentum = 120 kg m/s (in the other direction).

Her mass is 40 kg, so we can calculate her velocity:
$$\text{momentum} = mv$$
$$120 \text{ kg m/s} = 40 \text{ kg} \times v$$

Divide both sides by 40 kg: $v = \dfrac{120 \text{ kg m/s}}{40 \text{ kg}}$
$$= 3 \text{ m/s}$$

So the girl moves off in the opposite direction at 3 m/s.

Conservation of momentum in explosions and Newton's Third Law

Momentum is conserved in explosions because of Newton's Third Law. The two momenta are equal and opposite because the two forces acting (the force-pair) are equal and opposite. We can show this in the following way.

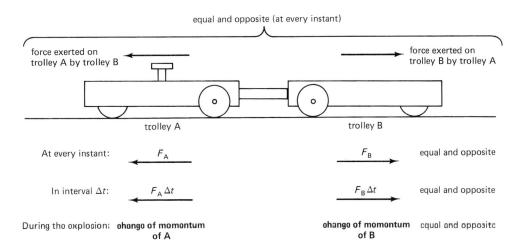

Imagine an explosion experiment with two trolleys. According to the Third Law, the force exerted on trolley **B** by trolley **A** is equal in size (at every instant) and opposite in direction to the force exerted on trolley **A** by trolley **B**. If we think of a tiny time interval Δt during the explosion, the impulse of the two forces ($F\Delta t$) will be the same for both trolleys. But we know that:

$$F\Delta t = \text{change of momentum}$$

So the change of momentum of one trolley in that tiny time interval must be equal and opposite to the change of momentum of the other trolley. This is true for each tiny time interval during the explosion. If we add all these up to find the total change of momentum for each trolley during the whole explosion, the same thing must apply. The change of momentum of one trolley is equal and opposite to the change of momentum of the other.

It doesn't matter what the masses of the trolleys are—conservation of momentum in explosions follows directly from the Third Law. Or we could put it the other way round: the experimental result that momentum is conserved in explosions is evidence that Newton's Third Law is true!

Rockets and jets

We can use the Third Law of motion and the idea of conservation of momentum to understand how rocket and jet engines produce their forward thrust.

Jet engine

In a jet engine, the compressor is like a large fan drawing in air and forcing it into the combustion chamber. Kerosene fuel burns in this compressed air, and the very hot gases force their way out through the blades of a turbine. As the drawing shows, a turbine is rather like a specially designed windmill, and works in much the same sort of way. Once the jet engine is started, the turbine can be used to drive the compressor also, but to get it all started, the compressor must be driven by a separate motor.

The exhaust gases emerge from the jet engine at high speed – more than 250 kg of air per second in a large jet engine. The force exerted on this gas by the engine is one force of a force-pair. The other force of the pair is the force exerted by the gas on the engine. This is equal and opposite, and so the jet engine experiences a forward force acting on it. The exhaust gases have been given momentum in the backwards direction. As total momentum is conserved, the engine (and the whole aeroplane) has an equal amount of momentum in the forward direction.

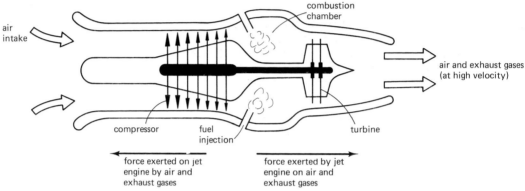

Rocket engine

Rockets are able to work in space where there is no air. A liquid-fuel rocket burns its fuel, forcing hot gases out of the rocket's exhaust at high speed. In this way the rocket exerts a force on the exhaust gases. The second force in this force pair is the forward force exerted by the gases on the rocket. This is what makes the rocket move forward. The momentum of the exhaust gases is equal and opposite to the momentum of the rocket.

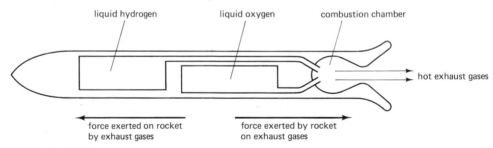

10.7 Collisions

Collisions are another example of an interaction between two bodies. Is momentum also conserved in collisions and can we use this to work out how fast an object will travel after a collision?

◁ A snooker player weighing up a shot is trying to decide exactly where the balls will go after they collide.

Policemen investigating a ▷ road accident try to do the same sort of thing in reverse. From the positions of the cars after the collision they can estimate how fast they were travelling before they collided.

Inelastic collisions

The easiest kind of collision to investigate experimentally is one where a moving body hits a stationary one, and the two stick together after the collision. This is called an **inelastic** collision. The experiment can be done with two trolleys on a friction-compensated runway (see section 8.1). When they collide, a pin fixed to one trolley sticks into a cork fixed to the other, keeping them both together. The velocities before and after collision can be measured by any of the methods described in section 5.3.

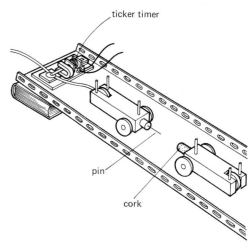

A ticker-tape shows clearly the change of velocity at the moment of collision.

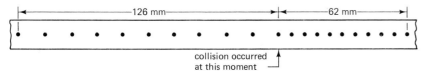

From the tape, it is easy to work out the velocity of the trolley before the collision and the velocity of the two combined trolleys after the collision. We measure the length of ten spaces shortly before and shortly after the point where the collision occurred. This is the distance travelled by the moving body in $\frac{10}{50}$ s (or $\frac{1}{5}$ s). For the sample tape shown in the drawing above, the results are as follows:

	Before collision	After collision
Length of 10-space section	126 mm	62 mm
Velocity of trolley(s)	$\frac{126 \text{ mm}}{0.2 \text{ s}}$	$\frac{62 \text{ mm}}{0.2 \text{ s}}$
	$= 630$ mm/s	$= 310$ mm/s

We can then repeat the experiment with two moving trolleys colliding with one stationary trolley; and then with three moving trolleys colliding with one stationary trolley.

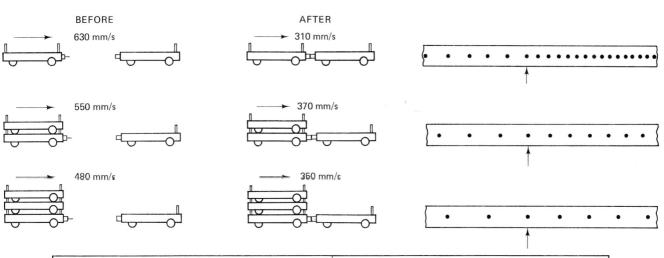

BEFORE COLLISION			AFTER COLLISION		
Moving mass (*m*)/units	Velocity (*v*)/(mm/s)	*mv*/units	Moving mass (*m*)/units	Velocity (*v*)/(mm/s)	*mv*/units
1	630	630	2	310	620
2	550	1100	3	370	1110
3	480	1440	4	350	1400

A typical set of results is shown above. We can see that the momentum (mass × velocity) is almost exactly the same before and after the collision. Within experimental error, the momentum is the same before and after. Momentum is conserved in an inelastic collision.

EXAMPLE 10.3

A fully loaded railway wagon **A** of mass 3000 kg moving at 8 m/s collides with a stationary empty wagon **B** of mass 1000 kg and both couple together. What is the velocity of the two wagons after the collision?

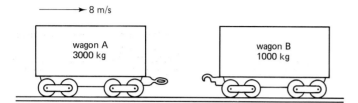
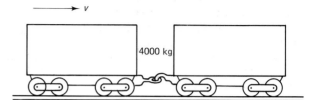

(i) Before collision:

momentum of wagon **A** = mass × velocity
= 3000 kg × 8 m/s
= 24 000 kg m/s

momentum of wagon **B** = 0

total momentum (of both wagons) = 24 000 + 0 kg m/s
= 24 000 kg m/s

(ii) After collision:

Momentum is conserved in a collision. So the total momentum of the system after the collision is also 24 000 kg m/s. The two coupled wagons have mass 4000 kg. Their momentum is 24 000 kg m/s.

momentum = mass × velocity

So $24\,000 \text{ kg m/s} = 4000 \text{ kg} \times v$

Divide across by 4000 kg: $v = \dfrac{24\,000 \text{ kg m/s}}{4000 \text{ kg}}$

= 6 m/s

The final velocity of the two wagons is 6 m/s.

Elastic collisions

In many collisions, the colliding objects do not stick together, but bounce apart. Some collisions of this sort can also be investigated using trolleys, provided we make sure that the direction of motion doesn't change after the collision. To make the trolleys bounce cleanly apart, one is fitted with a nose cone, and the other with an elastic band stretched between its posts. One trolley is stationary before the collision, but both will be moving separately afterwards. So we need two lengths of ticker-tape, one attached to each trolley. To keep the tape from the trolleys out of each other's way, we can fit brackets to hold the tape well clear of the moving trolleys themselves. We need only one ticker-timer however; by using two carbon discs, one timer can mark both tapes.

We can do a series of experiments using one, two and then three moving trolleys colliding with a single stationary one. In each case, the total momentum after the collision is calculated by working out the momenta of each trolley separately. These are then simply added, since the trolleys are all moving in the same direction.

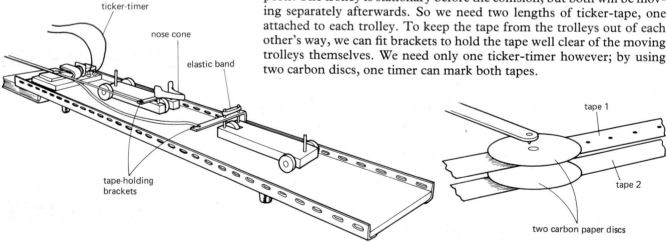

Again, the results show that the total momentum of both trolleys after the collision is equal (within the accuracy of the experiment) to the momentum of the moving trolley before the collision. Momentum is also conserved in collisions of this kind.

(Note: In a perfectly elastic collision, another quantity – the total kinetic energy – is also conserved. After you have read Chapter 11, use the data in the table to work out the total kinetic energy of the moving trolleys before and after these collisions. What do you find?)

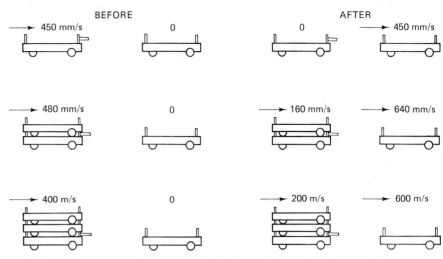

BEFORE COLLISION			AFTER COLLISION						
For trolley 1			For trolley 1			For trolley 2			TOTAL
Mass	Velocity /(mm/s)	Momentum	Mass	Velocity /(mm/s)	Momentum	Mass	Velocity /(mm/s)	Momentum	Momentum
1	450	450	1	0	0	1	450	450	450
2	480	960	2	160	320	1	640	640	960
3	400	1200	3	200	600	1	600	600	1200

Collisions in general

Both the trolley experiments described above deal with special examples of collision. In general, a collision might involve both bodies moving before the collision. They could be moving in the same direction, or in opposite directions. After the collision they might bounce apart in opposite directions also.

When two directions of motion are involved, it is important to remember that momentum is a vector quantity. When we are adding two momenta we must take their directions into account.

To investigate a more general type of collision, we can use an air-track and two photodetector timers.

This sort of measurement has become much easier since microcomputers became available. The microcomputer can time the gliders as they pass the photodetectors before and after the collision, and calculate their speeds. We can then analyse these results at leisure after the experiment is done. (For more information, see Appendix 2.)

We always find that, within experimental error, the total momentum after the collision is equal to the total momentum before. Although we cannot try out every possible collision, we can feel fairly confident that momentum is conserved in *all* collisions.

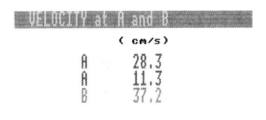

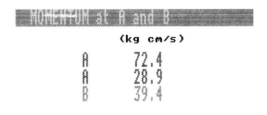

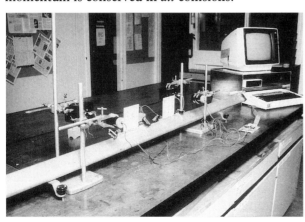

Conservation of momentum in collisions and Newton's Third Law

When two objects collide, the forces they exert on each other during the collision are equal and opposite at every moment they are in contact. According to the Third Law, the force on body **A** is equal and opposite to the force on body **B** at each instant during the collision. The forces on each body act for exactly the same time – the duration of the impact.

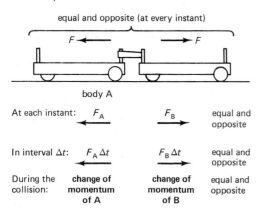

We know that: force × time = change of momentum

So the change of momentum of body **A** must be equal to the change of momentum of body **B**. The forces on the bodies are in opposite directions, so the two changes of momentum are in opposite directions also. They add to zero. The **total** momentum of the two bodies together does not change.

Momentum is conserved in any collision between bodies when no external forces are acting. The only force is then the pair of interaction forces between the colliding bodies.

10.8 Conservation of momentum: a reliable law?

We can test the law of conservation of momentum by doing a few collision experiments but we cannot possibly try all situations. So are we really wise to believe that momentum is always conserved in interactions between bodies? In fact, the law of conservation of momentum is one of the laws in which physicists have most confidence.

In the 1930s some results appeared from experiments which seemed to suggest that momentum might not be conserved in one particular situation. Some radioactive atoms (we will have a lot more to say about radioactivity later in the book in Chapter 33) are able to eject a tiny particle called a beta-particle. This is like an explosion on a tiny scale, and so the remainder of the atom recoils in the opposite direction. But when the exact calculations were done, it seemed as if the total momentum of the two objects after the 'explosion' wasn't the same as the momentum of the atom beforehand. The physicists involved had two options:

decide that they had found an exception to the law of conservation of momentum

OR

continue to believe in conservation of momentum, and find another way to explain away this peculiar result.

They decided that they would continue to believe in conservation of momentum. So they deduced that in the explosion there must be a third particle involved, which would account for the differences observed. From their calculations, this third particle had to have some very peculiar properties indeed: it had no mass, travelled at the speed of light, had momentum, and was spinning! They called it a **neutrino**.

For 20 years the neutrino was a particle which had been 'invented' to allow us to continue to believe in momentum conservation; but did it really exist? In 1956, two American physicists, Reines and Cowan, carried out an experiment in which they successfully detected a neutrino. It seems that it really does exist.

The faith of the earlier physicists in the law of conservation of momentum appears to have been justified. We have a 'hunch' that momentum conservation is one of the really fundamental laws of the physical universe.

Questions

1. What is the momentum of:
   (a) A hockey ball of mass 0.5 kg moving at 3 m/s?
   (b) A boy of mass 45 kg running at 5 m/s?
   (c) A lorry of mass 10 000 kg travelling at 15 m/s?
   (d) A car ferry of mass 20 000 000 kg moving at 10 m/s?

2. (a) A bus of mass 8 000 kg has momentum 96 000 kg m/s. What speed is it travelling at?
   (b) A cyclist and her bike have total mass 65 kg and momentum of 260 kg m/s. What is their speed?

3. A car, travelling at 15 m/s, has momentum 12 000 kg m/s. What is its mass?

4. A rock climber dislodges a 5 kg boulder, which falls down a vertical rock face. What is the momentum of the boulder:
   (a) after 1 s?
   (b) after 2 s?
   (c) when it reaches the ground 80 m below?

5. Neil kicks a football of mass 0.2 kg with an average horizontal force of 120 N.
   (a) If the ball was stationary before being kicked and moved at 10 m/s immediately afterwards, what is the change in momentum of the ball?
   (b) Calculate the time of contract between Neil's boot and the ball.

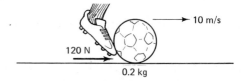

6. A rubber ball of mass 0.1 kg hits a wall at 15 m/s and rebounds at 10 m/s in the opposite direction.
   (a) If the ball is in contact with the wall for 0.1 s, what is the average force exerted by the wall on the ball?
   Imagine we now throw a ball of Plasticine, also of mass 0.1 kg, at the same wall. The Plasticine doesn't rebound but stops in 0.1 s also. Is the force at the wall bigger or smaller? Explain your answer.

7 A trolley of mass 4 kg, travelling at 3 m/s, collides with a stationary 2 kg trolley. The two stick together.
(a) What is the momentum of the moving trolley before the crash?
(b) What is the **total** momentum of both trolleys before the crash?
(c) Work out the speed of the two joined-up trolleys after they collide.

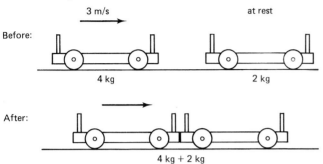

8 The diagram below shows apparatus used to investigate collisions.

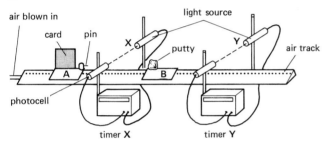

In one experiment, vehicle **B** is initially at rest and vehicle **A** is moving at a steady speed. After passing **X**, vehicle **A** collides with vehicle **B** and remains joined to it. The results are shown below.

Mass of vehicle **A**	= 0.10 kg
Mass of vehicle **B**	= 0.06 kg
Length of card on vehicle **A**	= 0.10 m
time to pass photocell **X**	= 0.50 s
Time to pass photocell **Y**	= 0.80 s

(a) Calculate the speed of vehicle **A** before the collision.
(b) Calculate the speed of vehicles **A** and **B** after the collision.
(c) Use these results to show that momentum is conserved in this collision. (SEB)

9 A truck is rolling freely along a horizontal surface at 2.0 m/s. A packing case is dropped on to the truck and both move on together. If the mass of the truck is 100 kg and the mass of the packing case is 50 kg, what is their velocity?

10 When the candle burns through the thread, the elastic catapult shoots the lead pellet (mass 0.001 kg) away at 10 m/s. If the mass of the balsa wood base is 0.05 kg, what velocity will it reach.

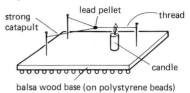

11 (a) In an experiment at an accident research unit, a moving car **P** was made to collide with a second stationary car **Q**. On impact, the two cars stuck together and afterwards skidded to rest.

The graph shows the velocity of car **P** throughout the experiment.

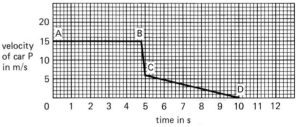

(i) Which section of the graph corresponds to the collision between the cars?
(ii) Which section of the graph corresponds to the cars skidding to rest?
(iii) State the velocity of car **Q** at the time corresponding to point **B** on the graph.
(b) A dummy of mass 70 kg was firmly fixed in car **P** using a seat belt. Information from the graph indicates a deceleration during the collision of 45 m/s².
  (i) Calculate the force exerted by the seat belt on the dummy.
  (ii) Describe carefully what would have happened to the dummy if it had not been wearing a seat belt. Explain your answer in terms of Newton's laws of motion.
(c) (i) Calculate the deceleration of the cars as they skidded to rest.
  (ii) The experiment was repeated with identical cars when the road was wet. When a graph was drawn of the velocity of the moving car, it was found to be identical to the above graph up to point **C**. What difference would you expect to find in the section CD of the graph? Explain your answer. (SEB)

INVESTIGATION

What is the time of contact between your boot and a football when you kick it? What is the average force on the ball during the kick? (Note: You could do this for other sporting 'collisions' if you prefer: a tennis ball and a tennis racket; a hockey ball and a hockey stick, etc.)

11: Work and Energy

A force acting on a body can make it speed up or can lift it up to a higher point. In other words, it can increase the kinetic energy or the potential energy of the body. In this chapter we will look at how we can **measure** the amount of kinetic or potential energy a body has. This turns out to be very useful and leads to the important ideas of **power** and **efficiency**.

11.1 Work

If we want to find a way to measure kinetic and potential energy, the starting point is the physical idea of **work**. In everyday talk, **work** can mean quite a lot of different things: digging the garden, doing the washing up, writing an answer to a physics problem, or even sitting reading a textbook.

In physics, **work** has a very particular and exact meaning. Work is done when a force makes something move. If you push a car to get it started on a frosty morning, you are doing work once the car starts to move. If the car doesn't move, then you are not doing work, no matter how tired it is making you! To lift some luggage up on to the roof rack of a car, you must do work; but if you are just standing holding the luggage, you are not doing any work.

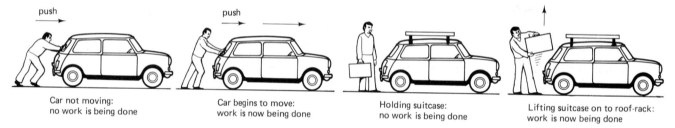

Car not moving: no work is being done

Car begins to move: work is now being done

Holding suitcase: no work is being done

Lifting suitcase on to roof-rack: work is now being done

Work and energy

There is a good reason for using the word **work** in this way. To understand the reason, let us look at some examples of work being done:

Imagine a trolley standing on a level floor. The trolley is at rest, so its kinetic energy is zero. If we apply a force, we can make the trolley move. We have increased the kinetic energy of the trolley. Since energy is always conserved, this extra kinetic energy must have come from somewhere. It has come from the chemical energy stored in our muscles. When we push the trolley, we transfer some of the chemical energy in our muscles into the kinetic energy of the trolley. The force we apply is part of the process of transforming some chemical energy into kinetic energy.

In the same way, if we lift a box up on to a shelf, we must apply an upwards force big enough to balance the force of gravity (the weight of the box). We transfer some chemical energy in our muscles into potential energy of the box. Again the force which moves the box is part of the process of transforming chemical energy into potential.

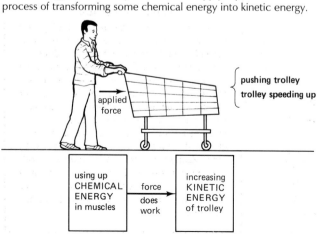

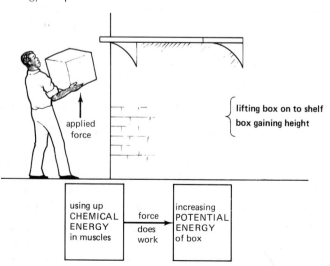

Of course we don't have to do all lifting jobs with our own unaided 'muscle power'. Often heavy lifting work is done by electrically-powered cranes. In this case, the crane is being used to transfer electrical energy into potential energy when it lifts the load. In order to carry out this conversion – to lift the load – the crane must apply an upwards force. A force is always involved when energy is being transferred.

As a final example, imagine pushing a heavy table across the floor. The speed of the table is constant, because there is a friction force equal to the force we apply, and so the total force on the table is zero. There is no *increase* in kinetic energy this time. The chemical energy in our muscles is being transferred into internal energy of the table legs and floor – they are getting hotter. The force is again part of the process by which energy is transferred. We will return later in the book (in Chapter 17) to look in more detail at heating and internal energy. In this chapter we will concentrate mainly on kinetic and potential energy.

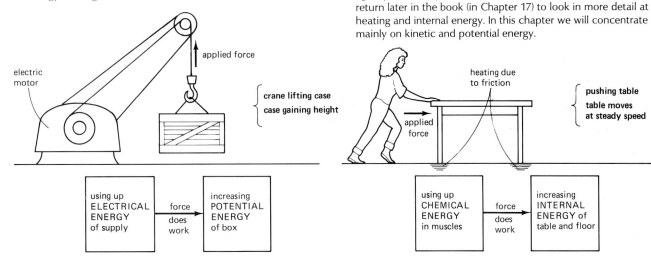

Summarising then:

When energy is transferred from one place to another, or transformed from one form into another, work is always done.

We can take this one stage further:

The *amount* of work done in any process is equal to the *amount* of energy transferred.

11.2 Calculating work

This last idea is a useful one. If we can measure the amount of work done, we have a way of measuring the amount of energy transferred. This gives us a way of measuring amounts of energy – provided we can find a way to measure amounts of work done!

How could we measure work? We have seen that work is done when a force makes something move. The amount of work done will depend on the size of the **force**, and on the **distance moved**. We define the amount of work done in the following way:

work done = force × distance moved in the direction of the force

In the SI system, force is measured in newtons and distance in metres, so work is measured in newton-metres, or joules (J) for short.

1 newton-metre = 1 joule

1 joule is the amount of work done when a force of 1 newton moves a distance of 1 metre in the direction of the force.

Table 11.1. Work done in different tasks (work = force × distance)

Task	Force	Distance	Work done
Opening a kitchen drawer	10 N	0.5 m	5 J
Closing a door	15 N	1.0 m	15 J
Putting luggage on a car roof rack	150 N	1.5 m	225 J
Putting a new light bulb in its socket	1 N	2 m	2 J
Pushing a supermarket trolley	20 N	200 m	4000 J

> **Work is a scalar**
> Although work is calculated by multiplying two vector quantities, force and distance moved in the direction of the force (displacement), work is a scalar quantity. It is not easy to see mathematically why this should be so. However, work is a measure of the amount of energy transferred in a process and it is pretty clear that energy must be a scalar quantity. Most forms of energy – chemical, internal, electrical – have no direction associated with them. They cannot be vector quantities. Work and energy are both scalars. We can add amounts of work or energy by ordinary arithmetic – there is no need to use vector addition.

11.3 Gravitational potential energy

Let us now see how we can use these ideas to measure gravitational potential energy. Let's imagine we give a box some more potential energy by lifting it up. To raise a box up to a height, we must do some work. The amount of potential energy the box gains is equal to the amount of work we do.

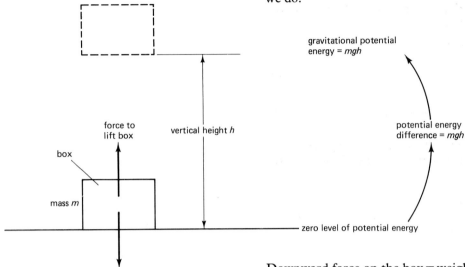

Downward force on the box = weight of box = mg newtons

So, force needed to lift box at a steady speed = mg newtons

Distance moved = h metres

Work done in lifting the box = force × distance moved
$$= mg \, \text{N} \times h \, \text{m}$$
$$= mgh \text{ newton-metres (joules)}$$

Therefore, the gain in potential energy of the box is mgh joules.

Usually we can regard the ground, or the bench, or some other convenient level as the zero level of gravitational potential energy. Then we can say that the potential energy of a body at height h above this level is mgh.

In symbols, **the equation for calculating gravitational potential energy is:**

$$E_p = mgh$$

Notice that energy, like work, is measured in joules.

EXAMPLE 11.1

A suitcase has a mass of 15 kg. It has to be raised by 2 m to put it into the luggage rack in a railway carriage. How much potential energy does it gain?

We use the equation for potential energy:

$$E_p = mgh$$

Substituting: $E_p = 15 \text{ kg} \times 10 \text{ N/kg} \times 2 \text{ m}$
$= 300 \text{ Nm (or J)}$

The suitcase gains 300 J of gravitational potential energy.

11.4 Kinetic energy

We can use the same sort of method to measure kinetic energy. This time, instead of using a force to lift a body, we apply a horizontal force for a certain distance to make a body move faster. Using the laws of motion we can work out how fast the body will travel.

Imagine a trolley of mass m at rest on a level bench. We push it with a constant force F for a distance s. At the end of this distance, the trolley has speed v.

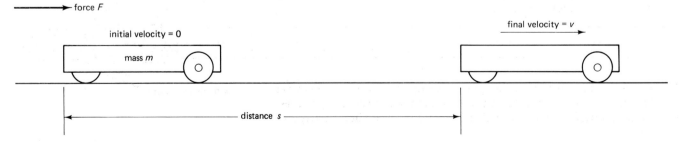

The trolley will have uniform acceleration a. To calculate v, we can use the equation of motion:
$$v^2 = u^2 + 2as$$

The trolley starts at rest, so $u = 0$. The equation of motion simplifies to:
$$v^2 = 2as \quad (1)$$

If friction is small enough to ignore, F is the total force acting. So we can calculate the trolley's acceleration using:
$$F = ma$$

Divide both sides by m:
$$a = \frac{F}{m} \quad (2)$$

Now substitute this into equation (1):
$$v^2 = \frac{2Fs}{m} \quad (3)$$

But Fs on the top of this fraction is (force × distance moved by the force) – it is the work done. We rearrange the equation (3) to get Fs on its own:

Multiply both sides by m: $\quad mv^2 = 2Fs$
Divide both sides by 2: $\quad Fs = \tfrac{1}{2}mv^2$

So the work done by the applied force is equal to $\tfrac{1}{2}mv^2$. This work done is equal to the kinetic energy gained by the trolley.

So, the gain in kinetic energy is $\tfrac{1}{2}mv^2$.

In other words, **a body of mass m travelling at a velocity v has kinetic energy $\tfrac{1}{2}mv^2$.**
$$\boldsymbol{E_k = \tfrac{1}{2}mv^2}$$

It follows from this equation that if we double the velocity of a body, its kinetic energy increases four times (2 squared). If we increase the velocity by three times, the kinetic energy increases nine times (3 squared).

Units of kinetic energy

The units of kinetic energy are the units of mass × (units of velocity)^2, that is: kg × (m/s)^2, or kg m^2/s^2.

We can show that these units are really equivalent to joules:

1 joule = 1 newton × 1 metre
 = 1 kg m/s^2 × 1 m = 1 kg m^2/s^2 = 1 kg (m/s)^2

So kinetic energy is measured in joules, as we would expect.

11.5 Checking the kinetic energy formula by experiment

The kinetic energy equation ($E_k = \tfrac{1}{2}mv^2$) can be investigated by experiment. We must investigate the two factors, mass and velocity, separately.

In the experiment, a body is given a certain amount of energy, and its resulting velocity is measured. The easiest way to do this is to use a catapult to set the body moving. When a catapult is stretched, it stores elastic potential energy. On release, this is transferred into kinetic energy of the moving body. The amount of kinetic energy is equal to the amount of elastic potential energy stored in the catapult, and this in turn depends on how far the catapult was stretched. If we keep constant the amount of stretch of the catapult, we can repeat the experiment giving the same amount of kinetic energy each time.

EXAMPLE 11.2

A bicycle and cyclist have a total mass of 80 kg. What is their kinetic energy when travelling at 5 m/s?

We use the equation for kinetic energy:
$$E_k = \tfrac{1}{2}mv^2$$

Substituting: $E_k = \tfrac{1}{2}mv^2$
$= \tfrac{1}{2} \times 80\,\text{kg} \times (5\,\text{m/s})^2$
$= (\tfrac{1}{2} \times 80 \times 25)\,\text{kg m}^2/\text{s}^2$
$= 1000\,\text{J}$

The bicycle and cyclist have 1000 J of kinetic energy.

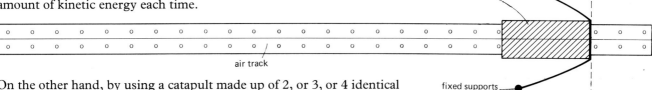

On the other hand, by using a catapult made up of 2, or 3, or 4 identical elastic cords, we can increase the amount of elastic potential energy stored by 2, 3, or 4 times, and so we can give 2, 3, or 4 units of kinetic energy to the body.

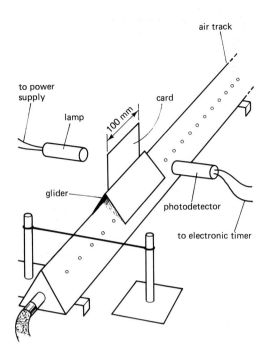

Kinetic energy and velocity

The first experiment investigates the relationship between kinetic energy and velocity. We use the same body throughout this experiment (so that the mass m is constant). The experiment can be done using a trolley on a friction-compensated runway, or using a glider on an air-track. The body's velocity (or speed, since they will be the same size) can be measured by any of the methods described in section 5.3 (ticker-timer, strobe photography, or electronic timer and photodetector switch).

On the air-track, a glider is catapulted along the track using one elastic cord stretched to a known mark. The velocity of the glider is measured by timing a 100 mm card as it cuts through the light-beam. The light-beam and photodetector are placed near the point where the glider leaves the catapult, so that any slowing down due to friction has little chance to have an effect. Results from a typical experiment are shown in Table 11.2.

Table 11.2

Kinetic energy of glider	Number of elastic cords			
	1	2	3	4
Velocity/(m/s)	0.20	0.27	0.36	0.41
Velocity²/(m/s)²	0.040	0.073	0.13	0.17

The graph of velocity against kinetic energy is not a straight line, but a curve. However, a graph of (velocity)² against kinetic energy is a straight line through the origin.

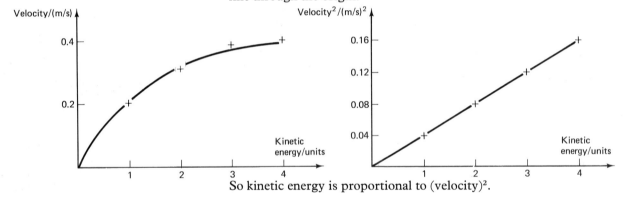

So kinetic energy is proportional to (velocity)².

$$E_k \propto v^2 \text{ (when mass } m \text{ is kept constant)}$$

Kinetic energy and mass

A second experiment can be carried out to check how kinetic energy depends on the mass of the moving body. One way to do this is to fire a glider of twice the original mass from a catapult with two elastic cords, a glider of three times the original mass from a catapult of three elastics, and so on. A typical set of results is shown in Table 11.3.

Table 11.3

Kinetic energy of glider	Number of elastic cords			
	1	2	3	4
Mass of glider	1	2	3	4
Velocity/(m/s)	0.20	0.21	0.19	0.20

Within experimental accuracy, the velocity is the same each time. The energy stored in the stretched elastic has gone up from 1 to 2 to 3 to 4 units; the mass has increased each time in the same way. The mass is proportional to the energy released by the elastics (the number of elastics). And the result is that the glider has the same velocity each time. So, at a constant velocity, the kinetic energy is directly proportional to the mass of the moving body.

$$E_k \propto m \text{ (when velocity } v \text{ is constant)}$$

Combining the two experimental results

From the first experiment we have: $E_k \propto v^2$ (when m is constant)

From the second experiment we have: $E_k \propto m$ (when v is constant)

In general, this means that:

$$E_k \propto mv^2$$

This agrees with the equation for kinetic energy that we worked out in section 11.4 above: $E_k = \tfrac{1}{2}mv^2$.

These two experiments do not, however, check that the factor $\tfrac{1}{2}$ in the kinetic energy formula is correct. We will see later (Example 11.5) how that can also be checked experimentally.

11.6 Energy transfer: kinetic and potential energy

Newton's First and Second Laws of Motion can be used to solve many problems involving motion. But there are some quite simple situations where it is either difficult or completely impossible to use $F = ma$ to get an answer. For instance, if we want to work out the speed which a parcel will reach in sliding down a smooth curved ramp, it is very difficult to do it using $F = ma$ (see (A)).

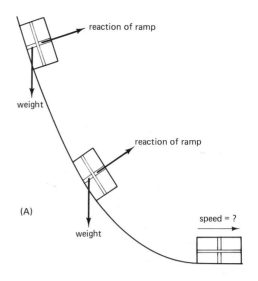

The size and direction of one of the forces on the parcel is changing all the time, so its motion is far from simple. A completely different approach to this problem is to use the idea of conservation of energy – the total amount of energy is the same at the end of the process as it was at the beginning. The parcel on the ramp has gravitational potential energy when it is at the top. As it slides down it loses this potential energy and gains kinetic energy. The total amount of energy stays constant. The kinetic energy of the parcel at the bottom is the same as the amount of potential energy which it had at the top (see (B)).

This gives us the general idea for tackling the problem. Let us see how it works for this particular parcel and ramp (see (C)).

The mass of the parcel is 3 kg. At the top of the ramp, its extra gravitational potential energy is:

$$\begin{aligned}E_p &= mgh \\ &= 3 \text{ kg} \times 10 \text{ m/s}^2 \times 5 \text{ m} \\ &= 150 \text{ J}\end{aligned}$$

At the top, the parcel has kinetic energy (E_k) of 0 J.
So the total energy of the parcel ($E_p + E_k$) is 150 J.

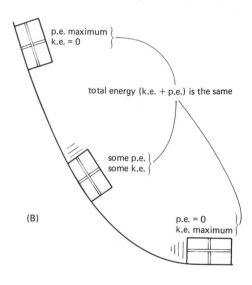

The parcel slides down the ramp. By the time it reaches the bottom, it has lost all its potential energy and gained kinetic energy. As energy is conserved, its kinetic energy must now be 150 J.

$$E_k = \tfrac{1}{2}mv^2$$

$$\begin{aligned}150 \text{ J} &= \tfrac{1}{2}mv^2 \\ &= \tfrac{1}{2} \times 3 \text{ kg} \times v^2\end{aligned}$$

Multiply both sides by 2: $\quad 300 \text{ J} = 3 \text{ kg} \times v^2$

Divide both sides by 3 kg: $\quad \dfrac{300 \text{ J}}{3 \text{ kg}} = v^2$

$$v^2 = 100 \text{ (m/s)}^2$$
$$v = 10 \text{ m/s}$$

The parcel has a speed of 10 m/s when it reaches the bottom.

By using the idea of energy conservation, we have been able to calculate the parcel's speed at the bottom of the ramp quite easily – much more easily than by trying to use $F = ma$ directly.

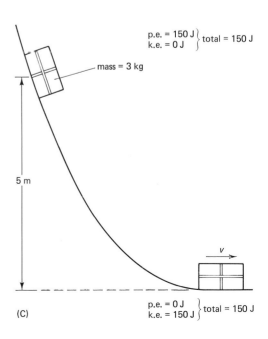

1. In the calculation on p. 97, the mass of the parcel doesn't make any difference to the final speed. Try repeating the calculation for a parcel of mass 5 kg. You should get the same answer: 10 m/s.

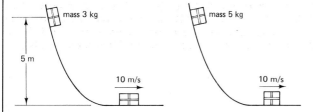

2. The exact shape of the ramp has no effect on the final speed. As long as the ramp is smooth, it is just the vertical drop that matters. The potential energy at the top depends only on the height of the ramp. And so the kinetic energy (and hence the speed) at the bottom is the same no matter what the shape of the ramp is.

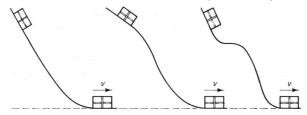

3. In reality, of course, ramps are not completely smooth. So some potential energy is transferred into internal energy of the ramp and parcel, because of friction. The ramp and parcel get hotter. As a result the kinetic energy at the bottom is a bit less than the potential energy at the top, the difference being equal to the increase in internal energy. And this may well depend on the shape and the length of the ramp.

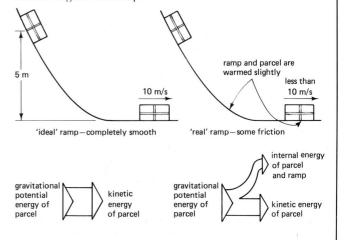

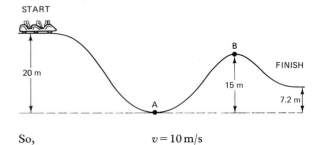

This idea of using energy conservation to solve problems is a useful one, and we will now look at a few more examples.

EXAMPLE 11.3

The diagram shows the heights above the ground of some points on the track of a fairground rollercoaster. The rollercoaster and its passengers have a total mass of 500 kg. What is the speed of the coaster at points **A** and **B**, and at the finish?

In this problem, we must assume that there is no friction on the track.

(a) To find the speed at **A**:

At the start, gravitational potential energy (relative to the ground) is

$$E_p = mgh = 500 \text{ kg} \times 10 \text{ N/kg} \times 20 \text{ m} = 100\,000 \text{ J}$$

When it reaches **A**, the vehicle has lost 100 000 joules of potential energy. So it has gained 100 000 joules of kinetic energy.

At **A**:
$$E_k = \tfrac{1}{2}mv^2$$
$$100\,000 \text{ J} = \tfrac{1}{2} \times 500 \text{ kg} \times v^2 = 250 \text{ kg} \times v^2$$

Divide both sides by 250 kg: $\dfrac{100\,000 \text{ J}}{250 \text{ kg}} = v^2$

$$v^2 = 400 \text{ (m/s)}^2, \quad \text{so,} \quad v = 20 \text{ m/s}$$

The vehicle has speed 20 m/s at **A**.

(b) To find the speed at **B**:

To do this, it is easier to consider the entire motion from the start to **B**. Over this stage, the drop in height is 5 m.

Between start and **B**, loss in potential energy is:
$$E_p = mgh = 500 \text{ kg} \times 10 \text{ N/kg} \times 5 \text{ m} = 25\,000 \text{ J}$$

So at **B**, the vehicle has 25 000 joules of kinetic energy.

At **B**:
$$E_k = \tfrac{1}{2}mv^2$$
$$25\,000 \text{ J} = \tfrac{1}{2} \times 500 \text{ kg} \times v^2 = 250 \text{ kg} \times v^2$$

Divide both sides by 250 kg: $\dfrac{25\,000 \text{ J}}{250 \text{ kg}} = v^2$

$$v^2 = 100 \text{ (m/s)}^2$$

So, $v = 10 \text{ m/s}$

The vehicle has speed 10 m/s at **B**.

(c) To find the speed at the finish:

This is just the same calculation as part (b) above:

Drop in height from start to finish = $20 - 7.2$ m
$$= 12.8 \text{ m}$$

Loss in potential energy: $E_p = mgh$
$$= 500 \text{ kg} \times 10 \text{ N/kg} \times 12.8 \text{ m} = 64\,000 \text{ J}$$

So at the finish, the vehicle has 64 000 joules of kinetic energy.

At finish:
$$E_k = \tfrac{1}{2}mv^2$$
$$64\,000 \text{ J} = \tfrac{1}{2} \times 500 \text{ kg} \times v^2 = 250 \text{ kg} \times v^2$$

Divide across by 250 kg: $\dfrac{64\,000 \text{ J}}{250 \text{ kg}} = v^2$

$$v^2 = 256 \text{ (m/s)}^2, \quad \text{so,} \quad v = 16 \text{ m/s}$$

At the finish, the vehicle has speed 16 m/s.

In fact, to do these three calculations, it is not necessary to know the mass of the roller coaster and passengers. We can simply call this m kg and do the calculation in the same way. The m's cancel out and we get the same answers each time – try it!

EXAMPLE 11.4

A child is sitting on a swing. To start her off, her mother pulls the swing back and lets go. At the start, the child is 0.45 m above the middle position. What is her maximum speed as she swings?

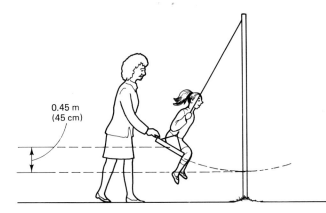

When you are swinging, your maximum speed is in the middle – at the lowest point. At the end of each swing, the child has extra gravitational potential energy and no kinetic energy – she is stopped for a moment as the swing changes direction. At the middle of each swing, all this potential energy has been transferred into kinetic energy.

To work out the maximum speed, we first calculate the amount of extra potential energy at the start. The child's mass is m kg.

$$E_p = mgh$$
$$= m \text{ kg} \times 10 \text{ N/kg} \times 0.45 \text{ m}$$
$$= 4.5m \text{ J}$$

At the middle of the swing, this has all been transferred to kinetic energy. The child has $4.5m$ J of kinetic energy. So we can work out her velocity:

$$E_k = \tfrac{1}{2}mv^2$$
Substituting: $\quad 4.5m = \tfrac{1}{2}mv^2$

Divide both sides by m: $\quad 4.5 = \tfrac{1}{2}v^2$
Multiply both sides by 2: $\quad v^2 = 9 \text{ (m/s)}^2$
So, $\quad v = 3 \text{ m/s}$

The child's maximum speed at the middle point of the swing is 3 m/s.

Pendulums

The child's swing (Example 11.4) is an example of a pendulum. The motion of a pendulum can most easily be understood from an energy point of view. When it is moved to one extreme position, the bob gains gravitational potential energy. When it is released, it loses potential energy and gains kinetic energy. At the mid-point of its swing, its potential energy is least; so its kinetic energy is largest. This is where the bob is moving fastest. As it swings on, the kinetic energy becomes less and the bob gains potential energy again. While it swings, the energy of the pendulum is continually being transferred from potential to kinetic and back to potential again. The pendulum gradually slows down and stops because friction and air resistance are always present causing a small amount of heating and resulting in some energy being transferred into internal energy (of the surrounding air mainly).

EXAMPLE 11.5

The following experiment is carried out to check the '$\tfrac{1}{2}$' in the kinetic energy equation $E_k = \tfrac{1}{2}mv^2$. A ball of diameter 5 cm (0.05 m) is dropped. After it has fallen exactly 20 cm (0.2 m), it cuts through the light beam of a photodetector switch. An electronic timer records a time of 25 milliseconds (0.025 s).

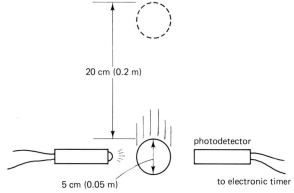

First, we calculate the speed of the ball as it cuts through the light beam:

Distance travelled through beam = 0.05 m
Time taken = 0.025 s

$$\text{average velocity} = \frac{\text{distance}}{\text{time}}$$
$$= \frac{0.05 \text{ m}}{0.025 \text{ s}}$$
$$= 2 \text{ m/s}$$

As the distance is very short, we can assume that this is a good estimate of the ball's instantaneous velocity.

Now we work out the gravitational potential energy lost by the ball as it falls. The ball's mass is m kg.

$$E_p = mgh$$
$$= m \text{ kg} \times 10 \text{ N/kg} \times 0.2 \text{ m}$$
$$= 2m \text{ J}$$

The kinetic energy of the ball must be equal to this. To check the equation, we will use $E_k = kmv^2$, and try to find k.

$$E_k = kmv^2$$
$$2m \text{ J} = k \times m \text{ kg} \times (2 \text{ m/s})^2$$

Divide both sides by m: $\quad 2 \text{ J} = k \times 4 \text{ J}$

Divide both sides by 4 J: $\quad \dfrac{2 \text{ J}}{4 \text{ J}} = k$

$$k = \tfrac{1}{2}$$

In this way we can complete the experimental findings discussed in section 11.5 and confirm that the kinetic energy equation is $E_k = \tfrac{1}{2}mv^2$.

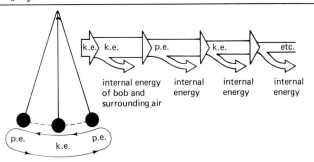

11.7 Elastic potential energy

There is a second type of potential energy – the energy stored in elastic objects when they are stretched or compressed. We can also measure amounts of elastic potential energy.

Imagine that we begin pulling on a spring to stretch it. As it stretches, the force we need to pull with gets larger. (Remember: Hooke's Law – section 3.2.)

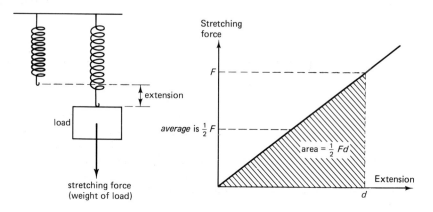

If it takes a force F to produce a total extension d, then the **average** force as we pulled was $\frac{1}{2}F$. The force has increased steadily from 0 to F – so its average value is half-way between: $\frac{1}{2}F$. We use this average force to calculate the amount of work we have done pulling the spring:

$$\text{work done} = \textbf{average force} \times \text{distance}$$
$$= \tfrac{1}{2} F \times d$$

The elastic potential energy stored in a stretched spring is $\frac{1}{2}Fd$. Notice that this is equal to the area under the force–extension graph.

If the spring constant for this particular spring is k, then F and d are also related. In fact,

$$F = kd$$

So the equation for elastic potential energy stored in the spring becomes:

$$\text{elastic potential energy} = \tfrac{1}{2}Fd \quad = \tfrac{1}{2}kd\,d \quad = \tfrac{1}{2}kd^2$$

$$E_e = \tfrac{1}{2}kd^2$$

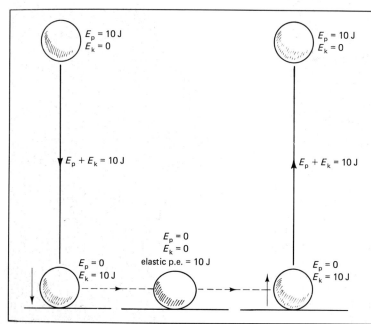

Bouncing ball

The motion of a bouncing ball involves kinetic energy, gravitational potential energy and elastic potential energy. When a ball is dropped from a height, its gravitational potential energy changes into kinetic energy as it falls. If we ignore the small amount of heating caused by air resistance, the kinetic energy gained is equal to the potential energy lost. Just before it hits the ground, the kinetic energy of the ball is equal to the potential energy it had originally. When it hits the ground, the ball is stationary for a very brief moment. At this instant, the ball is compressed, and the energy is now stored in the form of elastic potential energy. As this compression springs back, the ball bounces up again. If no energy is lost at this stage, the ball will rebound with the same kinetic energy (and hence with the same speed) as it had just before it hit the ground. It will rise until this kinetic energy has all been converted into potential energy again. So the ball rises to its original height.

EXAMPLE 11.6

A ball of mass 2 kg is dropped from a height of 1.25 m. What is its speed just before it hits the ground?

We can tackle this problem using energy conservation.

At the moment it is dropped, the ball has potential energy.

$$E_p = mgh$$
$$= 2\,\text{kg} \times 10\,\text{N/kg} \times 1.25\,\text{m}$$
$$= 25\,\text{J}$$

Just before it hits the ground, the ball has kinetic energy 25 J. So we can calculate its speed.

$$E_k = \tfrac{1}{2}mv^2$$
$$25\,\text{J} = \tfrac{1}{2} \times 2\,\text{kg} \times v^2$$
$$= 1\,\text{kg} \times v^2$$

Divide both sides by 1 kg: $v^2 = 25\,\text{J/kg}$ (or $(\text{m/s})^2$)
$$v = 5\,\text{m/s}$$

The ball has speed 5 m/s² just before it hits the ground.

Absorbing energy in deformation

To stretch a spring or any solid object, a force must be applied. As the stretching occurs, the force moves – some work is done and energy is stored in the stretched spring. When the spring is released, it springs back to its original position and the stored energy is released again. If a spring is stretched beyond its elastic limit, some of the stored energy is not recovered in this way. Some work has gone into causing the permanent stretch (or deformation) of the metal – in fact, it will have made the metal become slightly warmer.

This work done in deformation can be very useful. Here are two examples:

1. In a car collision, the kinetic energy of the moving car is dissipated as the car is suddenly stopped. An important safety feature is a 'crumple zone' at the front and rear of a car. The passengers are seated within a rigid metal frame, but the front (or rear) can deform in a collision. As it deforms, the metal is stretched beyond its elastic limit and this dissipates much of the car's kinetic energy. A rigid body cannot absorb energy in this way. (**Note**. The effect of 'crumple zones' can also be understood using the idea of impulse. This was discussed in section 10.4.)

2. A climbing rope has to hold a climber in the event of a fall. Climbing ropes are elastic. If a climber slips, he will fall until the rope becomes taut. The rope then begins to stretch, gradually absorbing the falling climber's kinetic energy. As a result, the maximum force which the rope has to bear is much less. If the rope was unable to stretch, it could not absorb the energy. The sudden force on the rope as it became tight would be very great and it would almost certainly snap. After a fall, a rope should be discarded and not used again – it has been permanently stretched and weakened by absorbing the energy of the fall. (**Note.** You can also use the idea of change of momentum [section 10.4] to explain why the force on the rope is smaller if it stretches. Can you see how to do this?)

11.8 Power

The two electric motors in the photograph can both lift a load of 0.1 kg from the floor to the bench. The difference is that the larger motor can lift the load more quickly. It can do the same amount of work in a shorter time.

mass = 1290 kg   0–96 km/h in 8.9 s

mass = 1310 kg   0–96 km/h in 5.3 s

The two cars can reach a speed of 96 km/h (60 m.p.h.). Both have approximately the same mass, so when they are travelling at this speed they have the same kinetic energy. But one can accelerate from rest to 110 km/h in a shorter time. Its engine can do the same amount of work in a shorter time.

The rate at which something can do work is called its **power**.

$$\text{power} = \frac{\text{work done}}{\text{time taken}}$$

When work is done, energy is transformed from one form to another. So power can also be written:

$$\text{power} = \frac{\text{energy transformed}}{\text{time taken}}$$

In the SI system of units, power is measured in joules per second (J/s), or watts (W), for short.

1 watt = 1 joule per second

Large powers are measured in kilowatts (kW) or megawatts (MW).

1 kilowatt = 1000 watts
1 megawatt = 1000 kilowatts = 1 000 000 watts

Table 11.4
Some typical values of power

	Power
Radio	3 W
Light bulb	60 W
Hi-fi amplifier	125 W
Food mixer	400 W
Electric kettle	3 kW
Car engine	50 kW
Lorry engine	200 kW
Jet aircraft engine	500 kW
Power station	2000 MW

EXAMPLE 11.7

A lift in a shop can raise 8 people through a height of 20 m in 5 seconds. The lift plus passengers has a mass of 800 kg. What is the power of the lift motor?

$$\text{Mass of the lift and passengers} = 800 \text{ kg}$$
$$\text{So, weight of lift and passengers} = (800 \times 10) \text{N}$$
$$= 8000 \text{ N}$$

To lift the load at steady speed, the lift must exert an upward force of 8000 N.

$$\text{Work done by the lift} = \text{force} \times \text{distance}$$
$$= 8000 \text{ N} \times 20 \text{ m}$$
$$= 160\,000 \text{ J}$$

This work is done in 5 s.

$$\text{Power of the lift} = \frac{\text{work done}}{\text{time taken}}$$
$$= \frac{160\,000 \text{ J}}{5 \text{ s}}$$
$$= 32\,000 \text{ W}$$

The lift motor has a power of 32 000 W (or 32 kW).

(**Note.** Another way to do this would be to calculate first the gain in potential energy of the lift and passengers using $E_p = mgh$. The result is 160 000 J. Power can then be calculated as:

$$\frac{\text{energy transformed}}{\text{time taken}}$$

The result is the same: 32 000 W.)

EXAMPLE 11.8 Measuring human power

One way to estimate your own power is to time yourself running up a flight of stairs. To do the calculation you need to know your mass, and the height of the stairs. Here is a typical set of results:

Mass of person = 50 kg
(Vertical) height of stairs = 12 m
Time taken = 15 s

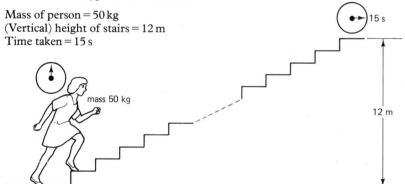

To calculate the power, we need to work out the energy gained per second.

$$\text{gravitational potential energy gained} = mgh$$
$$= (50 \times 10 \times 12) \text{J}$$
$$= 6000 \text{ J}$$

This is the amount of energy transferred from chemical energy in the person's muscles into gravitational potential energy.

$$\text{power} = \frac{\text{energy transformed}}{\text{time taken}}$$
$$= \frac{6000 \text{ J}}{15 \text{ s}}$$
$$= 400 \text{ W}$$

The person's power output during this task is 400 W.

104 Work and Energy

Hand drilling through wood. The handle is designed to provide the extra leverage required.

11.9 Machines and efficiency

A 'machine' is anything which helps us to do a job. For example, it is easier to drill a hole using an electric drill than by hand.

Using an electric drill to drill a hole in concrete.

Inside the electric drill is an electric motor. It transfers electrical energy from the mains (or batteries) into kinetic energy. All machines transfer energy. But in the process, some of the energy often goes into places and forms that we do not want. In particular, some energy is almost always wasted in heating up parts of the machine itself.

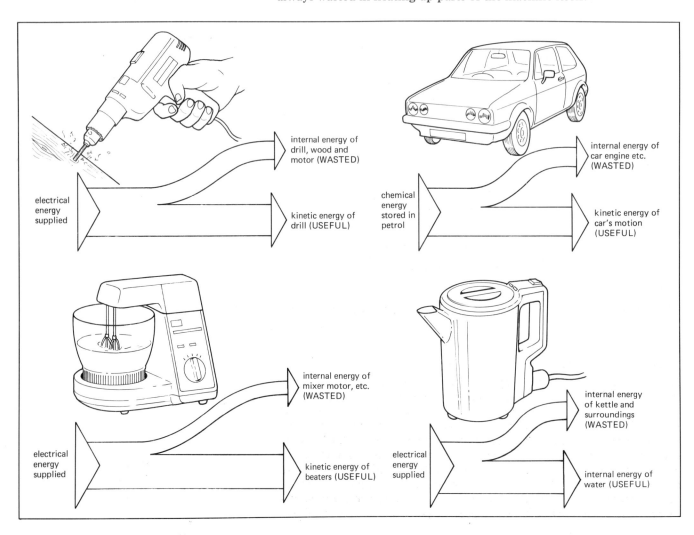

The **efficiency** of a machine is a measure of how much energy it transfers to where we want it to go. If a lot of energy is wasted doing things we don't want, then the machine's efficiency is low.

Usually the efficiency of a machine is calculated as a percentage:

$$\% \text{efficiency} = \frac{\text{useful energy output}}{\text{total energy input}} \times 100$$

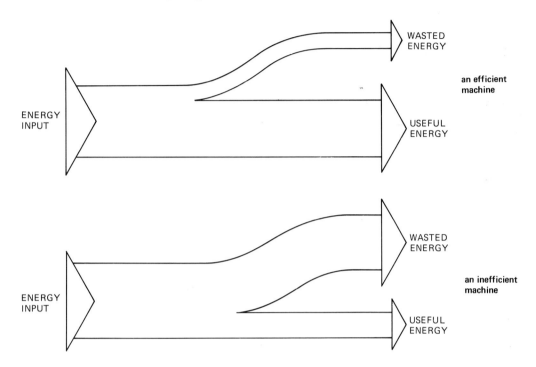

Sometimes it is more convenient to calculate the energy input or the energy output (or both) **indirectly** – by calculating the **work done**.

Work done at the machine's input side (the **work input**) is a measure of the energy input to the machine. Work done by the machine (its **work output**) is a measure of its energy output.

So other ways of calculating efficiency are:

$$\% \text{efficiency} = \frac{\text{useful work output}}{\text{total work input}} \times 100$$

or

$$\% \text{efficiency} = \frac{\text{useful energy output}}{\text{total work input}} \times 100$$

...and so on.

All of these are entirely equivalent. We just use the one that is the most convenient in each case.

Measuring efficiency

If we can measure the energy supplied *to* a machine and the useful energy output *from* the machine, we can calculate its efficiency. For the moment, we can only do the calculation for simple machines where kinetic and potential energy are the only types involved. Later in the book we will see how to measure internal energy and electrical energy; this will allow us to consider a wider range of machines. The same general idea of efficiency applies to all types of machines—from simple manual (hand-operated) machines like levers, pulleys and jacks to the most complicated motor-driven machine. When we talk about efficiency we are always asking: how much of the input energy (or work) goes to where we want it to go?

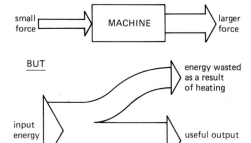

A smaller force is needed to push the barrow up the ramp

The force needed to lift the barrow straight up is so large that the man is unable to do it

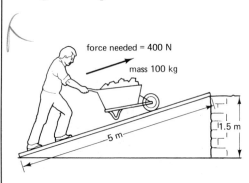

Ramp

Using a ramp to raise a heavy load is such a simple idea that you may not think of it as a machine at all. Yet it does all the things that a machine should do! If we want to get a barrow load of bricks up to a higher level, it is much easier to push the barrow up a ramp than to lift it straight up. In fact, since there is a limit to the force our muscles can exert we might be completely unable to lift the barrow directly. The ramp is a machine which allows us to do the task.

All machines have this property of allowing us to do a task using a smaller force than we would otherwise need. This is called the **mechanical advantage** of the machine.

$$\text{mechanical advantage} = \frac{\text{force needed to do task using the machine}}{\text{force which would be needed to do the task without the machine}}$$

So a machine **multiplies** the force we apply. Does this mean that we are getting something for nothing? Not quite! The barrow has to move a much longer distance if we use the ramp; the direct vertical distance is shorter. As a result, the amount of work (force × distance) which we have to put *into* the ramp is always greater than the amount of work we get *out*. Another way of putting this is to say that the **efficiency** of the ramp is less than 100%. All machines are less than 100% efficient, because of the way energy always spreads and some does not go into doing the job we want done.

Mass of barrow full of bricks = 100 kg
Vertical height raised = 1.5 m
Length of ramp = 5 m

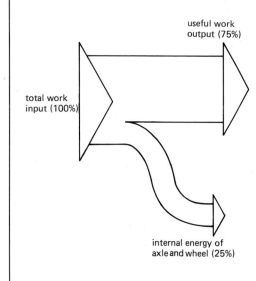

Here is a typical task involving a ramp By measurement, we find that the force needed to push the barrow up the ramp is 400 N. So what is the efficiency of this ramp?

First, note that the ramp **does** make the task easier: to lift the barrow straight up we would need a force of 1000 N (weight = mg). Using the ramp the force needed is only 400 N.

So, mechanical advantage of this ramp = $\dfrac{1000 \text{ N}}{400 \text{ N}}$ = 2.5

To calculate efficiency, we need to compare the work done pushing the barrow up the ramp with the actual potential energy gained:

Work done to push the barrow and bricks up the ramp
= force × distance
= 400 N × 5 m
= 2000 J

Gravitational potential energy gained by barrow and bricks
= mgh
= 1000 N × 1.5 m
= 1500 J

So, efficiency of the ramp = $\dfrac{\text{useful energy output}}{\text{total work input}}$ × 100%
= $\dfrac{1500 \text{ N}}{2000 \text{ N}}$ × 100%
= 0.75 × 100%
= 75%

The ramp has a mechanical advantage of 2.5 and an efficiency of 75%.

The ramp is only 75% efficient because one quarter (25%) of the work we put in is wasted overcoming friction as we push the barrow up the ramp. The axle and the wheel of the barrow get hot. The longer the ramp, the more energy is lost in this way. As we can never get rid of friction completely, a machine can never be 100% efficient.

Four more machines

Lever

We often use levers to make jobs easier. This 300 N load can be lifted using a force of just 60 N. The 60 N force is five times as far from the pivot, so the force is five times smaller. But the load only rises by 0.2 m for every metre we push down on the end. The efficiency is less than 100%.

The scoop of a mechanical digger is raised and lowered by a system of levers, operated hydraulically.

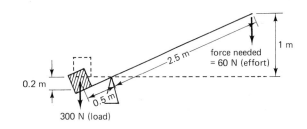

Pulley systems

Pulley systems are useful for lifting heavy loads. To lift out a car engine, a garage mechanic uses a block-and-tackle pulley system. In the diagram, the force needed to lift the engine is supplied by four ropes, so each rope provides just one-quarter of the lifting force – 250 N. The mechanic has to pull with a force of 250 N. If the car engine is lifted by 1 m, the whole of the lower pulley block will rise by 1 m, and each of the four sections of rope, **a**, **b**, **c** and **d**, will become 1 m shorter. So there will be 4 m of slack rope. We have to pull the rope down by 4 m to raise the engine by 1 m. Again the efficiency of the pulley system is less than 100%.

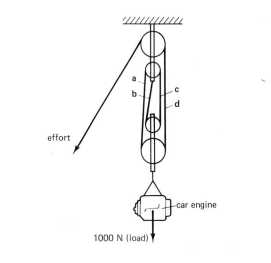

Jack

It would be impossible to lift a car without a jack. Each time the handle of the jack is turned, the car is lifted by a distance equal to the **pitch** of the screw. The effort force moves a lot further than the load and so it is much less than the load. The efficiency is quite low because there is a lot of friction at the screw thread. In fact this is an advantage of the jack! It prevents it unwinding when you let go of the handle.

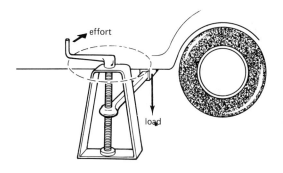

Gears

Gears are machines for magnifying turning forces (or couples). This gearing system has a gear ratio of 3:1. The small input wheel must turn 3 times to make the large wheel turn once. But the couple exerted by the output gear wheel will be almost 3 times larger than the couple applied by the input gear.

In a car gearbox, the gear change lever allows you to select which of a number of gear wheels are interlinked. In bottom gear, the input gear wheel (driven by the engine drive shaft) turns 4 or 5 times to make the output gear wheel (which drives the car wheels) turn once. The car moves relatively slowly, but the couple applied to the car wheels is large. This is necessary for starting off, or for going up a steep hill. In top gear, the ratio will be 1:1. The car goes much faster, but the couple applied to the car wheels is not so big.

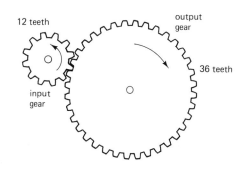

Questions

1. How much work is done in the following situations?
   (a) by a mother pushing a pram with a force of 40 N for a distance of 200 m;
   (b) by a workman pushng a barrow with a force of 60 N for a distance of 25 m;
   (c) by a student pushing a book with a force of 2 N for a distance of 1.5 m across the bench;
   (d) by a father pulling a child on a sledge with a force of 45 N for a distance of 150 m?

2. To start a car on a frosty morning, two men push it at a steady speed of 0.5 m/s. This requires a force of 400 N. How much work do they do in 10 s?

3. A boy pushes a box of mass 20 kg along a rough floor. He has to exert a force of 80 N to keep it moving.
   (a) How much work does he do in moving the box 20 m?
   (b) When he stops pushing, the box stops. Where has the work he has done 'gone'?

4. (a) A suitcase has mass 20 kg. How much work is done by a passenger lifting the suitcase to a height of 0.5 m off the ground? How much more work does he have to do to hold it here for 5 s?
   (b) A weightlifter raises a 55 kg bar off the ground and holds it above his head – a total distance of 2.5 m. How much work is done by the weightlifter in lifting the bar to this height? How much more work must he do to hold the bar steady above his head for 3 s?

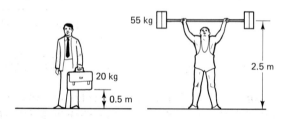

5. What is the gain in gravitational potential energy in each of the following situations?
   (a) when a child of mass 30 kg climbs a 10 m flight of steps;
   (b) when a high jumper of mass 50 kg raises her centre of gravity from 0.8 m (standing) to 2.0 m (as she crosses the bar);
   (c) when a car of mass 1200 kg is raised by 2 m on a garage ramp.

6. How high a mountain must a 75 kg man climb for his gravitational potential energy to increase by 1 MJ (1 000 000 J)?

7. What is the kinetic energy of:
   (a) a bicycle and rider of mass 80 kg, travelling at 5 m/s;
   (b) a volleyball of mass 800 g (0.8 kg), travelling at 4 m/s;
   (c) a baby of mass 12 kg crawling along the floor at 0.5 m/s?

8. (a) A racing car of mass 800 kg has kinetic energy 360 000 J. How fast is it travelling?
   (b) A fast bowler bowls a cricket ball of mass 0.2 kg with kinetic energy of 160 J. How fast does the ball move?
   (c) A golf ball leaves the club at a speed of 70 m/s. Its kinetic energy is 245 J. What is the mass of the ball?

9. A stone of mass 2 kg is dropped from the window of a high building, 45 m above the ground.
   (a) What is the initial gravitational potential energy of the stone (relative to the ground)?
   (b) What is the kinetic energy of the stone at the moment just before it hits the ground?
   (c) At what speed does the stone hit the ground?
   Would the result of calculation (c) have been any different if the stone's mass had been 3 kg? (If you are not sure, try the calculation again and see!)

10. If the stone in question 9, instead of falling freely, had been allowed to slide down a smooth ramp, how would its speed at the bottom compare with your answer to question 9?

11. A child slides down the park chute shown in the diagram. If the slide is very smooth, what will be her speed at the bottom?

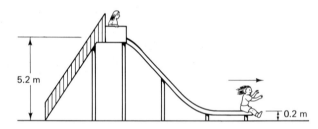

12. A trolley of mass 2.5 kg runs up a slope and stops at **B**.
    (a) How much gravitational potential energy does the trolley gain?
    (b) What is the minimum amount of kinetic energy it must have at **A** if it is to reach the top?
    (c) What is the trolley's minimum velocity at **A** to enable it to reach **B**?
    (d) Would a trolley of mass 5 kg need to be moving faster, slower, or at the same speed to reach **B**?

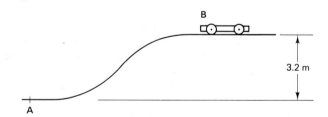

13. A family saloon car can accelerate from rest to 25 m/s (about 60 m.p.h.) in 12.5 s; a sports car can accelerate from rest to 25 m/s in 8 s. Both cars have mass 1000 kg.
    (a) How much kinetic energy is gained by each car as it accelerates?
    (b) What power is developed by the cars' engines during the acceleration?

14 What is the power of the following?
(a) the mother in question 1(a) if her journey took 160 s;
(b) the workman in question 1(b) if his tasks took 15 s;
(c) the student in question 1(c) if it took 0.5 s to pass the book;
(d) the father in question 1(d) if he pulls the sledge for 135 s;
(e) the two men pushing the car in question 2;
(f) the child in question 5(a) if she takes 12 s to climb the stairs;
(g) the high jumper in question 5(b) if the time from take-off until she clears the bar is 0.6 s;
(h) the garage ramp in question 5(c) if it takes 60 s to raise the car.

15 Photographs were taken of a karate blow breaking a piece of wood. By taking measurements on the photographs the speed of the fist could be found at various times during the blow and the graph below was plotted.

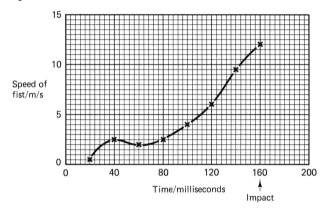

(a) (i) Use the graph to describe how the speed of the fist changed during the period 20 to 160 milliseconds.
(ii) Comment on any unexpected aspect of the way the speed changed.
(iii) What was the speed of the fist at impact?
(b) Calculate the kinetic energy of the fist when it first strikes the wood, assuming that the fist has a mass of 1 kg.
(c) Calculate the power of the blow assuming that the time of impact was 0.03 s. (AEB: SCISP)

16 A large packing case of mass 500 kg rests on the ground. A fork-life truck raises it 1.5 m, transports it at a steady speed of 2.0 m/s and deposits it on the loading platform of a lorry.
(a) (i) What is the minimum upward force exerted by the fork-lift?
(ii) How much potential energy is gained by the packing case?
(b) (i) Calculate the kinetic energy of the packing case while it is being transported at the steady speed.
(ii) What happens to this kinetic energy when the fork-lift truck stops?
(c) If the fork-lift uses energy at the rate of 25 kW and the lifting operation takes 3.0 s, calculate the apparent efficiency of this operation. (SEB)

17 The diagram below shows a machine in which a load at **P** can be raised or lowered by a force applied at **Q**. When this force is applied through a distance of 1.5 m the axle **Y** rotates once. Axles **X** and **Y** have the same diameter.

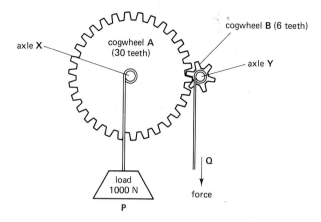

(a) (i) When the force is applied through 7.5 m how many times must cogwheel **B** rotate?
(ii) How many rotations will this cause in cogwheel **A**?
(iii) Through what distance will the load **P** have been moved?
(b) (i) Calculate the work done on the load.
(ii) Calculate the effort required to do this work.
(iii) What have you assumed about the machine in order to carry out this calculation?
(c) (i) Suggest and explain one alteration which could be made to the above machine in order that the force would raise the load by a greater distance.
(ii) Give one reason why your suggested alteration might not be helpful in practice.
(d) Is this machine a force multiplier or a distance multiplier? Explain your answer. (AEB: SCISP)

18 The operator of this lifting machine has to pull rope **P** a distance of 4 m to raise the 5 kg mass a height of 2 m. What is the **least** force he must exert on **P**?

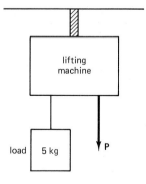

INVESTIGATION
What is the efficiency of a catapult made from an elastic band?

12: Pressure

Pressure diagrams

The area of each table leg is 10 cm². The weight of the table is 400 N. What is the pressure on the carpet?

Each foot has an area of 1500 cm². The weight of the elephant is 54 000 N. What pressure does the elephant exert on the ground?

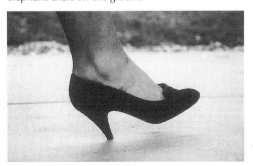

The woman weighs 600 N. The area of the heel is 3 cm². What pressure does the heel exert on the ground?

The thumb exerts a force of 20 N. The area of the head of the drawing pin is 1 cm². The area of the point of the pin is 0.002 cm². What pressure is exerted on the head of the pin? What pressure is exerted on the wooden board?

12.1 Force and pressure

Walking over deep snow is difficult and tiring because your feet sink in at every step. Skies or snowshoes make it much easier because they stop you from sinking in.

Your weight is a measure of the force you exert on the ground. Wearing skis or snowshoes doesn't make you weigh any less: the force exerted on the ground stays the same, but it is now spread over a larger area. So the force exerted on each square centimetre of the ground is less. We say that the pressure is less.

Pressure is calculated by dividing the force acting on a surface by the area on which the force acts:

$$\text{pressure} = \frac{\text{force}}{\text{area}}$$

If we increase the force while keeping the area the same the pressure gets larger; if we keep the force the same but make the area larger, the pressure goes down. Common sense tells us that this is what ought to happen, so the definition above is a sensible way to measure pressure.

In the SI system of units, force is measured in newtons and area in metres squared. So pressure is measured in newtons per metre squared (N/m²). This unit is also called the pascal (Pa).

$$1\,\text{Pa} = 1\,\text{N/m}^2$$

Sometimes it is more convenient to use N/cm² or N/mm² when the area involved is small. Some older (non-SI) units of pressure are still found in use, as we shall see later in this chapter.

◁ The diagrams on the left show some situations where pressure is exerted. The answers are given on page 121.

EXAMPLE 12.1

On a table is a full box of groceries which has a weight of 160 N. The area of the base of the box is 0.2 m². What is the pressure between the box and the table top?

By definition,
$$\text{pressure} = \frac{\text{force}}{\text{area}}$$
$$= \frac{160\,\text{N}}{0.2\,\text{m}^2}$$
$$= 800\,\text{N/m}^2 \text{ (or 800 Pa)}$$

The box exerts a pressure of 800 Pa on the table.

Pressure

12.2 Pressure in liquids

If you try to put your finger over the end of a tap when it is turned on, you can feel the pressure of the water in the pipe. This is just caused by the weight of the water in the pipes all the way back up to the cold water tank in the loft. The force of gravity acts on liquids, pulling them downwards into their container.

Pressure and depth

Pressure increases with depth in a liquid. The pressure at any depth is due to the weight of all the liquid above.

Aqualung divers do not normally go beyond a depth of 30 metres because the pressure of the water becomes too great for the human body to withstand. Small submarines are used for working at greater depths.

Pressure and liquid density

The pressure at all depths in sea water is slightly greater than in fresh water. This is because salt water is denser than fresh water, and so the mass of 1 m³ of sea water is greater than the mass of 1 m³ of fresh water.

Pressure and direction

Liquid pressure is caused by the weight of the liquid. Although weight is a force which acts downwards, the pressure acts in all directions. In fact, the pressure is the same in all directions. ▷

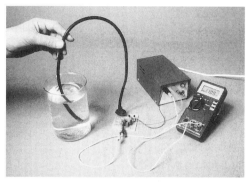

An electronic pressure sensor can be used to measure the pressure at different points in a liquid. The pressure increases with depth. At any given depth, the pressure is the same in whichever direction the open end of the tube points.

Pressure acts at right angles to surfaces

Liquids exert pressure at right angles (or *normally*) to any surface. If a crumpled metal can is punctured with a number of holes and then filled with water, the water spurts out of each hole at right angles to the surface of the can.

Pressure does not depend on the shape of the container

If a liquid is poured into a U-tube with one wide and one narrow arm, it will rise to the same height on each side. Even in a vessel of much more unusual shape, the liquid surface in all the arms is at the same height. A vessel like this is known as a Pascal vase, after the French mathematician and physicist, Blaise Pascal, after whom the pressure unit is named. Pascal used this apparatus to show that 'a liquid always finds its own level'.

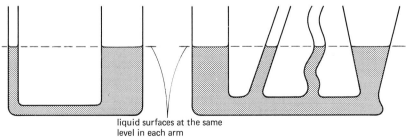

liquid surfaces at the same level in each arm

The pressure at a depth in a liquid depends only on the *depth* and not on the *amount* of liquid in each tube. The pressures at the bottom of each of the tubes must be the same; if they weren't, then liquid would flow from one to another until they became equal.

Liquid pressure

The molecules in a liquid are closely packed together but are not held tightly to their neighbours. A good model might be a pile of marbles in a tube.

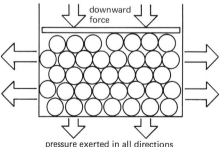

pressure exerted in all directions

If the pile is pushed downwards by exerting a force on the top, the marbles push out sideways as well as downwards. In the same sort of way, as gravity forces the liquid molecules downwards, this causes the liquid to push out sideways as well, and so the liquid exerts a sideways pressure too.

12.3 Calculating pressure in a liquid

The pressure at any point in a liquid depends on the depth and on the density of the liquid.

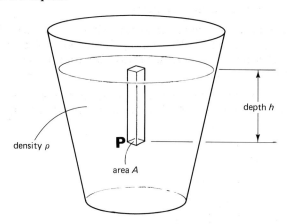

The point **P** is at a depth h. Imagine a small area A around **P**. The pressure at **P** is caused by the liquid standing on top of the area A.

Volume of liquid above A = area of base × height
$= Ah$
Mass of liquid above A = density × volume
$= \rho Ah$
Weight of liquid above A = mass × g
$= \rho Ahg$

The liquid exerts a force of size ρAhg on the base of the container. To calculate the pressure at **P**, we use:

$$\text{pressure} = \frac{\text{force}}{\text{area}}$$

$$= \frac{\rho Ahg}{A}$$

$$= \rho hg$$

So the pressure at a depth h in a liquid of density ρ is ρhg.

The pressure in a liquid depends only on the vertical distance below the liquid surface. The pressure at points **A** and **B** in the diagram would be the same. ◁

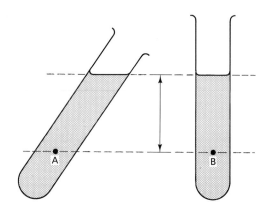

EXAMPLE 12.2

What is the pressure exerted by the water on the bottom of the deep end (2 m deep) of a swimming pool? The density of water is 1000 kg/m³.

Pressure at a depth in a liquid
$= \rho hg$
$= 1000 \text{ kg/m}^3 \times 2 \text{ m} \times 10 \text{ N/kg}$
$= 20\,000 \text{ N/m}^2$
$= 20\,000 \text{ Pa}$

The pressure caused by the water on the bottom of the pool is 20 000 Pa (or 20 kPa).

EXAMPLE 12.3

The surface of the water in a domestic cold water tank is 5 m above the downstairs cold water tap. What is the pressure of the water as it leaves the tap? The density of water is 1000 kg/m³.

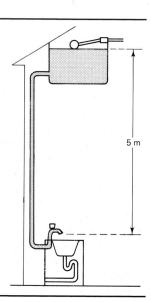

The pressure caused by the water in the tank and the pipes depends only on the vertical height (or depth, depending on how you look at it!).

Pressure at the tap
$= \rho hg$
$= 1000 \text{ kg/m}^3 \times 5 \text{ m} \times 10 \text{ N/kg}$
$= 50\,000 \text{ N/m}^2$
$= 50\,000 \text{ Pa}$

The pressure due to the height of the water is 50 kPa.

12.4 Hydraulic machines

A machine is any device which multiplies the force we apply and allows us to do a job more easily. Hydraulic machines use liquids under pressure. Liquids have two important properties:

1. They are almost completely incompressible. It is very difficult to squeeze a liquid into a smaller space (see section 1.12, part 5).
2. Changes in the liquid pressure are transmitted almost instantaneously to all parts of the liquid.

The drawing shows how a simple hydraulic machine works. A small force applied to the smaller piston can lift a greater load on the larger piston.

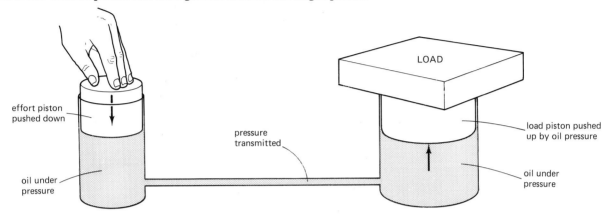

The 20 kg load on the larger piston has a weight of 200 N (weight = mg). To lift this load, an upward force of 200 N is needed. We can calculate the pressure on the base of the piston, which has an area of 0.1 m², which is needed to produce this force:

$$\text{pressure} = \frac{\text{force}}{\text{area}}$$

$$= \frac{200\,\text{N}}{0.1\,\text{m}^2} = 2000\,\text{Pa}$$

The pressure is the same everywhere within the liquid. In particular, the pressure under the smaller piston must also be 2000 Pa. This piston has an area of 0.01 m². So we can calculate the force which we would have to exert on the smaller piston to produce a pressure of 2000 Pa.

$$\text{pressure} = \frac{\text{force}}{\text{area}}$$

$$2000\,\text{Pa} = \frac{\text{force}}{0.01\,\text{m}^2}$$

Multiply both sides by 0.01 m²

$$2000\,\text{Pa} \times 0.01\,\text{m}^2 = \text{force}$$
$$\text{force} = 20\,\text{N}$$

This means that an applied effort force of just 20 N is enough to raise a 200 N load. The hydraulic machine has multiplied the force by 10.

As always, however, there are snags. If the load rises by 1 cm, some liquid must go into the larger cylinder to fill this extra space. Because of the difference between the cylinder sizes, the smaller piston would have to go down by 10 cm to push this amount of liquid through. So although the force needed is 10 times smaller, it has to move 10 times as far. In fact the work done by the effort is always more than the energy gained by the load: some energy is lost in friction. In other words, the efficiency of the hydraulic machine must be less than 100%. Even so, it is a useful device since the force needed to lift a heavy load (like a car) may be too big for our bodies to manage unaided.

The hydraulic jack

Hydraulic jacks are used to raise cars. To lift a heavy load with a small effort force, a small cylinder is used on the effort side and a much larger one on the load side: the load rises by only a small amount each time the small pump piston is pushed fully down. On the down stroke, the pump piston exerts pressure on the liquid which closes valve **A** and opens **B**. When the handle is raised again, valve **B** closes and **A** opens, drawing extra liquid into the system.

When the load has been raised, the high pressure of the liquid below the large piston keeps valve **B** closed and so the load stays up. To lower it again, a release valve is opened. This allows the liquid to flow back into the reservoir. This valve has to be closed again before the jack can be used to lift another load.

Any liquid could be used in a hydraulic jack. Oil is usually chosen because it helps to reduce the friction in the cylinders and prevent rusting of metal parts.

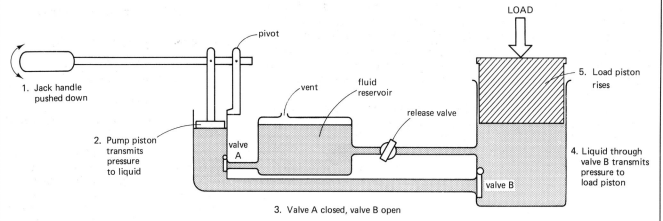

Hydraulic braking system

Car brakes use liquid pressure to transmit the pressure applied at the brake pedal to the brakes of all four wheels at the same instant. An advantage is that the connecting pipes can follow a winding route through the car without the difficulty you would have with a system of levers!

Disc brakes are usually fitted on the front wheels. Pistons **A** and **B** force two hard asbestos brake pads against the two sides of the disc attached to the wheel. The friction between the brake pads and the disc slows the car down.

Drum brakes are fitted on the back wheels of most cars. Each slave cylinder has two movable pistons. Pistons **C** and **D** push two semi-circular brake shoes apart. Their outer asbestos brake linings are pushed hard against the inner surface of the brake drums. These drums are attached to the car wheel. Again it is the friction between the brake shoes and the brake drum which slows the car down.

Disc brakes are usually more effective than drum brakes, so high performance cars often have disc brakes on all four wheels.

It is important to keep the brake fluid reservoir topped up. This prevents air from getting into the hydraulic brake system. Air is easily compressed by pressure (unlike liquids) and so an air bubble would prevent the pressure applied at the master cylinder being transmitted to the slave cylinders. This could result in one or more brakes failing to operate when needed. The process of removing air from the braking system is called 'bleeding' the brakes.

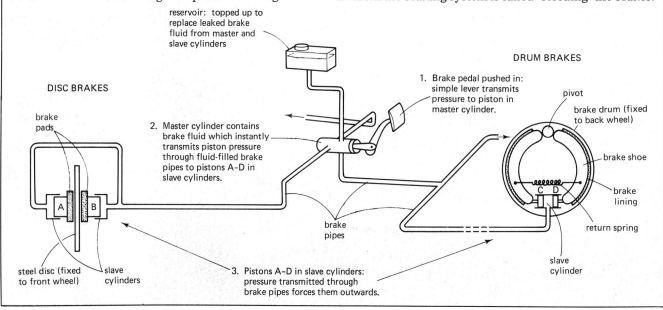

12.5 Air pressure

Air is a fluid. Just like a liquid, air exerts a pressure equally in all directions. The pressure of the atmosphere is due to the weight of air above us. At sea level this atmospheric pressure amounts to 100 000 Pa or 100 kPa. This is really quite a large pressure; it is equivalent to the weight of a double decker bus standing on one square metre!

Why are we not crushed by such an enormous pressure? Our bodies, of course, are made to cope with it. The pressure of the fluids inside our bodies exactly matches the atmospheric pressure outside. The only part of the body which is sensitive to changes in pressure is the ear-drum. It is connected to the back of the throat by two narrow tubes called the Eustachian tubes. Sometimes our ears 'pop' when we go up or down quickly (in an aeroplane, or in a lift, or even up a long hill in a car). This popping happens when air moves in the Eustachian tubes so that the pressure on the inner side of the ear-drum adjusts to match the new pressure on the outer side of the ear-drum. Swallowing helps to allow the pressures to equalise in this way.

One of the first people to realise that the atmosphere exerts a pressure was the Italian physicist, Torricelli. The idea came to him while he was trying to solve the problem of pumping water out of deep wells. He had been unable to pump water from a depth of more than 30 metres. Some thought that all that was needed was a better pump, but Torricelli became convinced that no pump, however good, would be able to do the job. We will see why later, in section 12.7.

In a very dramatic way, Otto van Guericke demonstrated the enormous size of atmospheric pressure. In Magdeburg in Germany he arranged for two large metal hemispheres to be made. The flat rims of the two hemispheres fitted closely together to form a sphere and the air inside was pumped out. Two teams of horses were then harnessed, one team to each hemisphere. The old engraving shows the scene. The horses were unable to pull the hemispheres apart! The diagram shows how the force caused by the atmospheric pressure holds the hemispheres together.

EXAMPLE 12.4

The pressure of the atmosphere at sea level is 100 000 Pa. The density of air is 1.3 kg/m³. From this information we can estimate the height of the Earth's atmosphere.

The pressure at a depth in a fluid is given by:

$$\text{pressure} = \rho h g$$

Substituting:

$$100\,000\,\text{Pa} = 1.3\,\text{kg/m}^3 \times h \times 10\,\text{N/kg}$$

Divide across by $(1.3\,\text{kg/m}^3 \times 10\,\text{N/kg})$:

$$\frac{100\,000\,\text{N/m}^2}{1.3\,\text{kg/m}^3 \times 10\,\text{N/kg}} = h$$

$$h = 7700\,\text{m}$$

From this calculation, the height of the Earth's atmosphere is 7700 m or 7.7 km.

This ignores one important difference between the atmosphere and a liquid. In a liquid, the density is the same throughout. The atmosphere, however, becomes less dense as we go upwards. The Earth's atmosphere is not a layer of air all of the same density, 7.7 km thick; it becomes gradually thinner as we go up and so it extends rather further than 7.7 km.

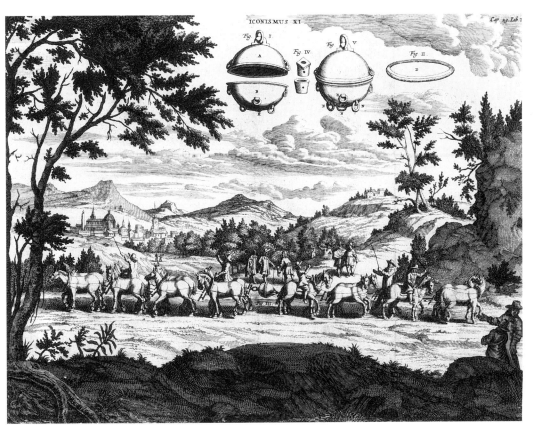

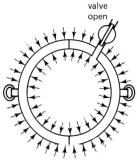

12.6 Effects of air pressure

Many everyday effects can be explained by atmospheric pressure. You may have heard explanations of some of them using the idea of 'suction', but from a physics point of view, there is no such force as 'suction'. To understand these effects we need to see how they are caused by atmospheric pressure.

The can crushing experiment

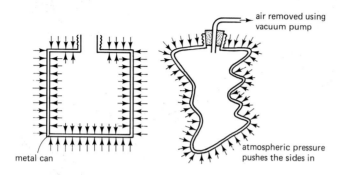

When the air in a metal can (such as a large oil can) is pumped out, its sides collapse inwards. The atmospheric pressure pushing on the outside of the can is quite large enough to bend and buckle the metal sides of the can. (The Magdeburg hemispheres would have collapsed in this way if they hadn't been very strong.)

Drinking straw

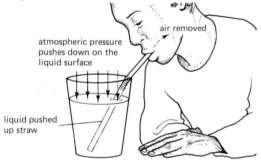

When you suck through a straw, what you are doing is lowering the air pressure inside the straw. Then the pressure of the atmosphere acting on the surface of the drink in the cup pushes the liquid up the straw and into your mouth.

It is interesting to consider what actually happens when you suck. You increase the volume of your lungs and this has the effect of reducing the pressure of the air inside them. This connection between volume and pressure will be discussed again more fully in section 15.3.

Syringe

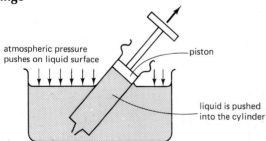

Pulling up the piston reduces the pressure inside the cylinder. The atmospheric pressure on the liquid surface then pushes the liquid up into the syringe.

If you then hold the plunger in place and lift the syringe out of the liquid, none will fall out. This is again due to atmospheric pressure.

The upturned milk bottle

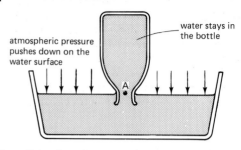

If you fill a milk bottle with water and then upturn it with its neck under the surface of water in a basin, the water will stay in the bottle.

The atmospheric pressure pushing down on the surface of the water in the basin is enough to hold up the water in the bottle. The extra pressure at **A** due to the height of water in the bottle is much smaller than atmospheric pressure, so the water is easily held inside the bottle.

Rubber sucker, e.g. for a kitchen hook

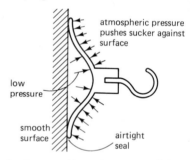

When the sucker is pressed into place, most of the air behind it is squeezed out. The sucker is held in position by the pressure of the atmosphere on the outside surface of the rubber. If the seal between the sucker and the surface is airtight, the sucker will stick permanently.

Larger suckers are used for handling large panes of glass while they are being installed in a building. Several suckers are attached at various points on the pane. When it has been fixed in position, a valve on each sucker is opened, to allow air inside again. The suckers can then be easily removed.

Vacuum cleaner

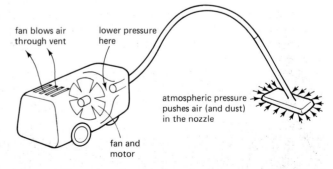

A 'vacuum cleaner' produces only a partial vacuum. The fan inside the cylinder blows air out of the vents. With less air inside, the air pressure there drops. The atmospheric pressure outside then pushes air up the cleaner hose, carrying dust and dirt with it.

Bicycle pump

Pumping up a bicycle tyre involves two valves, one in the pump and the other on the tyre. The tyre valve allows air to be forced into the inner tube, but prevents air escaping again. The pump valve is part of the piston and is simply made from a greased leather washer which fits inside the pump barrel.

When the pump handle is pushed inwards, the flap of leather around the washer presses against the inside of the cylinder, making a good fit. The pressure under the piston is increased and this forces the air through the tyre valve into the inner tube.

When the pump handle is pulled outwards again, the pressure beyond the piston becomes less and the valve flap opens. The air on the handle side of the piston is at atmospheric pressure, so it is forced past the piston. On the next downstroke this air will be forced into the tyre.

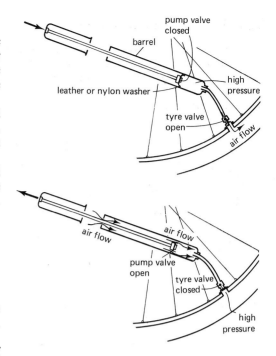

12.7 Measuring pressure

The manometer

A U-tube containing liquid can be used to measure pressure. This simple device is called a manometer. The height difference between the liquid levels is a measure of the pressure difference on the two sides. ▷ ▽

A manometer containing water can be used to measure the pressure of the gas supply. One arm of the manometer is connected by a rubber tube to a gas tap. When the gas is turned on, the height difference indicates how much the gas pressure is greater than atmospheric pressure. In a typical experiment the difference in levels might be 250 mm. Then the gas pressure is greater than atmospheric pressure by an amount equivalent to a 250 mm column of water. We can calculate how much this is in pascals:

$$\text{pressure} = \rho h g$$
$$= 1000 \text{ kg/m}^3 \times 0.25 \text{ m} \times 10 \text{ N/kg}$$
$$= 2500 \text{ N/m}^2 \text{ (or 2500 Pa)}$$

Atmospheric pressure is approximately 100 000 Pa, so the gas pressure is 102 500 Pa. This is not a lot bigger, but the difference is enough to make sure that the gas can flow out of the gas tap when needed.

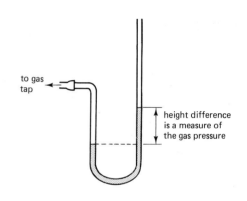

Bourdon gauge

The air hose on a garage forecourt has a pressure gauge to allow motorists to check their tyre pressures. This is not a manometer, but a gauge with a moving pointer. Smaller gauges of the same sort are found on top of cylinders of oxygen and other gases. These are called Bourdon gauges. They work on a very simple principle.

Inside the gauge is a hollow copper tube. When the gas pressure inside this increases, the tube straightens out a little. It behaves just like a stronger version of the 'party blower', which also uncoils when the pressure inside it is increased. The small movement of the copper coil is magnified by a system of levers, and the movement of the pointer is large enough to see and read off. Bourdon gauges are more suitable than manometers for measuring larger pressures. They have to be calibrated (that is, they have to have a scale marked off on them) before they can be used. ▷

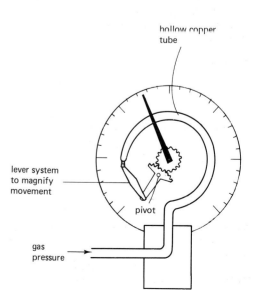

118 Pressure

If a manometer is used to measure larger pressure differences, it is usually filled with mercury rather than water. Mercury is 13.6 times more dense than water, so for the same pressure difference, the height difference of a mercury manometer is 13.6 times less than for a water-filled one.

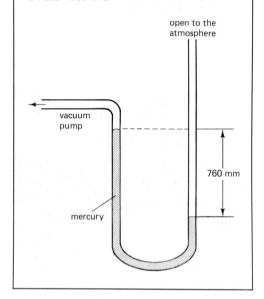

EXAMPLE 12.5

If water were used as the liquid in a barometer, how high would the water column be?

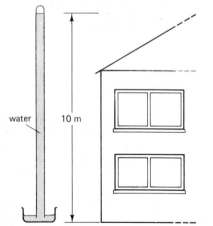

The water column would have the height needed to produce a pressure of 100 000 Pa at the bottom.

$$\text{pressure} = \rho h g$$
$$100\,000\,\text{N/m}^2 = 1000\,\text{kg/m}^3 \times h \times 10\,\text{N/kg}$$

Divide across by $(1000\,\text{kg/m}^3 \times 10\,\text{N/kg})$:

$$\frac{100\,000\,\text{Nm}^2}{1000\,\text{kg/m}^3 \times 10\,\text{N/kg}} = h$$

$$h = 10\,\text{m}$$

The atmosphere would therefore support a water column 10 m high! Water barometers are rarely used.

Measuring atmospheric pressure

Atmospheric pressure could be measured using a manometer by connecting one arm to a vacuum pump. The height difference would then be a measure of the atmospheric pressure on the other side.

With mercury in the manometer, the height difference is approximately 760 mm or 0.76 m. The density of mercury is 13 600 kg/m³, so atmospheric pressure can be calculated in pascals:

$$\begin{aligned}\text{pressure} &= \rho h g \\ &= 13\,600\,\text{kg/m}^3 \times 0.76\,\text{m} \times 9.8\,\text{N/kg} \\ &= 100\,000\,\text{Pa (to 2 s.f.)}\end{aligned}$$

(**Note.** The accurate value of g is used here.) However, atmospheric pressure is usually measured with a mercury **barometer**. The glass tube is filled with mercury and then up-ended in a mercury dish. The height of the mercury column does not depend on the diameter or the angle of the glass tube (pressure in a liquid depends only on depth). This is found to be approximately 760 mm.

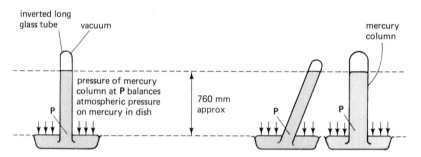

In SI units the pressure of the atmosphere is 100 000 Pa, but it is also sometimes given as 760 millimetres of mercury (mmHg) or as 1 atmosphere (atm).

The result of Example 12.5 means that water cannot be pumped up from a depth of more than 10 m in a single stage. With a perfect vacuum at the top of a pipe, water will rise just 10 m up the pipe. This explains why Torricelli had problems with pumping water which led him towards an understanding of atmospheric pressure (section 12.5).

Aneroid barometer

If you have a barometer on the wall at home, it is probably not filled with mercury. It is more likely to have a moving pointer and a round dial, with markings like 'very dry' and 'stormy' on it. Changes in atmospheric pressure can be used to try to forecast the weather. This sort of barometer is called an aneroid barometer. ▽

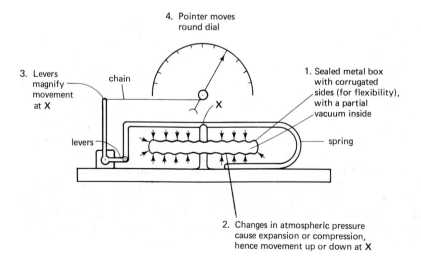

Altimeters

Atmospheric pressure is caused by the weight of air above. So the higher you go, the smaller the atmospheric pressure becomes.

A barometer can be used to measure height above sea level directly. An **altimeter** is an aneroid barometer with a scale marked in metres rather than pascals. The atmospheric pressure at ground level changes from day to day, so an altimeter must always be set at zero before taking it up to a height. Altimeters are fitted to aircraft; they are also used by parachutists and mountaineers to tell them how high they are.

Aircraft pressurization

Passenger airliners cruise at a height of around 15 km. The pressure at that height is about 30 kPa. Inside the cabin the passengers must be able to breathe air at normal atmospheric pressure (100 kPa). So the cabin is pressurised, and there is a large pressure difference between the inside and outside. The cabin walls, windows and doors must be designed to withstand the forces which this pressure difference causes. Specially strong plastics are used for aircraft windows, and doors fit like 'plugs' from inside, so that they cannot be forced out by the much higher pressure inside.

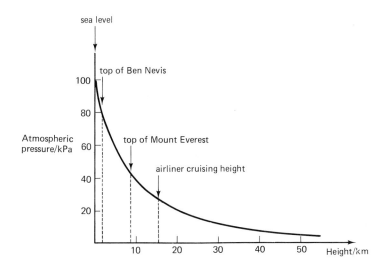

A doctor taking a patient's blood pressure. It is important that the cuff is placed at the top of the arm, level with the heart itself. The blood pressure at this point is the same as at the heart itself. (Remember: Pressure is the same at two points at the same level.)

Measuring blood pressure

A doctor uses a special mercury filled manometer (called a sphygmomanometer) to measure blood pressure – the maximum and minimum pressures of blood in the arteries as it is pumped from the heart.

An inflatable cuff is placed round the patient's arm and pumped up using a small hand pump. The mercury manometer indicates the pressure inside the cuff. The doctor pumps until the cuff is squeezing the patient's arm firmly enough to stop the blood flow in the arteries of the arm. Then she opens a valve and slowly releases the air in the cuff, at the same time listening for the sounds of blood flow in the patient's arm. When the pressure reaches the point where it is equal to the maximum blood pressure, she will hear the sound of a spurt of blood on every heartbeat. She takes one reading of the manometer at this moment. When the pressure has fallen until it is equal to the minimum blood pressure, the sounds disappear again. Another manometer reading is taken. Typical healthy values for maximum and minimum blood pressures are 120 mmHg and 90 mmHg. (What are these equivalent to in pascals?)

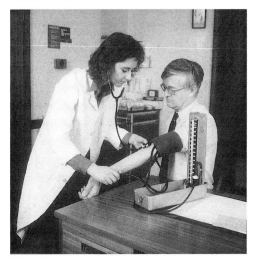

Pressure

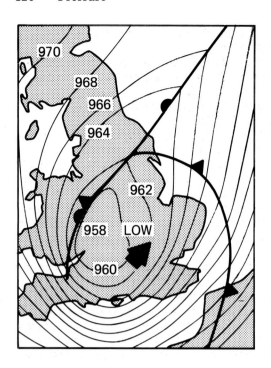

Weather changes

The words 'high' and 'low' on the television weather maps mean areas of high and low **atmospheric pressure**. Pressure is measured by a large number of weather stations and ships over the area of the map and the readings from each station are plotted on a map. Lines, called **isobars** are then drawn, joining up places where the pressure is the same. Isobars are usually labelled in units called bars, or millibars.

$$1 \text{ bar} = 100 \text{ kPa} = \text{normal atmospheric pressure}$$

$$1000 \text{ millibars} = 1 \text{ bar}$$

Our very changeable weather in Britain is the result of the constant mixing of cold air from the regions around the North Pole with warmer air from nearer the Equator. Cold air becomes denser and sinks towards the ground whereas hot air expands and rises. The dense and falling cold air causes an area of higher pressure, the rising hot air a region of lower pressure. Where the warm and cold air meets, water vapour carried in the warmer air condenses into clouds (like your breath condensing when it hits the cold air on a winter morning). This line of clouds where the cold and warm air masses meet is called a **front**.

To make things more complicated, all of this is going on on the surface of a globe which is rotating! The Earth's rotation makes the air flow round in curved paths and not in straight lines. The fronts become curved in shape also – as the weather map shows.

Questions

1. What pressure is exerted by each of the following?
   (a) a force of 200 N acting on an area of 4 cm²;
   (b) a force of 500 N acting on an area of 25 cm²;
   (c) a force of 1500 N acting on an area of 30 cm².

2. What force, acting on an area of 8 cm², would produce a pressure of:
   (a) 16 N/cm²;
   (b) 400 N/m²?

3. A floor surface is damaged by a pressure of more than 4 kN/m². Over what area must a force of 40 N be spread to keep within this limit?

4. A boy weighs 400 N. The total area of both his shoes is 200 cm².
   (a) What pressure (in N/cm²) does he exert on the floor?
   (b) What is this equivalent to in N/m² (or Pa)?

5. Would the table in the diagram make a deeper mark on the carpet if it is standing upright, or if it is turned upside-down? Explain your answer.

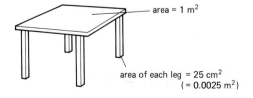

6. The flat head of a drawing pin has area 1 cm². The point of the pin has area 0.01 cm². A force of 15 N is needed to push the pin into a board.

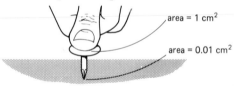

(a) What is the pressure between your thumb and the pin as you push?
(b) What is the pressure between the pin point and the board?

7. A rectangular block measures 8 cm by 5 cm by 4 cm, and has a mass of 1.25 kg.
   (a) (i) If the gravitational field strength is 10 N/kg, what is the weight of the block?
   (ii) What is the area of the smallest face of the block?
   (iii) What pressure (in N/cm²) will the block exert when it is resting on a table on its smallest face?
   (iv) What is the least pressure the block could exert on the table?
   (b) (i) What is the volume of the block?
   (ii) Calculate the density of the material from which the block is made. (Sp. NEA)

8. What is the extra pressure (i.e. above atmospheric) at a depth of 5 m in water? (The density of water is 1000 kg/m³.)
   At what depth in sea water (density 1050 kg/m³) would the pressure be the same?

9 Use the idea of pressure to explain each of the following:
(a) Rucksacks are made with padded shoulder straps.
(b) Cheese is most easily cut using a wire cheese-cutter.
(c) If a suitcase handle breaks and is temporarily repaired with string, the case will be much less comfortable to carry.
(d) The cold water storage tank is usually placed at the highest point in a house.
(e) When a dam wall is built to store water in a reservoir, the wall is always made much thicker at the bottom than at the top.
(f) To walk on a shed roof made of corrugated plastic sheeting, it is safer to lay large planks on the roof and walk on these.
(g) If someone faints, you should let them lie down and then raise their feet above the level of their head.

10 The diagram shows a car windscreen washer pump.

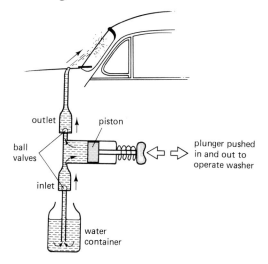

Study the diagram and write a short explanation of how the pump works. Use the ideas of pressure in liquids to explain when each valve is open and closed.

11 (a) The diagram shows a tall cylinder which is kept filled with water.
 (i) Copy the diagram and add to it the paths followed by the water passing through and away from the outlets **X** and **Y**.
 (ii) What does this experiment demonstrate about the pressure in a liquid?
 (iii) Explain how this simple experiment affects the design of the wall of a dam.
 (iv) If the experiment were carried out using a liquid of lower density than water, **state** and **explain** what difference would be observed.
(b) The level of water in the header tank of a house is situated 5 m above the hot water tap in the kitchen. The density of water is 1000 kg/m³ and the value of g is 10 N/kg.
 (i) What is the pressure of the hot water emerging from the tap, in Pa, and in kPa?
 (ii) The area of the tap where water emerges is 10^{-4} m², i.e. 0.0001 m². A person places his thumb over the outlet of the tap and stops the hot water flowing. Use your answer to (b)(i) to calculate the force exerted by the water on the person's thumb.
(JMB)

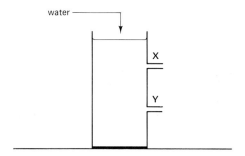

INVESTIGATION
What pressure do you exert when you stand on the floor?

The answers to the questions on page 110 are: table 10 N/cm²; elephant 9 N/cm²; heel 200 N/cm²; pin 20 N/cm², 10 000 N/cm².

13: Floating and Sinking

A heavy log of wood will float on water yet a light copper coin will sink. The property which determines whether an object floats or sinks is not its weight, but its **density**. Look at the density values in Table 2.1. Solids like foam polystyrene, cork, wood and ice with densities less than that of water will float on water. Oil is less dense than water so it floats on top of water. Polythene will float in water but would sink in ethanol.

But the situation starts to get more complicated when we consider objects of a more complicated shape. A ship with a steel hull can float on water but a solid block of steel sinks. What makes the difference? We will find in this chapter that the answer involves ideas about pressure which we learnt about in Chapter 12.

13.1 Upthrust

You may have noticed that objects seem less heavy if you lift them under water. Lifting a large stone from the bed of a stream is easier while it is submerged. It feels heavier once it is out into the air. When any object is under water, the water exerts an upwards force on it. This force is called an **upthrust**.

We can investigate this by weighing an object in air and then in water. The diagrams show the results of weighing a metal block in air and in water. The weight appears to be less when the block is immersed in liquid. The upthrust is equal to this apparent reduction in weight.

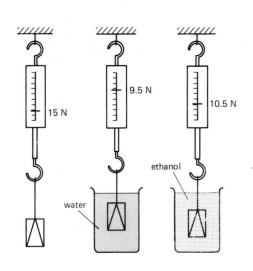

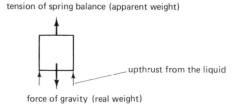

The Greek natural philosopher Archimedes (287–212 BC) was the first person to realise that the upthrust came from the liquid itself. He noticed that the water level rose in the container when a solid object was immersed in it. The solid takes up some 'space' where the water was previously; it displaces some water. Archimedes realised that the upthrust force was a result of this displaced liquid.

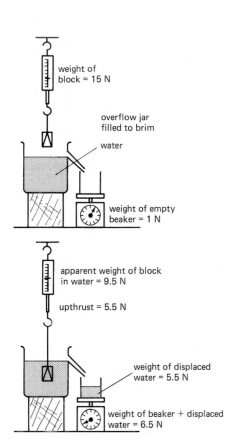

If we repeat the weighing experiment above, but this time arrange to collect the displaced water, we get an interesting and important result. This time the block is weighed in air and then lowered carefully into a displacement can filled to the brim with water. The water which is displaced is collected in an empty beaker. The weight of the displaced water is 4 N. This is the same as the apparent loss in weight of the block. In other words, the upthrust is equal to the weight of the water displaced. This is one example of a more general result known as **Archimedes' Principle**. It states:

When an object is immersed in a fluid (a liquid or a gas), the upthrust force on the object is equal in size to the weight of fluid displaced by the object.

Archimedes' Principle applies to all liquids, not just to water. It also applies to gases. An object's true weight should be measured in a vacuum. If it is weighed in air, the displaced air causes an upthrust and the apparent weight is less. Usually this is too small to make any significant difference. The upthrust on your body due to the air you displace is about 0.5 N (the weight of an average sized tomato).

Archimedes' Principle also applies to an object which is only partly below the liquid surface and to floating bodies. In both cases the upthrust is equal to the weight of liquid displaced.

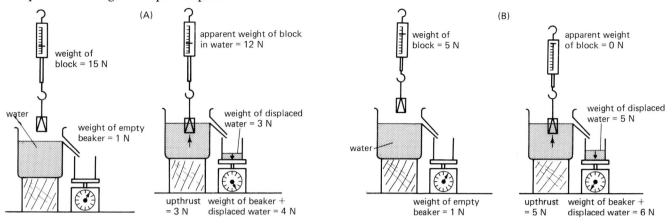

13.2 What causes the upthrust?

Upthrust is a result of the fact that pressure increases with depth. Imagine a brick hanging from a string, completely immersed in water. The water exerts a pressure on all the surfaces of the brick. The pressure on the sides of the brick causes equal and opposite forces, which cancel out. But the pressure on the bottom surface of the brick is greater than on the top surface, because it is deeper. So the force acting upwards on the bottom of the brick is bigger than the force acting downwards on the top of the brick. This difference is what causes the upthrust.

If you use $p = \rho h g$ and work out the pressures and forces on the top and bottom of the brick, you can prove that the upthrust force is equal to the weight of the liquid displaced by the brick. This is Archimedes' Principle.

13.3 Floating

Imagine two blocks of the same size and shape, one made of steel and the other made of wood, held below the surface of water and then released. The steel block will sink to the bottom and the wooden block will rise to the top. Both blocks displace exactly the same volume of water, so the upthrust force is the same in both cases. But their weights are different. The weight of the steel block is larger than the upthrust, so the block will accelerate downwards; the weight of the wooden block is smaller than the upthrust, so the block will accelerate upwards.

When it reaches the surface, the wooden block will float. The upthrust is now equal to the weight. The block is only partially submerged, so it is displacing less water than before. In fact the weight of water it is now displacing is equal to its weight. This is sometimes called the law of flotation, but it is really just a special case of Archimedes' Principle. The law of flotation states:

A floating object displaces its own weight of the fluid in which it floats.

Another way to look at this is to imagine a block of wood weighing 5 N being slowly lowered into water. As more and more water is displaced the upthrust gets gradually larger. When the upthrust is equal to the weight, the block floats.

If a steel block is gently lowered into water in the same way, it is not able to displace a weight of water equal to its own weight, even when it is fully immersed. So the steel block cannot float.

In general, if a block is made of a material which is less dense than water, it will be able to displace its own weight of water before it is fully submerged – so it will float. If it is made of a material which is denser than water, then it is unable to displace its own volume of water and will sink.

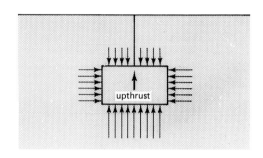

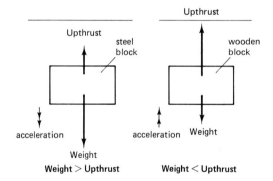

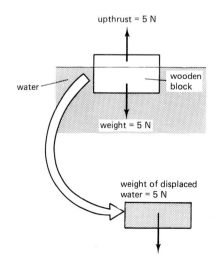

Floating and sinking

13.4 Ships

Steel is denser than water. Solid blocks of steel sink, yet steel ships can float. We are now in a position to understand how this is possible. The hollow shape of a ship's hull causes it to displace much more water than could the same mass of steel in the form of a solid block.

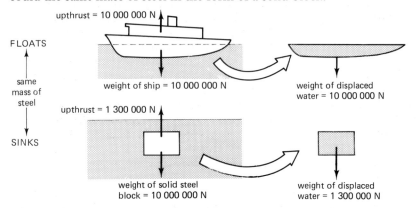

The ship's hull displaces a weight of water equal to the weight of the ship well before it is completely immersed. So the ship floats.

The most unlikely materials can be used for ship-building if the hull is shaped to displace enough water. In China in the 1970s the high cost of steel led to the making of concrete boats for use on rivers and canals – a good example of 'appropriate technology'!

Of course if a steel or a concrete boat capsizes it will sink, because the actual *material* of the boat is denser than water. On the other hand a capsized wooden boat will float.

13.5 Depth of floating

It is easier to float in salt water than in fresh water. The Dead Sea is so salty that you can even read the newspaper while floating! The thing that makes the difference is the density of the liquid. ▽

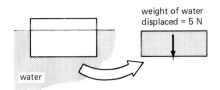

A 5 N wooden block floats in water. It sinks until it has displaced exactly 5 N of water.

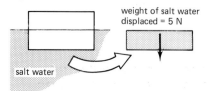

Salt water is denser than fresh water. The 5 N wooden block doesn't have to sink so far to displace 5 N of salt water. So it floats higher in the salt water.

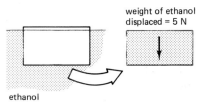

Ethanol is less dense than water. The block will have to sink further to displace 5 N of ethanol. So it floats deeper.

Note. In all three cases, the **weight** of liquid displaced is the same – 5 N. The **volume** displaced, however, is different each time as it depends on the density of the liquid.

Hydrometers

The depth at which an object floats depends on the density of the liquid it is floating in. An instrument to measure liquid density (called a hydrometer) is based on this idea. It has a glass bulb which contains some lead to make it float upright, and a long narrow neck with a scale marked on it. The drawings show a hydrometer floating in water and in a sugar solution. The sugar solution is denser than water so the hydrometer can displace its own weight of liquid without sinking so deep. The reading on the scale gives a measure of the density of the liquid. This is not a particularly accurate method of measuring density but it is quick and easy.

Hydrometers are used in home beer and wine-making. Wine is made from fruit juice which contains sugars. The juice is denser than water to begin with. Yeast is added and fermentation begins. The sugar is used up and alcohol (ethanol) is produced. The density of ethanol is less than that of water, so as more and more ethanol is produced the density of the liquid mixture gradually drops. By measuring the density you can estimate how far the fermentation has gone and so decide when to stop this stage of the process.

Another application is in checking car-battery acid. A car battery contains concentrated sulphuric acid. When the battery is fully charged the density of the acid should be about 1.25 times larger than the density of water (about 1250 kg/m³). When the battery is discharged, the acid will have a density almost the same as that of water. Measuring the density of the acid is an easy way to check if the battery is fully charged. Because concentrated acid can burn skin and clothes, a special pipette containing a small hydrometer is used to draw acid from the battery and measure its density in safety. ▷

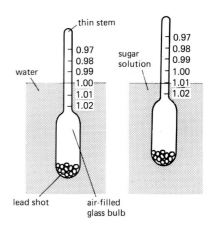

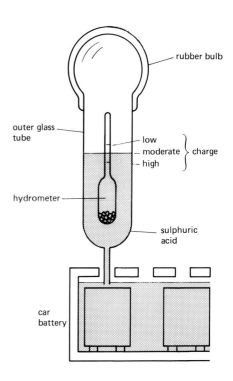

Submarines (and fish)

A submarine can sail on the surface or underneath it. Its outer hull is entirely watertight. Inside the hull are a number of tanks called ballast tanks. When these are full of air, the vessel can easily displace its own weight of water and so it floats on the surface like an ordinary ship. To make the submarine go under water, valves are opened which allow water into the ballast tanks. These are filled until the weight of the submarine is slightly greater than the weight of water it can displace and the submarine then slowly dives. It would be almost impossible to keep the weight and the upthrust exactly equal. Instead the submarine's weight is slightly greater than the upthrust when it is submerged. As it sails through the water, its short side fins act like the wings of an aircraft and produce some additional upward force. This provides the means of controlling the submarine's depth.

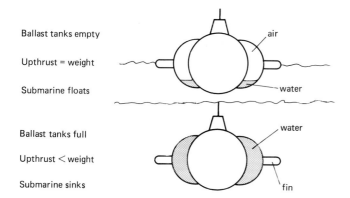

When it needs to surface again, water is pumped out of the ballast tanks. The submarine's weight is now less than the upthrust and so it rises to the surface.

A fish controls its buoyancy in a similar way through the use of an air-filled swim-bladder inside its body.

Floating and sinking

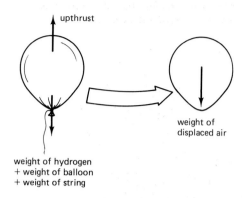

13.6 Floating in air

A hydrogen-filled balloon will float up into the air if it is released. The upwards force comes from the air. Air is a fluid and produces an upthrust.

Air is not very dense but, if a lot of air is displaced, the force can be quite large. The airship below displaces more than 5 tonnes of air. So it will float upwards if its total weight is less than this. Modern airships are filled with helium. This gas is denser than hydrogen but has the great advantage of not being inflammable. It is still much less dense than air. When the airship is just hovering, the weight of air displaced is equal to the weight of the helium plus the weight of the airship and the passengers.

Hot air balloons use a similar idea. Hot air is less dense than cold air. (We will study this in more detail in Chapter 15.) The weight of cold air displaced can be made equal to the combined weight of the hot air inside the balloon, the balloon fabric, the basket and passengers, provided a large enough balloon is used. Once it is airborne, the balloon's height can be controlled by turning the gas burner which heats the air on and off as needed.

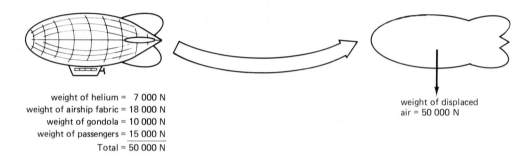

Questions

1. Write a paragraph as if you were explaining to your younger brother how it is possible for a boat made of Plasticine to float, yet a ball of Plasticine sinks.

2. Here are the densities of several solids and liquids:

Materials (in alphabetic order)	Density (g/cm³)
Cork	0.25
Mercury	13.6
Oil	0.8
Polythene	0.9
Steel	8.0
Water	1.0

A small cork, a polythene ball and a steel ball bearing are put into a tall jar and 25 cm³ each of mercury, oil and water are poured in. Draw a diagram to show what the jar will look like when its contents have had time to settle.

3. Explain why a hydrometer for measuring the density of homebrewed beer has the larger numbers at the bottom of the stem and the smaller numbers at the top.

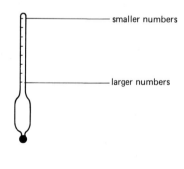

INVESTIGATION

A puzzle! Imagine you are in a rowing boat in a small pool. You have several concrete blocks in the boat. You drop these over the edge. Does the water level at the side of the pool rise or fall?

Try it with a model boat in a small tank and see if your prediction is right.

14: Thermal Properties of Matter

14.1 Temperature

We can tell whether something is 'hot' or 'cold' by touching it. We can judge its 'hotness' or **temperature**. Something which is hot has a high temperature; something which is cold has a low temperature. But our unaided senses are only satisfactory for *estimating* temperature. We cannot detect very small temperature differences. We are also influenced by what we were handling before. For instance, if you have been washing dishes in hot water and then pick up a mug of tea, the mug will feel quite cool. But if you had been outside on a frosty day without gloves on and then picked up the same mug of tea, it would feel hot.

To *measure* temperature, we need to find something which changes with temperature and which we can measure accurately. The commonest thing to use is the length of the column of mercury in a mercury-in-glass thermometer.

After your hand has been in cold water for a while, tepid water feels hot. After it has been in hot water, tepid water feels cool.

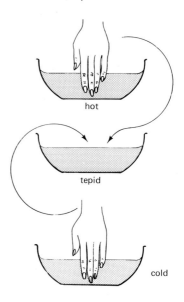

Marking a scale on the thermometer

The thermometer on the right has a scale marked on it. For the marking we need to have some agreement about which numbers to use for each level of 'hotness'. We pick two standard temperatures which we can easily set up and be sure of. These are known as **fixed points**.

The **lower fixed point** (the ice point) is the temperature at which pure water freezes.

The **upper fixed point** (the steam point) is the temperature of the steam above pure boiling water when the pressure is 1 atmosphere (standard atmospheric pressure, or 760 mmHg).

We can choose any two numbers we wish for these two fixed points. The most commonly used scale, and the one which all scientists have agreed to use for scientific work, is the **Celsius** scale. On the Celsius scale, the ice point is 0 degrees Celsius, or 0°C, and the steam point is 100 degrees Celsius, or 100°C.

This gives us a practical way of marking a scale on a thermometer. First we place it in melting ice and mark the position of the top of the mercury thread. This is 0°C. Then we place the thermometer in the steam from boiling water and mark the new position of the thread. This is 100°C.

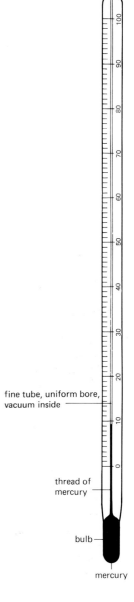

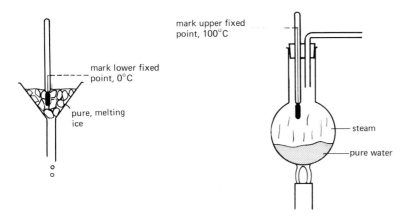

Then we divide the distance between the two marks into 100 equal divisions. Each of these is called one degree. The older name for a temperature scale like this is a 'centigrade scale', meaning 'one hundred divisions'. Marking the scale in this way is called **calibrating** the thermometer.

Some temperatures on the Celsius scale

°C	
6000	surface of the sun
4000	centre of the earth
1540	iron melts
900	bunsen burner flame
800	red hot
357	mercury boils
100	boiling water
58	record highest atmospheric temperature recorded
37	human body temperature
25	fine summer day
0	water freezes
−10	cold winter day
−39	mercury freezes
−88	record lowest atmospheric temperature recorded
−196	nitrogen gas turns to liquid

Is there a highest possible temperature or a lowest possible temperature? That is a question we will come back to later in this chapter. Another point to note is that temperatures above 357°C and temperatures below −39°C could not be measured with a mercury thermometer! A different type of thermometer would be needed.

14.2 What is temperature?

The mercury-in-glass thermometer gives us a way of measuring temperatures – between 357°C and −39°C anyhow! But what *is* temperature? What is the real difference between a hot object and a cold one? The answer lies in what the molecules in the object are doing. According to the molecular theory of matter (see Chapter 1), everything is made of molecules which are constantly moving around randomly. If an object is heated, its molecules move faster. A hot object is one where the molecules are vibrating fast or moving around fast – with a lot of energy. A cold object is one where the molecules are vibrating or moving around slowly – with less energy.

Temperature is really a measure of the average energy of the molecules in the material.

One prediction which follows from this is that there is a 'lowest possible temperature'. This is when the molecules are stationary and have no kinetic or potential energy. The material cannot be cooled any further. This temperature is called **absolute zero**.

14.3 Types of thermometer

Mercury-in-glass thermometers have a number of limitations. One is that they can only work between the freezing point and the boiling point of mercury. To go beyond this range, other liquids can be used. A common one is ethanol, usually coloured red to make the thread easier to see. It works between about 70°C and −117°C. This raises the question: Do thermometers with different liquids always agree? Do they give the same results when we use them to measure temperatures? The simple answer is that they don't quite agree but the differences are small. For most purposes the mercury thermometer is accurate enough.

However, all liquid-in-glass thermometers have limits since glass itself eventually begins to melt. Another problem is that they are not much use for measuring temperatures in inaccessible or dangerous places because someone has to be there to take the readings! For instance, you couldn't easily measure the temperature inside the cylinder of a car engine when it was running using a liquid-in-glass thermometer. Completely different types of thermometer, such as electrical thermometers, are needed. Although we have not dealt with electricity yet in this book, we will describe some electrical thermometers briefly here. You may find this section easier to understand if you come back and read it again after you have studied the ideas about electricity in Chapters 25 and 26.

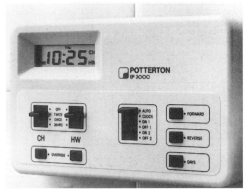

An electrical thermometer controlling central heating. Another advantage of electrical thermometers over the mercury-in-glass type is that temperatures can be read automatically and fed back to a device which switches the boiler on and off.

Resistance thermometer

These are based on the fact that the electrical resistance of a piece of metal increases as it gets hotter. It becomes more difficult for current to pass through the metal and so the reading on an ammeter drops. The meter can be calibrated to read temperatures directly in degrees Celsius. A film of platinum is often used for the resistance. Resistance thermometers can be used to measure temperatures from −200°C to over 1000°C. ▷

A platinum resistance thermometer: its resistance gets bigger as it gets hotter.

Thermistor thermometer

A thermistor is a small device made from semiconductor material (see Chapter 32). It behaves in the opposite way to a metal resistor – its resistance decreases as it gets hotter, so more current can flow through it.
▷

Some thermistors: their resistance drops as they get hotter.

Thermocouple thermometer

This is made from wires of two different metals. These wires are used to make two junctions. When the junctions are at different temperatures, the thermocouple behaves like a small electrical cell (or battery). It produces a small electric current which can be measured on a sensitive meter.

The junctions are very small and so thermocouples are very useful for measuring temperatures inside ovens and furnaces. They can be used over a wide temperature range from as low as −200°C to over 1500°C.

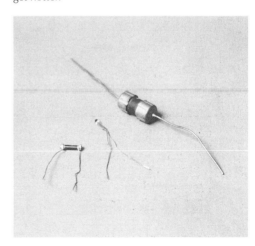

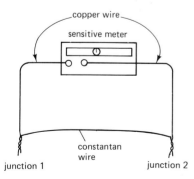

The meter reading shows the temperature difference between the junctions.

The ESMI meter uses a thermocouple to measure temperature.

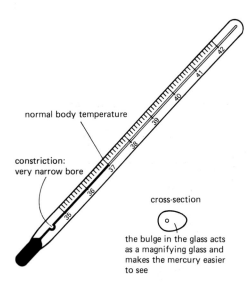

Clinical thermometer

◁ The thermometer used by a doctor to take a patient's temperature is a special type of mercury-in-glass thermometer. It works only over a small range of temperatures from 35°C to 42°C, but can read temperatures to an accuracy of $\frac{1}{10}$°C. It has a narrow constriction in the tube just above the bulb. When the thermometer is removed from the patient, the mercury begins to cool and the thread breaks at this constriction. The part of the thread in the tube then stays still and the doctor can read it at leisure.

(**Important note.** This makes a doctor's thermometer different from an ordinary mercury thermometer which **must** be read while it is still in contact with the object whose temperature you are measuring.)

After reading the temperature, the doctor shakes the thermometer to get the mercury thread back into the bulb. The thermometer is then ready to use again.

If you have ever looked closely at a clinical thermometer, you will have noticed that the tube is not exactly round, but slightly pear-shaped. This magnifies the very thin mercury thread and makes it easier to read the temperature.

14.4 Expansion of solids

The mercury thermometer works because mercury expands when heated. So do many other materials. Most solids expand only a little when heated but the effects can be shown easily enough. The metal bar of a bar and gauge fits snugly into the gauge when both are cold. But when the bar is heated, it no longer fits. The bar also expands in width and no longer fits into the hole in the gauge. Expansion occurs in all directions. The ball and ring shows the same thing. Solids expand slightly when heated.

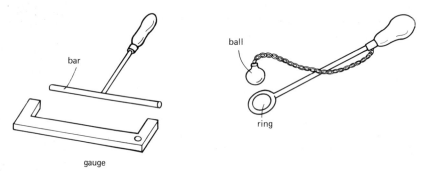

Although the expansion is small, the forces which result can be very large. The bar breaker apparatus shows just how big they are. The metal bar is heated to make it expand and then the nut is tightened to hold the cast-iron pin firmly against the pillars on the base of the apparatus. As the metal bar cools, it contracts, pulling the pin more and more tightly against the pillars. The force is eventually enough to snap the pin.

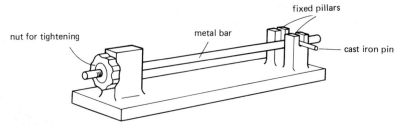

Molecular explanation

As solids are heated, their molecules vibrate more rapidly. This pushes the molecules further apart as their vibrations become more vigorous. As the solid cools again, the vibrations become smaller and the forces of attraction between the molecules pull them together again. The sum of all these little forces causes the large force which is enough to break the cast-iron bar of the bar-breaker.

Expansion and contraction: problems and applications

Although a short metal bar does not expand very much when heated, the expansion of a long metal beam will be much greater. And, of course, all solids expand, not just metals. For example, the steel or concrete beams used to make the deck of a bridge might be 100 metres long or more. On a hot summer day a beam like this would be about 5 cm longer than on a cold day in winter. As we have seen, the forces involved in expansion are very large. So room has to be allowed for expansion, especially where long lengths of material are involved.

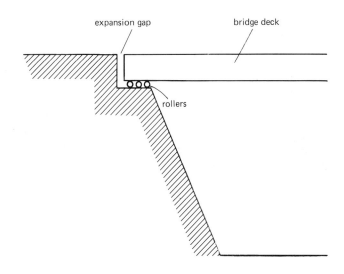

Bridges

The main span of a bridge is mounted on rollers which allow it to expand and contract without damaging the bridge supports. Expansion joints are left in the road surface to allow it to expand without buckling.

Railway tracks

Railway track used to be laid in lengths of about 20 metres with an expansion gap between each length. Nowadays, to give a smoother ride, modern track is welded together as it is laid, and each piece is about 1 km long. The problem of expansion and contraction from summer to winter is dealt with by heating the rail as it is laid so that it expands. It is then fastened to heavy concrete sleepers which hold the rail as it cools. These can take the large forces which are exerted without any damage. The railway track is always under tension – like a stretched piece of elastic. On a hot day, the tension is smaller than on a cold day. Between the lengths of track, a tapering joint allows for any expansion movement which does take place, and also lets the train cross smoothly from one length of track to the next.

Pipelines

Oil is often carried in metal pipelines many kilometres long. Similarly in chemical engineering works, long pipelines are often used to carry liquids or gases from one part of the plant to another. Unless some precaution is taken, these pipes will buckle in hot conditions due to expansion. The usual solution is to put a flexible expansion loop into the pipeline at regular intervals.

hearth for heating steel tyre
metal train wheel
axle
the tyre contracts to fit the wheel

Fitting tyres

Expansion can be made use of in fitting metal tyres. Train wheels have steel tyres which fit tightly on to the metal wheel. The tyre is heated in a furnace to make it expand. When it is red hot, it is fitted to the wheel and as it contracts the fit becomes very tight indeed. No extra nails or screws are needed. The same idea is used to put metal tyres on to wooden cart wheels, or metal hoops on to barrels.

Getting stubborn tops off bottles

If the screw top of a bottle or jar refuses to come off, you can use expansion to help you move it. The top is usually made of metal so if you hold it under a hot tap for a few seconds, the hot water will make the metal expand. This often loosens the top enough, and you can unscrew it easily.

Of course the hot water will make the glass expand as well. Why does this not stop the method working? The reason is that different materials expands by different amounts when heated. The metal top expands more; it also probably expands more quickly because it is being heated more directly.

14.5 Linear expansivity

When a solid is heated it expands in all directions. However if we know the amount by which the length increases we can calculate any other changes. This increase in length is called **linear expansion**. It depends on the original length of the solid and on the temperature rise.

increase in length = α × original length × temperature rise

The constant α is called the **linear expansivity** of the material. It has different values for different materials.

EXAMPLE 14.1

The main span of a road bridge is made of concrete and is 100 m long. How much longer is the bridge on a summer day when the temperature is 20°C than on a winter day when it is 0°C?

From Table 14.1, the linear expansivity of concrete is 0.000 011/°C; in other words, each metre of concrete expands by 0.000 011 m for every 1°C rise in temperature. We use the equation:

increase in length = α × original length × temperature rise
= 0.000 011/°C × 100 m × 20°C
= 0.022 m

The increase in length is 0.022 m, or 2.2 cm. This bridge is 2.2 cm longer on the summer day. This difference is large enough to see easily, and an expansion gap would have to be left to allow the bridge to expand safely.

Table 14.1

Material	Linear expansivity (/°C)
Aluminium	0.000 026
Brass	0.000 019
Steel	0.000 011
Concrete	0.000 011
Glass	0.000 009
Pyrex	0.000 003
Invar	0.000 001

Making use of the expansivities of different materials

Choosing the same expansivity

The linear expansivity of concrete is the same as that of steel. This means that steel rods can be used to reinforce concrete without any problems being caused by expansion. The steel is embedded in the concrete when it is being made. In the finished concrete beam, both the steel and the concrete will expand and contract together when the temperature changes. If they expanded by different amounts, this might crack the concrete. ▷

Another place where it is important to have two materials with the same linear expansivity is in making light bulbs. The metal wires from the bulb filament have to be sealed into the glass bulb. A light bulb is filled with an inert gas (not air) and so the seal must be extremely good to prevent the gas getting out or air getting in. The problem is that the bulb gets hotter when it is in use. The glass and the metal wires expand. An expensive platinum alloy with the same linear expansivity as glass is used for the part of the wire which makes the seal with the glass. ▷

Advantages of small expansivity

Pyrex glass has a much smaller linear expansivity than ordinary glass. This is why it is oven-proof and can stand large changes of temperature. Ordinary glass jars may crack if boiling water is poured into them. The inside of the glass is heated more quickly than the outside and starts to expand. This causes large stresses in the glass which may be enough to make it crack. Pyrex dishes are less likely to crack because they don't expand as much.

Materials with different expansivities: the bimetallic strip

A bimetallic strip is made of two thin strips of different metals, tightly joined together side by side. Two common metals to use are brass and invar (an alloy with a very small linear expansivity). When it is heated, the brass expands more than the invar and this makes the strip bend, with the brass always on the outside of the curve. (It is further round the outside of a curve than round the inside!) As it cools it bends back again.

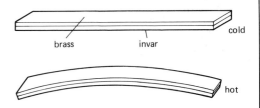

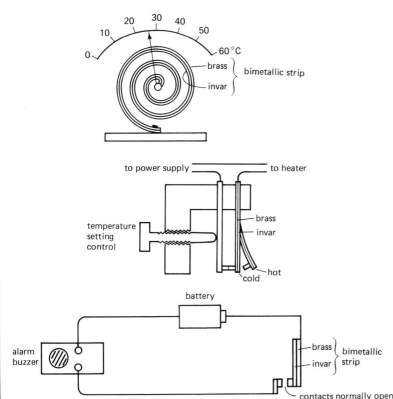

Some simple domestic thermometers use a long spiral bimetallic strip. As the temperature rises the spiral coils up more tightly and this moves the pointer over the scale. Thermometers like this are not very accurate but they are easy to read.

A room thermostat. When the room is warm, the bimetallic strip bends and the electrical contacts move apart. This switches the room heater off. As it cools down again, the bimetallic strip straightens and switches the heater on again. You can choose the temperature you want to have in the room by setting the control knob. If it is at a high setting (screwed in), the strip has to bend more before the heater is switched off. This means it has to reach a higher temperature. Ovens and fridges also have similar thermostats to keep their temperature steady – by switching on and off when the temperature is too low or too high.

A simple fire alarm uses a bimetallic strip. When the temperature rises, the bending of the bimetallic strip switches on an alarm.

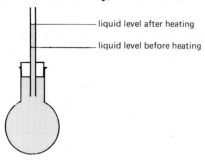

liquid level after heating
liquid level before heating

14.6 Expansion of liquids

Like solids, liquids expand when heated. It is the increase in **volume** which we consider when thinking about liquid expansion, and liquids have a much larger increase in volume than solids when they are heated. A flask like the one in the drawing can be used to demonstrate the expansion of a liquid. It is heated by placing it in hot water. If you look closely when doing this you may notice that the liquid level in the tube falls slightly at first. This is because the glass flask has heated first and expanded before the liquid begins to heat up. In due course, the liquid is heated and expands. As a result the level rises in the tube. The liquid expands much more than the glass. Liquid-in-glass thermometers work in exactly the same way, with a narrow tube to make the expansion of the liquid more easily seen.

If a liquid is kept in a closed container there must always be some space left for expansion. This is why bottles and cans of liquid are never filled completely.

◁ When a liquid is heated and expands this means that the same mass of liquid is now taking up a larger space. The density of the liquid is now smaller. The hot liquid will rise to the top of its container.

We will look again at the way hot liquids expand and rise when we consider the transfer of thermal energy in Chapter 16.

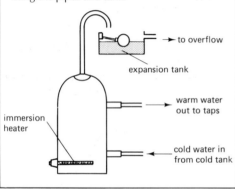

In a hot water tank with an immersion heater, the water close to the heater will be heated first. It will expand a little. This makes it less dense than the rest of the water and so it rises to the top. The pipe taking the hot water from the tank to the hot-water taps leaves from the top of the tank. Notice too the pipe from the very top of the tank; this is called the expansion pipe and allows the water inside the whole heating system to expand without bursting the pipes and tank.

14.7 The unusual behaviour of water

Most liquids expand steadily when heated and contract steadily when cooled. Water doesn't! If water is cooled from 100°C, it contracts until it reaches 4°C. But from 4°C to 0°C, it expands. When the water at 0°C freezes and forms ice, it expands considerably more again. If the ice is cooled further, it contracts just like any other solid.

The large increase in volume when water freezes is what causes burst water pipes in cold weather. The ice takes up more space and cracks the pipes. The increase also means that ice is less dense than water and floats on the surface.

The unusual behaviour of water on cooling has some very important consequences. It means that as a lake or pond freezes, the ice always forms first on the surface. It also has the result that the pond is less likely to freeze entirely solid, and so fish can survive. To see why this is the case, imagine a pond cooling down in cold weather.

Graph **A** shows the volume change of 1 kg of water as it is cooled from 100°C to 0°C. Graph **B** shows the changes around 4°C in more detail. The volume is smallest at 4°C. Water has its maximum density at this temperature.

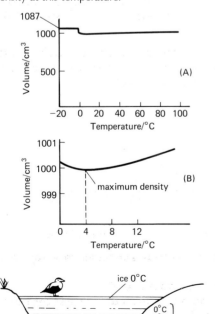

Stage 1: All the water cools to 4°C
The temperature of the air above the pond is below 0°C, so the water on the surface cools. This makes it contract and become denser, and so it sinks to the bottom. Slightly warmer, less dense water comes up to the surface to take its place. This goes on until all the water has reached 4°C and is at its maximum density.

Stage 2: When it freezes, the deeper water stays warmer
If the water on the surface is now cooled further, it actually expands slightly and becomes less dense! So it stays on the surface. Between 4°C and 0°C, the coldest water is nearer the surface. In due course ice begins to form on the surface. The water temperatures at different depths are as shown in the drawing. The denser water at 4°C is unlikely to freeze because it does not circulate upwards towards the top, and the ice and the layers of water above it insulate it from the cold air outside.

We are used to seeing ponds freezing in this way, so we take it for granted. However it is really quite unusual. If you cool and freeze a sample of any other liquid, it begins to go solid from the bottom, and rapidly freezes solid. The unusual behaviour of water makes it possible for fish and other animals to survive during the winter by living in the warmer water at 4°C near the bottom.

Questions

1 When you measure the temperature of water in a beaker, you must keep the thermometer bulb *in* the water while you take the reading. But when the doctor measures your temperature, he takes the thermometer bulb *out* from under your arm to read it! What is the difference between the two thermometers which means they can be used in these two different ways?

2 Imagine you are given a thermocouple thermometer like the one in the diagram. The meter reading is bigger when the thermocouple junction is hot. Describe what you would do to calibrate this thermometer. How could you then use it to measure the temperature of the room you are in?

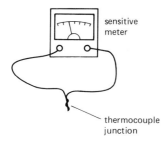

3 For both parts of this question, you need to know that the linear expansivity of steel is 0.000 011/°C.
(a) If steel railway lines are 30 m long and are designed to cope with a possible temperature difference of 50°C between the highest summer temperature and the lowest winter one, what gap for expansion would you need to leave between the rails?
(b) The Forth Railway bridge is made of steel and is 1200 m long. How much longer is the bridge on a hot summer day (30°C) than on a cold winter day (−10°C)?

4 The 1 litre cartons in which fruit juice is packed are made of thin card with a layer of aluminium foil on the inside. If a strip cut from a carton is thrown on to a fire, it curls up before catching fire. Explain why this happens.

5 Imagine that you have just invented the bimetal strip and you have designed a fire alarm based on it. Design an advertising poster for your new invention.

6 Water would not be a good liquid to use in a thermometer. Why is it unsuitable?

INVESTIGATION

How does an automatic toaster know when to pop up? (Safety note: do not take the toaster apart or poke anything inside it while it is plugged in.)

---

7 (a) A thermostat can be used to control a system. The diagram below shows how a bimetallic strip can be used in a simple fire alarm.

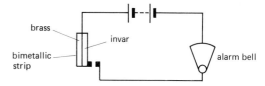

(i) Explain fully how a fire would cause the alarm to sound.
(ii) State **two** ways in which the thermostat could be altered to sound the alarm at a lower temperature.
(b) In an electric iron it is important to be able to set the temperature of the iron, which should then remain constant.

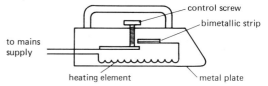

Describe how the working temperature of the iron is (i) set; (ii) kept constant.
(c) The graph shows how the temperature of an iron varied with time after the iron is switched on.

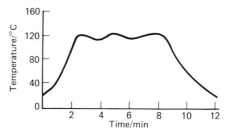

(i) What was the room temperature?
(ii) To what temperature was the iron set?
(iii) Explain why the temperature of the iron does not remain exactly constant. (Sp. LEAG)

8 An investigation was carried out on an electric iron to record the times when its heater switches on and off. The graph summarises the results.

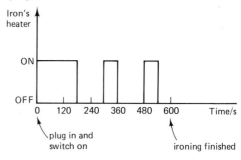

(a) For how many seconds altogether was the iron switched on?
(b) How long did the ironing take?
(c) For what percentage of the total time was the iron's heater switched on?
(d) Explain why the first period the iron was on was the longest.

15: The Behaviour of Gases

Like solids and liquids, gases also expand when heated. In fact gases increase in volume much more than solids or liquids for the same change in temperature. However there is another completely separate way of changing the volume of a gas – by changing its pressure. This might seem to make the behaviour of gases more complicated than solids and liquids, but in other ways they are easier to understand. By studying the behaviour of gases, we are able to improve our understanding of all the states of matter.

15.1 Expansion of gases

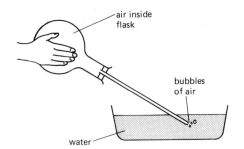

◁ If you cup your hands around a flask as shown, air bubbles begin to come out of the tube. By heating the air slightly with your hands, you have made it expand and some has been pushed out of the open tube. Of course, the glass must have expanded a little as well. But the expansion of the gas has been much greater.

To study the expansion of air more carefully, we use the apparatus ◁ illustrated. A small sample of air is trapped in a very fine capillary tube (this just means a tube with a narrow bore). The tube is sealed at the bottom and the air is trapped under a bead of concentrated sulphuric acid. The acid is used because it absorbs water and keeps the air sample dry. The air is free to expand when heated, but its pressure will stay constant. The pressure is just atmospheric pressure plus a tiny extra bit due to the small column of acid.

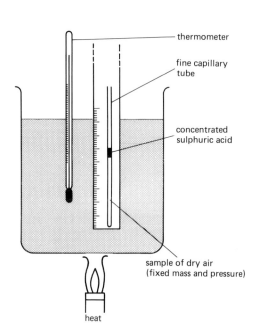

The air is heated by putting it into a large beaker of water which we can heat up. If the water is heated slowly and stirred well, the temperature of the trapped air should be the same as the temperature of the water. As the air is heated it will expand. We can measure the change of **volume** by measuring the **length** of the air column, because the width of the tube is constant. If the volume doubles, the length must double, and so on. The capillary tube is fixed to a ruler so that the length of the air sample can easily be read off the scale.

The water is then heated up slowly and the length of the air column is measured at different temperatures. The graph of volume (length of air column) against temperature is a straight line.

Temp./°C	Length/mm
0	36
10	37
20	39
30	40
40	41
50	43
60	44
70	45
80	47
90	48
100	49

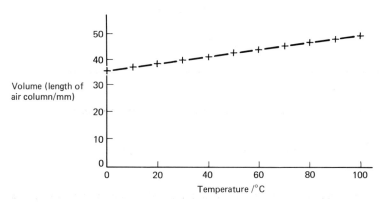

What does this mean? Strictly speaking, it means that the air expands in the same steady way as the mercury in the thermometer. The volume of the air sample increases steadily with temperature (measured on the mercury scale). This result was first discovered by the French scientist, Jacques Charles, in 1787 and is known as Charles' Law.

15.2 Temperature, volume and absolute zero

According to the results shown in the graph on the page opposite, the volume of air decreases steadily as the temperature drops. This makes us wonder what would happen if we were to cool the air below 0°C. We can redraw the graph on different axes and extend the straight line back until it meets the temperature axis.

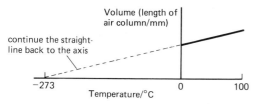
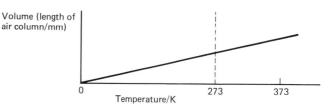

We find that the graph cuts the axis at around −273°C. At this temperature it looks as if the volume of the air has become zero – it has disappeared! It is always risky extending a graph like this well below the temperatures where we made the measurements. We cannot be sure that it will behave like this. In fact the air will become liquid before it reaches −273°C and will stop following the straight-line graph. However graphs of volume against temperature for all gases show the same behaviour – the graph is a straight line, and if it is extended back it meets the horizontal axis at −273°C. This temperature is known as **absolute zero**.

(**Note.** −273°C is just an approximate result. The most accurate measurements give a value of −273.15°C. For our work, however, −273°C will be accurate enough.)

If we were now to move the position of the axes of the graph so that the zero of the temperature scale is at absolute zero, the graph would be a straight line through the new origin. This temperature scale is called the kelvin scale. On the kelvin scale, absolute zero is 0 K. The freezing point of water (0°C) is 273 K; the boiling point of water is 373 K. The size of the kelvin is the same as the Celsius degree. To change Celsius to kelvin, you just have to add 273.

Notice that we write kelvin temperatures without the degree sign (°). The symbol is just K.

We can now state the Charles' Law result in a different way. From the graph, we can see that if the temperature measured in kelvin doubles, the volume of the gas will double, and so on.

The volume of a fixed mass of gas is directly proportional to its temperature (in kelvin) if the pressure is constant.

Writing this in symbols: $V \propto T$ (if p is constant)

$$V = \text{constant} \times T$$

$$\frac{V}{T} = \text{constant}$$

EXAMPLE 15.1

On a hot summer day, the temperature is 25°C. What is this in kelvin?

To convert from degrees Celsius to kelvin, we add 273:

So the temperature in question is $25 + 273\,\text{K} = 298\,\text{K}$

EXAMPLE 15.2

Mercury freezes at 234 K and boils at 630 K. What are these temperatures in degrees Celsius?

To convert from kelvin to degrees Celsius, we subtract 273:

So, the freezing point of mercury is $234 − 273°C = −39°C$
And the boiling point of mercury is $630 − 273°C = 357°C$

EXAMPLE 15.3

In an experiment, water is heated from 20°C to 50°C. What is the temperature **difference** in kelvin and in degrees Celsius?

In Celsius, the temperature rise is $50 − 20°C$, which is $30°C$.

In kelvin, the two temperatures are 293 K and 323 K. The temperature difference is $323 − 293\,\text{K}$, which is 30 K.

The size of the degree Celsius and the kelvin are the same. Temperature differences are numerically the same on both scales.

15.3 Pressure and volume

In the Charles' Law experiment, we had to keep the pressure of the air sample constant. This is necessary because changes in pressure also affect the volume of a sample of a gas. We can investigate the effects of pressure on volume directly using the apparatus shown.

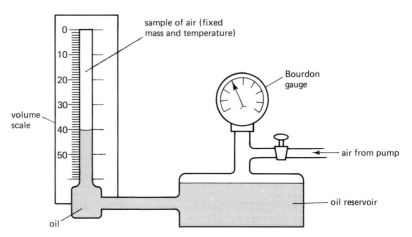

The air sample is trapped above a column of oil. If air from outside is pumped into the space above the oil reservoir, this forces oil up the glass tube and reduces the volume of the trapped air. The Bourdon gauge measures the pressure of the air pumped in, which is almost exactly the same as the pressure of the trapped air. (The difference is just the small amount due to the height difference of the oil in the two containers, and this can be ignored.) The gauge reads the actual pressure of the air, including atmospheric pressure.

Several measurements of the volume and pressure of the air are taken. Pumping up the air heats it up a little and so it is better to wait a few minutes each time to let the temperature settle down again. A table and graph of the results are shown below.

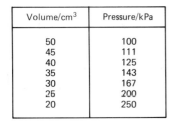

Volume/cm^3	Pressure/kPa
50	100
45	111
40	125
35	143
30	167
25	200
20	250

As we would expect, this shows that volume decreases as pressure increases. Looking more closely at the results, we see that when the pressure is doubled, the volume is halved; if pressure increases three times, volume falls to a third of its original value. This is called **inverse proportion**. This is confirmed by the fact that the quantity pressure × volume (pV) has the same value all the time; it is constant. The same result is obtained using any gas; it doesn't just apply to air. It was first discovered by Robert Boyle (1627–1691), who spoke of his experiments on 'the spring of the air'.

The result is known as **Boyle's Law** and can be stated:

The volume of a fixed mass of gas is inversely proportional to its pressure if the temperature is constant.

Writing this in symbols: $pV = $ constant (if T is constant)

15.4 Pressure and temperature

If we heat a gas but don't allow it to expand, its pressure will increase. We can investigate how the pressure changes with temperature when we keep the volume fixed. An easy way to do this is to heat up the air in a sealed flask and measure its pressure as it heats up.

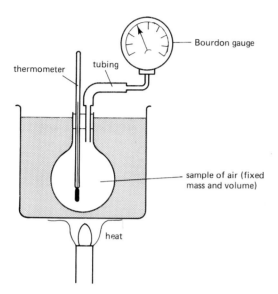

The results will not be quite accurate because the small amount of air in the tubing does not get heated. However, this is small compared with the volume of air in the flask. The water is heated up slowly so that the flask can keep at the same temperature as the water. Several measurements of pressure and temperature are taken. These are used to plot a graph.

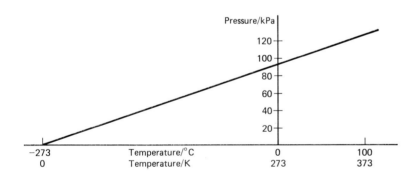

Temp/°C	Pressure/kPa
0	93
20	100
40	107
60	114
80	120
100	127

This has exactly the same form as the volume–temperature graph on p. 136. It is a straight line and if we continue it back, it meets the axis at $-273°C$ – absolute zero. If we measure temperatures on the kelvin scale, the graph is a straight line through the origin. The result is known as the **Pressure Law**. It can be stated:

The pressure of a fixed mass of gas is directly proportional to its temperature (in kelvin) if the volume is constant.

In symbols:
$$p \propto T \text{ (if } V \text{ is constant)}$$
$$p = \text{constant} \times T$$
$$\frac{p}{T} = \text{constant}$$

As with Charles' and Boyle's Laws, this result applies to all gases, and not just to air.

15.5 Kinetic theory of gases

The three gas laws are experimental results telling us how the pressure, temperature and volume of a sample of gas are related to each other. However the three laws have a very simple form, and this led scientists to think that there must be some explanation for them. They wondered what a gas must really be made of to behave in this way. Over two hundred years ago, the Swiss scientist Daniel Bernoulli suggested that gases might be made up of millions of very tiny particles moving around randomly at high speed. Later on, additional evidence to support Bernoulli's suggestion was produced (e.g. Brownian motion) and some of this is explained in Chapter 1. We now think of all materials (solids and liquids, as well as gases) as made up of tiny molecules. In gases these are in constant random motion. However, the original idea of the **kinetic theory of gases** was proposed to explain the gas laws. The basic assumptions are:

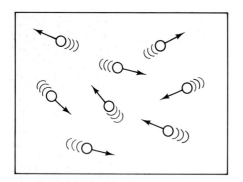

1. All gases are made up of a very large number of tiny molecules.

2. These molecules are constantly moving around randomly at high speeds.

3. The molecules collide elastically with anything they meet. If they hit the wall of the vessel containing the gas, they bounce off again at the same speed.

4. The molecules are so small and so far apart that they almost never collide with each other. They do not exert any forces on each other, but move independently.

5. The pressure of the gas is caused by collisions between the molecules and the walls of the container. Imagine a molecule of mass m approaching one wall with speed v. Its momentum is mv. It rebounds with the same speed in the opposite direction. Its momentum is now $-mv$. Its change of momentum is $mv - (-mv) = 2mv$.

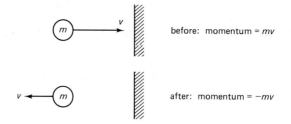

This change in momentum is caused by the force which the wall exerts on the molecule (see section 10.2). The molecule exerts an equal but opposite force on the wall (Newton's Third Law). The force exerted by one molecule is tiny, but there is a very large number of molecules. The *total* force exerted on unit area of the walls of the container is the gas pressure.

6. As a gas is heated, the molecules move faster. Their kinetic energy increases. In fact, the average kinetic energy of the gas molecules is proportional to the kelvin temperature. This is our understanding of what temperature **is** using the kinetic theory.

We can now use this theory to explain the three gas law results.

Pressure Law

The sample of gas is kept at constant volume. As it is heated up, its molecules move more rapidly. As a result, they collide with the walls at higher speed. The change of momentum when they collide with the walls is greater, and so the force they exert on the walls is larger. Because the molecules are moving around more rapidly in the same space, each molecule will also hit the walls of the container more often. This also increases the force exerted on the walls. For both these reasons the pressure increases.

i.e. **as temperature increases, pressure increases**

Boyle's Law

Imagine a sample of gas being compressed, with the temperature staying constant. The speed of the molecules stays the same but they are now confined to a smaller space. The molecules are squeezed closer together. As a result, the number of collisions every second with each unit area of the walls increases. So the pressure increases.

i.e. **as volume decreases, pressure increases**

Charles' Law

As the sample of gas is heated, the molecules move faster. They collide with the walls more frequently and at greater speed. So they exert a larger pressure on the walls of the container. As a result, the gas will expand if it is able to. This allows the molecules to spread out a little which reduces the number of collisions per second with each unit area of the walls. The gas continues to expand until the pressure is back to its original value.

i.e. **if temperature is increased but pressure stays the same, the volume must increase.**

We have only used the kinetic theory to show (in words) that the gas law results are reasonable. In fact it is possible to derive the gas laws mathematically from the kinetic theory, though that is too advanced for this book. However, the fact that it can be done makes us more confident that the kinetic theory of gases is a useful picture of gases and how they behave.

Absolute zero

If temperature is related to the kinetic energy of molecules, then we might expect that there would be a temperature where the molecules would be stationary and their kinetic energy would be zero. At absolute zero the kinetic energy of molecules is a minimum. No object can be cooled to a lower temperature than this.

In fact, the theory called **quantum mechanics**, which deals with the behaviour of very small objects like atoms and molecules, indicates that the molecules do have a small amount of kinetic energy left even at absolute zero. However their kinetic energy is then the minimum possible.

A model of the kinetic theory of gases: the moving beads ('molecules') exert pressure on the underside of the piston. By speeding up the motor we increase the 'temperature' of the gas. With this model we can demonstrate the three gas laws. Can you see how?

15.6 General gas equation

The results of the three gas laws can be summarised in a single equation:

For a fixed mass of gas, $\dfrac{pV}{T} = \text{constant}$

This is called the general gas equation.

A gas which obeys the three gas laws exactly is called an **ideal gas**. (Ideal gases do not exist in the real world, only in our imaginations!) Real gases obey the gas law approximately. If the gas is at relatively low pressure, it behaves very like an ideal gas. For much of our work, we can treat real gases as if they were ideal.

15.7 Calculations using the gas laws

Sometimes we want to use the gas laws to calculate the volume, or pressure, or temperature of a sample of a gas after a change has occurred. There is a way of rewriting the gas laws to make this easier.

Charles' Law can be written in the form:

$$\frac{V}{T} = \text{constant} \quad (T \text{ in kelvin})$$

This means the following: if we take a sample of gas and measure its volume as the temperature changes, the number we calculate if we divide the volume V by its corresponding value of T (in K) is the same each time. Imagine a gas sample of volume V_1 and temperature T_1. This is heated until its temperature is T_2; its volume is then V_2. From Charles' Law,

$$\frac{V_1}{T_1} = \text{constant} \quad \text{and} \quad \frac{V_2}{T_2} = \text{the same constant}$$

So it follows that, $\frac{V_1}{T_1} = \frac{V_2}{T_2}$ (T_1 and T_2 are in kelvin)

This is another way of writing Charles' Law.

In the same way, Boyle's Law can be written as: $p_1 V_1 = p_2 V_2$

and the Pressure Law as: $\frac{p_1}{T_1} = \frac{p_2}{T_2}$ (T_1 and T_2 in kelvin)

Finally, in a general situation where pressure, volume and temperature can *all* change, the general gas equation is required. It can be written:

$$\frac{p_1 V_1}{T_1} = \frac{p_2 V_2}{T_2} \quad (T_1 \text{ and } T_2 \text{ in kelvin})$$

EXAMPLE 15.4

A mixture of air and petrol vapour is drawn into the cylinder of a car engine when the cylinder volume is 120 cm³. Its pressure is then 1.0 atm. The valve closes and the mixture is compressed until its volume is 15 cm³. What is its pressure now?

This situation relates pressure and volume. Boyle's Law is the gas law to use:

$$p_1 V_1 = p_2 V_2$$

Using the information given in the question:

$p_1 = 1.0$ atm
$V_1 = 120$ cm³
p_2 is to be found
$V_2 = 15$ cm³

Volume has been reduced to almost one-tenth of its starting value, so we would expect pressure to have increased by almost ten times. Substituting into the equation:

$$1.0 \text{ atm} \times 120 \text{ cm}^3 = p_2 \times 15 \text{ cm}^3$$

Divide both sides by 15 cm³:

$$p_2 = \frac{1.0 \text{ atm} \times 120 \text{ cm}^3}{15 \text{ cm}^3}$$
$$= 8.0 \text{ atm}$$

The pressure of the gas mixture is 8.0 atm.

EXAMPLE 15.5

A motorist blows up her car tyres to a pressure of 2.7 atm on a cold morning when the temperature is $-3°C$. What will be the pressure in the tyres on a hot day if the temperature is 27°C?

This situation involves pressure and temperature. The Pressure Law relates these:

$$\frac{p_1}{T_1} = \frac{p_2}{T_2}$$

In this equation, T_1 and T_2 must be in kelvin.

So
$p_1 = 2.7$ atm
$T_1 = -3°C = 270$ K
p_2 is to be found
$T_2 = 27°C = 300$ K

Before substituting into the equation, it is useful to estimate what we expect. Pressure increases as temperature increases, so the result will be bigger than 2.7 atm. The temperature rises only from 270 K to 300 K so the increase in pressure will be quite small too.

Substituting into the equation:

$$\frac{2.7 \text{ atm}}{270 \text{ K}} = \frac{p_2}{300 \text{ K}}$$

Multiply both sides by 300 K:

$$p_2 = \frac{2.7 \text{ atm} \times 300 \text{ K}}{270 \text{ K}}$$
$$= 3.0 \text{ atm}$$

The final pressure of the air in the tyres is 3.0 atm.

Questions

1. A sample of gas has a volume of 100 cm³ at 20°C. To what temperature would you have to heat it if you wanted to double the volume to 200 cm³? At what temperature would its volume be halved, i.e. 50 cm³?

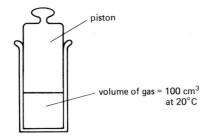

2. Make a copy of the temperature scale diagram on p. 128, this time showing all the temperatures in kelvin, not in degrees Celsius.

3. Convert the following to degrees Celsius:

   (a) 373K (b) 223K (c) 6000K
   (d) 546K (e) 100K (f) 4K

4. Some air at atmospheric pressure (10^5 Pa) is trapped inside the barrel of a bicycle pump by sealing the nozzle. The distance (**d**) from the end of the barrel to the washer is 24 cm.

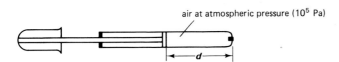

   (a) What is the pressure of the air if the handle is pulled out until distance **d** is 30 cm?
   (b) How far would you have to push the handle in to make the pressure three times atmospheric (3×10^5 Pa)?

5. The diagram shows the apparatus in an experiment to find how the volume of a fixed mass of air is related to its pressure when the temperature is kept constant.

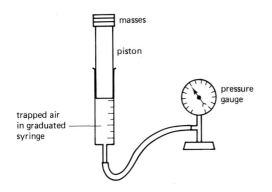

(a) The trapped air exerts an upwards pressure on the base of the piston. Explain how air molecules cause this pressure.
(b) The results of the experiment show that as the pressure increases, the volume of the trapped air decreases. Values of $\frac{1}{\text{volume}}$ are calculated and the appropriate points are plotted as shown

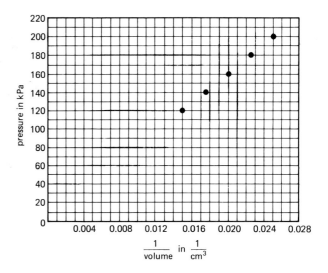

(i) A pupil looks at the results and concludes that the pressure varies directly as $\frac{1}{\text{volume}}$. Explain how you would show that the conclusion is justified from these results.
(ii) From the graph find the volume of the trapped air at a pressure of 80 kPa.
(c) All of the results of the experiments have been obtained at pressures **greater** than normal atmospheric pressure, which is approximately 100 kPa. How could a pupil adapt the apparatus to obtain results at pressures lower than atmospheric pressure?
(d) An astronaut, returning from a 'space walk', re-enters his spacecraft via a door into a vacuum chamber (marked **A**).

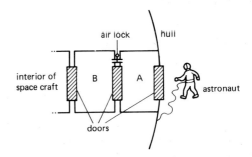

An air lock connects **A** to an inner chamber **B** which contains air. When the outer door is closed the air lock is opened. Air then rushes into compartment **A** from **B** until the pressure in both **A** and **B** is 100 kPa. The effective volumes of **A** and **B** are 4 m³ and 8 m³ respectively.
  Assuming there are no temperature changes, what must have been the pressure in chamber **B** before the air lock was opened?   (SEB)

6 (a) A pupil investigates the relationship between the pressure and temperature of a fixed mass of gas using the apparatus shown.

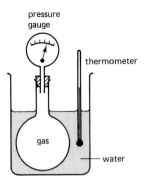

He heats the water continuously using a Bunsen burner and records the pressure and temperature readings every minute.
  (i) State two ways in which this experiment may be improved.
  (ii) Using the results from an improved experiment, describe how the relationship between pressure and temperature on the kelvin scale may be found.
(b) A cylinder of oxygen at 27°C has a gas pressure of 3×10^6 Pa.
  (i) Calculate the pressure of the gas if the cylinder is cooled to 0°C.
  (ii) Describe what happens to the gas molecules as the gas is cooled and indicate how this results in a reduction of pressure.   (SEB)

7 Use the pressure law (pressure $\propto$ temperature (in K)) to explain the following:
(a) Although the gas used in aerosol cans is not inflammable, cans carry a warning not to throw them on the fire when they are empty.
(b) A dented table-tennis ball can sometimes be fixed by putting it into hot water.
(c) Filament light bulbs are usually filled with inert gas at less than atmospheric pressure.
(Hint: When switched on, the lamp's temperature will be several hundred degrees Celsius.)

8 Explain each of the following in terms of the movement of the molecules of the gas:
(a) On a cold winter day a football feels softer than on a warm day.
(b) As time goes by, the pressure inside a leaky oxygen cylinder falls.
(c) When the air inside a bicycle pump is compressed, the air pressure increases.
(d) A party balloon hung above a hot radiator expands and becomes bigger than the other balloons.

INVESTIGATION

Does the 'bounciness' of a tennis ball change with temperature?

16: Heating and Cooling

16.1 Temperature and energy

If we heat something up, its temperature rises. But what is happening to the object while it is being heated? The drawing shows a small electrical immersion heater placed in a beaker of water. When the immersion heater is switched on, we notice that the temperature of the water begins to rise. The immersion heater is supplying energy to the water. As more energy is given to the water, its temperature rises.

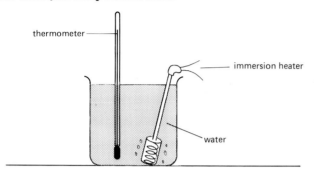

In everyday language, this energy is usually called 'heat'. Is **heat** the same as **temperature**? The answer is No, and it is important to try to understand clearly the difference between them. For example, consider a nice hot bath full of water, and a cup of boiling water. The temperature of the bath water is around 50°C and the temperature of the water in the cup is close to 100°C. But there is less than 0.25 litres of water in the cup and over 25 litres in the bath. There is more energy in the bath water than in the water in the cup.

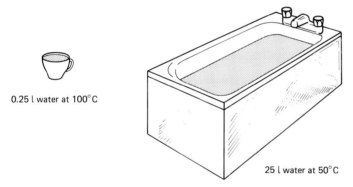

Using the kinetic theory of matter, we can see where this energy *is*. As a substance is heated up, its molecules move more rapidly. There is more energy in their movement (kinetic energy). In the case of solids and liquids, the bonds between the molecules are stretched more as the molecules vibrate more strongly, so they have more elastic potential energy. From outside the material, none of this extra movement or stretching of bonds can be seen. The energy in the movement and stretching does not show up in any movement or stretching of the material as a whole. It is all happening on a smaller-than-microscopic scale, *inside* the material. We call it **internal energy**, energy *inside* the object.

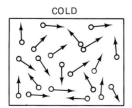
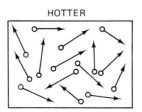

> ### Average and total energies
>
> In the kinetic theory, the temperature of an object is a measure of the **average** energy of its molecules. The internal energy is the **total** energy of the molecules. These are not the same thing. The water in the bath has a lower temperature than the water in the cup – the average energy of the water molecules is lower in the bath. But the total energy of all the molecules in the bath is much larger than the total energy of the molecules in the cup.

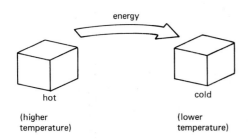

16.2 Internal energy on the move

We will return to some of these ideas in more detail in the next chapter. In this chapter, we will concentrate on one particular aspect of internal energy – its tendency to spread out. In the first diagram on p. 145, the energy from the immersion heater does not stay in the heater element or even in the water close to the heater. Instead it spreads throughout the water. But there is more happening than this. The beaker itself is also getting warmer. So is the table on which the beaker is sitting, and the air above the beaker. Energy is spreading out from the immersion heater into the water and its surroundings. In general, internal energy always seems to spread from an object at a higher temperature to an object at a lower temperature. If left to itself, this would go on until both were at the same temperature. Internal energy tends to spread and temperature differences tend to even out.

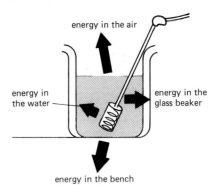

Several different 'methods of spreading' are involved and it is useful to think about these separately: they are convection, conduction and radiation.

16.3 Convection

When a liquid or gas is heated, it expands. So its density becomes less. (The same mass is now taking up more space, so mass/volume (= density) has become smaller.) The warmer material rises to the top, carrying its energy with it. This process is called **convection**.

By placing a crystal of a strongly coloured substance (like potassium permanganate) in one corner of a beaker full of water, and heating gently just below it, we can see the convection currents in the water. Warmer water rises and cooler water moves in to take its place. The circulation of water spreads the energy from the bottom to the top of the beaker.

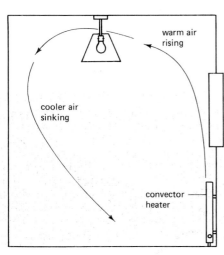

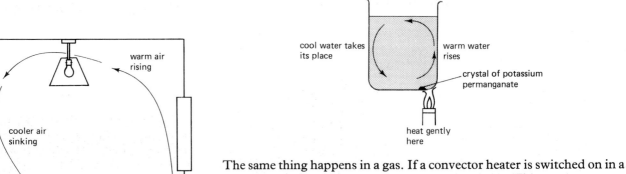

The same thing happens in a gas. If a convector heater is switched on in a room, the air above the heater will expand and rise. This movement can be detected by holding fine hairs or feathers above the heater. Cooler air moves in to replace the warmer air, and a convection current is set up in the room.

Convection is an important method by which energy spreads through a material. In convection it is the warmer material itself which moves, carrying its extra energy with it.

The domestic hot water system

The hot water system at home uses convection to circulate water through the pipes. The diagram shows a simplified version of how it is connected. Try to work out where the different parts are in your home. The hot water storage tank is usually in a cupboard (an airing cupboard) and the header tank in the loft.

Water is heated by a boiler or a fire. (If you use an electric immersion heater for hot water, the system is simpler – see p. 134.) It expands and rises into the top of the storage tank. This pushes cold water out of the bottom of the storage tank and into the boiler to be heated. Hot water collects in the storage tank from the top downwards. The hot taps are connected to the top of the storage tank.

The header tank (the cold water tank) has a number of jobs. It is higher than all the other parts of the system and the pressure due to this extra height of water pushes the water out of the taps when they are turned on. If the water level in the header tank falls, the ball-cock valve opens and lets more mains cold water into the tank. So it is always filled up to the same level. The expansion pipe allows for the expansion of the water when it is heated. It can overflow safely into the header tank. If steam or air bubbles ever form inside the system, these can also escape via the expansion pipe.

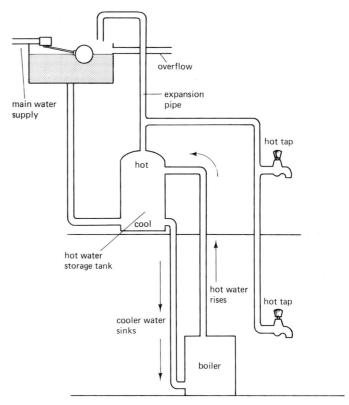

Convection currents in air

Most of our weather is the result of convection. Air over the polar regions is cooled, becomes more dense and sinks. Air over the warmer tropical regions is heated and rises. The currents set up by this, 'stirred up' by the Earth's rotation, cause the complicated pattern of highs, lows and fronts which we see on the weather forecast. ▷

On a smaller scale, convection causes the onshore and offshore winds which you may have noticed at the seaside. During the day, the sun heats up the land to a higher temperature than the sea. The air over the land is heated and rises, and cooler air from over the sea flows in to replace it – a sea breeze. At night the position is reversed. The land cools more quickly and becomes cooler than the sea. Warmer air rises over the sea and cooler air from over the land blows out to sea – a land breeze.

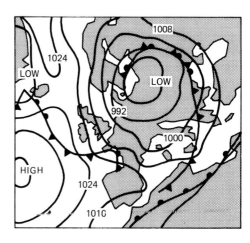

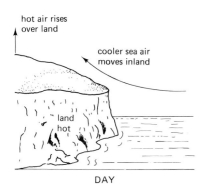

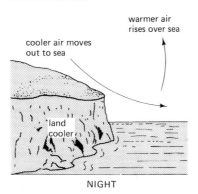

The rising currents of warm air are called thermals. These are much used by glider pilots to help them gain height. Glider pilots become very experienced at spotting thermals. There is usually a thermal over a built-up area caused by the warmer air rising from houses and factories.

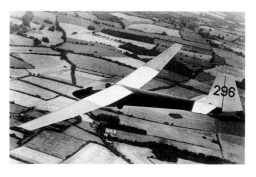

16.4 Conduction

Energy can also spread in solids. If you use a saucepan with a metal handle to boil some milk, you find that the handle also becomes hot – maybe even too hot to hold comfortably. The heat from the cooker ring has spread through the saucepan and along the handle. This cannot be due to convection – in solids the molecules are tightly bonded together and cannot move around inside the material, carrying energy with them. A different process is responsible, called **conduction**. Metals are good conductors – energy can flow easily through them.

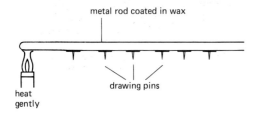

If we coat a metal rod with a thin layer of wax, we can use the wax to stick drawing pins to the bottom side of the rod. If one end of the rod is then heated gently, the drawing pints will fall off in turn, beginning with the one nearest the hot end. The energy from the bunsen flame has spread along the metal rod, melting the wax as it goes. The metal has **conducted** the energy along.

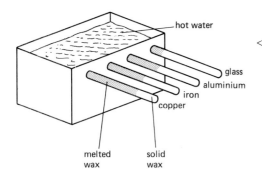

The speed at which the energy spreads depends on the material the rod is made from. The drawing shows one method of comparing different conductors. All the rods have to be the same thickness, and they are heated equally by the boiling water in the tank. After a few minutes, the length of melted wax shows which materials are the best conductors. Metals conduct well, with copper being a particularly good conductor. Non-metals like glass are poor conductors. Table 16.1 shows some examples of good and bad conductors. Bad conductors are known as **insulators**.

Table 16.1 Conductors and Insulators

Good conductors	Bad conductors/good insulators
copper aluminium iron brass silver gold	glass stone water plastics wood materials containing trapped air: wool, foam polystyrene, fibreglass

16.5 Conduction in water and air

Water is a poor conductor. If a boiling tube of water is heated gently at the top, as in (**A**), the water there will boil. Yet a thermometer with its bulb at the bottom of the tube shows that the temperature of the water there has hardly risen at all.

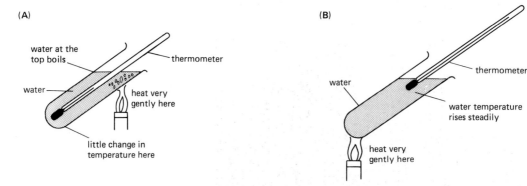

Of course, if we heat at the bottom and measure temperature at the top, as in (**B**), then the thermometer *will* show a rapid rise in temperature. This is due to convection, as hot water rises from the bottom of the tube. If convection is prevented (as in (**A**)), water can be seen to be a poor conductor.

Air is an even poorer conductor than water. A similar experiment shows this. If the tube is gently heated near the top, the temperature measured at the bottom scarcely changes at all. ▷

This is why many insulating materials are ones which contain many small pockets of trapped air. For example, wool, down, fibreglass and foam polystyrene are all good insulating materials (poor conductors). The air trapped inside them is a very poor conductor. Since it is trapped in small pockets, convection cannot take place. So the material is a good insulator.

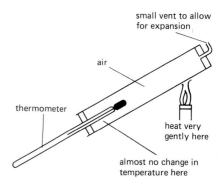

Conductors and insulators in everyday life

On a cold day, we put on more warm clothes. These don't actually supply energy, but they insulate our bodies from the colder air outside. We keep our body temperature at a steady 37°C which is usually higher than our surroundings. Clothes stop this energy spreading, and keep us warm. The warmest materials are things like wool and down. Clothes made from these contain pockets of air which is an excellent insulator. Birds and animals use the natural properties of down and fur to keep warm in cold weather. Birds can fluff up their feathers to trap extra air on cold days.

In the kitchen, we use good conductors (metals) where we want energy to spread quickly. Saucepans and pots are made of metal to transfer energy quickly from the hotplate or gas ring to the food being cooked. The handles, on the other hand, are usually made of plastic or wood which are poor conductors. ▽

Conduction is also often responsible for how warm or cold things *feel*. A tiled floor feels cold on bare feet but a woollen rug on the same floor feels pleasantly warm. Both are at the same temperature. The tiles conduct better than the rug, and allow energy to spread away from your foot into the floor. So the temperature of your foot drops – it feels cold. The rug is a good insulator (it has trapped air pockets) and a little energy spreads from your foot into it – it feels warmer.

 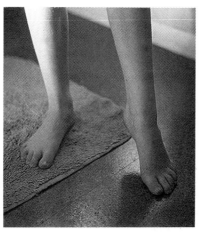

Heating and Cooling

Insulating your home

Heating a house costs money. We buy fuel (coal, or gas, or electricity) which we use up to heat the air in the house. If the house is not well insulated, much of our money is wasted. ▷

Let us look carefully at how insulation helps.

a. To begin with, the heating is off. There is no temperature difference between inside and outside.

b. The heating is then turned on. The house heats up. After a while, there will be a temperature difference between inside and outside. As a result, energy is lost from the house all the time – escaping to the outside.

c. The amount of energy lost every second depends on the temperature difference between inside and outside. As the house gets warmer, this temperature difference gets larger. Energy escapes faster.

d. Eventually the point is reached where the energy escaping is equal to the energy produced by the fuel. The inside temperature stops rising and remains steady.

e. If the house is insulated, less energy escapes every second for the same temperature difference between inside and outside.

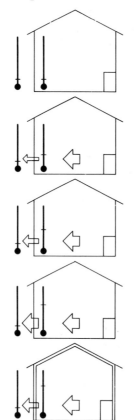

The great energy escape

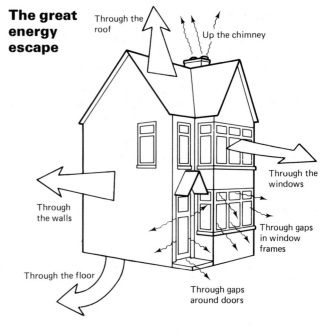

f. So the inside of the house reaches a higher temperature before the energy escaping equals the energy produced.

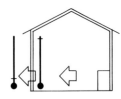

g. Of course, at home we don't want the rooms to become too hot. So we turn down the heating (or use it for shorter periods) and still have the steady temperature we want. With insulation, we can do this using less fuel – so we save on heating bills.

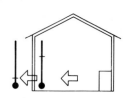

Insulation helps us save on fuel. But when we are deciding whether or not to install insulation, the actual cost of installing it is also important. How many years will it take before the money saved on fuel bills pays back the amount you spent on the insulation? Draught-proofing and loft insulation have the quickest 'pay-back time', insulating cavity walls comes next, with double-glazing taking longest to pay for itself. Of course people install double-glazing for other reasons as well as insulation.

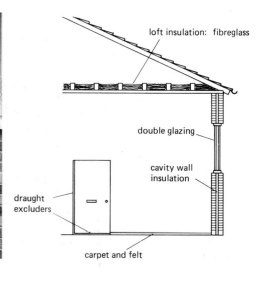

16.6 U-values

The biggest savings from house insulation come from insulating the places where most energy can escape. The rate at which energy is lost through the walls, windows and roof depends on:

- the area through which the energy can escape
- the temperature difference between inside and outside
- the material used (its type and thickness)

To calculate energy losses, builders use the **U-value** for the material. Some typical U-values are shown in Table 16.2.

Table 16.2 Some common U-values (in W/m² °C)

Walls	solid	2.5
	cavity: no insulation	1.5
	cavity: insulated	0.5
Floor	wooden	0.6
	solid	0.5
Roof/loft	no insulation	2.0
	2" insulation in loft	0.5
	6" insulation in loft	0.25
Windows	single glazed	5.0
	double glazed	3.0

For example, a brick cavity wall with no insulation has a U-value of 1.5 watts per square metre for every 1 degree difference in temperature between inside and outside. This means that:

- if the inside is 1 K (or 1°C) warmer than outside, 1.5 J escape through every square metre of wall every second.
- if the inside is 20 K (or 20°C) warmer than outside, 30 J escape through every square metre of wall every second.

If all four walls of a house have a total area of 100 m², then, when the inside is 20°C warmer than outside, 3000 J escape through the walls every second. In these conditions, it would take 3000 W (3000 J/s) to keep the house temperature steady. By installing cavity-wall insulation, the U-value is reduced to around 0.5 W/m² °C. It would then take only 1000 W to keep the house temperature steady under the same conditions.

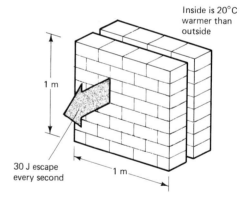

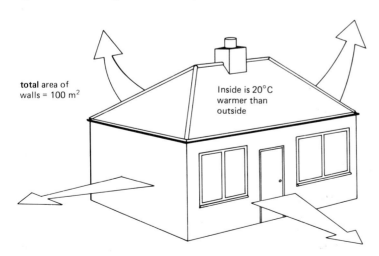

In general, the lower the U-value of a material, the better it is as an insulator.

16.7 How materials conduct

It is easy to understand how convection works, but can we understand conduction? In all solids, the molecules are held together by bonds which join them to their neighbours. When one end of a bar is heated, the molecules there will begin to move faster and vibrate more. They can pass on some energy to their neighbours. So the vibrations are passed on through the solid bar.

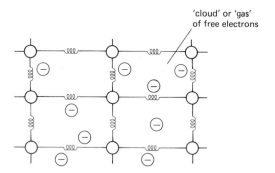

But this explanation would apply to all solids. It doesn't explain why metals are much better conductors than non-metals. The explanation above is a good model for the transfer of energy in non-metals. In metals something *extra* is involved. The special feature of metals (as we will see in Chapter 24) is that their atoms provide **free electrons** which can move anywhere within the solid.

These free electrons behave like a gas. Usually they move randomly through the metal. If one end of a metal bar is heated, the electrons there begin to move more quickly. When they collide with the atoms, they make them vibrate more. This raises the temperature of the metal. As the electrons are free to move, the energy can be transferred very quickly through the bar.

16.8 Radiation

We are warmed all the time by the Sun. But there are 150 million kilometres of empty space between us and the Sun. The Sun's energy cannot reach us by convection or conduction; there must be another method. Some of the Sun's energy reaches us in the form of **infra-red radiation**. As you might expect, this radiation is quite similar to the light which the Sun also provides. We will look at radiation in more detail in Chapter 22.

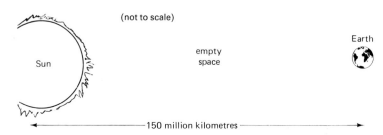

Questions

1. Write a few sentences explaining clearly the difference between **internal energy** (what we normally call 'heat') and **temperature**.

2. A drop of boiling water is at a higher temperature than a bath of warm water. But it takes more energy to heat up the bath than the drop!
Use the molecular model of matter to explain how it is possible for the drop to be at the higher temperature but still contain less energy.

3. Explain why:
   (a) The freezing compartment in a fridge is at the top.
   (b) A downie (a continental quilt) keeps you warmer than several blankets.
   (c) You often see dust marks on the wall above a central heating radiator.
   (d) A tiled shower feels colder to your feet than a cloth bath-mat.
   (e) Foam polystyrene blocks are used to fill the wall cavity in many buildings.
   (f) Saucepan handles are usually made of plastic (not metal).
   (g) The bases of goods saucepans are often made of copper or aluminium.

4. The drawing shows a pudding which has been cooked in an oven at 200°C for a time of fifteen minutes.

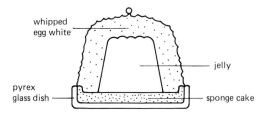

   Explain why the jelly has not melted. (Sp. NEA)

5. People insulate their houses to save money on their fuel bills and to make their houses more comfortable to live in. Use the information in the series of diagrams about house insulation on page 150 to write a paragraph explaining clearly how insulating a house works.

> **INVESTIGATION**
> Compare some insulating materials to find out which would be best for insulating a domestic hot water tank.

6. Here are some data for a typical detached house:

   Total area of all windows = 15 m²
   Total area of all walls = 80 m²
   Total area of roof = 25 m²
   Total area of floor = 20 m²

   The house has cavity walls with no insulation, single glazing, wooden floors and no loft insulation.
   (a) Using the data in Table 16.2, calculate how much energy is lost per second through the windows, walls, roof and floor on a day when the temperature difference between inside and outside is 10°C.
   (b) If there are no other energy losses, work out what percentage of the total energy is lost through each of these.
   (c) The householder is considering installing either loft insulation, cavity wall insulation or double glazing. What advice would you give her? Explain your answer.

7. The kitchen window in a house is 1 m high and 2 m wide. It is single glazed (U-value = 5.0 W/m² °C).

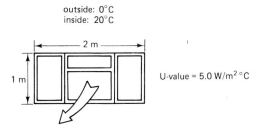

   (a) On a winter day, the temperature outside is 0°C whilst inside it is 20°C. How much energy is lost through the window every second?
   (b) How much energy is lost in one day?
   (c) Throughout the year, taking an average, the inside temperature is 8°C higher than outside. How much energy is lost through this window in one year?

8. The householder in question 7 replaces his kitchen window with a double glazed window (U-value = 3.0 W/m² °C).
   (a) Repeat the calculation of 7(c) for this new window to find how much energy is lost through the new window in a year.
   (b) How much energy is saved in a year by installing the double glazed window?
   (c) The house is heated by gas and 1 MJ (1 000 000 J) of energy costs 0.5 p. How much money is saved in a year by installing this double glazed window?

17: Heating and Internal Energy

Heating an object makes its temperature rise. This may seem too obvious to be worth saying! Looked at from an energy point of view, what it means is that when energy is given to an object, its temperature rises. In this chapter, we will look in more detail at the relationship between temperature and internal energy.

17.1 Heating materials

Temperature rise

If we put some water into a polystyrene cup, and heat it with a small immersion heater, we can measure its temperature every half minute. The heater supplies energy steadily, so the time the heater is on is a measure of the amount of energy supplied. We must stir the water to make sure it is evenly heated. If we do this, we find that its temperature rises steadily – the same rise in temperature in each same time interval.

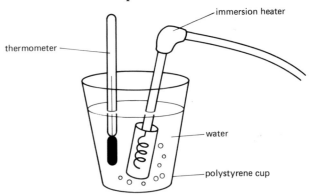

If we go on heating for too long, this simple result will break down, because energy starts to spread from the water into the surrounding air and table. Using a good insulating material for the cup helps to prevent this. The important thing is that the temperature rose steadily to begin with. Within the accuracy of the experiment we can deduce that:

The temperature rise of an object is proportional to the amount of energy supplied.

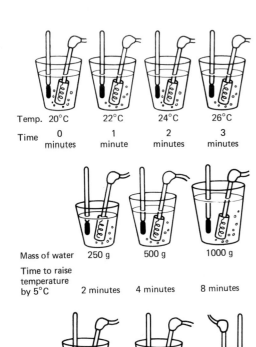

Mass

Clearly the amount of water (the mass) is also important in determining the temperature rise. If we have several polystyrene cups containing different masses of water, we can time how long it takes to raise the temperature of the water in each cup by the same amount (say, 5°C).

We find that the time we need (the amount of energy supplied) is proportional to the mass of water heated.

For a given substance, the energy needed to produce a given temperature rise is proportional to the mass of substance heated.

The substance used

The energy needed to produce a certain temperature rise also depends on what substance is being heated. We can time how long it takes to raise the temperature of the same mass of different substances by the same amount. Water takes longer than ethanol, and both take much longer than an aluminium block.

Water has a greater capacity than ethanol or aluminium to store internal energy. It can absorb more energy for the same temperature rise.

154

17.2 Specific heat capacity (s.h.c.)

These results don't just apply to water. *All* substances behave this way. If we want to do calculations on energy and temperature change, a useful quantity to know is the amount of energy needed to raise the temperature of a given mass (1 kg) of the substance by 1 K (or 1°C). This quantity is called the **specific heat capacity**, or **s.h.c.** for short.

The s.h.c. of water is 4200 J/kg K. This means that:

to raise the temperature of 1 kg of water by 1 K, you need 4200 J of energy.

From this we can calculate how much energy is needed for any mass of water and any temperature rise. For example:

to raise the temperature of 1 kg of water by 5 K you need 21 000 J of energy.

to raise the temperature of 3 kg of water by 1 K you need 12 600 J of energy.

to raise the temperature of 2 kg of water by 10 K you need 84 000 J of energy.

In general, **energy = s.h.c. × mass × temperature rise**
or, in symbols: $E = cm\Delta T$

(**Note:** ΔT means 'the change in temperature'.)

The same equation applies to both heating and cooling. As an object cools, its temperature falls and it loses internal energy. The amount of energy released is equal to (s.h.c. × mass × temperature drop).

Table 17.1 S.h.c. of some common substances

Substance	S.h.c.(J/kg K)
water	4200
ethanol	2500
ice	2100
aluminium	900
concrete	800
steel	500
copper	380
mercury	150

EXAMPLE 17.1

An aluminium saucepan is used to boil 0.5 kg water. The saucepan has a mass of 0.4 kg. The saucepan and water start at 20°C. How much energy is needed to heat: (a) the water, (b) the saucepan?

For both water and saucepan, the temperature change is from 20°C to 100°C, i.e. 80°C or 80 K.

We use the relation: energy = s.h.c. × mass × temperature rise
For the water, energy = (4200 × 0.5 × 80) J
                      = 168 000 J
For the saucepan, energy = (900 × 0.4 × 80) J
                        = 28 800 J

So, 168 000 J are needed to heat the water and 28 800 J to heat the saucepan. We have assumed that the whole saucepan reaches 100°C, which may not be quite true in practice.

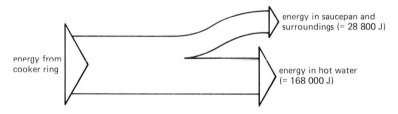

It is interesting to look at this situation in terms of efficiency. The task we want done is the water heating; the energy used to heat the saucepan is 'wasted'. We could work out the efficiency of the water heating:

$$\% \text{ efficiency} = \frac{\text{useful energy}}{\text{total energy}} \times 100$$

$$= \frac{168\,000 \text{ J}}{196\,800 \text{ J}} \times 100$$

$$= 85\%$$

All processes involving heating are bound to be inefficient for this sort of reason. The problem lies in the tendency of energy to spread out. It is hard to keep it where we want it!

For overnight energy storage, a large well-insulated water tank can be used, though it is more efficient if a group of houses share an even larger tank.

Making use of S.H.C.

As Table 17.1 shows, water has a very high s.h.c. This makes it very useful for storing energy. For a given rise in temperature, water can store more energy than most other substances; as it cools, it releases this energy again. This makes it a good substance to use in a central heating system. It is good at storing energy and carrying it from the boiler to the radiators; as it cools there, it releases a lot of energy into the room. It is fortunate that water is so cheap, safe and readily available!

The National Centre for Alternative Technology at Machynlleth in Wales also uses a large underground water tank to store energy for heating an exhibition hall throughout the year.

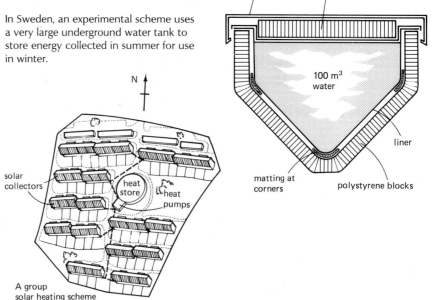

In Sweden, an experimental scheme uses a very large underground water tank to store energy collected in summer for use in winter.

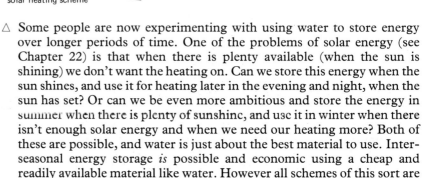

▷ Some people are now experimenting with using water to store energy over longer periods of time. One of the problems of solar energy (see Chapter 22) is that when there is plenty available (when the sun is shining) we don't want the heating on. Can we store this energy when the sun shines, and use it for heating later in the evening and night, when the sun has set? Or can we be even more ambitious and store the energy in summer when there is plenty of sunshine, and use it in winter when there isn't enough solar energy and when we need our heating more? Both of these are possible, and water is just about the best material to use. Inter-seasonal energy storage *is* possible and economic using a cheap and readily available material like water. However all schemes of this sort are worthwhile only if the houses they are heating are very well insulated.

◁ Electric storage heaters use large concrete blocks, rather than water, to store energy. They use electricity at night (when the price of electricity is lower because there is less demand) to heat the blocks. These then release energy during the day as they cool down. Although concrete has a lower s.h.c. than water, it is more dense and so concrete takes up less space than the same mass of water.

◁ In a car engine cooling system, water is circulated through pipes around the engine block to absorb energy from the hot engine and so to keep it cool. From the cylinder block, the water passes into the radiator. This is made of a honeycomb of fine tubes and pipes and is placed at the front of the car where air can blow through it. The water loses its energy to the air and circulates back towards the engine to cool it again. The high s.h.c. of water makes it ideal for this purpose. It can absorb and carry away a lot of energy each time it passes through the engine.

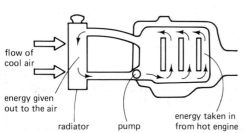

17.3 Measuring s.h.c.

The s.h.c. of a material is the amount of energy needed to raise the temperature of 1 kg of the material by 1 K (or 1°C). So, to measure s.h.c., we need to heat up a sample of the material, using a measured amount of energy, and find out how much its temperature rises as a result. The sample does not, of course, have to be exactly 1 kg and the temperature rise doesn't have to be just 1 K! If we know the energy supplied, the mass and the temperature rise, we will be able to calculate the s.h.c. (c):

$$\text{energy} = \text{s.h.c.} \times \text{mass} \times \text{temperature rise}$$
$$E = cm\Delta T$$

Divide across by (mass × temperature rise):

$$\text{s.h.c.} = \frac{\text{energy}}{(\text{mass} \times \text{temperature rise})} \qquad c = \frac{E}{m\Delta T}$$

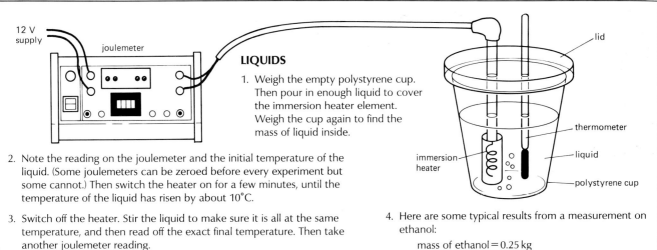

LIQUIDS

1. Weigh the empty polystyrene cup. Then pour in enough liquid to cover the immersion heater element. Weigh the cup again to find the mass of liquid inside.

2. Note the reading on the joulemeter and the initial temperature of the liquid. (Some joulemeters can be zeroed before every experiment but some cannot.) Then switch the heater on for a few minutes, until the temperature of the liquid has risen by about 10°C.

3. Switch off the heater. Stir the liquid to make sure it is all at the same temperature, and then read off the exact final temperature. Then take another joulemeter reading.

4. Here are some typical results from a measurement on ethanol:

   mass of ethanol = 0.25 kg
   energy supplied = 5000 J
   initial temperature = 18°C
   final temperature = 26°C
   temperature rise = 8°C (8 K)

   $$c = \frac{E}{m\Delta T} = \frac{5000 \text{ J}}{0.25 \text{ kg} \times 8 \text{ K}} = 2500 \text{ J/kg K}$$

   So the s.h.c. of ethanol measured in this experiment is 2500 J/kg K.

SOLIDS

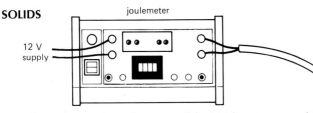

1. Note the reading on the joulemeter and the initial temperature of the block. Then switch the heater on for a few minutes, until the temperature has risen by a few degrees.

2. Switch off the heater. With a solid block, it is not possible to stir! So after the heater is switched off, keep watching the thermometer until the temperature stops rising any further. Metals are good conductors and the energy should soon spread evenly throughout the block. The highest temperature which the thermometer reaches is taken as the final temperature of the block.

3. Here are some typical results from a measurement on aluminium:

   mass of aluminium block = 1 kg
   energy supplied = 11 400 J
   initial temperature = 19°C
   final temperature = 31°C
   temperature rise = 12°C (12 K)

   $$c = \frac{E}{m\Delta T} = \frac{11\,400 \text{ J}}{1 \text{ kg} \times 12 \text{ K}}$$
   $$= 950 \text{ J/kg K}$$

   The s.h.c. of aluminium measured in this experiment is 950 J/kg K.

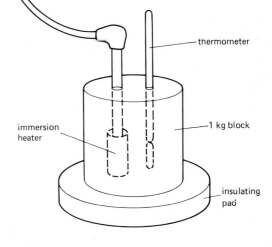

(**Note:** if no joulemeter is available for these experiments, the energy supplied can be calculated by multiplying the power rating of the heater by the number of seconds it is switched on for: energy (in J) = power (in W) × time (in s).)

EXAMPLE 17.2

A hot water bottle is filled with 0.75 kg of water at 80°C. By morning, it has cooled down to 30°C. How much energy has it given out?

The temperature change in this situation is a drop of 50°C (or 50 K).

We use the equation,

energy = s.h.c. × mass × temperature rise

energy given out = (4200 × 0.75 × 50) J
= 157 500 J

The bottle gives out 157 500 J during the night. If we take this to be roughly 8 hours (8 × 60 × 60 seconds), this is equivalent to 5.5 J every second, i.e. 5.5 W. The hot water bottle is, on average, a 5.5 W heater. Of course, it gives out energy more rapidly at first (when it is hotter), so it isn't a steady 5.5 W heater.

Experimental errors

The measured values of s.h.c. on p. 157 are both higher than the accurate values in Table 17.1. Experiments to measure s.h.c give only approximate results. Most of the errors are because some of the energy supplied escapes from the liquid or solid being heated and goes into the surroundings – the air or the bench. We can reduce these energy losses, but we cannot get rid of them completely. Some ways to reduce them are listed below.

In the experiment on liquids:
– use a cup which has a very small mass, so that the cup itself doesn't absorb much energy.
– use a cup made of good insulating material, so that only a little energy escapes into the bench or out through the sides of the cup.
– put a lid on the cup to reduce energy losses by convection.

In the experiment on solids:
– place the block on an insulating mat, to cut down the amount of energy escaping into the bench.
– wrap the sides of the block in a good insulating material to reduce energy losses to the surrounding air.

Even so, there will still be energy losses. For example, some energy is lost in heating up the immersion heater element itself! Can you see why these errors all tend to make the measured value of s.h.c. too high? It is because energy losses mean that the temperature rise is always slightly less than it really should be. So when we do the calculation, we get too large a value for s.h.c.

17.4 Melting and boiling

When we heat an object, we supply energy to it and its temperature usually rises. However, in some situations, it is possible to go on supplying energy to a substance and find that its temperature stays steady. For example, if we put some crushed ice into a beaker and surround the beaker with hot water, we find that the temperature of the ice remains steady at 0°C, even though the ice is constantly absorbing energy from the hot water. Only when all the ice has melted does the temperature rise above 0°C. Energy is needed to change the state from solid (ice) to liquid (water). This energy cannot be *seen* in the form of a temperature rise – it is 'hidden'. The energy needed is called the **latent heat of fusion**.

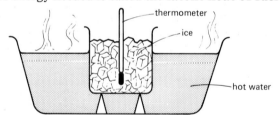

Something similar happens when we boil water in a kettle. The temperature rises steadily as the water comes to the boil, but once it reaches 100°C it stays steady even if the kettle element is kept switched on. Indeed, it will take a longer time to boil away all the water than it took to heat the water from 20°C to 100°C in the first place. A lot of energy is needed to change the state from liquid (water) to gas (steam). Again this energy does not show itself in the form of a temperature rise. It is called the **latent heat of vaporisation**.

17.5 Latent heat and the kinetic theory

The kinetic theory of gases can help us to understand what happens during changes of state, and what 'latent heat' is. In a solid, the molecules are linked to their neighbours by forces of attraction, which we can think of as rather like springs. As the solid is heated, the molecules vibrate more strongly. When the solid reaches its melting point, the vibrations have become so strong that the links begin to give way. Extra energy is needed to overcome these forces and separate the molecules. This is called the **latent heat of fusion**.

When a solid melts, the links are only partially disrupted. Molecules in a liquid are free enough to slide around and change neighbours, but they are still almost as close to each other as in a solid. The links are weaker but still effective. As the liquid is heated further, the kinetic energy of the molecules increases more. At the boiling point, the molecules break free from each other and become a gas. Energy is needed to overcome the remaining links. This is called the **latent heat of vaporisation**. ▷

When a vapour condenses to liquid, bonds are formed and energy is released again. The 'latent heat' which was required for vaporisation is given out again during condensation. This is why, for example, a burn from steam at 100°C is more severe than a burn from the same amount of water at 100°C. Energy is also released when a substance solidifies. As bonds form and the substance changes from liquid to solid, energy is given out.

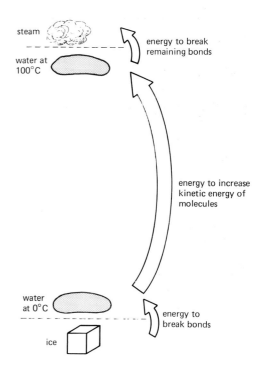

17.6 Cooling curves

A cooling curve is a graph showing the temperature of a sample of material at different times as it cools. If the material changes state during this cooling, the curve has a distinctive shape.

Graph **A** shows a cooling curve for a small sample of lead which was heated strongly in a crucible until it melted, and then allowed to cool down (see **B**). The temperature was measured using a thermocouple thermometer. Now compare this with the cooling curve for a block of copper, heated to the same initial temperature in an oven and then allowed to cool. (Graph **C**). ▷

The graph for copper is a smooth curve, showing that the hot metal cools rapidly at first, and then more slowly as its temperature gets closer to that of its surroundings. The cooling curve for the lead, however, has a distinctive flat section. This happens at the melting point of lead.

To explain why there is a flat section, it is easier to think about what the 'heating curve' would look like in the same experiment. It is almost impossible to do a 'heating curve' experiment because of the difficulty of supplying energy to the material at a steady rate, but we can imagine easily enough what it would look like.

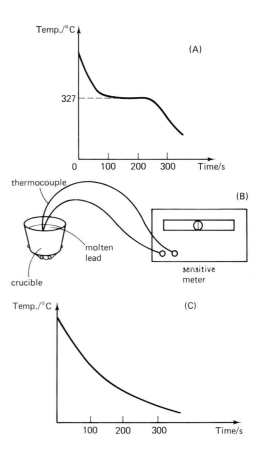

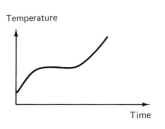

The flat section of the heating curve occurs because energy is needed at the melting point to overcome the forces which link the solid molecules tightly together. So the energy being supplied is used to break bonds and does not show as a rise in temperature. As the material melts, energy continues to be supplied but there is no temperature rise. The energy supplied during this period is the latent heat.

The cooling curve has a flat section because latent heat is released again during cooling. At the melting point, attractive forces link the molecules again and energy is released. The material is constantly losing energy because its temperature is well above room temperature, but its temperature does not drop. It loses its latent heat but the temperature stays steady. Only when all the material has solidified does the temperature begin to fall further.

17.7 Specific latent heat (s.l.h.)

The energy needed to change a substance from solid to liquid or from liquid to vapour depends on the mass of the substance. A useful quantity to know is the amount of energy needed to melt or vaporise 1 kg of the substance. This is called the **specific latent heat of fusion**, or the **specific latent heat of vaporisation**.

The s.l.h. of fusion of a substance is the amount of energy needed to change 1 kg of the substance from solid to liquid at its melting point, without any change of temperature.

The s.l.h. of vaporisation of a substance is the amount of energy needed to change 1 kg of the substance from liquid to vapour at its boiling point, without any change of temperature.

The energy needed to melt or vaporise any other mass of the substance can then be calculated using:

$$\text{energy} = \text{s.l.h.} \times \text{mass}$$
$$E = lm$$

To measure s.l.h., we have to heat a sample of the material, supplying a measured amount of energy, and find how much is melted or turned to vapour as a result.

FUSION

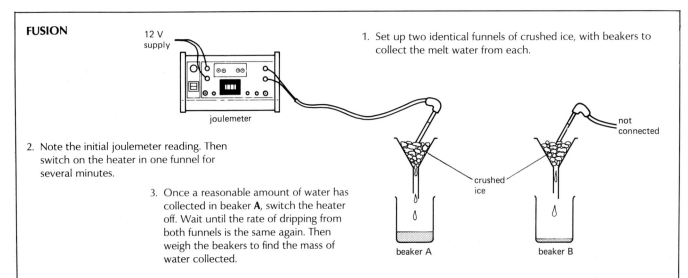

1. Set up two identical funnels of crushed ice, with beakers to collect the melt water from each.

2. Note the initial joulemeter reading. Then switch on the heater in one funnel for several minutes.

3. Once a reasonable amount of water has collected in beaker **A**, switch the heater off. Wait until the rate of dripping from both funnels is the same again. Then weigh the beakers to find the mass of water collected.

4. The **extra** water collected in beaker **A** comes from the ice which has been melted **by the heater**. Some will just have melted by sitting in the room. Beaker **B** lets us estimate how much this is.

5. Here are some typical results:
   mass of water collected in beaker **A** = 35 g
   mass of water collected in beaker **B** = 10 g
   energy supplied (joulemeter reading) = 8000 J

The mass of ice which has been melted **by the heater** is 35 g − 10 g = 25 g (0.025 kg). It has taken 8000 J of energy to do this.

$$\text{energy} = \text{s.l.h.} \times \text{mass}$$
$$8000 \text{ J} = \text{s.l.h.} \times 0.025 \text{ kg}$$

Divide both sides by 0.025 kg: $\text{s.l.h.} = \dfrac{8000 \text{ J}}{0.025 \text{ kg}} = 320\,000 \text{ J/kg}$

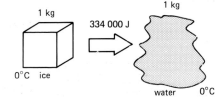

Errors

This is only an approximate result. There are several sources of inaccuracy in the experiment. For example, the contact between the heater and the ice may not be good and some energy may escape into the air. Also because the ice has to be in chunks which are big enough to stay in the funnel, small pieces of partly melted ice may come through the funnel. This can happen in either funnel. These errors might make the measured result too high or too low — we cannot predict. A more accurate measurement gives the value ◁ 334 000 J/kg for the s.l.h. of fusion of ice.

VAPORISATION

1. Place a large beaker of water on one pan of a two-pan balance. Clamp a mains immersion heater (500 W) with its heating element in the water. This has to be done carefully so that the beaker is free to move up and down on the scale pan without touching the heater, yet the heater element must stay immersed in the water all the time.

2. Put some weights on the other pan until the beaker is almost balanced. Leave it so that the beaker is slightly heavier and its side of the balance is down.

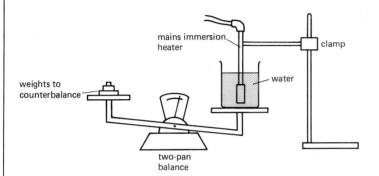

3. Then switch on the heater and bring the water to the boil. As the water boils away, the beaker becomes gradually lighter. Just as the pointer of the scales reaches the balance position, place a 20 g mass on the pan beside the beaker. At the same moment, start a stopclock.

4. After a short time, the scale pointer will come back to balance. This happens when a further 20 g of water has been boiled off. Stop the clock and note the time.

5. Here are some typical results:
Power rating of mains immersion heater = 500 W
Time taken to boil off 20 g = 90 s

The heater supplies 500 J/s.
So: energy supplied = (500 × 90 J)
= 45 000 J

This energy vaporises 20 g (0.02 kg) of water.
energy = s.l.h. × mass
45 000 J = s.l.h. × 0.02 kg

Divide both sides by 0.02 kg: s.l.h. = $\frac{45\,000 \text{ J}}{0.02 \text{ kg}}$

= 2 250 000 J/kg

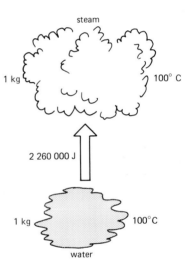

Errors

This is just an approximate method of measuring s.l.h. Errors arise because of splashes of water from the beaker as the water boils vigorously, and from drops of water condensing higher up on the immersion heater and clamp and dripping back into the beaker. Energy is also lost to the surroundings all the time. A more accurate measurement gives a value of 2 260 000 J/kg for the s.l.h. of vaporisation of water. ▷

17.8 Evaporation

A liquid doesn't change into a vapour *only* at its boiling point. Wet roads dry after a shower of rain. If a cup of water is left in the kitchen for a few days, the water in it will have gone. This process is called **evaporation** and it is going on all the time. Some factors can speed up evaporation:

1. Increase the temperature: hot liquids evaporate faster than cold ones. A hair-dryer makes use of this to speed up the drying process. ▷

2. Increase the surface area: liquids evaporate more quickly if the liquid surface is large. The two beakers **A** and **B** contain the same volume of water to start with, but the water in beaker **A** will evaporate faster. Wet clothes also dry faster if they are spread out, rather than rolled into a ball! The surface area for evaporation is larger. ▷

3. Pass air over the liquid surface. Clothes dry better on a breezy day; a hair dryer works better than a heater on its own. The moving air helps evaporation to occur.

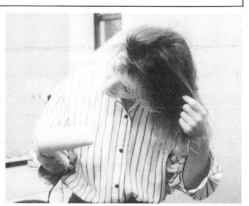

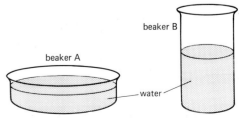

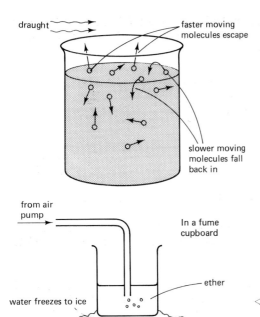

How can we explain evaporation? According to the kinetic theory, the molecules in a liquid can move around to some extent within the liquid but they bump into each other quite a lot. As a result of these collisions, at any moment some molecules will happen to be moving faster than the average (because they have just had a bump which has speeded them up), and some will be moving slower than average (because they have just had a bump which has slowed them down).

The temperature of the liquid depends on the **average** kinetic energy of the molecules – this doesn't mean that all the molecules have the **same** kinetic energy. If one of the faster moving molecules happens to be near the liquid surface, it may escape from the liquid altogether. It then becomes a molecule of water vapour. The more surface area there is, the more likely this is to happen. Of course, it may immediately fall back in again! But some molecules will escape, particularly if there is a draught of air to carry them away from the water surface. So the kinetic theory makes sense of the three factors listed above.

Evaporation and cooling

There is one other consequence of the process of evaporation. If the faster-than-average molecules are the ones to escape, it follows that the average kinetic energy of the remainder will get less. This is another way of saying that the remaining liquid will get colder. This is why wet clothes feel cold. A patch of water on the skin feels cold as it evaporates. If you use a liquid which evaporates better than water (like ethanol or ether), this cooling is even more noticeable.

There is another way to look at this cooling process which is sometimes useful. For a liquid to evaporate, it requires energy – its latent heat of vaporisation. It gets this energy by absorbing it from its surroundings. So when a drop of water evaporates from your hand, it absorbs the necessary latent heat of vaporisation from your hand – which has lost energy and feels colder as a result.

Waterproof clothing is essential for hill walking. If your clothes get wet through the wind will speed up evaporation, making you very cold. This can cause **exposure**, when the body temperature falls below 37°C.

Using evaporation: the refrigerator

The cooling effect in most fridges is caused by the evaporation of a liquid called Freon. Freon is a volatile liquid – it evaporates easily. In a fridge, the Freon is pumped round a closed circuit of pipes. There are several loops of this pipe around the freezing compartment inside the fridge. Here the Freon liquid evaporates, and takes its latent heat from the food inside the fridge. The vapour then passes outside the fridge where it is compressed by a compressor pump. This helps the Freon vapour to condense again. As it does so, it releases energy (its latent heat). But it is now outside the fridge. Here the pipe circuit is usually arranged in a series of loops with an array of fins to help this energy escape into the air by convection. The liquid Freon can then be pumped back into the freezing compartment to repeat the process.

A fridge is rather like a pump, pumping energy from the inside to the outside. It must, of course, be well insulated to prevent the contents simply heating up again immediately. Although the food inside is cooled, the room where the fridge sits will be heated. Indeed, the fins behind a fridge can get quite hot. Now that you know how a fridge works, have a closer look at yours at home. Feel the warm air rising from the fins and see if you can identify the different parts.

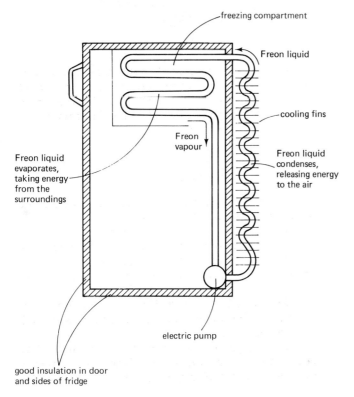

Questions

In these questions you may need to know that:
- the s.h.c. of water is 4200 J/kg K
- the s.l.h. of fusion of ice is 3.34×10^5 J/kg
- the s.l.h. of vaporisation of water is 2.26×10^6 J/kg

1. How much energy must be applied to a 0.25 kg block of aluminium to raise its temperature from 10°C to 60°C? (The s.h.c. of aluminium is 880 J/kg K.)

2. A heater supplies 10 000 J of energy to heat 0.1 kg of ethanol. This liquid has a s.h.c. of 2500 J/kg K. By how much will its temperature rise?

3. A heater supplies 13 500 J of energy to a copper block. The s.h.c. of copper is 450 J/kg K and the temperature of the block rises from 25°C to 85°C as a result of the heating. What is the mass of the block?

4. An immersion heater heats 0.3 kg of water for 126 s and the temperature rises by 25°C. What is the power rating of the heater?

5. The temperature of 0.6 kg metal block is raised from 18°C to 38°C in 110 s by a 50 W heater. What value does this experiment give for the s.h.c. of the metal? Do you think this answer will be greater or less than the accurate value of the s.h.c. measured by a more precise experiment? Explain why.

6. The graph shows the cooling curve of a substance between 250°C and room temperature.
   (a) In what state is the substance at 250°C?
   (b) What is the boiling point of the substance? What is its freezing point?
   (c) What is happening over the region **DE**?
   (d) Over which sections of the curve is the substance giving out energy to its surroundings?
   (e) For this substance, which is greater: the s.l.h. of vaporisation of the s.l.h. of fusion? Explain how you can tell.

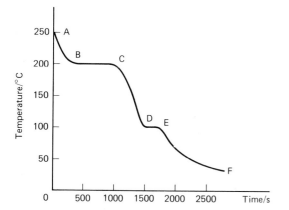

7. 0.5 kg of a solid is heated by a 100 W heater. The graph shows how the temperature of the substance varies with time.

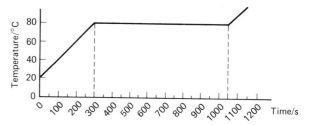

   (a) How long does it take for the solid to reach its melting point?
   (b) Use the information given above to calculate the specific latent heat of fusion of the solid. (SEB)

8. A heater supplies 200 J per second. How long will it take to vaporise 0.1 kg of water at 100°C? How much ice at 0°C could it melt in the same time?

9. (a) A man uses the steam wallpaper stripper, shown below, to help him prepare his living room for re-decoration.

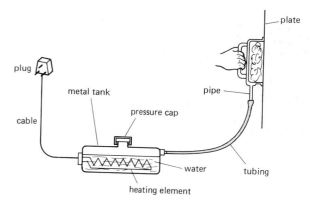

When the appliance is used, water is heated until it boils and produces steam. The plate is then held against the wall and the steam released from the pipe slackens the paper. The following information is shown on the appliance:

Power rating   2.5 kW
Voltage   240 V
Capacity   20 litres

Tap water at 20°C is used to fill the tank. One litre of water has a mass of one kilogram.
   (i) What mass of water fills the container?
   (ii) Calculate how much energy must be absorbed by the water to raise it to boiling point.
   (iii) Calculate the time taken for the water to begin to boil.
   (iv) State whether, in practice, it takes a longer or shorter time to raise the water to boiling point and explain why.
(b) The stripper is used until 50% of the water has been turned into steam. How much energy is required to bring about this change of state? (SEB)

10 Brian sets up the following apparatus to find the specific latent heat of ice.

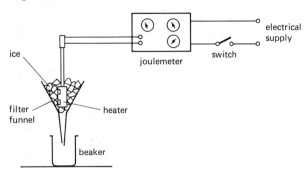

He takes large lumps of ice and places them in the filter funnel around the heater. He puts a beaker under the filter funnel and switches on the heater. A short time later he switches off the heater and removes the beaker. During the experiment he takes the following readings.

Initial reading on joulemeter = 24 000 J
Final reading on joulemeter = 34 000 J
Mass of empty beaker = 60 g
Mass of beaker + water = 110 g

(a) What value for the specific latent heat of ice do these results give?
(b) Jean tells Brian that his answer is inaccurate because some of the ice was melted by heat from the air.
  (i) Describe how Brian could find out how much ice was melted in this way.
  (ii) How would he use this result to get a more accurate value for the specific latent heat of ice?
(c) Explain why smaller lumps of ice would have improved the experiment.
(d) The value for the specific latent heat of ice listed in the data book is 3.3×10^5 J/kg. State in words what this means. (SEB)

11 An immersion heater is immersed in a large beaker full of water at an initial temperature of 10 °C. The heater supplies heat at a constant rate. The heater is switched on and the temperature of the water is taken every 30 seconds, the water being stirred each time before reading the thermometer. The results are:

Time/s	0	30	60	90	120	150	180
Temperature/°C	10	27	42	56	69	79	87

(a) Draw a graph of temperature (y-axis) against time (x-axis). From this graph estimate how long the water takes to reach its boiling point from the time when the heater was switched on. Show on your graph how you made your estimate.
(b) Explain why the graph is **not** a straight line.
(c) Which part of the graph would you use in order to estimate as accurately as possible the rate at which heat is given out by the heater? Explain your choice. (SEB)

12 A heating engineer is asked to find out how much energy can be saved by using a lagged hot-water tank instead of an unlagged one. He has two similar tanks, one of which is lagged and the other unlagged. Each tank is fitted with a 5 kW immersion heater.
(a) (i) If each tank is filled with 90 kg of water at 10°C, **estimate** the time needed to heat the water to 50°C.
  (ii) State any assumption you have made in your calculation.
(b) Each tank has a thermostat attached to it which switches on the heater when the temperature of the water falls to 40°C, and which switches off the heater when the temperature of the water rises to 50°C.
  The temperature–time graphs, (**A**) and (**B**), show one cooling–heating cycle for each tank.

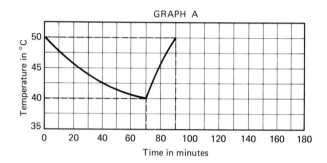

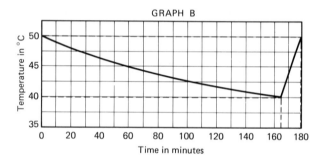

  (i) From the data given in the graphs, calculate the total time for which each heater would be switched on during a 24-hour period.
  (ii) Hence estimate the energy which would be saved during a 24-hour period by using a lagged tank instead of an unlagged one. (SEB)

12 A car of mass 850 kg is brought to rest from a speed of 20 m/s by applying the brakes. Each of the four steel brake drums has a mass of 2.5 kg and the s.h.c. of steel is 440 J/kg K.
(a) Draw an energy arrow diagram for this situation.
(b) Estimate the temperature rise of the brake drums.

INVESTIGATION

What is the power output of a candle?

18: Light and Reflection

We are so used to light that we often take it for granted. Without light, you wouldn't be able to read these words. But what *is* light? How do we see? These questions have interested people for thousands of years. The study of light has led to useful inventions like spectacles to correct poor eyesight, and telescopes and microscopes to see things which are very far away or very small. Its study also led to ideas about wave motion which link light with many other areas of study.

18.1 The nature of light

Rather than concern ourselves with what light **is**, let us begin by looking at some of the things we *can* say about light.

Light and energy

We can look at light from an energy point of view. The vanes inside the Crookes' Radiometer spin round when light shines on it. Without enough light, it stops. Light carries energy, and this makes the vanes turn.

A more dramatic demonstration that light carries energy was the flight of the Solar Challenger, which flew across the English Channel using sunlight as its fuel! Its wings were covered by solar cells which used the energy of the light to produce electricity to turn the propeller.

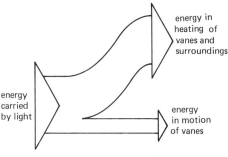

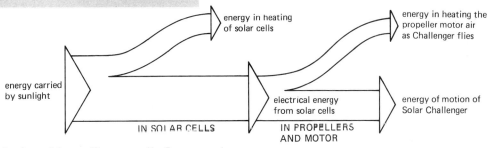

Sources of light

Light is **emitted** or given out by hot objects, like a candle flame, or the glowing filament of a light bulb, or the Sun. Some light sources are cooler, such as the screen of a television, or a fluorescent tube (strip light). An object which produces its own light is said to be **luminous**. ▷

Most of the objects we can see do not produce their own light – they are **non-luminous**. What we see is the light reflected off them. When light strikes an object, some is reflected and some is absorbed. Some surfaces are better reflectors of light than others. White walls help to make a poorly lit room brighter by reflecting most of the light.

However, all surfaces absorb some light. This is why your bedroom goes dark immediately you switch off the light at night. Light travels at an enormous speed – 300 000 000 metres per second. That means that light can travel across a typical sized bedroom and back about 15 million times in a second! Even if the walls absorb only 1% of the light each time, it will all be absorbed within a tiny fraction of a second.

Some materials like glass are **transparent**, which means that they allow light to pass through, though some is also reflected and absorbed at the same time.

165

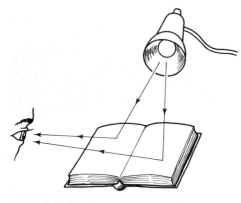

Light and seeing

We see objects when light from them enters our eye and strikes the retina, the light-sensitive 'screen' at the back of the eyeball. The light triggers off chemical reactions in some special cells in the retina, which then pass information along the optic nerve to the brain. The brain makes sense of these messages, and we 'see' the objects.

18.2 Light rays

In the diagram, we are really using a model, or a picture, of what light is. We have marked arrows on the lines which represent the light. Of course, we don't actually see arrows! What we see is something like the photograph on the left. Light seems to travel in straight lines.

A picture of light as **rays** travelling in straight lines from a luminous source and bouncing off objects, or being absorbed or transmitted by them, is a very useful one. We can think of a light ray as a very thin beam of light. A real beam contains many rays. But by drawing just a few of them we can get a clear idea of what is happening to the whole beam.

This picture of light rays can explain how one very simple device works – the pinhole camera. A pinhole camera consists of a tube or box, usually made of cardboard, with a pinhole at one end and a tracing paper screen at the opposite end.

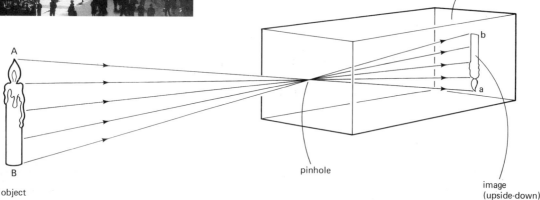

When the pinhole is pointed towards a bright object, an upside-down **image** appears on the screen. The size of the image depends on how close the camera is to the object. The diagram shows how we explain this by drawing light rays. Only a few of the rays are drawn. Every point on the object gives out light in all directions – we know this because we can see it from many different places. The only ray from point **A** which helps to produce the image is the one which sets off travelling in the direction of the pinhole. It travels straight through and hits the screen at **a**. One ray from every other point on the object also goes through the pinhole to the screen. The diagram shows only a few of these rays, and all the other rays which don't go towards the pinhole have been left out completely to make the diagram clearer.

We can see why the image is upside-down. It is also easy to see why it gets bigger if the camera is moved nearer the object. If we were to make a second pinhole in the front of the camera, we could predict that two separate images will appear on the screen.

The pinhole camera image is rather dim because very little light can get through a small pinhole, but if the hole is enlarged, the image becomes blurred as well as brighter. The larger hole is like several pinholes close together. They produce several images which are not quite in the same place. The result is a blurred image of the object.

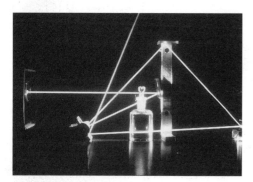

A laser beam shows clearly that light travels in straight lines.

18.3 Shadows

Shadows are cast where an opaque object stops light from reaching a screen. The shape of a shadow can be predicted using ray diagrams. The diagram shows the shadow formed when a ball is placed between a flat screen and a small light bulb. The bulb is small enough to be thought of as a **point source** – that is, the light from it comes almost from a single point. ▷

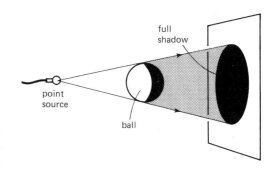

Light travels in straight lines from this point to the screen, and there is a circular area of shadow where no light can reach the screen. There is a sharp edge between the dark shadow and the bright region of the screen.

If we use a larger (or **extended**) light source, like an ordinary pearl light bulb, the shadow is more complicated, but we can still understand how it is formed by drawing rays. Light now comes from all the points on ▷ the bulb, and not just from a single point. There is still a circular region on the screen where no light can reach. And there is an outer region on the screen where all the light can reach. But in between is a ring-shaped region which receives light from only some parts of the bulb. This is in partial shadow. Imagine what you could see from behind the screen through a tiny hole at different points on the screen. If the hole were in the full shadow, you wouldn't be able to see any of the bulb at all. If it were in the outer bright area, you would be able to see all of the bulb. If it were in the area of partial shadow, you would be able to see part of the bulb but not all of it – the ball would be partly in the way.

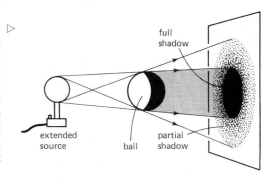

18.4 Eclipses

Eclipses of the Sun and Moon are special examples of casting shadows. By a strange chance, although the Sun is 156 million kilometres away from Earth and the Moon is only 400 000 kilometres away, the Sun is bigger than the Moon by just the right amount to make both of them look almost exactly the same size from Earth. The disc of the Moon looks almost the same size as the disc of the Sun. There is no particular reason why this should be so!

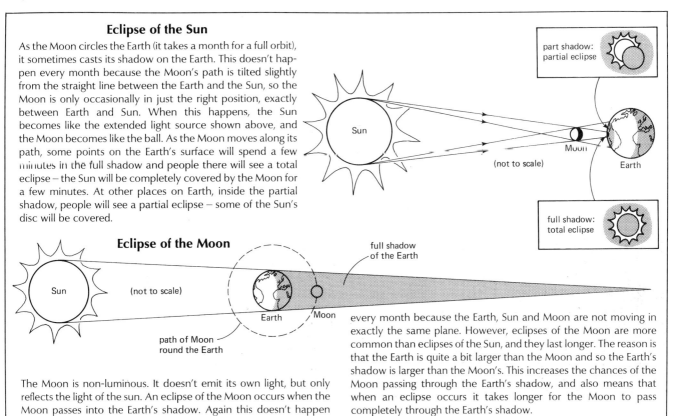

Eclipse of the Sun

As the Moon circles the Earth (it takes a month for a full orbit), it sometimes casts its shadow on the Earth. This doesn't happen every month because the Moon's path is tilted slightly from the straight line between the Earth and the Sun, so the Moon is only occasionally in just the right position, exactly between Earth and Sun. When this happens, the Sun becomes like the extended light source shown above, and the Moon becomes like the ball. As the Moon moves along its path, some points on the Earth's surface will spend a few minutes in the full shadow and people there will see a total eclipse – the Sun will be completely covered by the Moon for a few minutes. At other places on Earth, inside the partial shadow, people will see a partial eclipse – some of the Sun's disc will be covered.

Eclipse of the Moon

The Moon is non-luminous. It doesn't emit its own light, but only reflects the light of the sun. An eclipse of the Moon occurs when the Moon passes into the Earth's shadow. Again this doesn't happen every month because the Earth, Sun and Moon are not moving in exactly the same plane. However, eclipses of the Moon are more common than eclipses of the Sun, and they last longer. The reason is that the Earth is quite a bit larger than the Moon and so the Earth's shadow is larger than the Moon's. This increases the chances of the Moon passing through the Earth's shadow, and also means that when an eclipse occurs it takes longer for the Moon to pass completely through the Earth's shadow.

Measuring the brightness of light coming directly from the lamp.

Measuring the brightness of light reflected from the page of the book.

Table 18.1

Location or activity	Recommended brightness/lux
Living room	120–250
Kitchen	250–500
Classroom	250–500
Art or CDT room	500–1000
Shop	500–1000
Shop window	1000–2000
For fine work (sewing, drawing, etc.)	1000–2000
Very detailed work (watch repairing, engraving, etc.)	2000–5000

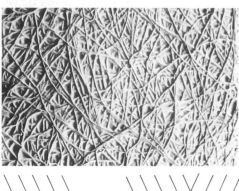

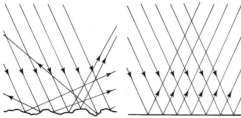

Diffuse reflection (from most objects)

Regular reflection (from a mirror)

18.5 Brightness of light

It is sometimes important to know how bright light is. To read without eye-strain, the light reflected from the pages of our book needs to be bright enough. Work which involves close attention to detail needs particularly good light levels. Light brightness is measured in units called lux, and Table 18.1 shows the recommended light levels in lux in different situations. In fact, our eyes are poor judges of light brightness because we can adjust them to compensate when there is less light. For reliable measurements, we need to use a lightmeter, or luxmeter.

Light spreads out from a source. The light is brightest close to the source and becomes steadily less bright as we go further away. If we take measurements with a luxmeter, we find that if we go twice as far away, the light level drops to one-quarter of its previous value. If we go three times as far away, it falls to one-ninth.

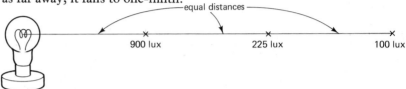

The relation is an inverse one: the greater the distance from the source, the lower the brightness. Because doubling distance reduces brightness by four times (2 squared), and trebling the distance reduces brightness by nine times (3 squared), and so on, the brightness is said to be linked to distance by an **inverse square law**.

Inverse square laws are quite important in physics because they crop up in several different places. This is probably the easiest inverse square law to understand. If we think of energy spreading from the light source, we can see that when it has gone twice the distance, the energy is spread over the surface of a sphere with double the radius. The surface area of this sphere is four times greater. The inverse square law is just a result of this spreading out of energy.

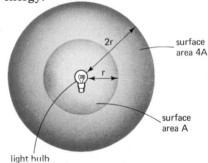

18.6 Reflection of light

We have said earlier in this chapter that most of the objects we see are non-luminous. We see them because of light which is reflected off them into our eyes. We have also said that the colour of walls can affect the amount of light reflected and make a room seem bright or dark. Usually we think of reflection as something special which happens at mirrors, not at all objects. What is the difference? The answer lies in how smooth the surfaces are. The surfaces of most objects, even though they look smooth, are very irregular when seen under a microscope. The photograph on the left shows the surface of the 'smooth' page of a book under high magnification. Light hitting this sort of surface is scattered in all directions. This is called **diffuse** reflection. It is what happens at most surfaces. On the other hand, the surface of a polished sheet of glass or metal can be made extremely smooth. The reflection from this is then **regular**.

Mirrors are made by putting a very thin coating of metal on to a flat piece of glass. Often this coating is put on the back surface of the mirror so that the glass protects it. Good mirrors reflect well over 90% of the light that reaches them, with only a small amount being absorbed.

Laws of reflection

We can use a plane (flat) mirror to investigate reflection in more detail. An easy way to do this is to place the mirror upright on a sheet of paper and to shine a narrow beam of light on to it. This is best done in a partly darkened room. ▷

We see a beam reflected off the mirror at the sort of angle we might expect. It looks rather as if the light has bounced off the mirror like a ball hitting a wall. We can check the angles by doing a few measurements. If we rule a line on the paper to show where the mirror is, we can shine the original ray (called the **incident** ray) on to the mirror at a number of different angles. The position of the reflected ray is then marked by putting a pair of pencil marks on the paper. These are joined by a ruler to show the reflected ray position. If we measure the angles carefully, we find that the reflected ray makes the same angle with the mirror as the incident ray.

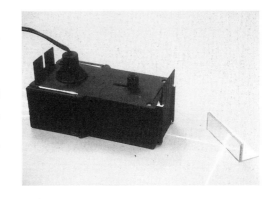

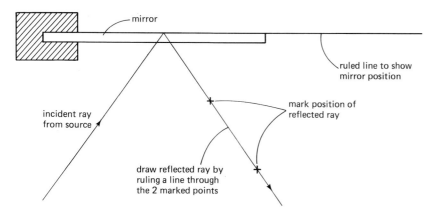

Usually these angles are measured, not from the mirror, but from a line at right angles to the mirror, called the **normal**. The **angle of incidence** is the angle between the incident ray and the normal; the **angle of reflection** is the angle between the reflected ray and the normal. Measurements on all shapes of reflecting surfaces (curved ones as well as plane ones) show that there are two laws of reflection:

1. The angle of incidence is equal to the angle of reflection.
2. The incident ray, the normal and the reflected ray all lie in the same plane. This is true in the simple reflection experiment described above, as we can draw all three on the same flat sheet of paper.

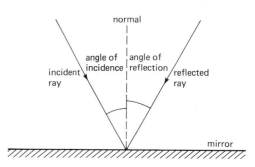

Since the angles of incidence and reflection are equal, it follows that light could equally well travel along the opposite path. If we shone an incident beam along the reflected beam's direction, then the new reflected beam would be exactly where the original incident beam was! All ray diagrams, no matter how complicated they are, can be reversed in this way. Light could travel equally well in the opposite direction.

18.7 Images

The most obvious thing about a mirror is that we can 'see ourselves' in it. There appears to another person behind the mirror and we sometimes talk about 'seeing our reflection', but what we are really seeing is an **image**. This image is just the result of the reflection of rays of light at the mirror.

To see how the image is formed, imagine two rays of light, spreading out from the same source, hitting a mirror. Each ray has simply been reflected at the mirror and for each, the angle of reflection is equal to the angle of incidence. The image is the same distance behind the mirror as the object is in front. ▷

The sort of image produced in a plane mirror is called a **virtual** image. The light rays do not really come from the image; they only appear to do so. **Real** images *do* exist, as we will see later (section 20.1).

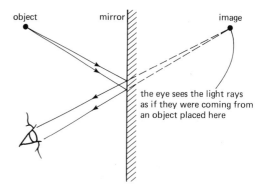

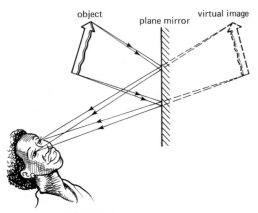

Images of real objects

When we look at ourselves in a mirror, we appear to be 'the wrong way round'. The watch on our left wrist appears to be on the 'image person's' right wrist, and so on. It is not the same as looking at a photograph of ourselves. The diagram shows how an image of a whole object is formed. Of course, every point on the object will emit light rays in all directions. In the diagram, only two rays from each end of the object are shown because this is enough to let us find where the image is.

At first sight, it is not at all obvious that this image *is* the wrong way round – it seems to point in the same direction as the object. But remember that we are seeing the *back* of the object when we look at the image. The real view of the object from behind (looking at it from the mirror) *does* look the opposite way round to its image. The image is said to be **laterally inverted**.

Laterally inverted writing – so that the mirror image is the right way round.

Images in plane mirrors

We can summarise what we have found about images in plane mirrors:

1. The image is the same size as the object.
2. The image is the same distance behind the mirror as the object is in front.
3. The image is virtual.
4. The image is laterally inverted (the wrong way round).
5. A line joining any point on the object to the corresponding point on the image meets the mirror at right-angles.

Using reflection

Accurate meters often have a mirror under the pointer to help you to read it correctly. For an accurate reading, your eye must be directly over the pointer. To use the meter, you move your eye until you cannot see the image of the pointer – it is directly below the pointer itself. Then your eye is in the correct position.

A periscope is a device for helping you to see over the top of an obstacle. It can be made from two plane mirrors. These are mounted in a tube, with both mirrors parallel but at an angle of 45° to the incident light rays. Light from the object is reflected at both mirrors into the eye. The eye sees a virtual image of the object, which is the right way up. The large periscopes used in submarines have prisms instead of mirrors for reflection (see chapter 22.5), but the principle they use is just the same.

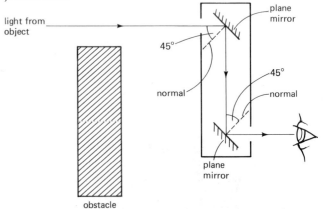

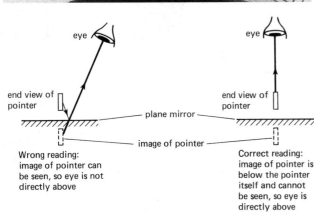

Wrong reading: image of pointer can be seen, so eye is not directly above

Correct reading: image of pointer is below the pointer itself and cannot be seen, so eye is directly above

18.8 Curved mirrors

The same laws of reflection apply when light is reflected at a curved mirror. There are two types of curved mirror:

Concave: the reflecting surface curves inwards at the centre. ▷
Convex: the reflecting surface bulges out in the centre. ▷

Concave mirrors

Perhaps the commonest everyday example of a concave mirror is a shaving mirror. This is a **spherical** concave mirror, because the curved reflecting surface is part of a large sphere. For simple experimental work it is easier to use a **cylindrical** concave miror – the reflecting surface is part of a large cylinder.

When parallel rays of light strike a concave mirror, they are reflected to meet at a point. The incident rays in the diagram travel parallel to the **principal axis** of the mirror. The point **F** where the reflected rays meet is called the **focus**, and the distance from **F** to the mirror is called the **focal length** of the mirror. The more curved the mirror, the shorter its focal length. ▷

If we look closely at what is happening to all the rays, we find that the simple law of reflection – angle of incidence is equal to angle of reflection – still holds at every point. A spherical concave mirror does the same thing in three dimensions, with a beam of light focused to a sharp (and bright) point at the focus.

Parabolic reflectors

The simple result in the diagram above applies exactly only if we have a narrow incident beam close to the principal axis. If a broader beam is used, the focus is no longer a sharp point. The solution to this problem is to use a mirror whose cross-section shape is not exactly circular. Instead, a shape known as a **parabola** is used. (A parabola is the path which a ball follows if you throw it into the air.) A parabolic reflector focuses a parallel beam to a sharp focus. ▷

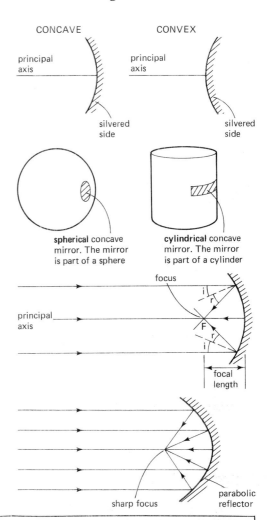

Parabolic reflectors in use

Parabolic dishes can be used to focus sunlight at a point – producing a very high temperature. On a small scale, this can be used to make a small solar cooker, or for heating water. ▷

On a large scale, the French solar power station at Odeillo in the Pyrenees has more than 60 mirrors arranged over a hillside to reflect the Sun's rays on to a 42 m diameter parabolic reflector. This then focuses the radiation on to a furnace which is in the central tower, in front of the reflector. The temperature here can rise as high as 3000 °C. The power station can generate 1 MW of electrical power. ▽

The commonest everyday application of parabolic reflectors is in producing a strong beam of light. If a light source is placed at the focus of a parabolic reflector, the reflected rays will form a parallel beam. Parabolic reflectors are used in car headlamps. The bulb is at the focus. Light ▷ leaves the bulb in all directions, but a lot of it is reflected forwards by the reflector to produce a strong parallel beam. Bicycle headlamps also use parabolic reflectors. Have a look at yours and see.

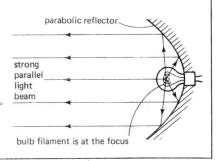

Questions

1. How would you go about demonstrating each of the following to someone who didn't believe them?
   (a) Light carries energy.
   (b) Light rays travel in straight lines.
   (c) Light is not only reflected from mirrors but from any surface.

2. Here is a set of objects which we can see because of the light coming from them: candle, the Moon, TV screen, filament lamp, cyclist's fluorescent safety armband, cats' eyes (on the road), stars, a gas fire.
   (a) List all the members of the sub-set of **luminous** sources – those which give off their own light.
   (b) List all the members of the sub-set of **non-luminous** sources – those which give off reflected light.

3. A worker is standing in front of a workbench, with a lamp above and behind him as shown.

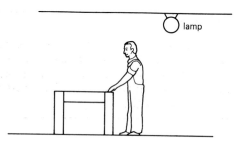

Copy the two diagrams below and draw **four** rays on **each**, to compare the shadows produced by the filament lamp and by the fluorescent lamp. In each case shade the area of bench in total shadow.   (JMB)

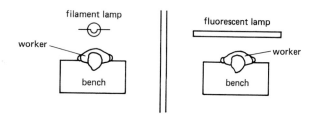

4. Three rays from a point on a lamp filament reach a plane mirror.

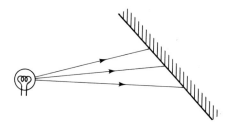

   (a) Draw a diagram to show the paths of the three rays after they are reflected.
   (b) Explain what you would see if you looked into the mirror along the direction of the reflected rays.
   (c) Where would the image of the lamp seem to be?

5. A narrow beam of light from a torch is shone on to a white wall in a dark room. The white patch on the wall can be seen from anywhere in the room. **But** if the torch beam is now shone on to a mirror hanging on the wall, the mirror looks dark from most directions. Only by looking straight into the mirror can the torchlight be seen.
   Explain these observations.

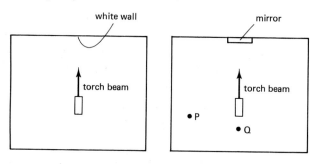

White patch on the wall can be seen everywhere in the room.

From P, the mirror looks dark; only from position Q can the reflected light be seen.

6. (a) Which letter in the diagram shows the position of the image of the girl's feet? Which shows the position of the image of the top of her head?

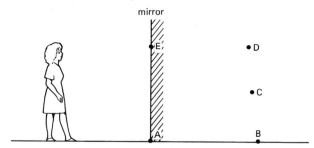

   (b) Copy the diagram and draw lines to show the direction in which she would have to look to see the image of her feet and the top of her head.
   (c) If she is 1.60 m tall, what is the shortest mirror which will allow her to see her complete image?
   (Hint: draw a scale diagram and mark the direction of the rays from her feet and from her head to her eyes.)

7. Copy and complete the diagram to show the paths of the two rays after reflection at the mirror.

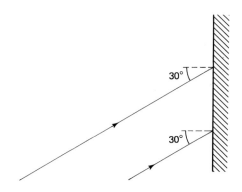

8 Complete the diagram to show the path of the light ray. Suggest one application which uses two mirrors arranged like this.

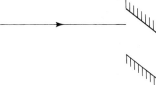

9 Draw a diagram to show the paths of rays of light from the lamp filament in this bicycle headlamp.

Make a list of other devices which use a concave mirror either to produce a beam of light or to focus light at a point.

10 Explain why:
(a) Accurate pointer meters often have a mirror under the pointer.
(b) The word FIRE is often written ɘЯIᖴ on the front of fire engines.
(c) If you look closely at an image in a mirror, you can see two images close together.

11 A car headlamp has two filaments, **A** for the main beam and **B** for the dipped beam. **A** is exactly at the focus of the parabolic mirror. Explain why the lamp gives a strong beam when filament **A** is on, but a weaker beam when filament **B** is on.

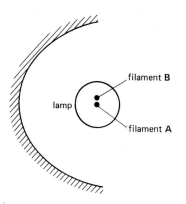

INVESTIGATION

What percentage of the light which shines on a surface is reflected? Does the colour or the shade make a difference? Are 'fluorescent' (bright yellow or orange) cycling harnesses or armbands better than most other materials at reflecting light?

19: Refraction

A swimming pool always looks shallower than it really is. A straw dipping into a drink looks as if it is bent at the surface. Both effects are caused by the same property of light – **refraction**.

19.1 Bending at a boundary

When a light beam shines on one side of a rectangular glass block, some light is reflected at the side of the block. But the strongest and most noticeable beam is the one which goes on into the block, at a slightly different direction from the original beam.

Light rays bend when they pass from one material (or **medium**) into another. This effect is called **refraction**. When light travels from air into glass, the angle of refraction is smaller than the angle of incidence. Notice that both these angles are measured between the ray and the **normal** to the boundary.

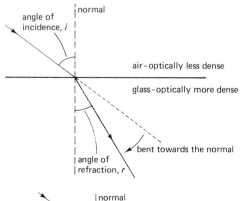

Compared with its original direction, we say that the light has been refracted **towards the normal**. When this happens, the medium the light is travelling into is said to be **optically more dense** than the medium it has left. On the other hand, if light travels from glass into air, it will be refracted **away from the normal**. It is now travelling into a medium which is **optically less dense**. Optical density is just a way of describing materials; it has nothing to do with 'ordinary' density. Table 19.1 summarises what happens to light at boundaries.

Table 19.1

increasing optical density →				
vacuum	air	water	glass	diamond

Light rays going from an optically *less* dense to an optically *more* dense medium are bent towards the normal.

Light rays going from an optically *more* dense to an optically *less* dense medium are bent away from the normal.

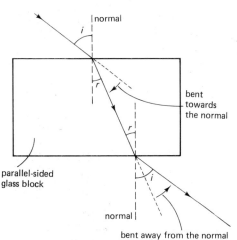

If light travels all the way through a rectangular glass block, we can see both directions of bending. As the light enters the glass, it is bent towards the normal. Where it re-emerges from the glass, it is bent away from the normal. This particular block has parallel sides, and the final ray travels in the same direction as the original ray. The only change is that it has been displaced slightly to one side.

19.2 Refraction and speed

Light changes direction at a boundary because the speed of light is different in different media. The change of speed when light crosses a boundary makes it change direction.

To see how a change of speed can cause bending, imagine a platoon of soldiers marching across a parade ground and on to a rather muddy grass field! It is harder to walk on the muddy field so each soldier slows down when he reaches it. The ones on the right-hand end of each row reach the field first, and slow down earliest. This makes the platoon change direction. (In reality, of course, they might compensate to allow for this slowing down!)

In the same way, a beam of light crossing a boundary into an optically more dense medium will be slowed down, and will bend towards the normal. One side of the light beam meets the boundary first and is slowed down. In the diagram, the light travels a distance **AD** in the glass **in the same time** as it travels the distance **CB** in the air.

The amount of bending depends on the change of speed. For example, the speed of light in vacuum (or air, since it is almost the same) is 300 000 km/s; in glass it is around 200 000 km/s. The **refractive index** for glass is defined as follows:

$$\text{refractive index for glass} = \frac{\text{speed of light in vacuum}}{\text{speed of light in glass}}$$

$$= \frac{300\,000 \text{ km/s}}{200\,000 \text{ km/s}}$$

$$= 1.5$$

The bigger the refractive index, the more bending occurs at the boundary.

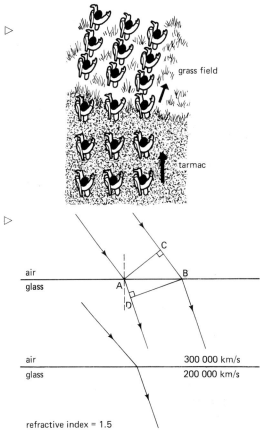

Refraction phenomena

Why does a swimming pool look shallower than it really is? If we think of the path which light takes to reach our eyes from the bottom of the pool we can see why.

At the water–air boundary, light is refracted away from the normal. The diagram shows how two rays from a point on the bottom of the pool change direction at the boundary. As a result, your eye sees the light as though it came from a point higher up – the **image** of the point on the pool bottom. The apparent depth is less than the real depth.

Another well-known effect due to refraction is that a straight object placed in water looks bent at the surface. Light from a point at the bottom of the straw is refracted at the water–air boundary and bends away from the normal. The image of this point on the straw will appear to be higher up. In fact, the images of **all** the points on the straw under water will appear to be higher than they really are. As a result, the straw appears bent at the surface (look at the top photo on p. 174).

Here is a trick you can try which uses refraction. Find a tall straight-sided mug and place a coin on the bottom. Get a friend to look into the mug at an angle such that the coin is just hidden. Tell her you will now make the coin appear without touching anything. To do it, you just pour water gently into the mug. As you do, the coin will appear!

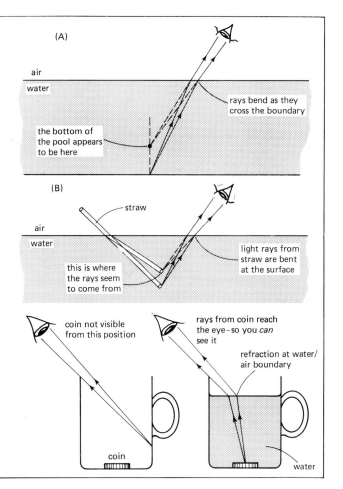

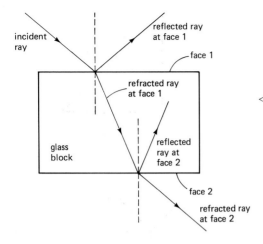

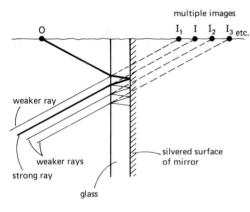

19.3 Internal reflection

If you look closely at the photograph on p.174 which shows light passing through a glass block you can see a reflected beam at both surfaces, where the light enters and leaves the block. At the second boundary, the light hits the boundary from inside the glass; the reflection here is called internal reflection. This is what causes the multiple reflections you see if you look closely in a mirror. In addition to the main image you should be able to notice some fainter images close to it. These are caused by extra light reflections at the two surfaces of the glass.

This is normally not a serious problem. But for work of high accuracy (for example, in telescopes and other optical instruments), multiple images are avoided by putting the silver reflecting coating on the front surface of the glass rather than the back. Of course it is then not so well protected and the mirror must be handled with great care.

Total internal reflection

In some situations internal reflection can become more important. We have already seen that light travelling from an optically more dense medium into air is bent away from the normal. Imagine what would happen if we placed a waterproof torch under water and shone a beam upwards towards the surface.

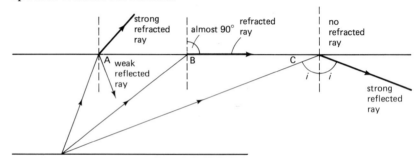

Where the beam hits the surface at **A**, a strong beam is refracted away from the normal and travels on into the air. The angle of refraction is greater than the angle of incidence. If we now shine the torch beam at a bigger angle, we will reach a point where the angle of refraction in the air is almost 90°. What happens if we now make the angle of incidence in the water even bigger still? The answer is that the refracted beam will disappear (the angle of refraction cannot become bigger than 90°), and all the light will be internally reflected. This is called **total internal reflection**. For rays coming from below, the water surface is behaving like a perfect mirror. The same thing happens for rays passing from any optically more dense medium into air.

Critical angle

The angle of incidence in the medium which results in an angle of refraction of 90° in air is called the **critical angle**. For angles smaller than the critical angle, the strongest beam is the refracted one leaving the medium. For angles larger than the critical angle, the light is totally internally reflected. The critical angle is the cut-off point between these two.

Measuring the critical angle

One way to measure the critical angle is to use a semi-circular glass or perspex block. A ray is shone through the curved side of the block, directly at the mid-point of the straight side.

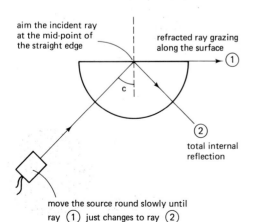

The ray passes straight through the curved side without bending because it hits the boundary straight on (the angle of incidence is 0°). By moving the source around, we can find an angle at which the ray just emerges from the straight side of the block into the air. This incident ray direction is marked, and the critical angle is measured.

With a perspex block, a typical value for critical angle is 42.5°.

Using prisms

The critical angle for glass is around 42°. This means that a block of glass in the shape of a 45–45–90° triangle can be used to reflect light. The angle of incidence in the glass is 45° which is larger than the critical angle, and so the light is totally internally reflected. ▷

A glass block like this is called a **prism**. A prism can be used to reflect light through 90° or through 180°. The great advantage over using mirrors is that there is no silvered surface to deteriorate as the prism gets older. It will always reflect totally – all the light is reflected. Of course some is lost by absorption in passing through the prism and by unwanted reflections at the other boundaries, but the prism is a more reliable reflector than a mirror and has a longer lifetime.

Prisms which reflect through 90° are used in periscopes, in exactly the same way as the mirrors in the drawing of the periscope on p. 170. Each half of a pair of binoculars also has two prisms which reflect light through 180°. The diagram shows how the prisms are arranged. By bending the path of light back on itself, the binoculars can be made more compact than a telescope. Another effect of the two prisms is to invert the image (upside down *and* right-to-left). This is useful because the lenses in the binoculars also invert the image and so the prisms put it back the right way again!

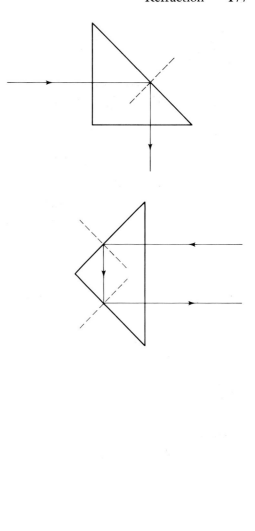

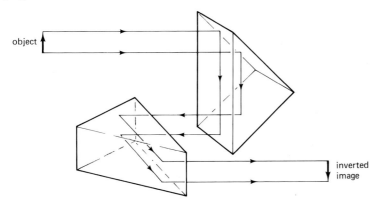

Optical fibres

Optical fibres are at the forefront of the modern revolution in communications – thousands of telephone conversations carried along a single glass fibre the thickness of a hair! An optical fibre depends very simply on total internal reflection. Light can travel along a glass fibre, even following bends in the fibre, being reflected every time it hits the surface. Once inside the fibre, the light cannot escape until it reaches the other end (see the front cover). The angle of incidence of the light ray inside the fibre is always greater than the critical angle and no light escapes. ▷

Optical fibres are normally coated with a second transparent layer to prevent the outer surface becoming scratched as this *does* lead to light losses. Communications cables can be made of bundles of tiny fibres, each of which is completely independent. When used for communications, the brightness of the light is varied to carry a message in the form of a code. The code consists of sequences of: light on/light off/light on, and so on.

Compared to metal wires, optical fibres are lighter and cheaper to produce (because glass is plentiful and cheap). Another advantage is that optical signals are not affected by interference in the same way as electrical signals in wires.

Another application of an optical fibre bundle is for looking at inaccessible places and even taking photographs of them. Doctors use optical fibre instruments for seeing inside the body. A fibre bundle is used to carry light into the body, and the image is viewed through a second fibre bundle. Light travels along each fibre independently without getting mixed up, and a clear image is visible at the outer end of the fibre. ▷

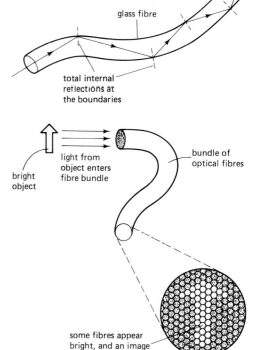

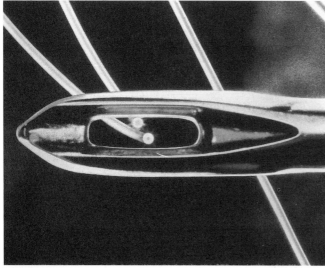

Optical fibres are very fine. This picture shows two fibres and the eye of a needle.

This amazing photograph of a 5-month-old human embryo inside the womb has been taken using fibre optic photography. ▷

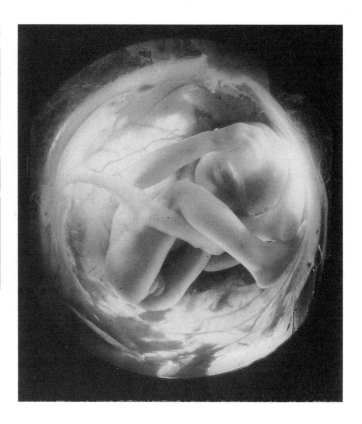

Mirages

You don't have to go to the desert to see a mirage. On a hot summer day, a distant road or runway will appear to have pools of water lying on the surface. What you are really seeing is a reflection of the sky. The air close to the ground is heated by the sun and expands. It is now less dense than the air higher up. Less dense air is also optically less dense. Although air has a very small refractive index, the variations in the refractive index of the air layers above the ground are enough to cause the mirage.

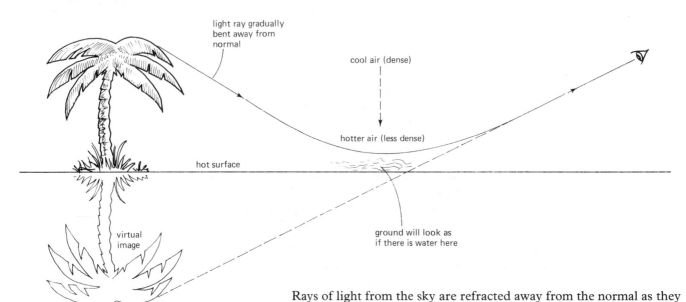

Rays of light from the sky are refracted away from the normal as they pass into the steadily less dense air close to the ground. At a very shallow angle, total internal reflection occurs. There are no sharp boundaries between the air layers, but the ray follows a curved path. You see an image of the sky appearing on the surface. In fact, the angles of bending are very small (much smaller than the diagram suggests) and mirages are seen only if you look at the ground in the distance, close to the horizon.

19.4 Prisms and dispersion

When light passes through a parallel-sided glass block, its direction isn't changed (see the photograph and diagram on p. 174). However, if the glass block does *not* have parallel sides, some other interesting things happen.

A 60° prism can be used to refract light. Light is bent towards the normal as it enters the prism, and away from the normal as it leaves it. The direction is changed. The prism has caused **deviation** of the light.

But there is something else which is even more noticeable. The incident beam is ordinary white light, but the beam coming out of the prism is coloured – with the colours of the rainbow. The range of colours produced is called a **spectrum** (see the photograph on the back cover).

Where do the colours come from? Newton carried out a famous series of experiments on prisms and spectra. He noticed that white light from a circular hole in a blind produced an oval shaped spectrum – as though the light beam was being stretched in one direction. He used slits to separate off a single colour from the refracted ray and tried to break it into even more colours using a second prism. He was unable to, but he *could* use a second prism to change the coloured spectrum back into white light.

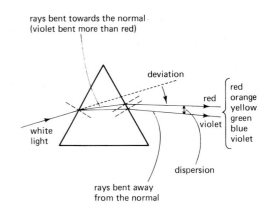

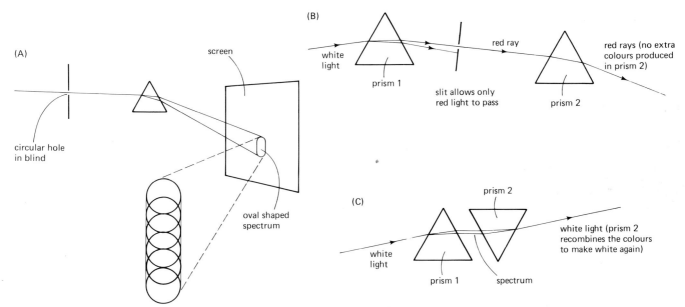

As a result of these experiments he convinced himself that the colours came from the white light itself. It took him some time to convince other scientists of his day (some of them thought the prism was changing the light and making it coloured), but eventually Newton's ideas were accepted!

White light is made up of light of all the colours of the spectrum. Light of each different colour has a slightly different refractive index and is bent by a slightly different angle. Violet light is bent more than red light. The differences are small, but they are enough to produce the spectrum. The effect is called **dispersion**.

The spectrum of white light (from a lamp or from the Sun) is a **continuous spectrum**. It has no gaps or breaks in it, but a continuous range of colours from the red end to the violet end. Some sources of light (for example, the light from yellow sodium street lights) contain only a limited part of the spectrum. This is called a **line spectrum**. Every source has its own line spectrum, so it is a sort of 'finger print' which can identify the source. Some examples of line spectra are shown on the back cover. Around the year 1900, new theories about how line spectra are produced led to important discoveries about the structure of atoms. Line spectra are not just pretty patterns, they are also useful evidence when trying to understand atoms and light. We will look at this again later in Chapter 33.

Rainbow

If a white light spectrum is produced by refraction in a prism, what makes a rainbow? It is caused by refraction of sunlight in tiny droplets of water. The light is refracted as it enters the drop, then totally internally reflected, then refracted again as it leaves. Violet light is refracted more than red. The angle between the white light and the red ray is always the same for every drop; so is the angle between the white light and the violet ray.

To see a rainbow you need sunshine and a rain shower at the same time. You always have your back to the Sun when you are looking at the rainbow. Red light reaches your eye from all the water droplets which are at just the right position in the sky, so you see red light coming from all the drops in this curve. A different set of drops sends your eye violet light. They lie on a slightly different curve. So what you see is a set of curved bands of colour – a rainbow.

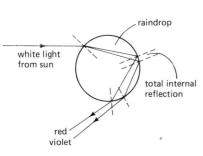

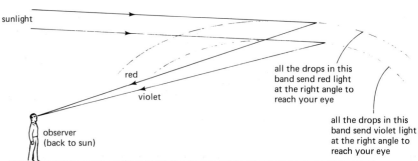

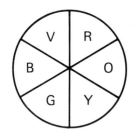

19.5 Colour vision

If white light is really a mixture of colours, why do we see it as white? We do not need a prism to demonstrate that white is made from colours. A ◁ Newton's disc has sectors of each colour of the spectrum. If it is spun quickly, it looks white.

But there is something even more surprising about colour mixing. We don't need *all* the colours of the spectrum to see white. If we shine red, green and blue light from three projectors on to a screen, the patch in the middle where all three overlap is white (see the back cover).

The explanation for this lies in the way the retina of the human eye is made. The retina at the back of the eyeball contains special cells which are sensitive to light. When each cell picks up light, it triggers and sends an electrical impulse 'message' to the brain. We now know that there are three different types of colour-sensitive cell – one which responds to red light, one to green and one to blue. White light makes all three of them trigger at once. So does a mixture of red, green and blue lights. As far as the brain is concerned, the message is the same. We 'see' the same thing.

In fact, we see *all* the different colours depending on how strongly they trigger the three types of sensor in the retina. We can produce any colour effect we want by changing the mixture. This is how colour television works. If you look closely at the screen, you will see that it is made of thousands of tiny red, green and blue patches arranged in groups of three.
◁ At a distance the eye cannot see the red, green and blue spots separately and so they become mixed to produce a single colour from each group of three. The separate spots can all be lit up to different brightnesses, producing any colour we wish. (Chapter 31 has more information on how television works.)

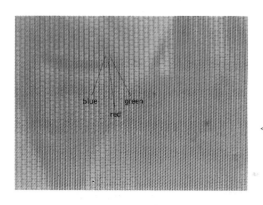

Seeing colours

Red, green and blue are known as the **primary colours** of light, because they can be mixed to produce white. Mixing any two of them produces a **secondary colour**. The diagram also shows what these are: yellow, cyan (turquoise) and magenta.

◁ Yellow is a rather surprising result from mixing red and green. After all yellow is also a member of the spectrum of white light. In fact, yellow light from the white light spectrum is completely different from a mixture of red and green. By using a prism, we could tell them apart quite easily. *To the eye*, however, they look exactly the same. The same cells are triggered in the retina by the red–green mixture and by the pure yellow light. So the brain gets the same message, and we 'see' the same thing. But here we are beginning to leave the realm of physics and enter that of physiology!

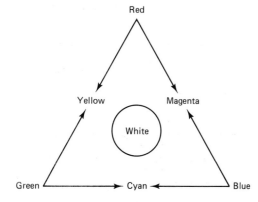

19.6 Colour filters and coloured objects

Colour filters

The easiest way to produce coloured light is not with a prism and spectrum. It is to use white light and a colour filter. A filter contains a substance which transmits only one colour of light and absorbs the rest. Some examples are shown here.

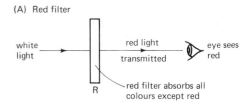

(A) Red filter

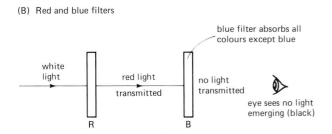

(B) Red and blue filters

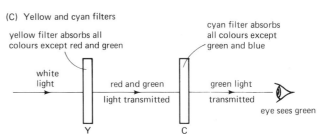

(C) Yellow and cyan filters

Real filters, especially cheap ones, are imperfect and allow small amounts of other colours through in addition to the colour we want to get. So experiments with colour filters sometimes don't work very convincingly. However, from the results you get, you should be able to work out which 'extra' colours are getting through.

Seeing coloured objects

Most objects do not give out their own light. Instead they reflect light from another luminous source. Objects look coloured if they reflect only some colours of light and absorb the rest. When we look at an object using white light, it is fairly easy to see why it has the colour it does. When we are using coloured light, we can explain what we see by thinking of white light as a mixture of red, green and blue lights. Some examples are shown in the margin. Again, these experiments may not work too well if you try them, because the colour of most objects is not 'pure' enough. It is really a mixture of several colours. So is the coloured light which is shining on them, if we use ordinary filters. This complicates the simple results shown on the left. Some objects look quite a peculiar colour in coloured light – but the theory of colour mixing can help you to explain what you see!

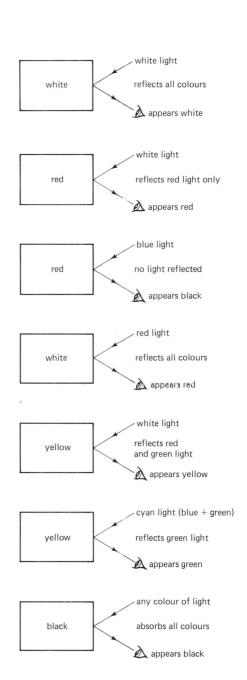

Mixing paint

To an artist, the three 'primary colours' are *not* red, green and blue, but red, **yellow** and blue. With these, any other colour of paint can be mixed. Why is this different from mixing coloured lights? The answer is that paints are made to absorb certain colours and reflect others; when we add two paints, it is their absorbing properties that we add. Take a common example: mixing yellow and blue paint gives green. Yellow paint reflects yellow light, but it will also reflect some of the neighbouring colours of the spectrum – orange and green. It absorbs the others. Similarly, blue paint reflects blue light, and also some of the neighbouring colours – violet and green. It absorbs the others. If yellow and blue are mixed, between them they absorb all the colours of the spectrum except green. Green is reflected. So the mixture looks green.

So mixing paints actually depends on the fact that none of the colours is 'pure'! They all reflect a group of colours. If coloured paints were pure, can you work out what colour we would get by mixing yellow and blue? Perhaps it is a good thing that the colours are impure!

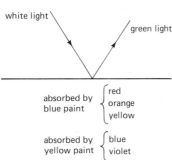

182 Refraction

Questions

1. Copy and complete these diagrams to show the path of the ray after it has crossed the boundary.

2. Use a protractor to measure the angle of incidence and the angle of refraction of the light ray entering the glass block. Copy and complete the diagram to show the path of the ray as it emerges from the first block and passes through the second block.

3. The diagrams show rays of light crossing boundaries between different materials. Arrange the materials in order, from the one in which light travels fastest to the one in which light travels slowest.

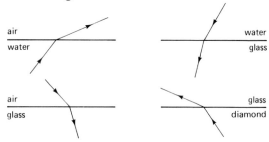

4. (a) Diagram (A) shows a long block of glass over an object **O**. Light from **O** reaches the top surface of the glass at **X**, **Y** and **Z**.

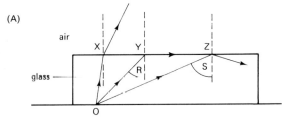

   (i) What is the name given to the bending of the light at **X**?
   (ii) Copy the following sentence, filling in the missing words.
   At **Z** light is _____ _____ reflected.
   (iii) Give the name of the angle marked **R**.
   (iv) Explain why light is reflected as shown at **Z**.

   (b) Diagram (B) shows two 45° 45° 90° glass prisms with two rays of light incident on a face of one of them.
   (i) Copy diagram (B) and complete the path of both rays through both prisms.
   (ii) Give a practical use for such a device. (NEA)

5. A ray **X** of red light is directed towards a glass prism.

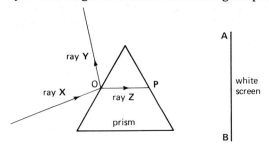

   (a) State what happens to the ray **X** after it strikes the glass at **O**.
   (b) Comment on the brightness of ray **Z** compared to the brightness of ray **X**.
   (c) Copy the diagram and show how the light leaves the prism at **P**.
   (d) Describe the changes which take place in (i) the wavelength, (ii) the speed of the red light as it enters the prism at **O**.
   (e) The ray **X** of red light is now replaced by a ray of white light. Describe in detail what is now seen on the screen **AB**. Explain how this happens. (SEB)

6. (a) The critical angle for glass is approximately 42°. Explain why any light ray which enters the end of a thin glass fibre cannot escape again through the side.
   (b) Could an optical fibre be made from a material whose critical angle was 48°?

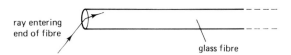

7. (a) Diagram (A) shows a beam of white light shining on a red filter.
   (i) Copy and complete the diagram to show what you would expect to see on the screen.
   (ii) Explain what the red filter does to the light.
   (b) Diagram (B) shows a prism producing a spectrum of white light.
   (i) Draw an accurate diagram to show what you would see if you placed a red filter in the path of the light beam between the prism and the screen.
   (ii) How could you use this experiment to argue against someone who said that 'a red filter produces red light by dyeing the white light as it passes through'?

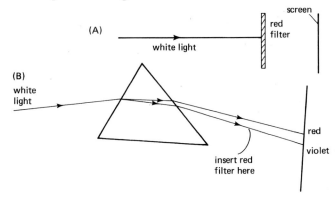

8 Here are two ways of using 45° glass prisms to reflect light. In each case, is the image upside-down or the right way up? Explain your answer with a diagram.

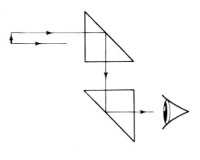

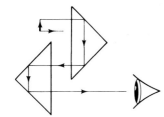

9 A bicycle reflector produces a red reflected beam when white light from a car's headlights shine on it. Using the diagram, explain: (a) how a reflected beam is produced; (b) why the reflected beam is red.

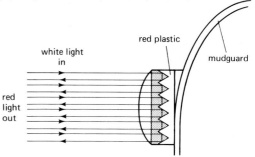

INVESTIGATION

Is there any difference (apart from brightness) between the light emitted by a candle, a filament bulb and a fluorescent tube?

20: Lenses and Optical Instruments

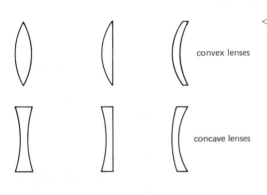

This is the earliest known picture of a person using spectacles. It was painted by Tomasso da Modena in 1392.

convex lenses

concave lenses

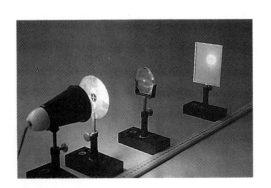

Probably the most important practical application of refraction is in lenses, and instruments which use lenses. The use of ground and shaped pieces of glass for aiding eyesight is recorded in the Middle Ages (around the year 1100). They were particularly useful for monks who spent many hours every day in copying and writing manuscripts, before the invention of printing.

Nowadays we still use lenses in spectacles (glasses), and also in cameras, telescopes and microscopes. The eye also contains a lens.

20.1 Lenses

There are many different types of lens. All have curved surfaces. In general, if a lens is thicker in the middle than at the edges, it is **convex**. If it is thinner in the middle than at the edges, it is **concave**. The simplest kind of convex lens has two equal convex surfaces; a simple concave lens ◁ has two equal concave surfaces. Other lenses have more complicated shapes.

Most lenses are **spherical** – each surface is shaped like part of a sphere. For simple experiments, it is sometimes easier to use a **cylindrical** lens, whose curved surfaces are shaped like part of a cylinder. These are really 2-dimensional versions of the 3-dimensional spherical lens.

Images in lenses

The really crucial property of lenses, and the thing which makes them so useful, is that a lens can produce an image. In a darkened room, a convex lens can cast an image of the objects outside the window on to a screen inside the room. The photograph (below left) shows a convex lens producing an image on the screen of the cross-wires of the ray box.

Let us think carefully about what is involved here, for it is really quite remarkable. Every point on the object is emitting light in all directions. Rays are leaving every single point and travelling in every possible direction. A small fraction of these rays from one point on the object will happen to be travelling in the right direction to hit the lens. *All* of these rays are bent by the lens *by just the right amount* to meet up again at a single point on the screen! The point where they meet is the image of that point on the object. This would be remarkable enough, but the lens is doing a lot more than that – it is doing the same thing for all the rays from *every point* on the object. Every point on the object emits many rays, and these are brought back together at the corresponding point on the image. A ray diagram like the one shown below is simplified to make it clearer, and shows only a few of the rays involved in forming the image.

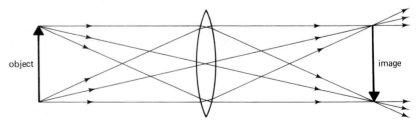

The ability of lenses to form images was probably the first thing discovered about them. We can imagine that some craftsman found – perhaps by accident – that a piece of curved glass could produce an image. This is what made it worthwhile to continue making and studying lenses. The fact that lenses produce images is still what makes it worth studying them in detail.

20.2 Focusing by lenses

Convex (or **converging**) lenses bend parallel rays of light so that they meet at a single point. The incident rays shown in the diagrams below are travelling parallel to the **principal axis** of the lens. The point where the refracted rays meet is called the **principal focus** of the lens (or simply the focus, for short), and the distance from the focus to the **optical centre** of the lens is called the **focal length**.

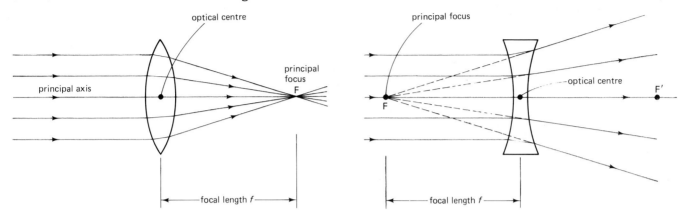

A concave (or **diverging**) lens bends a parallel beam outwards, so that it diverges. The rays do not meet at a point, but they look as if they were diverging from a single point. This point is again called the focus, and the distance from the focus to the lens is its focal length. A concave lens has a **virtual** focus.

A lens will also focus an incident beam which is not parallel to its principal axis. The rays meet at a point in the **focal plane** of the lens (a plane at a distance f from the lens), but not at the principal focus. It is easy enough to locate the point, because the ray which hits the optical centre of the lens goes straight on and is not bent.

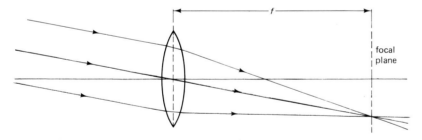

One way to see how lenses bend light the way they do is to imagine a lens as a series of prisms placed end to end.

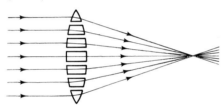

Light will be refracted towards the normal as it enters each prism, and away from the normal as it leaves the prism. Each prism will cause deviation of the light. The faces of the prisms near the edges are at a greater angle, so they deviate the light more. As a result, the parallel rays are brought to a single focal point.

All these results are approximate and hold only for thin lenses, and for parallel rays close to the principal axis of the lens. There is also the extra complication that different colours in the white light will be refracted by different amounts. This produces an effect called **chromatic aberration**: the edges of the image are tinged with red and blue light. You may have noticed this when using cheap binoculars or a cheap telescope.

In the remainder of this chapter, we will assume that the lenses are thin, and that the different bending of different colours of light can be ignored.

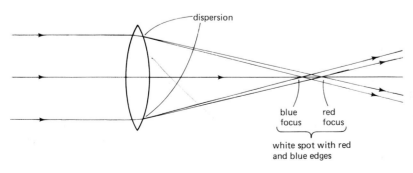

186 Lenses and Optical Instruments

20.3 Measuring the focal length of a convex lens

Cylindrical lenses

It is easy to measure the focal length of a cylindrical convex lens. Place the lens on a sheet of paper and shine parallel rays along its principal axis. Mark the position of the lens and the paths of the refracted rays on the paper. The focal length can then be measured with a ruler.

The more curved the lens, the shorter its focal length. Thick, strongly curved lenses bend the light more than thinner lenses.

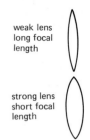

weak lens
long focal length

strong lens
short focal length

Spherical lenses

1. Rough method

An approximate way of finding the focal length of a cylindrical lens is to use it to form an image of a distant object. For instance, if we point the lens at a window, we can form an image of the window frame on the wall opposite or on a screen held behind the lens. The distance from the lens to the screen is (approximately) equal to its focal length.

2. More accurate method

A more accurate method is to use a convex lens and a plane mirror to cast an image back on to the object itself. The lens is mounted in a stand which can be moved along a ruler directly away from the source. A plane mirror is held behind the lens. As the lens is moved back, a point will be found where a sharply focused image is cast back on to the front of the ray-box, beside the cross-wires. The distance from the cross-wires to the lens is now equal to its focal length.

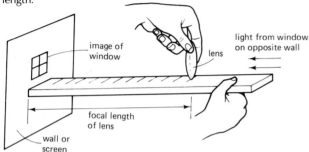

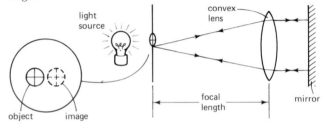

The method works for the following reason. When an object is really distant, the rays coming *from any one point on the object* are almost parallel when they hit the lens. These parallel rays are brought to a point at the focus. So the image of the distant object is formed at the focus of the lens (or, more correctly, in the focal plane). If we measure the distance from the lens to the screen, this is equal to the focal length.

Rays from a point on the cross-wires spread out to hit the lens. Because they came from the focus of the lens, the rays are parallel after refraction. When these parallel rays hit the mirror they are reflected straight back. They strike the lens again as a parallel beam and are brought to a point at the focus of the lens, i.e. back on the ray-box. This only happens if the distance from the ray-box to the lens is exactly equal to the focal length of the lens.

20.4 Images formed by convex lenses

The position of the image in a lens can be found by drawing an accurate scale drawing. At first sight, it might seem that this would be impossible. After all, every point on the object emits light rays in all directions, and it would become impossibly complicated to draw all of these and try to work out what happens to them when they are refracted at the lens. However, because we *know* that a lens produces an image, we can get away with drawing just enough rays to find where the image must be. To do this we use two special rays whose behaviour is easy to predict:

1. A ray of light which hits the centre of the lens passes straight on without any bending.
2. An incident ray of light which is parallel to the principal axis of the lens is refracted to pass through the principal focus.

A third ray is sometimes also useful (as a check on the others):

3. An incident ray which passes through the principal focus will leave the lens parallel to the principal axis. (This is just ray 2 in reverse.)

These diagrams have been slightly simplified. A light ray is really bent at *both* surfaces of the lens – on passing from air into glass and on passing from glass back into air. It is usual when drawing ray diagrams to draw the bending *as if* it all happened at one place – the centre line of the lens. This is done to make the drawing clearer.

Objects beyond the focus

Let us then see how we can use these special rays (plus our knowledge that an image *does* exist) to find where the image is. Each time we find the image by drawing two rays from the top of the object:

- a ray which goes straight through the centre of the lens
- a ray which goes parallel to the principal axis of the lens and is refracted to pass through the focus.

The image of the top of the object must lie where these two rays cross. The image of the bottom of the object lies on the principal axis.

In these three cases the image is inverted – it is upside down. It is also a **real image**. This means that light rays really do go to the image (unlike the image in a mirror which is virtual, and where light only *seems* to come from). A real image is one which can be cast on a screen.

Notice that the further the object is from the lens, the closer the image is to the lens. As the object moves closer, the image moves further away. It also gets bigger.

Close objects

If the object is closer to the lens than the focus, something different happens. We can try to draw the ray diagram by drawing rays 1 and 2 in the same way as before. Instead of coming to a point, these rays diverge after passing through the lens. However, if we imagine looking into the lens from the right hand side, the light will appear to be coming from a point behind the object. This is where the image of the top of the object appears to be. It is a **virtual image** – the light doesn't really come from it, and it couldn't be cast on a screen – but there *is* an image. It is erect (the same way up as the object) and magnified. When used like this, a convex lens works as a magnifying glass. We will have more to say about magnifying glasses later (in section 20.9).

Two special object positions

A special situation occurs if the object is a very long way from the lens. Light from a point on a very distant object reaches the lens as a parallel beam. So rays from the top of the distant object will be refracted to meet in the focal plane. Similarly rays from the bottom of the object also reach the lens as a parallel beam, but at a different angle. After passing through the lens, these rays meet at the focus. The image of the distant object is formed in the focal plane.

Another special situation occurs if the object is placed at the focus. The two special rays are now parallel after passing through the lens; they don't meet anywhere. The image is a long way from the lens (at infinity). It makes more sense to think of this image as a virtual one. If we look into the lens, we would see the image as if we were looking at a very distant object. The image is not only very distant, it is also very large! Think of where the rays would be when they did meet! So we *will* be able to see the image. We say it is 'at infinity'.

Although these two cases may seem to be rather special ones, they turn out to be important in understanding how microscopes and telescopes work.

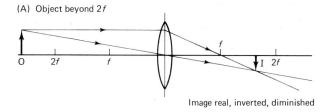
(A) Object beyond 2f

Image real, inverted, diminished

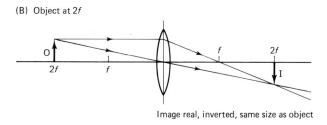
(B) Object at 2f

Image real, inverted, same size as object

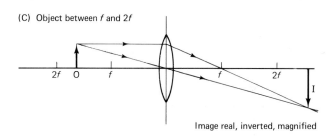
(C) Object between f and 2f

Image real, inverted, magnified

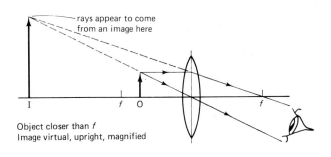
Object closer than f
Image virtual, upright, magnified

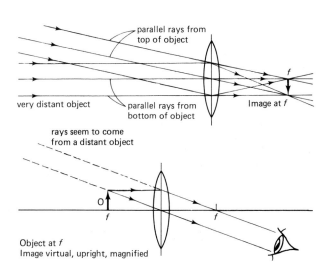
Object at f
Image virtual, upright, magnified

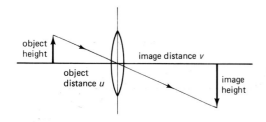

20.5 Magnification

From the ray drawings in section 20.4, we can see that images are sometimes magnified, sometimes diminished. In fact, there is a very simple relationship between the object and image distances, and the magnification.

The diagram on the left is a simplified ray diagram. The ray through the centre of the lens goes straight on. The two triangles are similar, so their sides are in proportion. If the image distance is twice the object distance, it follows that the image height is twice the height of the object. This is called the **linear magnification**.

$$\text{linear magnification} = \frac{\text{image height}}{\text{object height}} = \frac{\text{image distance}}{\text{object distance}}$$

However, linear magnification is not quite the same thing as magnification in a telescope or microscope, as we will see later. The size which something appears to us depends both on how big it really is (linear magnification) *and* on how far away it is.

20.6 The eye

Parts of the eye

We have already mentioned the eye at several points in this chapter and in chapters 18 and 19. Let us now take a fuller look at the structure of the eye, as we need to know some facts about if we are to understand how optical instruments work. In any case, the eye itself is a very interesting and important 'optical instrument'!

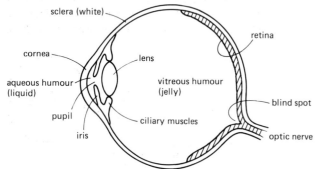

The **eyeball** is almost spherical in shape. It is made of a very tough white material called **sclera**, except for a transparent bulging disc at the front, called the **cornea**. This lets light enter the eye. At the back of the eyeball is a layer of light-sensitive cells which form a 'screen' called the **retina**. The cells in the retina are sensitive to light. Some, as we have seen earlier, are sensitive to different colours of light; some are particularly sensitive to very dim light and help us to see in the dark. These two sorts of cells are known as **rods** and **cones**. Separate nerves lead from every light-sensitive cell to a central nerve called the **optic nerve**. 'Messages' carried by nerves are like little electrical impulses, which travel along the optic nerve to the brain. Part of the brain works to interpret the nerve messages. In fact, it is impossible to say exactly where the eye ends and the brain begins! In a way, we also 'see' with our brain.

The amount of light entering the eye is controlled by a ring of muscles called the **iris**. This is the part that is brown or blue in most people's eyes. The black circle in the centre of the iris is really a hole (called the **pupil**) which allows light into the eye. If you are in a bright room, your pupils will be small. If you then go into a darker room, or if the light goes out, the iris contracts and the pupil expands, to let more light in. It takes a short time for the iris to adjust to changes in brightness. That is why it always seems so bright when you are suddenly woken up on a sunny morning; and why after a little time on a dark night, or in a dark room, you begin to see quite well.

To see clearly, the eye must produce a focused image on the retina of what we are looking at. Most of the focusing is done by the curved cornea and the clear liquid (**aqueous humour**) behind it. These form a liquid lens with a fixed focal length. The **eye lens** can make small adjustments to the focusing, letting us focus on close objects or on distant objects, when we want. This process is called **accommodation**. The lens is naturally thick and has to be pulled outwards to make it thinner. The thickness of the lens is controlled by the ring of ciliary muscles which hold it round the edges. When we are looking at something far away, the eye lens is thinnest, and the **ciliary muscles** are relaxed. To look at something close, the lens must become fatter, and the muscles are contracted. This is why it is tiring to concentrate for a long time on something close to the eye (see below).

The space between the lens and the retina is filled with a transparent jelly-like substance called **vitreous humour**. It also helps in the focusing process.

The image on the retina will, of course, be inverted (upside down), and diminished. The brain interprets this image so that we see things the right way up.

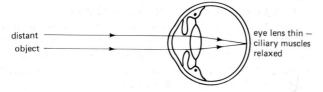

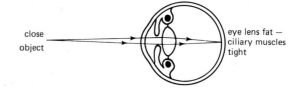

How large do objects look?

The size which an object appears to us depends on two things: how big it is, and how far away it is.

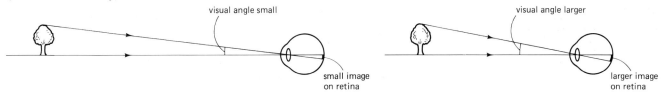

What *really* determines how big things appear to us is the size of the image they make on the retina. The figure above shows this clearly. When the tree is further away, its image covers a smaller part of the retina. Notice how we can find the position of the top and bottom of the image on the retina by using just one ray from the top and bottom of the object – the rays which go through the centre of the lens. The important factor in determining apparent size is the *angle* which the object makes at the eye. This is known as the **visual angle**.

It is not possible to make objects appear as big as we want simply by bringing them very close to the eye making the visual angle very large. We cannot focus sharply on objects which are closer to our eyes than the **least distance of distinct vision**. This differs from person to person and gets larger as you get older. You will probably find that you can focus for a short time on objects as close as 10 cm, but closer than that it becomes impossible.

Some optical instruments (microscopes and telescopes) are designed to make objects which normally look small appear bigger. They work by increasing the visual angle (see sections 20.9 and 20.10).

Defects of vision

Some people's eyes are not able to accommodate (change the shape of the eye lens) to see both close and distant objects clearly. They must then wear glasses or contact lenses.

Short-sighted people can see close objects but distant ones look blurred. The reason is either that the eyeball is slightly too long, or that the eye lens is too strong. In either case, the light from a distant object is brought to a focus in front of the retina.

Long-sighted people can see distant objects sharply but cannot focus on closer ones. Here the problem is either that the eyeball is slightly too short, or that the eye lens is too weak. Rays of light from close objects cannot be bent enough and would only meet beyond the retina.

To correct short-sight, we need a concave lens in front of the eye. This makes light from a distant object diverge, and the eye lens can then focus it exactly on the retina. Long sight is corrected by using a convex lens to help the eye lens bring the diverging rays from a close object to a focus on the retina. The diagram below shows how these work.

(A) Correcting short-sight

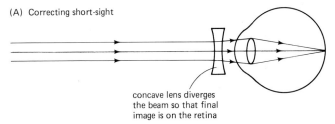

concave lens diverges the beam so that final image is on the retina

(A) Normal eye

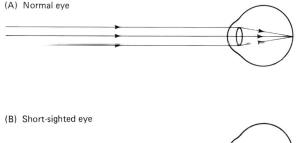

(B) Short-sighted eye

image formed in front of retina

(C) Long-sighted eye

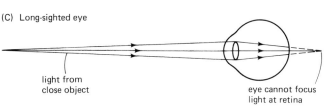

light from close object — eye cannot focus light at retina

(B) Correcting long-sight

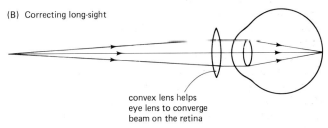

convex lens helps eye lens to converge beam on the retina

As people get older, their eyes may lose the ability to accommodate as well as they did when younger. They cannot see distant objects or close objects clearly because the eye lens cannot adjust quite enough in either case. They need glasses with convex lenses for seeing close-up, and glasses with concave lenses for seeing distant objects. Sometimes these are made as the upper and lower halves of a single pair of glasses – called **bi-focals**.

Other defects of vision are more complicated. **Astigmatism**, which is quite common, is caused by having a cornea which isn't exactly spherical in shape, but is slightly cylindrical (or 'barrel shaped'). A lens of the right shape can also compensate for this.

20.7 The camera

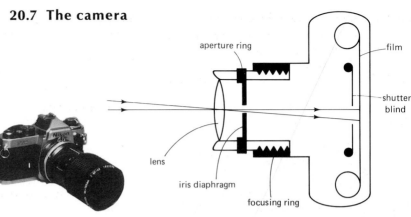

There are many different types of camera but all have a lot in common. A camera uses a convex lens to focus a real image on the film. A film is made of paper coated with chemicals which are sensitive to light. Normally it must be kept in total darkness. A shutter mechanism is used to allow light to reach the film when you want to take a photograph. The shutter in some cameras is a circular hole which is briefly uncovered; in others, it is a blind with a slit in it which moves rapidly across the film. On many cameras the shutter speed can be varied from several seconds to $\frac{1}{500}$ s or less. The film needs only a very short time to respond to the light. It is then treated with chemicals (developer) to make the picture appear. The image cast on the film will, of course, be upside down, but this doesn't matter as the developed photograph can easily be turned upright!

We may want to take photographs of objects or people close up, or of distant objects. We need to be able to adjust the lens so that either of these can be focused on the film. The lens is made of glass and cannot change its thickness, so the adjustment is made by moving the lens slightly closer to the film, or slightly further away. Usually the lens is mounted on a screw thread, and adjusted inwards or outwards by turning it.

We may also want to take photographs on a very bright sunny day, or in duller conditions. So cameras have an aperture control to adjust the amount of light reaching the film. This is a ring of sliding plates. The hole which lets the light through can be made either large or small. On more expensive cameras this is labelled with f-numbers. The larger the f-number, the smaller the aperture. $f/2.8$ means a large hole to let a lot of light in (for photographing indoors or in very dull weather); $f/22$ is a small hole for photographing on bright days.

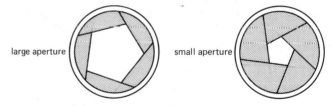

Of course, the shutter speed also provides a way of adjusting the amount of light reaching the film. So why do we need an aperture control as well? One reason is that it gives you more control over what your photograph will look like. Look at the two photographs on the left. In one, the girl is sharp but the background is blurred. In the other, both girl and background are sharp. The first photograph was taken with a large aperture and a faster shutter speed. The second used a smaller aperture and a slower shutter speed to let the same amount of light through. This one is said to have more **depth of field**. Sometimes we want one effect, sometimes the other. So we can choose.

With a large aperture, the lens has to be able to focus over a wide diameter without distortion. Lenses which can do this are usually made as combinations of several separate lenses. They are expensive to produce, and so good quality cameras are expensive.

Lenses and Optical Instruments 191

Comparing the eye and the camera

In many ways, the eye and the camera are similar. Both form a real image of objects on a screen. The table on the right lists the similarities and differences between the eye and a camera.

SLR camera

Table 20.1 The eye and the camera: similarities and differences

Eye	Camera
Convex lens system produces a real image on the retina	Convex lens system produces a real image on the film
Light falling on the retina stimulates cells to send messages to the brain	Light falling on the film causes chemical changes
Iris controls the amount of light entering	Aperture rings control the amount of light entering
Can focus on objects between least distance of distinct vision and infinity	Can take sharp photos of objects at any distance from a few centimetres up to infinity
To focus, the eye lens changes thickness	To focus, the lens-to-film distance is changed

Many keen photographers use a type of camera called a single-lens reflex (SLR) camera. The big advantage of an SLR camera is that when you look through the viewfinder to see what you are going to photograph, you are actually looking through the camera lens. Your eye sees exactly what the film will 'see'. As you adjust the lens on its screw thread, the view you see goes in and out of focus. You don't need to guess about what the depth of field will be with the aperture you have chosen – you can see for yourself. How does the SLR camera do this?

Inside the camera is a mirror normally set at 45°. This reflects light upwards into a special five-sized prism called a **pentaprism**. Total internal reflection in the pentaprism reflects the light into your eye. At the moment you press the shutter release, the mirror flips briefly out of the way, and a shutter blind moves across in front of the film. The photograph is taken.

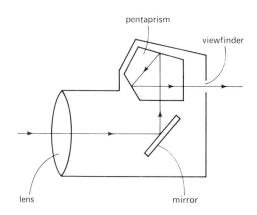

20.8 Slide projector

A projector for showing slides or a movie film uses a lens to produce a real image on a screen. In this case, the image is bigger than the object. The basic slide projector is really very simple. It has a light source and a single convex lens. The slide is between f and $2f$ for the lens, and so a magnified and inverted image is produced on the screen. So that we see the picture the right way up, the slide is placed in the projector upside down.

You could make a simple projector like this for yourself. If you did, you would find that the picture you project is much brighter at the centre than near the edges. A real projector improves this by using condenser lenses to concentrate more light on the slide. These are two plano-convex lenses (i.e. flat on one side and convex on the other) between the source and the slide. Another improvement is a concave mirror to reflect light back towards the slide. Concentrating the light like this means that the slide will get quite hot while being projected, so an energy absorbing filter is added. Even so, it is better not to keep a slide in a projector for too long.

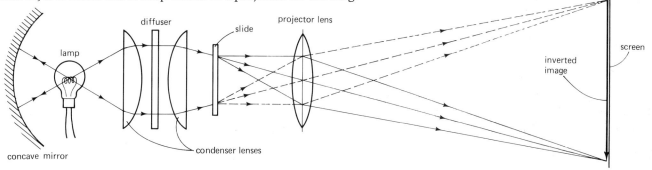

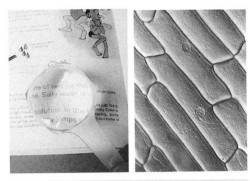

20.9 Magnifying glass and microscope

A microscope allows us to look at things which are too small to see clearly with the unaided eye – like the small print of a book, or the cells in a piece of onion skin.

A single convex lens, used as a magnifying glass, is the simplest form of microscope. The first microscope, invented around 1590 by a Dutch lens-maker, Anton van Leeuwenhoek, was made simply from a tiny polished grain of sand as the lens. Van Leeuwenhoek's enormous skill was his ability to polish something as tiny as a sand grain until it was a perfectly smooth sphere. It was then mounted in a tiny hole in a metal plate. The thing you wanted to observe was placed on a pin, and you looked through the tiny lens by putting your eye right up against the metal plate. With this, van Leeuwenhoek became the first man to see micro-organisms (microbes) in water.

Simple microscope

How does a single convex lens magnify things? To understand this, we need to remember what it is that makes objects look large or small. **An object looks large if it makes a large visual angle at the eye; it looks small if the visual angle is small** (see the diagram at the top of p. 189). To make something which is small appear larger, we need to increase the visual angle. One way to do this is to bring it closer, but there is a limit to this. When it is at the least distance of distinct vision, this is the largest visual angle possible.

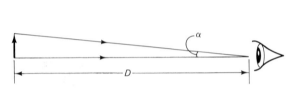
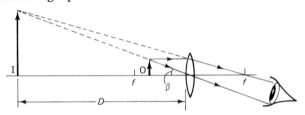

However, if we put the object in front of a convex lens (closer than the focus), an upright, magnified, virtual image is produced. Imagine your eye placed right behind the lens, looking through it at this virtual image. The visual angle of the rays now coming to your eye *from the image* is β. This is larger than the visual angle α without the lens. The object will *appear* larger. This is how the magnifying glass produces its magnifying effect.

Compound microscope

There is a limit to how much magnification you can get with a single lens. More powerful microscopes use two lenses.

1. The objective lens (convex) produces a magnified, real image of the small object.

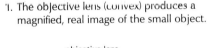

2. The eyepiece lens (also convex) is just a simple magnifying glass looking at this real image.

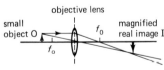

3. Putting these diagrams together:

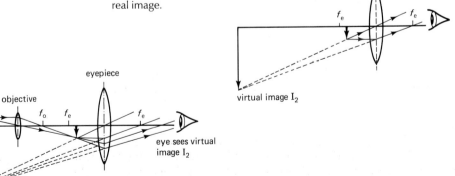

The final image is inverted. For the sorts of things we look at with a microscope this usually doesn't matter. But you will have noticed this inversion if you have ever tried to move a microscope slide whilst looking down the microscope.

20.10 Telescope

A telescope is also used to look at objects which appear small. But this time, they appear small not because they *are* small, but because they are far away. We cannot move the object – it has to be a long way from the telescope lens, well beyond $2f$.

The simplest telescopes are astronomical telescopes, used to look at the planets, stars, and galaxies. The final image is inverted, but this does not matter for astronomical work.

1. The objective lens (convex) of a telescope produces a real image of the distant object.

2. The eyepiece lens (convex) acts as a magnifying glass for looking at this small real image. It is usual to have the real image exactly at the focus of the eyepiece lens. This is called **normal adjustment**.

3. Putting the two diagrams together:

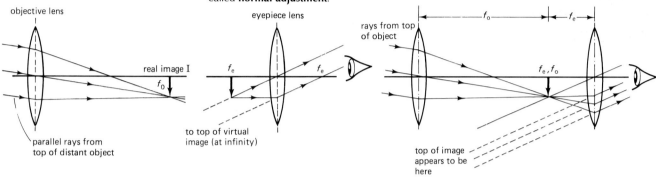

4. The telescope magnifies because the visual angle made by the final image is larger than the visual angle made by the object alone.

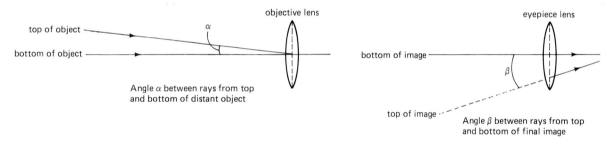

5. The magnification will be larger if the objective lens has a longer focal length. This is simply because the longer its focal length, the larger the real image of the distant object will be.

6. The eyepiece is a magnifying glass and so it magnifies more if it has a short focal length. So for high magnification we want an objective lens with long focal length, and an eyepiece with short focal length. The total length of the telescope (is $(f_o + f_e)$, so this puts a limit to how long we can make the focal length of the objective – otherwise we would have a very long telescope!

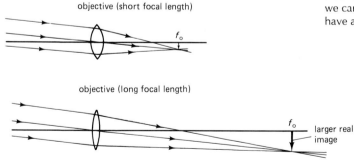

Binoculars

An astronomical telescope is not much use for looking at distant objects on Earth. It is long and cumbersome, and the image is upside down. Binoculars make clever use of two 45° reflecting prisms to shorten the instrument, and to turn the image right way up at the same time.

The prisms fold the light path back on itself. By having the second prism set at 90° to the plane of the first, the two reflections invert the image in both directions (upside-down and left-to-right – see page 177).

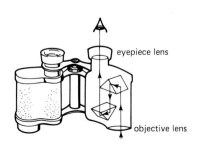

194 Lenses and Optical Instruments

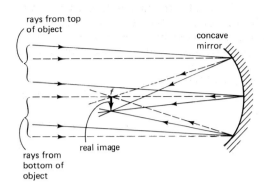

Reflecting telescopes

The largest **refracting** telescope in the world is at the Yerkes observatory in Wisconsin, USA. Its objective lens is 1 metre in diameter. To see very faint distant objects, a telescope needs to have a large diameter to collect as much light as possible. But large lenses are very heavy and difficult to mount; they are also difficult to make accurately. All the really large optical telescopes are **reflecting** telescopes. A concave mirror focuses the beams of parallel rays from the distant object to produce a real image. The eyepiece lens (convex) is used as a magnifying glass to look at this real image.

One problem is where to place the convex lens and the person using the telescope so as not to block the incoming light beam! Newton solved this difficulty and made the first reflecting telescope by using a small plane mirror, placed just before the focus of the concave mirror, to reflect the light beam sideways. The eyepiece and the astronomer can then be at the side of the telescope, clear of the incident beam.

The Newtonian reflecting telescope

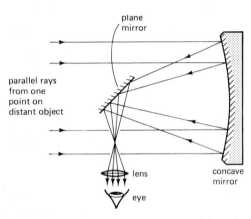

The concave mirror is the most important part of the reflecting telescope. It must be made with the correct smooth curvature over its entire surface. However, because nothing has to go behind the mirror, it can be supported here to take its great weight. The Royal Observatory's new William Herschel telescope on La Palma in the Canary Islands is a reflector with a mirror of diameter 419 cm. The site on La Palma is much better than any site in Britain because of the clear skies there, with fewer cloudy days and nights, and much less atmospheric pollution (dust and smoke). Even so, the atmosphere distorts light by refracting it slightly as it passes through, and so the idea of putting a telescope into orbit outside the Earth's atmosphere is an attractive one to astronomers.

Questions

1 Copy and complete these diagrams to show where the light rays go in the glass and after they have passed through.

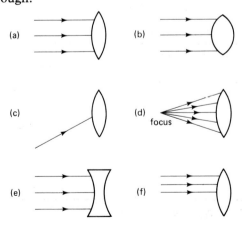

2 If you came across an old convex lens in a drawer at home, describe a method you could use to find out its focal length.

3 The diagrams show what happens when rays of light are affected by some objects. The objects have been hidden behind screens.

(a) Study the way the light is affected in each diagram and then draw on each of the diagrams what you think is behind the screen.

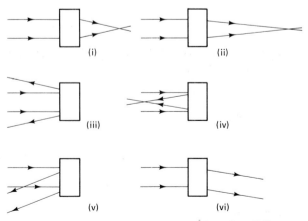

(b) Explain carefully the reason for the difference between diagrams (i) and (ii). (Sp.NEA)

4 A convex lens casts a sharp image on the screen. Someone now holds a piece of black card to cover half of the lens. What would you now expect to see on the screen? Explain your answer.

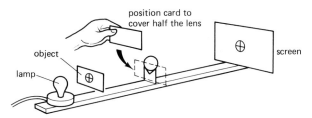

5 The diagram shows three light rays from the top of an object to a convex lens.

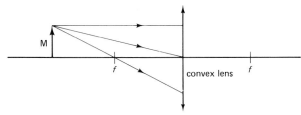

(a) Copy and complete the diagram by showing where the rays will go after they pass through the lens.
(b) The three rays should meet again at a point. What is at this point?
(c) On the same diagram, draw three rays which start from the **middle** of the object and pass through the same three points on the lens.

6 'As an object comes closer to a convex lens, its image moves further away from the lens and gets larger.' By drawing several ray diagrams, show that this statement is true, so long as the object doesn't come closer to the lens than its focus.

7 (a) Which position of the photographic enlarger would produce the larger photograph?

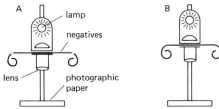

(b) Which position of the slide projector and screen will give the larger picture on the screen?

8 Imagine that the slide projector in diagram (D) of the previous question is adjusted so that the picture on the screen is sharply focused. To get a larger picture, the projector is then moved back (as in diagram (C)). Which of the following would you now have to do to focus the picture again?
(a) nothing – it will be in focus;
(b) move the lens outwards so that it is further from the slide;
(c) move the lens inwards so that it is closer to the slide.
Explain your answer.

9 (a) Copy and complete the entries in this table:

Part of the eye	Corresponding part of a camera	What the part does
Retina		
	Lens	
		Controls the amount of light getting in

(b) The eye and the camera have two rather different methods of adjusting their focusing so that either close or distant objects can be sharply focused. Explain what the two methods are and how they differ.

10 Copy and complete the table by writing either: **microscope**, **telescope**, or **both** beside each entry.

For looking at small objects	
For looking at distant objects	
Has two convex lenses	
The lens nearest the eye is strong (has a short focal length)	
The lens furthest from the eye is weak (has a long focal length)	
The final image is upside down	
The final image is virtual	
Forms an intermediate image	
The intermediate image is a real image	

INVESTIGATION

Is there any connection between the amount of magnification produced by a magnifying glass and the distance you hold it from the object you are looking at?

21: Waves

It may not be immediately obvious why we are going on from the study of light to look at waves. The connection will become clearer as we go on; we will see that light behaves as though it is a form of wave motion. Waves are worth studying in their own right. As well as the waves we see on water, studying waves helps us to understand light, radio, sound, and other phenomena.

21.1 Transverse waves

All waves have one thing in common: they transfer energy from one place to another. If we drop a stone into a pond, the kinetic energy of the stone makes the water surface move up and down near where the stone lands; ripples spread outwards and a cork floating on the water some distance away will start to bob up and down. The cork now has kinetic energy. The original energy of the moving stone has been transferred to the cork by the wave motion on the water surface.

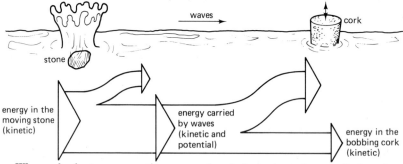

▽   We can look at wave motion more closely by using a long coiled spring. The spring is laid out on a long bench or on the floor. One end is held fixed and the other can be moved from side to side. If the free end is moved once from side to side, a single pulse goes down the spring. The pulse first stretches the coils at one end to one side. Each coil then pulls on its neighbouring coil moving it aside, and so on along the spring. As the coils move sideways, the 'springiness' of the stretched spring tends to pull them back into line. This makes a pulse pass along the spring. If the free end is moved continuously from side to side, a train of pulses travels along the spring, making a continuous wave. In practice, the pulses are reflected back from the fixed end and this makes the moving wave effect rather harder to see.

There are some useful things we *can* observe. If we mark one coil of the spring, we can watch it carefully as the wave pulses pass. The pulses travel *along* the spring, but the coil itself just moves from side to side. The movement of each coil is an **oscillation** – a movement to and fro. The direction of movement of the coils is at right-angles to the direction of movement of the wave. A wave like this is called a **transverse wave**. The spring is the **medium** which carries the wave. **In a transverse wave, the particles of the medium move at right angles to the direction of motion of the wave.**

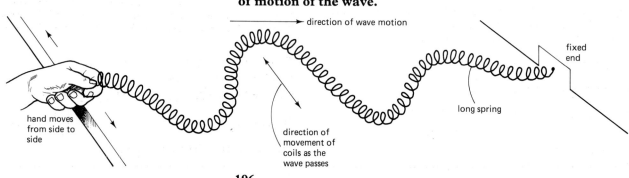

Some definitions

A number of terms are useful for describing waves:

1. The **amplitude** of a wave is its 'height' – the maximum distance which each point in the medium moves from its rest position as the wave passes (measured in metres, m, or millimetres, mm).

2. The **wavelength** is the length of one complete wave. For example, it is the distance from one wave crest to the next, or from one wave trough to the next (measured in metres, m, or millimetres, mm).

3. The number of complete waves leaving the wave source every second is called the **frequency** of the wave (measured in hertz, Hz). If the source produces 5 complete waves per second, the wave frequency is 5 Hz.

4. **Wavespeed** is the speed at which the waves move through the medium (measured in metres per second, m/s, or millimetres per second, mm/s). It is *not* the same thing as frequency. The frequency depends on the source; the wavespeed depends *only* on the medium the waves are travelling through. (For example, with the long spring, the tighter the spring is stretched, the faster the waves move.)

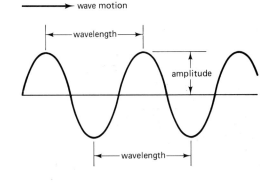

21.2 The wave equation

There is a connection between the wavelength and frequency of a wave and the speed at which it travels. Imagine a source which produces waves at a frequency of 5 Hz. That is, it produces 5 complete waves per second. These have a wavelength of 2 m in the medium they are travelling through. It is fairly clear that the wave will move forward by 5×2 m in one second. Its speed is 10 m/s.

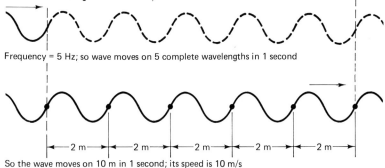

In general,

wavespeed = frequency × wavelength

This equation applies to all waves. In symbols, the wave equation is usually written:

$$v = f\lambda \quad (\lambda \text{ is the Greek letter 'lambda'.})$$

EXAMPLE 21.1

A pupil moves the end of a long spring from side to side 3 times per second. The wavelength of the wave on the spring is 0.5 m. With what speed do the waves move along the spring?

We use the wave equation: $v = f\lambda$

In this example: $f = 3$ Hz, $\lambda = 0.5$ m

Substituting into the equation: $v = 3 \text{ Hz} \times 0.5 \text{ m}$
$= 1.5 \text{ m/s}$

The wavespeed is 1.5 m/s.

EXAMPLE 21.2

A stone is dropped into a pond and waves travel outwards. By doing measurements we find that the speed of the waves is 0.8 m/s and their wavelength is 0.2 m. What is the frequency of these waves?

We use the wave equation: $v = f\lambda$

In this example: $v = 0.8$ m/s, $\lambda = 0.2$ m

Substituting into the equation: $0.8 \text{ m/s} = f \times 0.2 \text{ m}$

Divide both sides by 0.2 m: $f = \dfrac{0.8 \text{ m/s}}{0.2 \text{ m}} = 4 \text{ Hz}$

The frequency of the waves is 4 Hz.

21.3 Waves on water

Waves on a spring can move in only one direction. Transverse waves on a water surface can spread out in two dimensions. So we can find out more about how waves behave by studying water waves. In the laboratory, we use a ripple tank. A shallow tray of clear plastic holds the water, and a light above the water surface projects the wave patterns on to a sheet of paper on the bench below the tank. The legs of the tank can be adjusted to make the water surface exactly level. Plane waves are produced by a straight bar which hangs by two elastic bands from supports near one end of the tank. Circular waves are produced using dippers fixed to the bar. Waves can be produced by hand (by touching the hanging bar, or touching the water surface), or by a small motor mounted on the bar. The motor has a little weight fixed to its spindle in an off-centre position (an eccentric) so that when it runs, it makes the bar vibrate, producing a steady stream of waves.

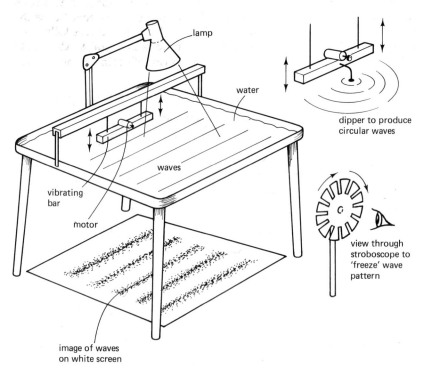

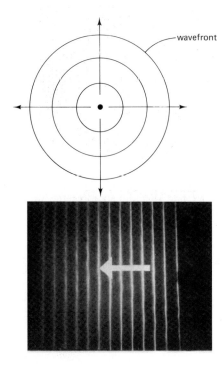

When a continuous stream of waves is used, it is sometimes easier to see what is going on by using a **stroboscope** to 'freeze' the wave pattern. A simple stroboscope (or strobe, for short) is a disc with slits. If you turn it as you look through it, you see the pattern only when a slit is exactly in front of your eye. Another type is the flashing light stroboscope, which has to be used in a darkened room. Again you see the pattern only when the light flashes on. Both allow you to see the wave pattern only intermittently. If the waves have moved by exactly one wavelength each time you see them, then the pattern will appear stationary.

Wave motion and wavefronts

◁ The first thing to try with a ripple tank is to see how waves travel on the water surface. If we touch the water surface in the middle of the tank, we see a pattern of circles spreading outwards.

One important result follows from this: the waves travel at the same speed in all directions. In the diagram, the circular lines represent the **wavefronts**. These lines show where the crests of the waves are at one particular instant. Waves always move at right angles to the wavefronts.

◁ The plane wavefronts in the photograph on the left are produced by touching the straight bar which dips into the ripple tank. The arrow shows the direction the waves are moving.

21.4 Wave properties

Four wave effects can be studied using the ripple tank. These are: **reflection**, **refraction**, **diffraction** and **interference**. We will look at each of these in turn.

Reflection

If a barrier is placed in the ripple tank, a circular spreading wave will be reflected from it. The incident waves come from the point **A**. The reflected waves look as if they are spreading out from the point **B**. Notice that **B** is the same distance behind the barrier as **A** is in front of it. This is just the same as the position of the image produced by light reflected in a plane mirror. ▷

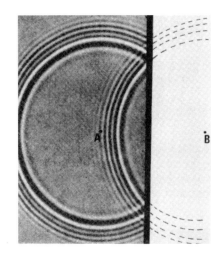

If plane waves strike a straight barrier, reflected plane waves are produced. This often shows up more clearly with single waves produced by touching the hanging bar, rather than with continuous waves from the motor. The angle at which the waves are reflected depends on the position of the barrier. By sketching the wave directions on the paper below the tank, we can measure these angles. We find that water waves follow the simple law of reflection: **the angle of incidence is equal to the angle of reflection**. At a straight barrier, water waves reflect in the same way as light at a mirror.

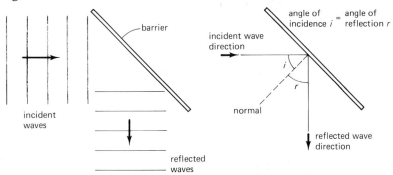

Plane waves are also reflected and focused by a concave barrier.

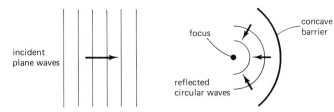

As with light, we can reverse this. If circular waves are generated at the focus, the reflected waves are plane waves. This is shown rather nicely in the photo. The circular waves in the lower photo are part of the original wave; the plane waves are the reflections from the curved barrier. ▷

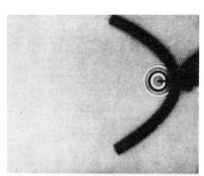
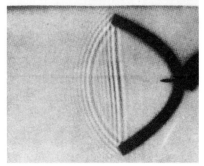

Refraction

Water waves travel more slowly in shallow water than in deeper water. The ripple tank has to be set up with a lot of care to show this effect. We can make a shallow region by placing a flat perspex plate on the bottom of the tank. The water over the perspex must be as shallow as possible, so that there is a big difference between the depths in the two different parts of the tank. So the tank must be very level, and the perspex shape must be cleaned with detergent to remove any grease.

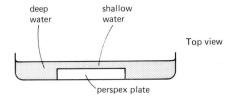
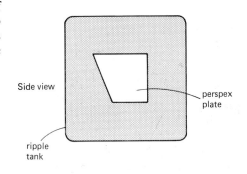

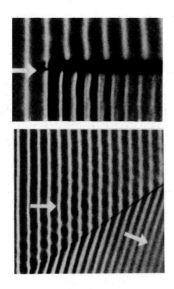

Plane waves in the shallow region are closer together. Their wavelength is less. The waves have 'bunched up' because they are slowed down as they enter the shallow region. The frequency of the waves in both regions is the same because frequency is determined by the wave source (the hanging bar and motor).

If the waves meet the boundary between deep and shallow water at an angle, then the direction of the waves changes. This is refraction, just like the bending of light as it crosses a boundary between two media.

If we think of what is happening to each wavefront, we can see why the bending occurs. For example, in diagram (A) below, one end of the wavefront enters the shallow region first. This end is slowed down. The remainder of the wavefront is still in the deep water and is travelling faster. The difference in speed across the wavefront causes the change in direction (see the drawing of the troops marching into the muddy field on p. 175). **When water waves travel from deep to shallow, they slow down and are bent towards the normal; when they travel from shallow to deep, they speed up and are bent away from the normal.**

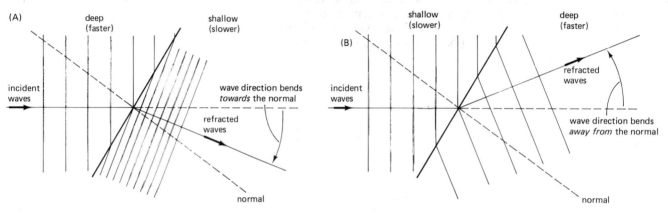

EXAMPLE 21.3

A perspex lens shape is placed in the bottom of a ripple tank. What will happen to plane wavefronts which pass over it?

Think first about what happens when the waves cross into the shallow region. The middle part of the wavefront enters the shallow water first. It is slowed down, whilst the outer ends move faster. The wavefront bends.

At the second boundary, the outer ends of the curved wavefronts emerge first into the deeper water. They are speeded up first. This makes the wavefronts curve even more. Remember that the direction of motion of the wave is at right angles to the wavefronts. The waves move inwards to a focus on the right hand side. The perspex shape acts as a lens, focusing the water waves.

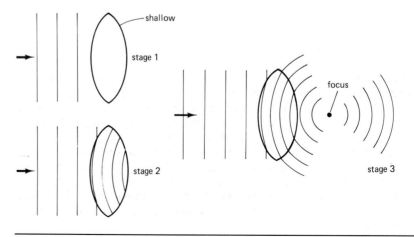

EXAMPLE 21.4

A ripple tank is placed on a slope, so that the water becomes gradually shallower towards one end. What will the wave pattern on the surface look like?

As the water gets gradually shallower, the wave speed gets gradually slower. The wavelength will also get steadily smaller towards the shallow end of the tank.

One result of this is that there is almost no reflection from a gradually shallowing 'beach'. This is why the sides of a ripple tank tray have a sloping region – to prevent reflections from the sides which would make the waves harder to see.

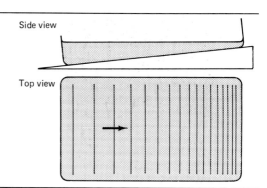

Diffraction

When plane water waves hit a short barrier, the 'shadow' region behind the barrier does not have perfectly sharp edges. The waves tend to bend around the obstacle a little. Indeed if the barrier is very short (about one wavelength of the waves), there is almost no shadow region at all. The waves pass the barrier as though it wasn't there. The bending of waves at the edges of obstacles is called **diffraction**. It is completely different from bending by refraction. One important difference is that the wavelength doesn't change during diffraction.

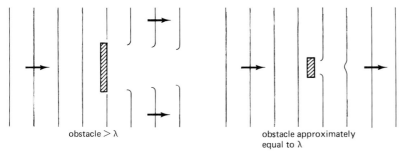

Diffraction can also be seen when waves pass through gaps between obstacles. If the waves pass through a gap which is long compared with their wavelength, the emerging waves are bent only at the edges. If the gap is similar in width to the wavelength of the waves, the emerging wavefronts are semi-circular.

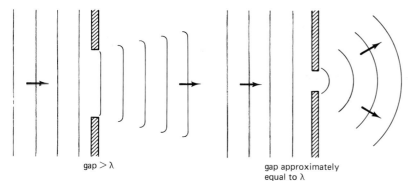

The photo on p. 196 shows diffraction of water waves as they enter the narrow gap of a harbour mouth. Why do waves bend around obstacles in this way? This question cannot really be answered – diffracting round corners is just something which waves do! The Dutch physicist, Christian Huygens (1629–1695) suggested that it is useful to think of the wave moving in stages – the first wavefront behaves like a new 'source' which then causes the next ripple, and so on. Looking at it like this, when the first wavefront reaches the gap, we have a 'source' which is just the line across the gap. Waves will spread out from this 'source' in all directions. So there is bending at the edges. You may find this a useful way to think of diffraction.

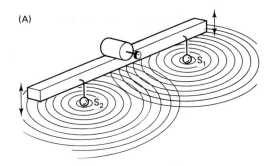

Interference

When two separate sets of waves meet in the ripple tank, their effects add together. If both waves have the same frequency, they combine in a particularly interesting way. The result is called **interference**.

Pictures (A) and (B) show what we see in a ripple tank when two circular waves are produced by two dippers operated from the same motor. This is known as an **interference pattern**. Along some lines the waves cancel each other out and there is calm water (**destructive interference**). Along other lines the waves add to produce larger ripples (**constructive interference**).

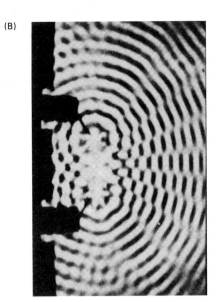

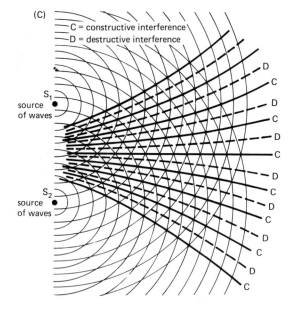

If we look carefully at diagram (C) above, we can see how interference occurs. Look at any point on a line of constructive interference. The two waves reaching this point from the two dippers are in step. When they add, they produce a larger wave. Now look at any point on a line of destructive interference. Here the two waves are exactly out of step. When they add, they cancel each other.

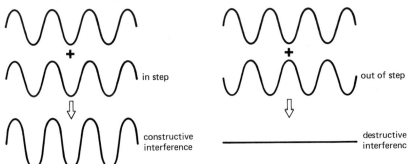

Of course, diagram (C) only shows the wavefronts at one instant. However, we see the interference pattern because it stays steady in the tank. If both waves have the same frequency (and hence, the same wavelength), the waves will *always* be in step at points along the solid lines; and they will *always* be exactly out of step at points along the dotted lines. Interference patterns are produced only when the two waves have the same frequency.

This is the reason why we use two dippers driven by one motor. Two separate dippers with two motors wouldn't work – it is impossible to adjust them accurately enough to make their speeds identical. Another way to produce an interference pattern is to pass a plane wave through two small gaps between barriers. The wave diffracts at both gaps and the two semi-circular waves interfere.

21.5 What IS a wave?

If we are thinking of water waves, this may seem a silly question. We can **see** what a wave is. But we also talk of light waves, radio waves, sound waves. It is less obvious why these are called 'waves'. What is really distinctive about a wave?

Of the four wave properties, reflection and refraction are not special wave properties. A billiard ball hitting a cushion will be reflected and the angle of reflection will be equal to the angle of incidence.

Particles can be refracted too. One example is a stream of marbles rolling at an angle down a slope which suddenly gets steeper. Seen from above, the marbles will change direction – they will bend towards the normal.

It is difficult, however, to imagine particles showing diffraction or interference. These two properties seem to apply only to waves. They are our evidence for wave motion. If we can produce diffraction and interference with whatever we are investigating, then we can deduce that it is some sort of wave.

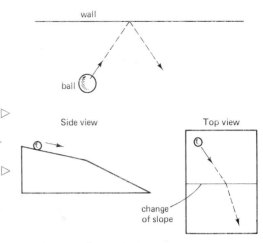

Particles *can* reflect and refract, but only waves can diffract or interfere.

21.6 Light waves

Ever since the time of the ancient Greeks, people have wondered about what light is, and about how we see. By the 18th century, some scientists thought that light was made of particles which were emitted by the light source; others thought light was a form of wave. Newton believed that light was a stream of particles. He thought that refraction in glass was due to the particles speeding up – like the marbles mentioned above. Newton was such a famous and well respected scientist that his view was accepted by many people. Then in 1801, Thomas Young did a famous experiment in which he produced an interference pattern with light.

What would an interference pattern with light look like? If we look back at diagram (C) on the page opposite we can see that if light from two sources interferes, this will produce lines of constructive interference (brighter light), and lines of destructive interference (darkness). If the light falls on a screen, we would expect to see alternate patches of bright – dark – bright – dark – and so on.

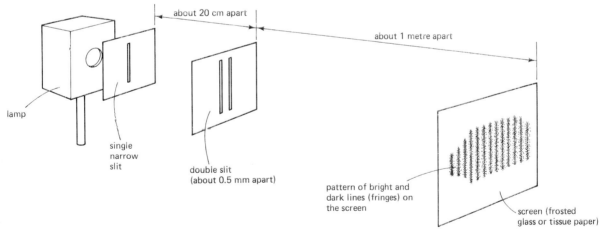

Young used two narrow slits in a slide lit by a single lamp as his two sources of light. In this way, he made sure that both sources had exactly the same frequency and were in step. The diagram above shows what he saw on the screen. The bright and dark bands on the screen are known as **interference fringes**.

For ideal results in Young's experiment, we need a light source which produces light of a single frequency (or wavelength). No light source can do this, though a laser or a sodium lamp comes close enough.

Young's experiment is strong evidence that light behaves as a wave. Only waves can show interference; if light was a stream of particles it would be very difficult to explain how we get an interference pattern.

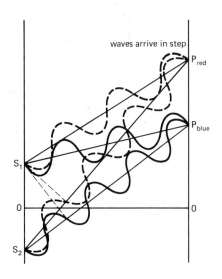

waves arrive in step

21.7 Wavelength and colour

We can obtain another important and useful result from Young's double-slit experiment by using different colours of light. If a white light source is used with a red and a blue filter in turn, we find that the red fringe pattern is more widely spaced than the blue one.

This suggests that the red light may have a longer wavelength than blue light. To see how this explains the different separations, look at the diagram. Imagine two waves of red light leaving the two slits. They are exactly in step when they leave the slits. There will be a bright patch (constructive interference) at the mid-point of the screen because the two waves are in step here. If we move upwards along the screen, the *next* bright patch will occur when the two waves are in step again. This happens at P_{red}. With blue light, the wavelength is less, and so we don't have to go as far up the screen to find the first point where the blue waves are first in step, P_{blue}.

Young's experiment can be used to *measure* the wavelength of light. When we do this we find that red light has a wavelength of 0.000 7 mm or 0.000 000 7 m (7×10^{-7} m). About 1400 of these waves would fit into 1 mm! The wavelength of blue light is even shorter, about 0.000 4 mm or 0.000 000 4 m (4×10^{-7} m). The other colours of the spectrum have wavelengths between red and blue. So although we often refer to light of a particular **colour**, it is really **wavelength** which determines the colour we see. Colour is an effect produced by our eye and brain; light of different wavelengths produces different effects which we see as different colours.

Light of just one wavelength, and therefore of a single colour, is called **monochromatic** light. No real light source can produce perfectly monochromatic light. Even a laser produces light with a group of similar wavelengths.

21.8 What IS light?

Young's experiment is strong evidence that light behaves as a wave. But what sort of a wave is it? Another piece of evidence comes from experiments with polaroid film – the sort that is used for polaroid sunglasses. If we look through two pieces of polaroid, we can alter the amount of light getting through by rotating one of the polaroids. In one position, the light is quite bright; turn one polaroid through 90° and almost no light comes through!

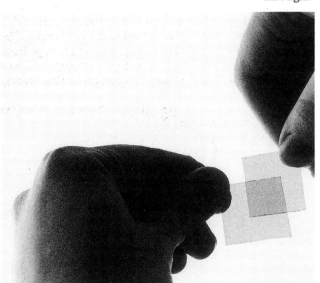

One explanation is that light is a **transverse** wave of some kind, and the polaroid allows waves through only if they are vibrating in one particular direction. If both polaroids are in line, then light gets through. If the polaroid directions are crossed, no light gets through.

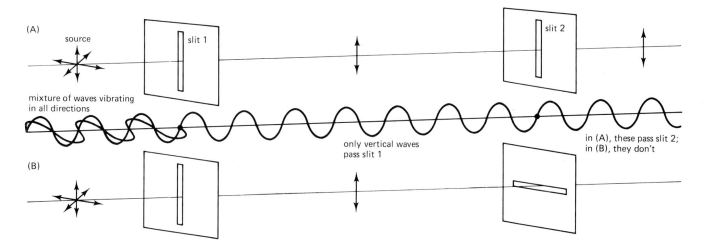

So we have evidence that light is a transverse wave. But what is the medium that this wave travels in? To make a transverse wave, we usually need a 'springy' medium of some kind; yet we know that light can travel through a vacuum (for example, between the Sun and the Earth).

There is no easy answer to this. We believe that light is an **electromagnetic wave**. It consists of a vibrating electric field and magnetic field moving along together. We have discussed fields in Chapter 1, and we will look in more detail at electric and magnetic fields in Chapters 24 and 29. A wave like this can travel in a vacuum. Indeed vacuum *is* the medium for an electromagnetic wave! Perhaps we are wrong to think of a vacuum as containing *nothing*. There are **fields** in the vacuum, and these can transmit the effects of gravity, magnetism and electrostatics. A vacuum is not empty – it contains fields. And so it can allow an electromagnetic wave to travel through it.

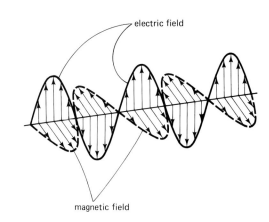

Questions

1 The diagram shows a wave travelling along an elastic cord (real size).

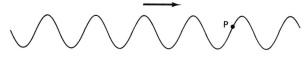

(a) Measure the wavelength of the waves.
(b) Measure the amplitude of the waves.
(c) The waves are moving from left to right. In which direction does the knot at **P** move?

2 A cork is floating on the surface of a pond. Someone throws a stone into the pond and it lands at **X**.

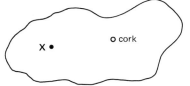

(a) Draw a diagram showing the shape and direction of motion of the waves made by the stone.
(b) In which direction will the cork move?

3 (a) Describe an experiment you could set up to demonstrate that water waves carry energy.
(b) One idea for getting power from the waves is the Salter's duck. The duck has an axle running through it and its 'nose' bobs up and down as the waves pass. The duck shape is particularly good at taking energy from the waves. If you watched a duck in operation, how could you tell that it was good at extracting energy from the waves?
(Hint: what would you notice about the waves on either side of the duck?)

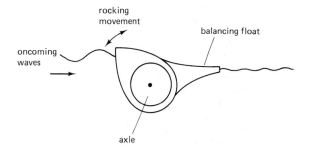

4 A ripple tank vibrator produces 8 waves every second. The wavelength of the waves in the tank is 6 cm. What is the speed of the waves?

5 What is the wavelength of a water wave of frequency 15 Hz, travelling at 60 cm/s?

6 A wave on a string travels at 2.4 m/s. Its wavelength is 0.6 m. What is the frequency of the wave?

7 Waves on a pond travel 1 m (100 cm) in 2.5 s. The wavelength of the waves is 4 cm.
(a) What is the wave speed?
(b) What is the frequency of the waves?

8 Copy and complete these diagrams, showing the behaviour of waves in a ripple tank.

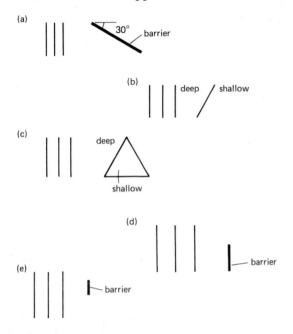

9 Draw diagrams to show two different ways of changing plane waves into circular waves.

10 The dipper S_1 is operated by a small motor.

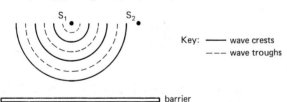

(a) Copy and complete the diagram showing more incident waves and some of the reflected waves from the barrier.
(b) Mark, with an **x**, three places where there will be constructive interference. Mark, with an **o**, three places where there will be destructive interference.
(c) The barrier is removed and a second dipper operated by a second motor is started at S_2. Will this produce an interference pattern? Explain your answer.

11 (a) A barrier is placed in a ripple tank in which plane waves are being generated.

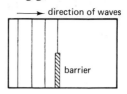

Copy the diagram and complete it to show the pattern of waves produced in the right hand side of the tank.
(b) The wavelength of the waves is reduced.

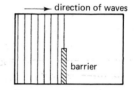

Copy and complete this diagram.
(c) A range of hills lies between a radio and television transmitting station and a house.

The house lies well within the normal reception range of both the radio and television transmissions from this station. Explain why a television set in the house cannot receive television transmissions from the station while a radio set can obtain good reception.
(SEB)

12 (a) Newton believed that light was a stream of particles emitted from a source. Now we think that light travels as a wave. Write a short paragraph explaining to someone who didn't know any physics why we think light is a wave and giving some evidence.
(b) After reading your explanation, they write back and say that light cannot be a wave because it can travel through a vacuum 'where there is nothing that can wave'. How would you answer this objection?

INVESTIGATION

Take four equal lengths of sewing thread, thin string, thick string and rope. Knot them together and investigate what happens to transverse waves at the knots. Are they reflected or do they continue on? Does it matter which direction the wave is travelling? Make careful notes of what you observe, and look for patterns in your results.

22: The Electromagnetic Spectrum

Light is just one of a group of electromagnetic waves, which together make up the electromagnetic spectrum. Other electromagnetic waves have important uses and applications. Radio waves, infra-red radiation ('heat' radiation), ultra-violet radiation and X rays are also members of the electromagnetic 'family'.

22.1 Beyond the visible spectrum

Chapter 21 ends with the idea that light is an electromagnetic wave – a vibration of electric and magnetic fields. We have also seen that white light is a mixture of lights of different colours – different wavelengths. The spectrum of white light stretches from red light of wavelength 0.000 7 mm to blue light of wavelength 0.000 4 mm. The human eye is sensitive to any electromagnetic waves with wavelengths in this range. But there are also electromagnetic waves with longer and shorter wavelengths, which we cannot see. The diagram on the right shows a very simple circuit with a light-dependent resistor. When the resistor is in the dark, there is almost no current. In the light, the resistance of the light-dependent resistor falls and there *is* a current. The interesting result is that if the light-dependent resistor is placed in the dark region just beyond the red end of the spectrum, we get a current. Some kind of radiation which we cannot see must be reaching the light-dependent resistor.

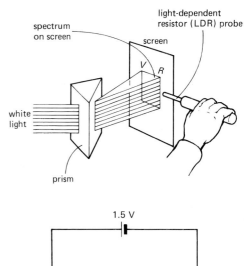

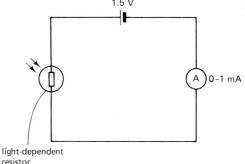

22.2 The electromagnetic spectrum

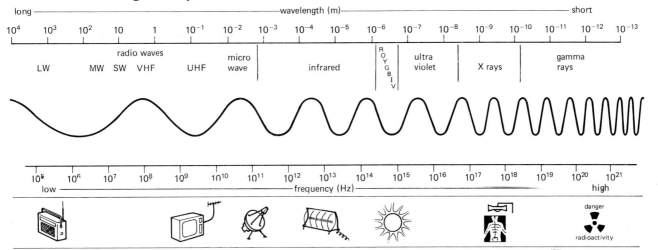

The diagram above shows all the members of the electromagnetic family of waves. All of these waves have some features in common:

1. They all transfer energy from one place to another.
2. They are all transverse, electromagnetic waves.
3. They can all travel through a vacuum.
4. They all travel at a speed of approximately 300 000 kilometres per second, or 300 000 000 metres per second. (3×10^8 m/s). This speed is almost too fast to imagine. According to Einstein's theory of relativity, it is the fastest speed possible; nothing can travel any faster than the speed of light in vacuum!
5. They all show the wave properties: reflection, refraction, diffraction and interference.
6. They all obey the wave equation $v = f\lambda$.

It is from properties 5 and 6 that we deduce that all the members of the electromagnetic spectrum *are* waves. The different types of electromagnetic radiation have very different properties because of their differences of wavelength. This ranges from many metres (in the case of some radio waves) to less than a picometre – a million millionth of a metre (in the case of some gamma rays). Notice that as the wavelength gets less, the frequency gets greater, since the speed ($v = f\lambda$) is the the same for all.

The electromagnetic spectrum is a continuous spectrum, with no gaps. The diagram above shows sharp boundaries between the different types of radiation, but in fact these really merge gradually into each other.

208 The Electromagnetic Spectrum

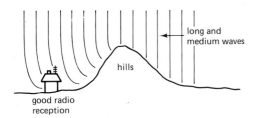

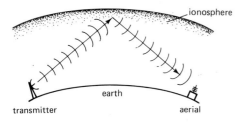

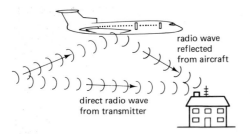

22.3 Types of electromagnetic radiation

Radio waves

◁ Radio waves have the longest wavelengths in the electromagnetic spectrum, ranging from several hundred metres (long wave, LW) to a few centimetres (UHF). They are used in radio and television to transmit sound and picture information over long distances. Radio waves are produced by oscillating electric currents in a transmitting aerial (or antenna). They are picked up by another aerial at the receiving end. Electromagnetic waves travel in straight lines, but long and medium wave radio transmissions will diffract round obstacles like large buildings and hills, so radio reception on long and medium wave is quite good in hilly areas. They also diffract round the Earth's curvature and are
◁ reflected off a layer of charged particles in the upper atmosphere, called the **ionosphere**. As a result, good reception of long and medium wave signals is possible a long way from the transmitter.

VHF (very high frequency) waves (used for stereo radio) and UHF (ultra high frequency) waves (used for television) have shorter wavelengths than LW and MW. They do not diffract as easily and so, for good reception, a much better aerial is needed. It also helps to be closer to the transmitting aerial and to have a fairly clear path to it.

The diffraction of radio waves is part of the evidence that they *are* waves. Radio waves can also interfere. People who live near an airport may notice 'interference' on their radio and TV reception when a low-
◁ flying aircraft is nearby. This is caused by interference between two radio waves: the direct wave from the transmitter to their aerial, and the wave reflected off the aircraft's metal body.

Microwaves

Microwaves are very similar to UHF radio waves. They have wavelengths of a few centimetres. Microwaves are produced by electrical oscillations in special valves. They are used for satellite communication.
◁ Some telephone links now use satellite microwave stages for part of the connection. Although the satellite is quite a long way away, the signal can travel in a straight line to the receiver. This is a parabolic dish which collects the weak radiation and focuses it on the detector. Microwave transmitters are also dish-shaped with the transmitter at the focus.

Nowadays perhaps the commonest use of microwaves is in microwave ovens. These work because microwaves of a particular wavelength are very strongly absorbed by water. The energy carried by the microwaves heats the water up. As almost all foods contain a lot of water, they can be heated and cooked very quickly in a microwave oven. Because it works in this way, the microwaves heat the food throughout its volume, not just from the outside as in a normal oven. This is why microwave ovens are so good for defrosting frozen food, and also why they can cook food so quickly.

◁ Microwaves can be investigated in the laboratory. 3 cm microwaves from a transmitter are reflected by a metal plate, and refracted by a large prism filled with paraffin. The real evidence for their wave nature comes from experiments to show diffraction and interference. When 3 cm microwaves pass through a narrow gap between two metal plates, they can be detected in all directions beyond the gap. The microwaves are diffracted by the narrow slit. With two gaps, an interference pattern is produced. The detector shown in the lower photograph is connected to an amplifier and loudspeaker. Together they convert the energy of the microwaves into sound. As the detector probe is moved across behind the plates, the loudness of the sound varies up and down. This is due to interference of the microwaves (not of sound!).

Infra-red radiation

The name 'infra-red' means 'beyond red'. These are waves just beyond the red end of the spectrum. The diagram at the start of this chapter has already shown one way to detect infra-red radiation.

In fact, *all* objects emit infra-red radiation. Hot bodies emit more infra-red radiation than colder ones, and of different wavelengths. The higher the temperature, the shorter the wavelengths in the infra-red radiation. The graph below shows this. At relatively low temperatures, all the radiation is infra-red. At 800°C, the main part of the radiation is still infra-red, but a little bit of visible light is also present, from the longer wavelength end of the visible spectrum – the red end. The hot object is now 'red hot'. It glows a dull red colour. If we heat it further, its colour will become orange, then yellow, and eventually white. By around 1200 °C, it is said to be 'white hot'.

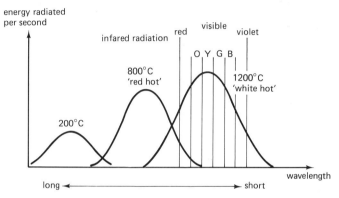

When objects absorb infra-red radiation, they become hotter. Infra-red 'heat' lamps have a hot filament which emits some visible light, and a large amount of infra-red radiation. The radiation coming from them feels warm.

Our eyes are sensitive only to the visible part of the spectrum and not to the infra-red region. But an infra-red camera can be used to 'see' the infra-red radiation from an object. The hottest parts of the picture emit more infra-red radiation, and of a shorter wavelength. This is converted by the camera into bright and dark areas, or sometimes into 'false colour' to show up the detail more clearly.

This infra-red photograph lets us 'see' the energy losses from a house. The hot spots are the light areas: the windows and the glass door panel. A lot of energy is being lost through these. Is the roof well insulated?

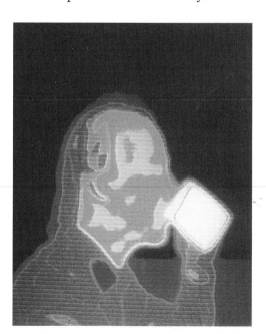

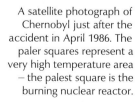

Having a hot drink. The hot liquid shows up as a light area in the infra-red photograph. Photographs like this can also be used in medical diagnosis.

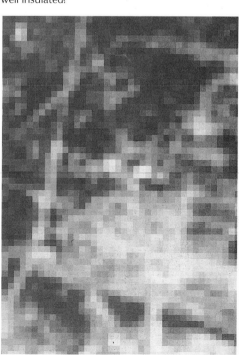

A satellite photograph of Chernobyl just after the accident in April 1986. The paler squares represent a very high temperature area – the palest square is the burning nuclear reactor.

We will take a further look at infra-red radiation and its properties in section 22.4.

210 The Electromagnetic Spectrum

The glass in this helmet protects the welder's eyes from ultra-violet radiation

Ultra-violet

There is also radiation beyond the violet end of the visible spectrum, called ultra-violet radiation. Although we cannot see it with the eye, this radiation can affect photographic paper and cause chemical changes. It is produced by any object heated to a very high temperature ('white hot'). The Sun emits ultra-violet radiation, which causes sun-tanning when it causes changes in some chemicals in the skin. Too much ultra-violet radiation is harmful and can damage the retina of the eye, so it is always unwise to look at an ultra-violet source. Glass stops most ultraviolet radiation; this is why you cannot get a suntan by sunbathing inside. Special protective glass can be used if more complete protection from ultra-violet radiation is needed.

Some chemicals **fluoresce** when ultra-violet radiation falls on them; they absorb the ultra-violet and emit visible light which makes them glow. Fluorescent paints and dyes work in this way. Some washing powders contain chemicals which fluoresce. In sunlight, these give your white clothes a 'whiter than white' appearance – or, at least, that is the idea! Fluorescent chemicals are also used to coat the inside of fluorescent light tubes (strip lights). An electric current in the gas inside the tube causes it to emit ultraviolet radiation. This is then absorbed by the chemical coating and visible white light is emitted instead.

X rays and gamma rays

X rays and gamma rays both have very short wavelengths. There isn't a single wavelength where X rays end and gamma rays begin. The difference is where the rays come from. X rays are produced in X-ray machines by making a beam of electrons collide at high speed with a metal target. The very sudden acceleration of the electrons (a negative acceleration – their velocity is suddenly reduced to zero) results in very short wavelength electromagnetic waves. They were first discovered by Wilhelm Röntgen in 1896.

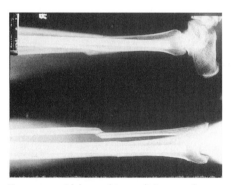

X rays are widely used in medicine to take photographs of the inside of the body. As X rays pass through the body, bone absorbs more than flesh. A 'shadow photograph' of the bones is produced on the photographic film. This can be a great help in diagnosing and setting fractures.

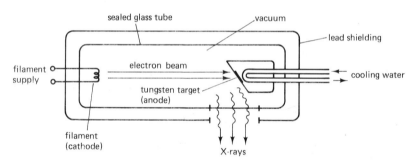

Gamma rays are emitted from some radioactive materials (see Chapter 33). They are bursts of very short wavelength radiation emitted continually by the atoms of these materials.

Both types of radiation have similar properties. Both are able to pass through many solid materials, including metals, with very little being absorbed. Both affect photographic plates and this is one method of detecting them. Both are also examples of **ionising radiation**, that is, they cause changes (ionisation) in the materials they pass through. For this reason X rays and gamma rays are harmful to humans, and we need to take careful precautions in using them. We will discuss ionisation in more detail in Chapter 24.

In taking medical X-ray photographs, the radiation dose to the patient must be kept as low as possible. The radiographer who operates the equipment is in particular need of protection because she or he is working with X rays every day. Shields of lead and concrete are used, as these materials are both good absorbers of X rays.

The uses and properties of gamma rays and the precautions needed when working with ionising radiation are discussed more fully in Chapter 33.

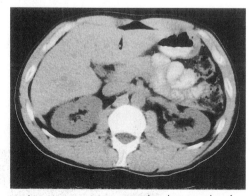

Modern X-ray scanners can take photographs of soft tissue inside the body also, helping in all kinds of medical diagnosis. A computer is used to produce the detailed image shown in this photograph from a CT scanner.

22.4 Radiation and heating

When electromagnetic radiation of any sort is absorbed by an object, it gets hotter. The energy of the electromagnetic wave makes the atoms of the material vibrate more vigorously, and its temperature rises. All electromagnetic waves do this, but infra-red radiation is best known for the heating effect it produces. All objects emit infra-red radiation, and all objects absorb it. Some materials, however, are better emitters and absorbers than others.

Good and bad absorbers

For absorbing radiation, it is the *surface* rather than the *material* the object is made of which is important. A quick way to make a rough comparison is to use a radiant heater between two metal plates with different surfaces. A drawing pin is fixed with wax to the back of each plate. The drawing pin attached to the plate with the black inner surface falls off first. ▷

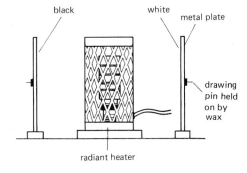

In general, matt black surfaces are the best absorbers of radiation, and shiny silvered surfaces are worst. White surfaces also absorb little radiation. For this reason, black cars are likely to get hotter than white cars when sitting in the sun. White clothes are cooler than dark clothes on sunny days.

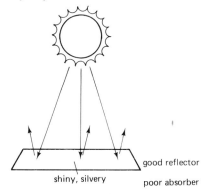
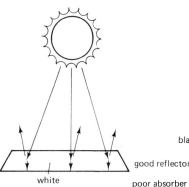
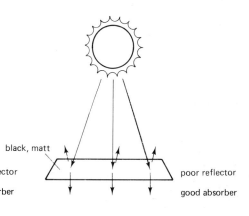

Good and bad emitters

Like absorption, emission depends on the surface of the emitter. The drawing shows a method of making a quick comparison between two surfaces.

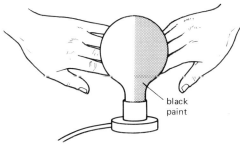

One side of the light bulb is painted matt black, the other is left white. If you hold the backs of your hands the same distance away from the two halves of the bulb, you should feel the hand nearer the black side getting hot quicker. The black side is emitting more infra-red radiation than the white.

So good absorbers are also good emitters. Black surfaces both absorb and emit a lot of radiation; shiny silvered surfaces absorb and emit much less. The cooling fins on motor-cycle engines are often painted matt black so that they emit more radiation and cool the engine more effectively. You may also have noticed black cooling fins attached to some electronic components. These are used where particular components are liable to heat up to too high a temperature. The fins keep the component cooler by radiating some of the energy away. ▷

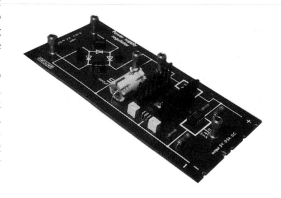

The greenhouse effect

Greenhouses help plants to grow by providing a warmer environment, and higher air temperatures. In winter, a greenhouse heater may be required, but in summer they can trap enough of the Sun's radiation to keep very warm inside.

The Sun's radiation contains a wide range of wavelengths of the electromagnetic spectrum, from infra-red right through to ultraviolet. Glass allows visible light to pass through it, and also the shorter wavelength infra-red, which is close to the red end of the visible spectrum. This is absorbed by plants and the soil inside the greenhouse, and heats them up. They are however much cooler than the Sun, and the infra-red radiation which they emit is at much longer wavelengths (look back at the diagram on p. 209). This radiation cannot pass through glass. The energy has been trapped inside the greenhouse. This is called the **greenhouse effect**.

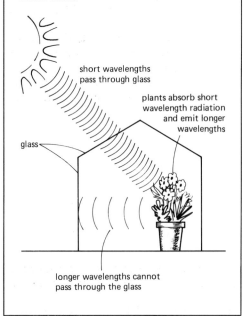

Collecting the energy carried by the Sun's radiation

Almost 85% of the energy we use in our homes is for heating the rooms and providing a supply of hot water. As fuels become scarcer and more expensive, people are turning to 'solar energy' as a way of heating water. One method is to use a solar panel on the roof. This has a black metal plate to absorb the Sun's radiation. Inside the plate, or fastened to it, is a circuit of pipes filled with water. As the plate heats up, the temperature of the water rises. This water is circulated by a small pump, and used to heat up the domestic hot water.

Notice that the collector plate is covered with a sheet of glass or clear plastic. This makes the solar panel into a 'mini-greenhouse' – it uses the greenhouse effect to increase the temperature of the air over the collector plate.

You may think we don't get much sun in Britain, but if every house had around 5 m² of solar panel on the roof, this could supply about half of the hot water required. Of course, one disadvantage is that we would get most hot water on hot days when we may not need it so much. This is why a method of storing the energy collected is so useful (see section 17.2).

Another way to collect the Sun's radiation in countries with a lot of direct sunlight is to use a parabolic reflector to focus the radiation at one spot – producing a very high temperature there. This has been described in section 18.8.

The vacuum flask: stopping energy losses by conduction, convection and radiation

A vacuum flask reduces energy losses from a hot drink in several ways:

1. The plastic stopper prevents convection.
2. The stopper and the plastic mounts are made of materials which are poor conductors. This reduces energy losses.
3. The flask itself is made of a double-walled glass bottle. Between the walls is a partial vacuum. Conduction and convection need a medium and so a vacuum prevents these losses.
4. Both glass surfaces are silvered on the inside. Shiny silvered surfaces reflect infrared radiation, and so the radiation from the hot drink will be partly reflected back into the flask. This slows down radiation losses.

Of course, the same features are just as good at preventing conduction, convection and radiation in the opposite direction also. So the flask is equally good at keeping cold drinks cool.

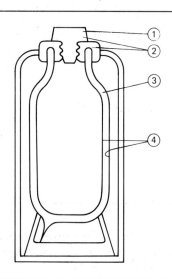

Questions

1 Use the information in Chapter 22 to complete a table with the following headings.

Type of radiation	How it is produced	How it is detected	What it is used for

2 (a) What types of radiation are **A** and **B**?

Radio	Microwaves	A	Visible	Ultra violet	B	Gamma rays

(b) Which type of radiation is used for cooking food rapidly? Give one more application for this type of radiation.
(c) Which type of radiation has a shorter wavelength than type **B**? How is this radiation produced? What can be used to detect it?

3 All electromagnetic waves travel at a speed of 300 000 km/s in vacuum (or in air).
(a) What is the time taken for a radio signal to travel from Britain to New Zealand (18 000 km)?
(b) How long does it take for a radio message to travel from Earth to an astronaut on the Moon (320 000 km)?
(c) How long does it take for light to reach Earth from the Sun (150 000 000 km)?
(d) When the space probe Voyager 2 sent back radio signals as it passed close to the planet Uranus, it was then about 2 700 000 000 km from Earth. How long would these radio signals take to travel from Voyager to ground control?
(e) The nearest star is around 4 light years away, i.e. the distance light travels in 4 years. How many kilometres is this?

4 A spectrum of white light is allowed to fall on a sheet of photographic paper in a darkened room. The position of the spectrum is marked in diagram (A). When the photographic paper is developed, a black area appears as shown in diagram (B).

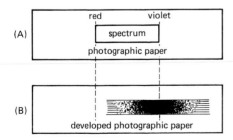

(a) What conclusion can be drawn from the fact that the blackened area on the developed photographic paper is much longer than the area of the spectrum?
(b) Explain how this experiment illustrates that it is quite safe to develop this photographic paper in a room illuminated only by red light. (SEB)

5 (a) In an experiment to show interference of microwaves, the transmitter is placed behind aluminium screens with two narrow gaps between them. A reading on the meter shows that microwaves are reaching the detector.

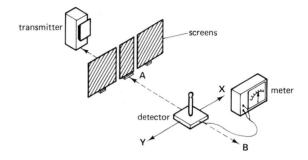

The detector is moved along the line **AB** and **constructive interference** is found. When moved along the line **XY**, a series of positions where **constructive** and **destructive interference** occurs is detected.
(i) Explain what is meant by each of the terms: 'constructive interference' and 'destructive interference'.
(ii) Describe how the meter reading would show these forms of interference when the detector is moved along **AB** and then along **XY**.
(b) The diagram shows part of a coastline with two land-based radio navigation stations **A** and **B** which are a large distance apart.

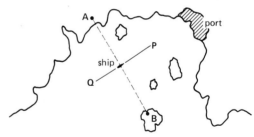

Both stations continuously transmit radio signals with the same amplitude and the same frequency of 1.5 MHz.
(i) Calculate the wavelength of the radio signals used.
(ii) The ship is heading for the port along route **QP**. When it is at a position exactly midway between **A** and **B**, it detects a signal whose amplitude is twice that of the signal from either station alone. What does this suggest about the signals transmitted from **A** and **B**?
(iii) Describe how you could use these radio signals to navigate to port in foggy conditions. (SEB)

INVESTIGATION

Is a light-dependent resistor equally sensitive to all the colours of the visible spectrum? Is it sensitive to any radiations beyond the visible spectrum?

23: Sound

We live in a world filled with sound. It is almost never completely silent. Even in the quietest place, you can usually hear the sound of the wind, or birds singing, or distant traffic. We use sound to communicate with each other, and for our entertainment. Sound is a form of wave motion, but sound waves are very different from light waves.

23.1 Longitudinal waves

We saw in section 21.1 how we could make a transverse wave on a long spring. If we use a spring with loose coils – a 'slinky' spring – we can produce a different kind of wave motion.

As before, one end of the spring is fixed. Instead of moving the free end from side-to-side, a pulse is produced by moving it in-and-out in the direction of the spring. This produces a small compression pulse, which then travels along the slinky. If we move the free end in and out continuously, a train of **compressions** and **rarefactions** travels along the slinky, as a wave. A compression is a region where the coils of the slinky are squeezed closer together than normal; a rarefaction is a region where the coils are stretched further apart than normal.

If we watch any individual coil of the slinky carefully, we see that it just oscillates briefly as the pulse goes past. The movement of each coil is *in the same direction* as the movement of the wave. A wave like this is called a **longitudinal wave**.

The wave equation applies to longitudinal waves also:

$$\text{speed} = \text{frequency} \times \text{wavelength}$$

$$v = f\lambda$$

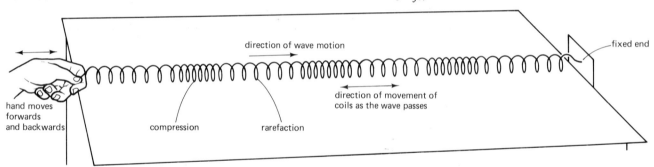

The **wavelength** of a longitudinal wave is the distance between the centre of one compression and the next, or between the centre of one rarefaction and the next. The **frequency** of the wave is controlled by the movements of the wave source.

The **wavespeed** depends on the medium the wave is travelling through. If the slinky is stretched so that it becomes tighter, the speed of the waves along it increases.

The **amplitude** of a longitudinal wave is a measure of the 'strength' of the wave. It is the maximum distance which any point in the medium has moved from its normal position. In a longitudinal wave, the amplitude is a measure of how compressed the compressions are. In the diagram above, the coils mid-way between a compression and rarefaction are the ones which have moved furthest from rest. Their displacement is the amplitude of the wave.

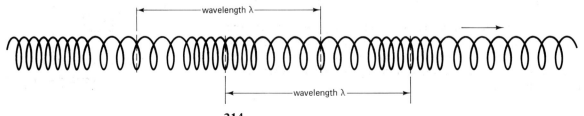

23.2 The nature of sound waves

Producing sounds

Sounds are caused by vibrating objects. A cello string vibrates to produce a sound; the skin of a drum vibrates when it is struck. The prongs of a sounding tuning fork are vibrating – you will see this very clearly if you dip the fork into a glass of water!

When a loudspeaker is producing sound, its cone is vibrating. As the cone vibrates, it is continually compressing and stretching the air next to it. This makes a series of compressions and rarefactions travel through the air, away from the loudspeaker. The movement of the air near the cone can be seen by placing a candle in front of the speaker. The flame flickers back and forward. All vibrating objects produce longitudinal waves in the air in this way. When the waves reach the ear, they cause small vibrations of the ear-drum. This causes the sensation we call 'sound'.

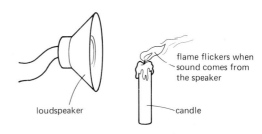

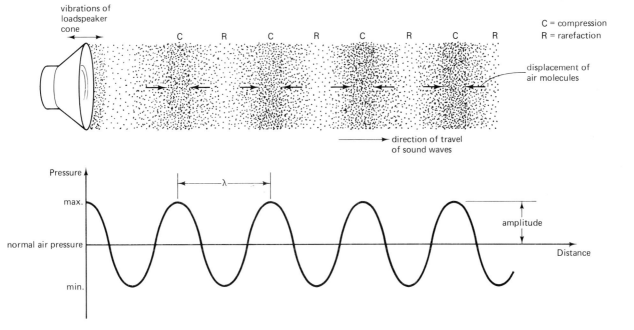

Sound needs a medium

Sound waves need a medium to travel through. Compressions and rarefactions need a material which can be compressed and stretched. Sound can travel through gases, but not through a vacuum. If an electric bell is placed inside a glass bell-jar, the sound becomes fainter and fainter as the air is pumped out. The bell can still be seen ringing, but no sound is heard.

If you try this experiment, you may still hear a very faint sound even when all the air is pumped out, because sound *can* travel through solid materials. In fact, solids and liquids are good media for sound. You can hear sounds through a wall, especially if you put your ear close to it. Sounds can also be heard underwater.

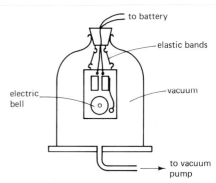

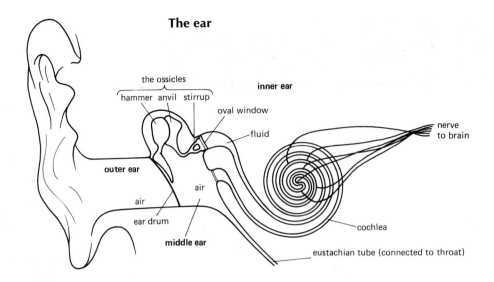

The ear

The human ear is able to hear a very wide range of sounds from very quiet to very loud. The outer ear – the part we can see from outside – collects sound waves and reflects them into the **auditory canal** which leads to the **ear-drum**. This is a stretched membrane of skin which vibrates when sound waves strike it.

These vibrations are very small. They are amplified by three small bones, called the **ossicles**, which are connected in a lever arrangement. The bones are called the **hammer**, the **anvil** and the **stirrup** because of their shapes; they are the smallest bones in the body.

The stirrup touches against a second membrane of skin, called the **oval window**. Beyond the oval window is a spiral-shaped organ, filled with fluid, called the **cochlea**. The vibrations of the oval window set up longitudinal waves in this fluid, and these changes of pressure are detected by nerve endings in the cochlea. The nerves carry messages to the brain.

So the ear really consists of three separate compartments: the outer ear, the middle ear and the inner ear. The middle ear has to have a passage to the outside to allow the pressure inside to adjust to changes in pressure outside. The **Eustachian tube** connects the middle ear to the back of the throat. This is why our ears sometimes 'pop' when we swallow, after going suddenly uphill or downhill. The 'pop' is the pressure beyond the eardrum becoming equal again to the pressure outside. The middle ear also contains two **semi-circular canals** which have nothing to do with hearing. These are like small spirit levels, and we use them to keep our balance.

23.3 Speed of sound

Sound travels quickly, but not nearly as fast as light. If you have ever watched someone hammering from some distance away, you will have noticed that you see the hammer land and then there is a short delay before you hear the blow. The sound takes a little time to reach you. The speed of sound waves depends on the medium. Sound travels fastest in solids, and slowest in gases.

Here are some typical values:

speed of sound in air	330 m/s
speed of sound in water	1400 m/s
speed of sound in steel	6000 m/s

When we talk of 'the speed of sound', we usually mean the speed of sound in air. 330 m/s is around 1200 km/h. A fast car might travel at 125 km/h; a jumbo jet at 800 km/h. The bar graph shows how some speeds compare with the speed of sound. The speed of light is too large to show on the same diagram – light travels almost one million times faster than sound!

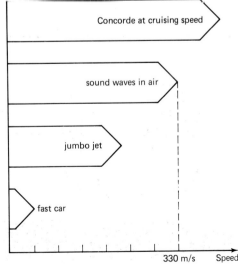

Measuring the speed of sound in air

One simple method of measuring the speed of sound in air is by using echoes. An echo is caused by the reflection of sound waves from a wall, or some other flat surface. There is always a delay between a sound and its echo; this is the time taken for the sound to travel to the reflecting wall and back again.

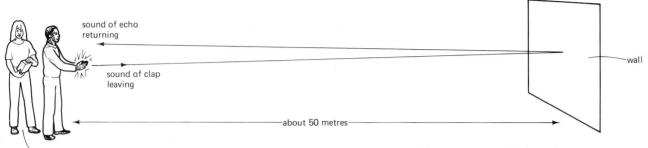

To do the measurement, you need to find a fairly large flat wall. Then measure a distance of 50 m from the wall. If you stand at this point and clap your hands, you will hear an echo. There is a very short delay between the clap and echo, and it would be difficult to measure this time accurately. But it is fairly easy to clap at a steady rhythm so that the time between clap-echo-clap-echo... is the same. Once you have got this even rhythm going, your assistant times you for twenty claps using a stop watch. (Take care here: you want to time twenty **intervals** between claps – the time from clap 1 to clap 21.) Do this several times to get an average. It is then easy to divide this time by 20 to find the interval between claps, and then divide this by 2 to find the interval between clap and echo. This is the time taken for sound to travel from your hands to the wall and back again.

Here are some typical results:

Distance from wall = 50 m
Time for 20 intervals between claps = 12 s

$$\text{We see that time between claps} = \frac{12\,\text{s}}{20}$$
$$= 0.6\,\text{s}$$

$$\text{So, time between clap and echo} = \frac{0.6\,\text{s}}{2}$$
$$= 0.3\,\text{s}$$

In this time, sound travels 100 m to the wall and back. So, speed of sound is:

$$\text{speed} = \frac{\text{distance}}{\text{time}} = \frac{100\,\text{m}}{0.3\,\text{s}} = 333\,\text{m/s}$$

The speed of sound is approximately 330 m/s in air.

Echo-sounding

In the clap-echo method, we time an echo in order to find the speed of sound in air. This method can be used the other way round. If we know the speed of sound in a medium, we can time echoes in order to measure distances. Ships use echo-sounding equipment to find how deep the water is. The time interval is measured between a pulse of sound and its echo from the sea bed. Usually **ultrasound** is used. Ultrasound means sound waves of a frequency that is too high to be heard by the human ear. By using one particular frequency of ultrasound and tuning the detector to pick up only this frequency, the echo-sounder is not confused by any other underwater sounds. If the echo sounder measures an interval of 0.4 s, and the speed of sound in water is 1500 m/s, the depth can be calculated as follows:

$$\text{distance travelled by pulse} = \text{speed} \times \text{time}$$
$$= 1500\,\text{m/s} \times 0.4\,\text{s}$$
$$= 600\,\text{m}$$

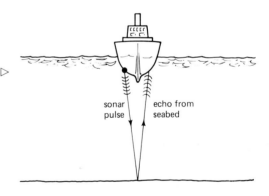

This is the time taken for the pulse to travel down to the sea bed and back, so the depth of water is 300 m.

In modern fishing trawlers, echo-sounding equipment is used to detect shoals of fish. The equipment can detect the reflected pulse from the shoal, and work out its position and depth. Geologists also use echo sounding methods to locate boundaries between rock layers far below the surface. This is useful, for example, in finding places where it might be worth drilling for oil. A small explosion on the surface sends a sound wave downwards into the Earth, and some of the sound is reflected from the boundaries between rock layers. The detector measures the time interval between the sound and the echoes. From this, the depth of the rock layers can be measured.

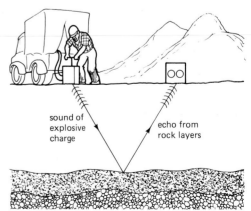

23.4 'Seeing' sounds

◁ The apparatus shown in the photograph produces a 'picture' of a sound wave.

The large instrument is called a cathode ray oscilloscope (or CRO, for short). We will look in more detail at how it works in Chapter 31. A microphone is connected to the input of the oscilloscope. Inside the microphone is a flat sheet of metal (a diaphragm) which vibrates when the compressions and rarefactions of a sound wave strike it, in much the same way as the ear-drum vibrates. The microphone changes these vibrations into electrical oscillations. The way a microphone works is explained in Chapter 30. These electrical oscillations make a spot of light move up and down on the oscilloscope screen.

The **waveform** which we see on the oscilloscope screen is really a graph showing how the air pressure in front of the microphone diaphragm changes with time. The crests of the wave on the oscilloscope screen correspond to compressions of the sound wave (where pressure is highest) and the troughs correspond to rarefactions (where the pressure is lowest). By allowing us to 'see' the sound wave, the oscilloscope lets us investigate which properties of the sound wave correspond to the different characteristics of the sounds we hear.

Amplitude and loudness

If you pluck a guitar string gently, the note is played quietly. If you pluck the string harder, the sound produced is louder. The loudness of the note depends on the **amplitude** of vibration of the string. The same is true for other musical instruments also. The diagram below shows the waveforms produced by a quiet note and an identical loud note on an oscilloscope. This confirms that the amplitude of a sound wave is what determines the loudness of a sound.

quiet sound

loud sound

Pitch and frequency

Another characteristic of sounds is their pitch. The bass string of a guitar produces a note of low pitch; the thinnest string produces a note of high pitch. The diagram below shows the waveforms produced by two notes of equal loudness but different pitch. The frequency of the sound waves is different. High frequency sound waves are heard as high pitched sounds; low frequency sound waves are heard as low pitched sounds.

low pitch
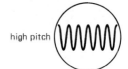
high pitch

You can actually see the difference in frequency if you look at the top and bottom guitar strings vibrating when they are played. The heavier bass string vibrates at a lower frequency. Another way to confirm the connection between pitch and frequency is to use a toothed gear wheel and a piece of card. As the wheel turns, it vibrates the card and produces a sound. The faster the wheel turns, the higher pitched the sound. You can produce much the same effect using a piece of cardboard and the spokes of ◁ your bicycle.

Sound and noise

Our ear can also tell the difference between musical sounds and noises. When we look at the sound waves on an oscilloscope, we find that the waveforms produced by musical instruments are regular, whilst those produced by noises are jagged and irregular. There is no definite frequency present.

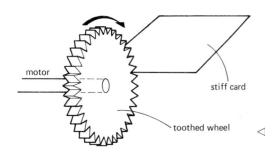
a musical sound   a noise

23.5 Wave properties of sound

Wave equation

The wave equation also applies to sound waves:

$$\text{speed} = \text{frequency} \times \text{wavelength}$$

or in symbols, $v = f\lambda$

Reflection

Echoes are caused by reflection of sound. It is possible to investigate the laws of reflection for sound using two long cardboard tubes and a quiet sound source, like a ticking watch.

One tube with the watch at one end is pointed towards a wall. The second tube is moved round until the watch can be heard most clearly. The barrier is used to prevent sound travelling directly from the watch to your ear. It is much more difficult to do this experiment with sound than with light, and the results are not so accurate. (We will see why later in this section.) Even so, we can confirm that the usual law of reflection holds: the angle of reflection is equal to the angle of incidence. ▷

Reverberation

Reflections from walls and ceilings of buildings cause changes in the sounds we hear. Try making a cassette recording of someone playing a musical instrument outside, and then inside a room. Can you hear any difference when you replay the recordings?

In a large empty hall with many flat hard surfaces, sounds will reflect many times before all the energy in the wave has been absorbed. As a result, sounds take a long time to die away – sometimes as long as several seconds. The effect is called **reverberation**. If an actor is speaking, or an orchestra playing, this means that the echoes of earlier sounds get mixed with the next sounds produced. The result can be an unclear sound which is difficult to make out.

Most halls have some walls of sound absorbent material, such as curtains, and carpeted floors to absorb the energy of the sound wave and make the reverberation time shorter. However, it is possible to go too far in the opposite direction also. A room which absorbs sound very well and cuts down reverberation to a minimum can sound 'dead'; the sounds are muffled. In some halls, special reflectors are used to increase the reverberation. The sound absorbing and reflecting characteristics of a room are called its **acoustics**.

EXAMPLE 23.1

A middle C tuning fork is marked 256 Hz. The speed of sound in air is 330 m/s. What is the wavelength of the sound from the tuning fork?

In this example: speed = 330 m/s
frequency = 256 Hz

Substituting in the equation: $v = f\lambda$
330 m/s = 256 Hz × λ

Divide both sides by 256 Hz:

$$\frac{330 \text{ m/s}}{256 \text{ Hz}} = \lambda$$

$$\lambda = 1.3 \text{ m}$$

The wavelength of the sound from the tuning fork is 1.3 m. This means that the compressions (and the rarefactions) sent out from the fork are 1.3 m apart.

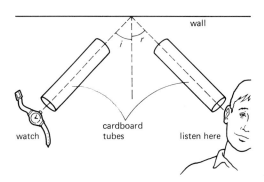

The curved reflectors hanging from the roof improve the acoustics of the Royal Albert Hall.

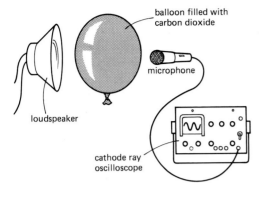

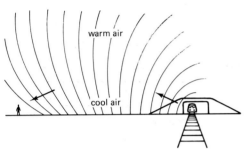

Refraction

◁ Sound waves can be refracted by a balloon filled with carbon dioxide. This gas is heavier than air, and sound travels slower in it. A balloon of carbon dioxide acts like a lens, focusing the sound wave. The microphone is moved around beyond the balloon to locate the position where the sound wave is strongest and produces the largest waveform on the oscilloscope screen. If the balloon is now removed, the amplitude of the waveform drops. A louder sound is detected with the balloon present.

Refraction of sound also explains why distant sounds often appear louder and clearer at night, though it is hard to know how much of this is simply due to the fact that it is quieter! At night, the air near the ground is often cooler than the air higher up, because it is close to the cold ground. The sound waves travel faster through the warmer, less dense layers of
◁ air, and this refracts the sound wave back towards the ground.

Diffraction

With light, it is quite difficult to see diffraction effects. With sound, there is so much diffraction that it makes the other effects hard to notice! Sound diffracts very readily. This is why we can hear sound round a corner, or behind an obstacle. The reason is that sound waves have long wavelengths in air, ranging from a few centimetres up to several metres. As we have already seen, long wavelength waves diffract more readily than those with short wavelengths.

Interference

A **signal generator** is an instrument which produces electrical oscillations of different frequencies. If its output is connected to a loudspeaker, sounds of different frequencies can be produced. When two loudspeakers are connected to the same signal generator, the two sound waves have the same frequency and produce an interference pattern. If you walk along a line across the front of the speakers, you can hear the loudness of the sound increase and decrease. At the loud places, the two sound waves are interfering constructively to produce larger compressions and rarefactions, so the sound appears louder. At the quieter places, there is destructive interference of the two waves.

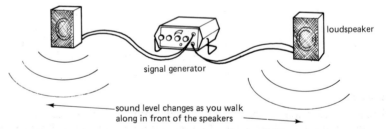

Inside a room, sound interference is sometimes difficult to demonstrate convincingly. This is because of reflection of sound from the walls, ceiling and floor. The interference effects can often be heard more clearly if the experiment is done outside, where the only reflector is the ground.

Demonstrating interference with sound is important because it provides us with good evidence that sound *is* a type of wave motion.

Comparing light and sound

Similarities:	Both are waves, so both have wavelength, frequency, amplitude etc., and both show reflection, refraction, diffraction and interference.	
Differences:	**Sound**	**Light**
	Longitudinal wave	Transverse wave
	Needs a medium	Vacuum is the medium
	Speed = 330 m/s	Speed = 300 000 km/s
	Mechanical vibrations	Electromagnetic vibrations
	Typical wavelength = 1 m	Typical wavelength = 0.000 5 mm

23.6 The spectrum of sound

With a signal generator and loudspeaker, we can produce sounds of a wide range of frequencies. Some of these sound are of too high a frequency for the human ear to hear. We are able to hear sounds with frequencies from around 20 Hz to almost 20 000 Hz. This upper limit falls as we get older to 15 000 Hz or even lower.

Frequencies lower than 20 Hz are sometimes 'felt' as low vibrations, particularly if the sound is loud, but we cannot really hear them. Sounds above the upper hearing limit are called **ultrasound**. Although we cannot hear them, an oscilloscope can confirm that there *is* a sound wave present, and other animals can hear sounds in this range. Dogs, bats and dolphins are all known to be able to hear sounds whose frequency is well above the limits of human hearing.

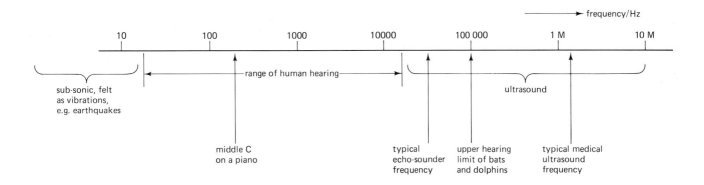

Using ultrasound

Using ultrasound is really a special example of echo-sounding. It has the advantage that the ultrasound source and detector can be tuned to exactly the same frequency, so that the detector only picks up signals which have come from the source and other sounds do not get in the way. Ultrasound has many important uses in medicine and in industry.

Medicine

Ultrasonic scanning in medicine involves sending ultrasound waves into the patient's body and detecting the echoes which come back. This can be used, for example, to see the position of an unborn baby inside its mother's womb, and to check that the baby's development is progressing satisfactorily. The whole process is completely painless and, as far as we know, harmless. It is certainly much safer than using X rays.

Doctors also use ultrasonic scanning to investigate the abdomen and for studying heart conditions.

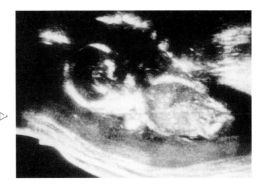

Industry

In industry, ultrasonic scanning is used to detect cracks in metal structures. This is how aircraft parts are checked for hidden cracks which might prove dangerous later. In routine checks, important parts of the aircraft are tested. Cracks inside the metal reflect the ultrasound and produce an echo which can be detected. This can then be investigated.

British Rail use ultrasonic scanning in much the same way to check for cracks in track.

A rather different application of ultrasound is in cleaning! The high frequency vibrations caused by ultrasound can actually vibrate the dirt off an object. Dentists, for example, use ultrasonic probes to vibrate hard tartar off the surface of teeth and so prevent gum disease.

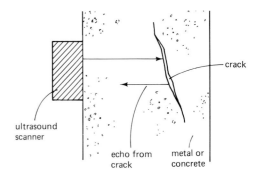

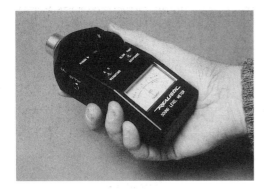

23.7 The loudness of sounds

Loudness depends on the amplitude of a sound wave. A sound level meter, or dB meter, has a microphone to pick up the sound. Specially designed electrical circuits convert the electrical signals from the microphone into a direct reading of loudness on the scale. The meter is marked in units called **decibels** (or dB, for short).

◁ The dB scale is a slightly unusual one. The zero on the decibel scale is the threshold of human hearing. Above that, an increase of +10 dB is heard as a doubling in the loudness of the sound. So, if we are listening to a sound of loudness 60 dB, an increase to 70 dB will make the sound seem twice as loud; a drop to 50 dB will make it seem half as loud. A change of 10 dB from any starting point means doubling or halving the loudness we hear. The smallest change in loudness that our ears are able to detect is around 3 dB.

Noise pollution

Too much noise of the wrong sort – especially very loud noise – is a form of pollution.

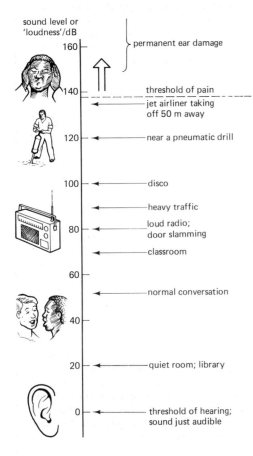

Noise pollution affects workers who have to spend their working day in a noisy environment – noise from pumps, fans, drills, saws and so on. Doctors believe that working for long periods in sound levels over 80 dB can permanently damage your hearing. Many industrial workers are partially deaf as a result of high noise levels at work. Some companies set their own noise limits for their workers.

Loud music can also be a problem. The sound level in a disco may be over 100 dB. Near the speakers at a rock concert, it can be as much as 120 dB. Recently doctors have become concerned about the effects of listening to very loud sound levels on personal stereo headphones. Although the sound may not seem loud to anyone else, the pressure changes produced by headphones at the eardrum can be very large, and can damage hearing.

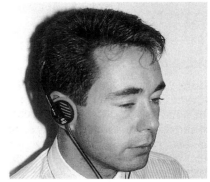

23.8 Sound and music

Stringed instruments

In a stringed instrument like a guitar or a violin, the sound is produced by vibrations of a tightly-stretched wire or string (made of gut or nylon). When the string is plucked or bowed, a transverse wave travels along it and is reflected at the end. The two waves travelling in opposite directions along the string interfere and form a **standing wave pattern**. Some points on the string, called **nodes,** are at rest; others, called **antinodes,** are vibrating with maximum amplitude.

The simplest standing wave pattern is called the **fundamental**. It has nodes at the ends and an antinode in the middle.

By plucking the string a quarter way along, we can set up the second **harmonic**. Its wavelength is half that of the fundamental. So its frequency is twice the fundamental.

The third harmonic would look like this.

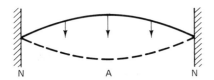
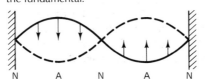
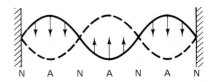

When we play a guitar or violin normally, the note we produce is a mixture of the fundamental and several harmonics. The exact mixture differs from one instrument to another. This is how we are able to recognise musical instruments. The same note sounds different on different instruments.

The frequency of the fundamental note on a stringed instrument depends on:
- the **length** of the string: shorter strings produce higher pitched notes.
- the **tension**: the tighter the string, the higher the note.
- the **mass per unit length**: heavy strings produce lower notes than lighter strings.

If you play a stringed instrument you will know this already. The highest notes on a guitar come from the thinnest string. To play different notes, you press the string against the fretboard which makes it effectively shorter. To tune the instrument, you tighten or slacken the strings. ▷ △

Wind instruments

Wind instruments depend on the idea of **resonance**. Longitudinal pressure waves can be set up in the air inside the instrument. This column of air has its own natural frequencies at which it can vibrate. When we blow, we use the mouthpiece (or reed, if there is one) to start some vibrations. Those which happen to match exactly the natural frequencies of the instrument are picked out and magnified.

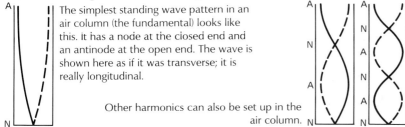

The simplest standing wave pattern in an air column (the fundamental) looks like this. it has a node at the closed end and an antinode at the open end. The wave is shown here as if it was transverse; it is really longitudinal.

Other harmonics can also be set up in the air column.

Notes from different wind instruments are mixtures of the fundamental and various harmonics.

The frequency of the fundamental note depends on the length of the air column. We can alter this by opening and closing holes along the instrument (flute, recorder, clarinet), or by opening and closing valves to include extra lengths of tubing (trumpet, horn), or by simply extending the column telescopically (trombone). In some instruments, the air column is straight, in others it is bent and folded. What matters is its total length.

Frequency spectra: the 'fingerprints' of notes

A microcomputer interfaced to an analogue-to-digital converter can be used to sample a sound signal and display it on the screen like a CRO. A program can then calculate which frequencies are present in the note. We can obtain a graph showing how much of each harmonic is present. This is called the **frequency spectrum** of the note and is like a 'fingerprint' of the note.

The diagram on the right shows the same note (G) played on a piano and on a clarinet, along with their frequency spectra. Although the fundamental frequency (the largest peak) is the same, the amounts of the different harmonics present are different. This is one of the ways in which we recognise the different sounds of the piano and clarinet when they are playing the same note.

Resonance

All vibrating objects have their own natural frequency of vibration. If something else happens to be vibrating at exactly this frequency, it can set the object vibrating very strongly. This is called **resonance**.

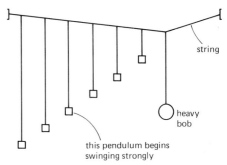

If the heavy pendulum is started swinging, it forces all the lighter pendulums to swing at the same frequency. But the one which is the same length as the heavy one will swing with a much bigger amplitude. The natural frequency of a pendulum depends only on its length and this one has the same natural frequency as the forcing vibration.

You may have noticed how something loose inside a car can vibrate noisily at just one particular speed. This is also due to resonance. At this speed, the frequency of the engine's vibrations exactly matches the natural frequency of the object and so they set it vibrating strongly.

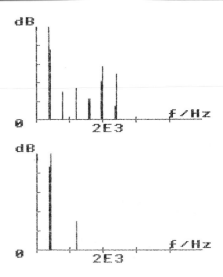

Questions

In some of these questions you may need to know that:
- the speed of sound in air is 330 m/s
- the speed of light in air is 300 000 km/s

1 (a) Use the information in this chapter to complete the following table:

Similarities between sound and light	Differences between sound and light

(b) Describe an experiment to demonstrate one of the differences.

2 A coastguard sees an emergency flare explode out at sea. 4.0 seconds later he hears the sound produced by the exploding flare.
(a) Explain why there is a delay of 4.0 seconds.
(b) Estimate the distance to the flare using the information given above.

3 A new bridge is being opened by a Member of Parliament. The opening ceremony is being broadcast worldwide on radio and is also being broadcast to spectators by loudspeakers. A spectator, 1 kilometre away at the opposite end of the bridge, hears the opening announcement on his transitor radio 3 seconds before he hears the sound of the same announcement coming to him directly from the loudspeakers.

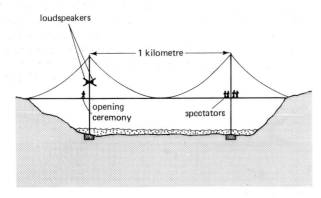

(a) Explain why this spectator hears the announcement of the opening of the bridge on his radio before he hears the same announcement directly from the loudspeakers.
(b) A man in Australia is also listening to the live broadcast of the opening ceremony. If the radio waves travel a distance of 18 000 kilometres to him, calculate whether he will hear the opening announcement on his radio before or after the spectator at the other end of the bridge hears the sound directly from the loudspeakers.
(SEB)

4 The frequency limits of human hearing are around 15 Hz and 20 000 Hz. What are the wavelengths of the longest and shortest sound waves you can hear?

5 The drawing shows a beam of sonar waves sent to a shoal of fish directly underneath a fishing boat.

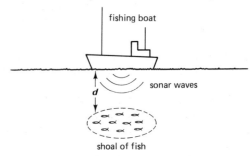

(a) The speed of the sonar waves in water is 1400 m/s and the echo returns after 0.1 seconds. Use the equation

distance = speed × time

to calculate the depth d of the shoal of fish.
(b) Explain why the returning pulse lasts for a longer time than the pulse sent out. (Sp.NEA)

6 A geophysicist searching for oil sets off an explosive charge, and 0.4 s later his detector receives an echo from a rock band in the Earth's crust.
(a) If the speed of the sound waves through the surface layer is 4000 m/s, how deep is the top of the rock band?
(b) A second reflected pulse is picked up 0.1 s later still. Where has this come from? How thick is the rock band?

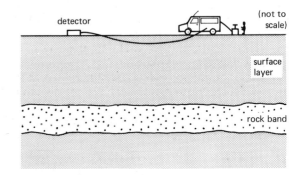

7 A noise level meter is used to make measurements at different distances from a pneumatic drill. The results are as follows:

Distance from drill/m	5	10	20	40	80
Noise level/dB		90	70	50	30	10

(a) What pattern do you notice in this data? How does the loudness change with distance?
(b) Plot a graph of loudness against distance.
(c) From your graph, estimate the noise level at 50 m from the drill.

8 Two loudspeakers, in a large open space, were connected to the same audio frequency generator as shown in diagram (A) below.

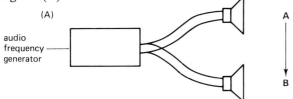

A person walking from **A** to **B** in front of the loudspeakers found that the sound level varied in a regular pattern.
(a) (i) What name is given to the effect causing this variation in sound level?
 (ii) Explain, using a labelled diagram(s), how this effect occurs.
 (iii) Suggest **two** reasons why a domestic 'stereo' system does **not** usually show such regular changes in sound level.
(b) Diagram (B) below shows part of an experimental system designed to reduce the noise from an engine.

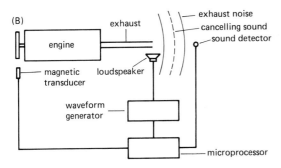

For each complete cycle of the engine the same waveform is produced by the exhaust. The loudspeaker emits a sound each time the engine fires. This sound cancels out the exhaust noise.
 (i) Name a device that could be used as a sound detector in this system.
 (ii) Diagram (C) shows the amplitude of the sound waveform emitted by the exhaust during one cycle of the engine. Sketch (C) and on it draw the waveform that would be emitted by the loudspeaker during that cycle. (AEB: SCISP)

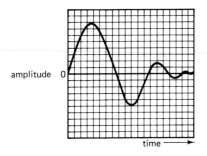

9 The graph below shows the percentage sound wave absorption at different frequencies for various everyday furnishings and for a window pane of 4 mm thickness. Use it in answering the questions below.

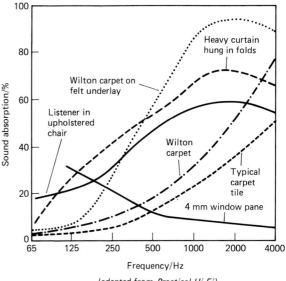

(adapted from *Practical Hi-Fi*)

(a) For the **furnishings** shown, which frequencies are least absorbed?
(b) Which two lines on the graph show the most similar pattern of sound absorption? Describe this pattern.
(c) Give **two** effects, other than absorption, which these materials could have on the sound waves.
(d) If the speed of sound in air is 330 m/s, calculate the wavelength of sound waves of frequency 990 Hz.
(e) Suggest **two** differences which would be apparent in the sound reproduction produced by two identical domestic stereo systems, if one were placed in a fully furnished room and the other in an empty greenhouse.
(f) Apart from the quality of the stereo system and the nature of the furnishings, suggest **one** feature of a room which might affect listening. Explain your answer.
(AEB: SCISP)

INVESTIGATION
The same note sounds different when played on different musical instruments (this is how we can recognise each instrument). But what differences can you actually observe and measure between the waveforms of the notes?

24: Electrostatics

Most of the electrical devices we come across in everyday life are made with wires and connectors, cables and plugs, batteries, and so on. But this kind of electricity – current electricity – was not the first to be discovered and investigated. In the 17th century, the first electrical experimenters worked with what we now call **static electricity** or **electrostatics**. Nowadays electrostatics has many applications, including photocopiers and improved crop-sprayers.

24.1 Static electricity

Have you ever pulled off a pullover over a nylon shirt or blouse and heard little crackles? Try it in a dark room and you will see that the crackles are really tiny sparks. This is **static electricity**. It can also produce sparks on a larger scale – it is what causes lightning!

Static effects are easy to produce. A balloon rubbed on a woollen pullover will 'stick' to the wall; rub a plastic comb on your sleeve and it will attract and pick up light pieces of tissue paper or feathers.

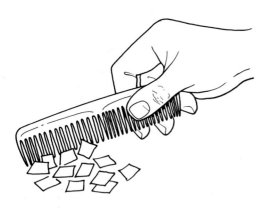

Static effects of this kind have been known for a very long time. The ancient Greeks knew before 500 BC that the substance **amber** would produce tiny sparks if you rubbed it (the Greek word for amber is 'elektron'). But the study of static electricity really only began in earnest in the 17th and 18th centuries.

In this book, we will also begin our study of electricity by looking at electrostatics.

24.2 Charging by rubbing

Take a polythene rod, rub one end of it with a cloth and then hang it in a stirrup. Then rub a second identical polythene rod and bring the two rubbed ends close together. We find that they **repel** one another; they push each other apart. The balanced rod moves away.

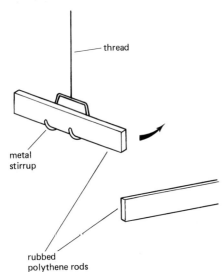

If we use two perspex rods, the same thing happens – the rubbed rods repel. But when we do the experiment with one polythene rod and one perspex rod, we find that the rods **attract** one another.

Something happens to the rods when they are rubbed to make them behave in this way. They are now able to influence other rubbed rods nearby. We say that the rods have become **charged**. Two charges of the same type repel one another; but two different charges attract.

What would happen if we now took a third rod of another different material? Perhaps when it is rubbed, it will be able to repel both polythene and perspex rods. In fact, no rod has ever been discovered which can do this. It seems that there are just two types of charge. These have been given the names positive (like the charge on a perspex rod) and negative (like the charge on a polythene rod). The table summarises the rule we have discovered.

Charge on one rod	Charge on other rod	Result
Positive (+)	Positive (+)	REPEL
Negative (−)	Negative (−)	REPEL
Positive (+)	Negative (−)	ATTRACT
Like charges repel	**Unlike charges attract**	

The force between the charged rods is quite small. They must be brought quite close together if we are to observe it. The size of the force depends on the distance between the rods; it gets smaller still as the two rods are moved further apart.

24.3 Where does charge come from?

When a plastic rod is rubbed with a cloth it becomes charged. But what *is* charge and where has it come from? In fact we think that charge is a basic property of matter – it is always present in matter. Of course, most objects are not charged most of the time. This is because they contain equal amounts of positive and negative charge mixed up together.

Scientists have gradually built up a picture or **model** of matter as made out of tiny atoms. Chapter 1 described some of the experimental results which can be explained using the idea of atoms. Now we need a model of what the atom itself is like. Later, in Chapters 31 and 33, we will see some of the evidence for this model of the atom. For the moment, we will simply use the model to help us understand electricity.

A model of the atom

At the centre of each atom is a tiny **nucleus**. The nucleus itself is made up of even smaller particles called **protons** and **neutrons**. Surrounding this nucleus are some very much lighter particles called **electrons**. The electrons can be thought of as circling constantly around the nucleus, rather like the planets in the solar system going round the Sun. The particles inside the atom are electrically charged:

Electrons have a negative (−) charge.
Protons have a positive (+) charge.
Neutrons have no charge.

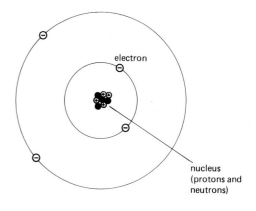

The negative charge on an electron is the same size as the positive charge on a proton. Every atom has the same number of electrons and protons, and so their charge cancels out. The atom as a whole is neutral (uncharged).

Explaining 'charging by rubbing'

When a polythene rod is rubbed with a cloth, some electrons may be pulled off the atoms of the rod, and stick on to the atoms of the cloth (or vice versa). The charge on polythene when it is rubbed is negative. Electrons are negative charges, so the polythene must have picked up some extra electrons from the cloth. The cloth will be positive as a result.

Perspex acquires a positive charge when it is rubbed, by losing some electrons to the cloth, which this time becomes negative.

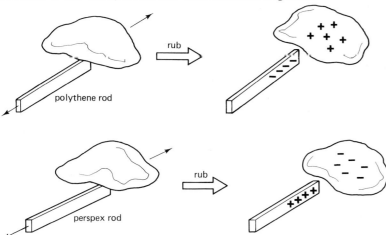

Rubbing two materials together doesn't *make* charge. It just separates some of the charge which is already there. The atoms of some materials have a stronger hold on their electrons than other materials. When we rub two different materials together, the direction in which the electrons go depends on which two materials we use.

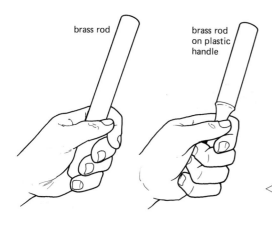

24.4 Conductors and insulators

Some materials seem not to become charged when they are rubbed. If we rub a metal rod, it seems to have no effect on a charged polythene or perspex rod. However, if the metal rod has a plastic handle, it *can* be charged. Metals can be charged by rubbing, but if we hold the metal directly in our hand, the charge escapes as quickly as we produce it.

This is an important difference between metal rods and plastic rods (like polythene and perspex). When we rub a polythene rod and electrons are transferred to it from the cloth, the electrons stay put; they don't escape. But when a metal rod is rubbed and electrons are transferred to it, they *do* escape. The route by which they escape is through the hand and arm of the person holding the rod, and into the Earth.

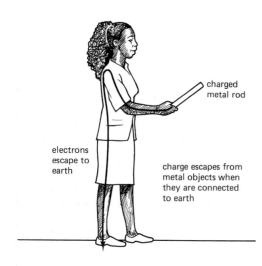

Before the electrons can escape in this way, they must be able to move through the rod itself to reach the hand. Electrons can move through metal rods, but in polythene rods they are unable to move. Materials in which electrons can move around freely are called **conductors**. Metals are conductors, but so too are many non-metallic materials (like the human body, for example!). Materials in which electrons are unable to move around are called **insulators**. Plastics are insulators, as are glass, ceramic and rubber. The charging experiments with rods work only because air is a good insulator.

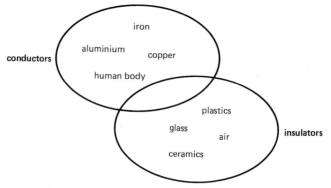

What materials do you think could be put in the overlap? We talk about these later on, on p. 320.

EXAMPLE 24.1

Explain how the electrons move when someone touches a positively charged metal rod.

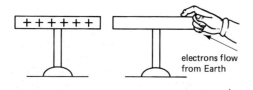

The rod is positively (+ve) charged. It has lost some electrons. When someone touches it, their body forms a connection between the rod and Earth. Electrons will be attracted by the positive charge on the rod and will move *from* Earth *into* the rod to cancel out the positive charge.

If a conductor does become charged, the charge will spread evenly over it because charge is free to move and like charges repel each other. So they will try to move as far apart as possible, and this will make them spread all over the conductor.

The electron model of the atom helps to explain *why* conductors and insulators are different. In an insulator, the electrons are tightly held to their own atoms and are not free to move along. In a conductor some of the electrons are not so tightly bound to their own atoms, and can drift through the material, hopping from one atom to the next (see the diagram on the left below). Indeed, a good model of a metal is to think of it as an array of atoms – each one having lost one of its electrons – 'embedded' in a sea of free electrons. Normally the electrons move around randomly, rather like the molecules of a gas. (Note: we have already used this model to explain conduction of energy in section 16.7.)

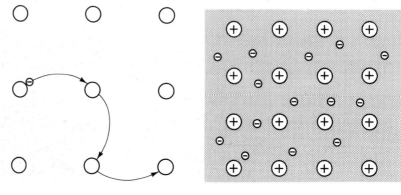

24.5 Attracting light objects

A charged rod will attract or repel other charged rods. But it also attracts light objects like feathers or small pieces of tissue paper. How does this happen?

If a charged polythene rod (negative) is held near a piece of paper, the electrons in the paper will be repelled by the rod (remember: like charges repel). Paper is a good enough conductor to allow the electrons to move and so they are pushed towards end **B**, leaving positive charge behind at end **A**. There is a force of attraction between end **A** and the charged rod, and a force of repulsion between end **B** and the rod. However, **A** is closer to the rod and so the attraction is stronger. The paper is pulled towards the rod.

The paper would also be **attracted** by a positively charged rod. Check, by drawing another diagram similar to the one on the right, that you can explain this too.

There is not a permanent charge on the piece of paper. When the rod is taken away the charges in the paper will move together again and it will be uncharged. The temporary charges on ends **A** and **B** while the rod is nearby are called **induced** charges.

The attraction between a rod and a fine stream of water can be explained in the same way. The force of attraction is even greater in this case because the repelled electrons can flow through the water and the tap to Earth, leaving only positive charge on the water stream. The effect is quite dramatic, as the photograph below shows. Try it!

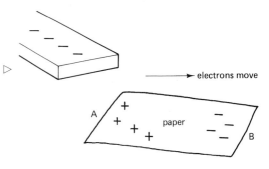

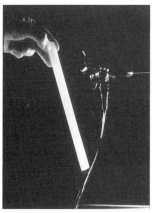
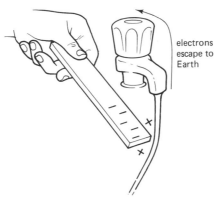

Static problems

The electrostatic attraction of light objects can sometimes be quite a nuisance.

Some carpets, especially those made from certain man-made fibres, become charged as people's shoes rub over them. They then attract dust, making them hard to clean. Trying to remove the dust by brushing often only makes the problem worse as it charges the carpet more!

Records become charged as they are taken from their sleeves and attract dust on to their surfaces. Special anti-static cleaning fluids and cloths are sold for keeping records free of dust.

24.6 The van de Graaff generator

When a van de Graaff generator is running, its dome is continually being charged up. The charge on the dome quickly becomes much larger than you can produce by rubbing a plastic rod with a cloth.

If you stand on an insulating support (a large block of foam polystyrene, for example) and touch the dome of the van de Graaff generator, the charge on the dome spreads over your body as well. As you charge up, your hair starts to stand on end. Each hair has the same charge, and like charges repel. So every hair is trying to get as far as possible from all the other hairs. The effect is 'hair raising'! ▽

The diagram below shows how a van de Graaff generator works. The lower roller is turned either by a handle or by a small motor. This drives a rubber belt which also passes over the top roller. There is always some slipping between the rollers and the belt and so the rubber belt gets charged by rubbing as the generator runs. The charge collects on the dome.

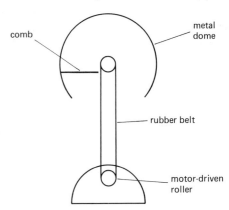

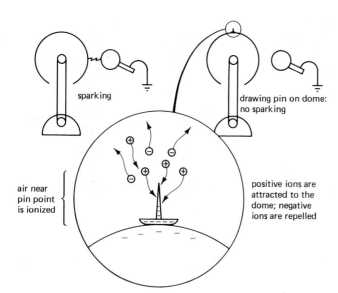

24.7 Sparks

Sometimes a second metal sphere is placed near the dome of a van de Graaff generator as it charges up. After a short time, large sparks begin to jump across the gap between the two metal spheres. What causes these sparks?

The explanation is that as the charge builds up on the dome, it begins to have an effect on the air nearby. Imagine an atom in the air close to the dome. The strong negative charge on the dome attracts the positive nucleus of the atom and repels the negative electrons. If the forces are large enough, they may even pull one electron completely off its atom. This leaves behind a positive 'charged atom' which has lost one of its electrons. The electron itself will probably join on to another atom making it a negative 'charged atom'. The proper name for a 'charged atom' is an **ion**. The process is called **ionization**. Once an ion is formed, it will be attracted strongly towards the dome. It moves rapidly towards the dome and bumps into other atoms in the air as it goes. These collisions are violent enough to knock electrons off those other atoms, forming more ions. And so the process goes on – it is an **avalanche effect**.

The spark we see is a sudden avalanche of ions. They move at such a high speed that they emit energy in the form of light and sound as they go.

The effect of sharp points

Around a sharp point or edge on a conductor, the charge can become quite concentrated, simply because of the shape. As a result, ions are more likely to form in the air near a point. If a drawing pin is placed on the top of a van de Graaff dome, it will stop it sparking to a nearby sphere. Ions are forming continually in the air around the pin point. These carry the charge away from the dome, so that a charge strong enough to cause a spark never builds up. At the end of the experiment, when the generator stops, its dome is uncharged.

The lightning conductor

Most tall buildings have a lightning conductor. A thick strip of copper runs from a pointed metal post on top of the building down into the ground.

Thunderclouds are electrically charged. The water droplets in the cloud are moving around all the time because of convection currents (section 16.3) inside the cloud, and they become charged up by 'rubbing' against the air as they move. A large thundercloud will **induce** charge on the ground and any buildings under it. There is a danger of a large spark jumping from the cloud to the nearest point on the ground. The lightning conductor reduces the risk of this happening in two ways:

1. The lightning conductor itself has become charged (with induced charge) by the cloud overhead. Near the sharp point of the conductor, ions will form in the air. The stream of ions may cancel out some of the charge on the cloud and make a lightning strike less likely.

2. If lightning does strike, then it is more likely to strike the highest point. The copper strip provides a conducting path for electrons to pass harmlessly into the ground without damaging the building. The Earth is so large that it can absorb extra electrons without any difficulty.

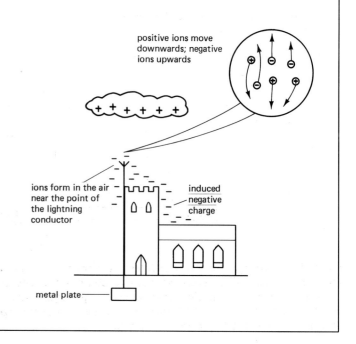

24.8 Electric fields

In Chapter 1, we looked at the strange phenomenon of 'action at a distance'. What causes the force between two rubbed rods? How does one rod 'know' that the other rod is there? How is the force carried from one rod to the other? It is really only in the last 50 years or so that theoretical physicists have begun to find some possible answers to questions like these. Their answers are extremely complex and involve rather advanced mathematics!

To begin to understand 'action-at-a-distance', we use the idea of a **field**. This doesn't solve the problem, but gives us another way of looking at it, which turns out to be very useful. Around a charged object there is a region where electrostatic forces can be felt. We say that the charged object has set up an **electric field** around itself. Any other charged body which comes into the field will experience a force. The electric field is strongest near the charged object; the force is strongest there. The field gets weaker as you move further away.

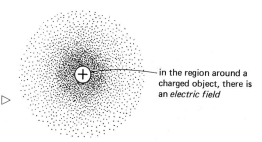

in the region around a charged object, there is an *electric field*

Field diagrams

Rather than simply describing in words the field close to a charged object, we can draw a 'map' of the field. The map is a series of lines (called **field lines**) which show at each point the direction of the force which would be felt by a small positive test charge.

We can observe electric field patterns experimentally using the apparatus in the diagram below.

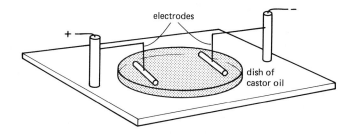

Metal electrodes dip into a shallow dish of castor oil, which is an insulating liquid. One electrode is connected to the dome of a van de Graaff generator and the other is earthed. When the van de Graaff generator is switched on, the electrodes are charged. Some semolina is sprinkled on the oil, and the grains line up in the direction of the electric field lines.

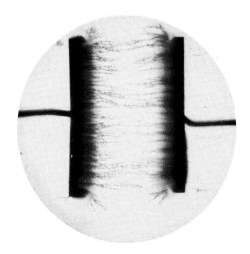

The photograph shows the electric field pattern between two parallel charged plates.

The diagrams below show the electric field patterns around some simple arrangements of charges.

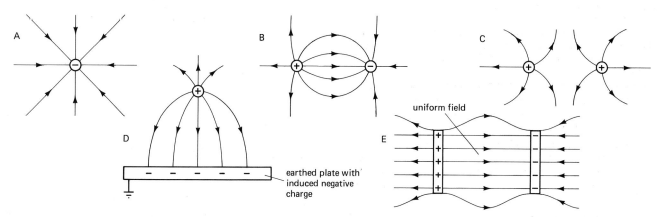

The field is strongest where the lines are closest together. Field lines never cross over. They start on positive charges and end on negative charges. In diagram **E**, notice that there is a large region between the two parallel plates where the field lines are parallel and evenly spaced. The field is said to be **uniform** in this region.

232 Electrostatics

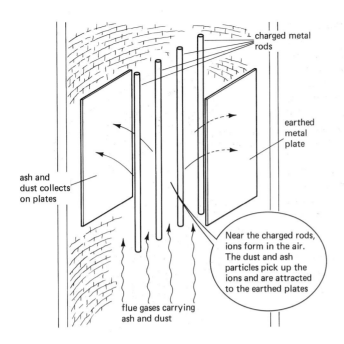

Applications of electrostatics

One very important industrial application of electrostatics is in removing fly-ash and other dust from the waste gases in power station chimneys. An electrostatic precipitator inside the chimney consists of two flat metal plates with a number of wires running vertically between them. The plates are earthed but the wires between are kept strongly charged (negative). There is an electric field in the region between the wires and the plates.

Around the wires, ions form in the air. Positive ions are attracted back to the wires but the negative ions are picked up by the tiny particles of ash and dust. The charged dust particles then move towards the metal plates where they are collected. This method can collect as much as 99.5% of the ash and dust in the chimney gases – up to 50 tonnes every hour in a large power station! The fly-ash can then be used for road building.

Electrostatic precipitation cannot remove unwanted gases from the chimney, so the sulphur dioxide and nitrogen oxides which cause acid rain are not removed by this process. Other more expensive methods must be used if we want to clean the waste gases further.

Electrostatic paint spraying is widely used nowadays. Paint droplets from an aerosol usually become charged by rubbing against the nozzle of the spray. If a car body is earthed during spraying, the paint droplets will be attracted on to the metal body, giving a more even coating, and ensuring that the paint reaches even the most inaccessible parts. ▽

New crop sprayers are using the same idea. A ▷ strongly charged metal wire charges tiny drops of pesticide as they leave the nozzle. When a drop approaches a plant leaf, it induces an opposite charge on the leaf and is attracted to it. Notice how the paths of the droplets are the same as the electric field lines in diagram **D** on p. 231. The charge also makes the drops stick to the leaves of the plant. The drops all have the same charge, so they repel in the spray and spread out more evenly as a result. All these electrostatic effects have the same advantage for the farmer – less spray is needed to treat the crop effectively. In some Third World countries this can make the difference between a good harvest and a famine.

Office photocopiers are based on electrostatics. Many photocopiers use the **Xerox** process. This makes use of a rather unusual property of the metal selenium. It is a **photoconductor** – it conducts when it is in the light, and is an insulator when it is in the dark! Inside the copier is a drum coated with a thin layer of selenium. First the whole surface of the drum is charged by rotating it near a highly charged wire. The ions produced in the air near the wire are picked up by the drum and charge it. When a printed page is photocopied, light is reflected off the page on to the drum. Some parts of the drum receive a lot of light from the white parts of the page; these areas become conducting and lose their charge. Other parts of the drum correspond to the black parts of the page and receive no light at all. They remain insulating and hold on to their charge. So the drum ends up with a pattern of charge which is an exact copy of the pattern of printing on the original page. Fine particles of powdered ink (toner) are then attracted to the charged areas of the drum and are printed on to the photocopy as the drum rotates and presses against the copy paper.

The final stage is to heat the paper to melt the toner powder and fix it to the paper surface. You may have noticed that copies coming from a photocopier are often electrically charged – the sheets stick together. Perhaps now you can understand why this is so.

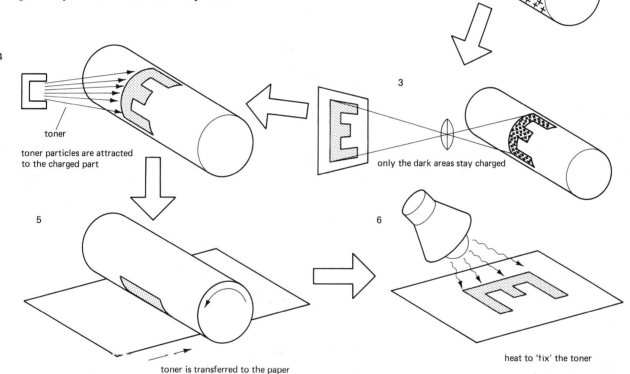

Electrostatic hazards

Many man-made fibres used in clothing (like nylon, acrylic, and so on) are good insulators and easily become charged. This is why a nylon shirt or blouse becomes charged when you pull off a pullover over it – charging by rubbing. People may also pick up charge as they walk on carpets or other flooring materials made of man-made fibres. In some situations it is necessary to take precautions against getting charged up.

Some electronic components are very sensitive to static charge and can be damaged by it. Chips of the CMOS type (complementary metal oxide semiconductor), which are quite common in many electronic circuits, are an example. When working with them it is wise to keep oneself earthed, either by working on an earthed pad (e.g. a metal sheet connected to a water pipe) on the workbench, or standing on a conducting (rather than an insulating) floor so that charge cannot build up on the body.

This worker has an earth lead attached to his wrist.

Questions

1 Imagine that you are:
(a) an electron on the surface of a perspex rod which is being rubbed with a cloth;
(b) an electron on a negatively charged conductor which is then connected to Earth.
Describe what happens to you!

2 A strip of polythene from a carrier bag is charged by rubbing it with a cloth. When it is then hung over a string, the ends push apart. Explain why this happens.

3 (a) When you rub your comb on your sleeve you find that it becomes charged. Explain clearly how you could test it if it is charged positively or negatively. (Assume you have a polythene and a perspex rod available.)
(b) The charged comb attracts small pieces of paper. Explain, in terms of the movement of electrons, why there is a force of attraction between the comb and the paper.

4 An aircraft flies just below a negatively-charged thunder cloud. Movement of free electrons causes electrostatic charges to be induced in the aircraft.
(a) Draw the diagram and mark on it the positions and signs of the induced charges on the aircraft.

(b) Explain, in terms of the movement of electrons, the distribution of the charges you have shown.
(c) What will happen to the induced charges when the aircraft flies away from the cloud? (Sp.LEAG)

5 After cleaning the windows, they still seem to have small pieces of dust sticking to them. Rubbing harder just makes it worse. Suggest a reason for this.

6 In one method of making sandpaper, a roll of paper passes through nylon friction pads. A spray of fine droplets of glue is produced from an aerosol. These stick evenly to the paper. The sticky paper then passes over a flat table covered with sand grains.

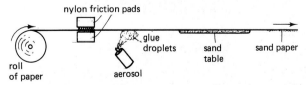

Explain how electrostatics is used in this process:
(a) to make sure the glue is spread evenly on the paper;
(b) to produce an even coating of sand on the paper.

7 When a valuable ore is mined it is often mixed with waste particles of rock and sand. One method of separating the ore involves crushing the material and then allowing it to slide down a plastic chute and fall between two charged plates. Look at the diagram and explain how this method might work.

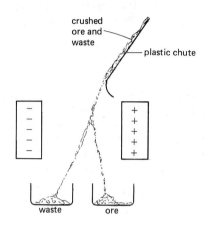

8 A light metal-covered sphere is suspended between two parallel plates, as in diagram (A). The plates are connected to a d.c. supply set at 2 kV.

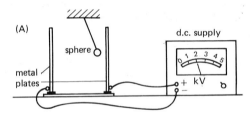

(a) What is the sign of the charge on the sphere?
(b) When the plates are moved to the right (diagram (B)), the sphere is seen to stay in the same position.

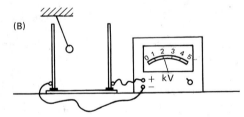

What does this tell you about the electric field between the plates?
(c) Draw lines to represent the electric field between the plates. Indicate the direction of the field.
(d) Give one method of increasing the deflection of the sphere from the vertical. (SEB)

INVESTIGATION

A metal rod can be charged by rubbing if it has an insulating handle. Which materials are good enough insulators to be suitable as handles?

25: Current Electricity: Basic Ideas

We use electricity every day. Try to make a list of all the things you use electricity for in your home. But how do electric circuits work? In this chapter, we will look at four electrical ideas which help us to understand and predict what happens in electric circuits. They are: **charge**, **current**, **potential difference** (or **voltage**) and **resistance**.

25.1 Current and charge

Current

A few years before 1800, an Italian professor of anatomy, Luigi Galvani, was dissecting the leg of a dead frog. He noticed that the leg muscle twitched when he touched a nerve with two dissecting instruments made of different metals. He thought (wrongly) that the effect was due to some property of the nerve and muscle.

His fellow-countryman, Alessandro Volta, took a different view. He believed that it was really caused by the two different metals. He set out to test his idea. Volta made a pile of pairs of silver and zinc discs separated by pieces of felt soaked in brine (salt water). This is now called a **Voltaic pile**. When he touched the top and bottom discs at the same time, he got an electric shock. Indeed, if the pile was large enough, he was able to get sparks from it.

The importance of Volta's experiment is that it meant that electricity could easily be produced in the laboratory. The batteries (or 'dry cells') we use nowadays are really just a development and improvement of Volta's original idea (see Chapter 28).

Alessandro Volta and Luigi Galvani

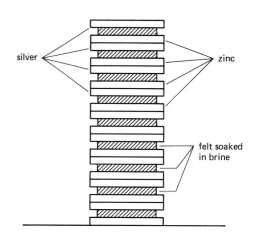

'Moving charge' equals 'current'

At first sight, this sort of electricity seems very different from electrostatics. However, we can show that the two are closely related in the following way.

If a van de Graaff generator is switched on for a few seconds, the dome will become charged.

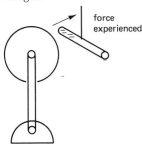

But if we begin by connecting the dome to an earthed point (e.g. a metal water pipe) — using a connecting wire, and *then* run the generator, we find that the dome does not become charged. The only difference between (a) and (b) is the connecting lead. It seems that the charge has been able to escape from the dome along this route to Earth.

We now take a sensitive meter and put it into this connecting link. When the van de Graaff generator is switched on, the pointer of the meter moves. The meter reading is caused by the movement of charge from the dome to Earth.

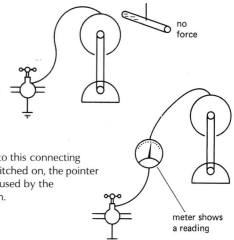

Current electricity and static electricity are closely related.

An electric current is a movement of charge.

You can do an experiment similar to Volta's in the laboratory. Take an iron nail and a piece of thick copper wire and stick them into a lemon. Then connect a sensitive meter between the two pieces of metal. It will indicate that electricity is being produced.

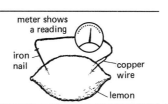

25.2 A working circuit

Closed loops

One modern development of Volta's pile is the torch battery (more correctly called a 'dry cell'). Using a dry cell, a torch bulb, and a few pieces of wire, it is quite easy to obtain the results shown in the diagram below.

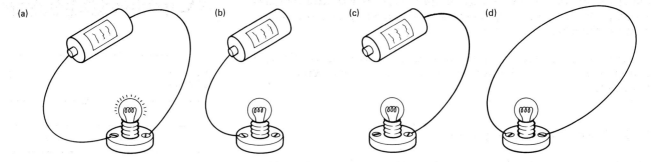

No matter how we connect the wires, we can never get the lamp to light without using a cell. The cell is the energy source which makes things happen. We also find that the lamp lights only when there is a wire from the cell to the lamp and another wire back to the cell again (see diagram (a) above). Unless the circuit is in the form of a closed loop, the lamp is off. When the lamp lights, we say that there is a **current** in the electrical circuit.

How can we picture what is going on here? Something seems to be travelling from the cell to the lamp (to make it light up), but the second wire back to the cell is necessary, forming a closed loop. One way to make sense of this is to imagine that something is **flowing** along the wires, like water in a pipe. The cell is like a pump, pushing the water along.

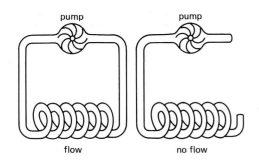

If there *is* a closed loop, then water can flow round; but if the loop is blocked at any point, then no water can flow anywhere! Compare the diagrams on the left with their electrical equivalents (diagrams (a) and (b) above). There is a current only if the cell, lamp and wires are connected in the form of a closed loop.

Circuit symbols and diagrams

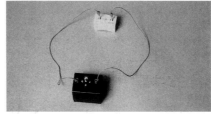 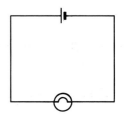

The photograph above shows a lamp connected to a cell by wires. An arrangement like this is called an electric **circuit**. It is much easier to use symbols when we want to draw an electric circuit, rather than having to draw a 'picture' of the circuit as we did in the torch battery example above. We use these symbols:

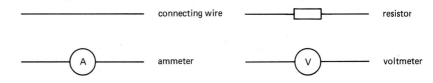

Circuit diagrams are drawn with the connecting wires shown as neat straight lines, with sharp corners. The circuit would not look exactly like this if it was actually constructed, but the important thing is that the connections are the same. The photograph above shows a real circuit alongside the corresponding circuit diagram; the diagram does not show the shape of the actual circuit, but it *does* show which points are connected.

Current Electricity: Basic Ideas

Conductors and insulators

We have seen that we need a closed loop for an electric circuit to work. But if we put together the circuit shown in the diagram on the right, we find that the lamp lights when some materials are used to connect between **A** and **B**, but does not light for other materials.

To test a particular material, we connect it between **A** and **B**. If the lamp lights up, the material is a **conductor**. If it is an **insulator**, the lamp will stay off. The results of testing some common materials in this way are shown in the table.

A **conductor** is a material which an electric current can flow through; an **insulator** is a material which does not allow an electric current to flow through it.

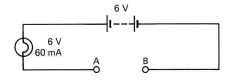

Material	Lamp
Wood	Off
Iron	On
Plastic	Off
Steel	On
Paper	Off
Carbon	On
Glass	Off

25.3 Closed conducting loops

The idea of closed loops of conductors is a useful one. If you can trace out a closed loop from the cell to a lamp, through the lamp and back to the other end of the cell, then the lamp will be lit.

These three circuits work in exactly the same way. With the switch open, the loop is not complete – both bulbs are off.
With the switch closed, there is a complete loop – both bulbs come on.

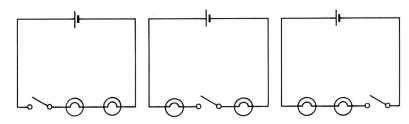

These two circuits are equivalent. When switch **1** is closed, there is a complete loop through bulb **A** – it comes on.
When switch **2** is closed, there is a closed loop through bulb **B** – it comes on.
The two bulbs can be switched on and off independently.

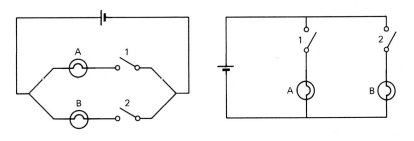

Can you work out what will happen when the switch is closed in this circuit?

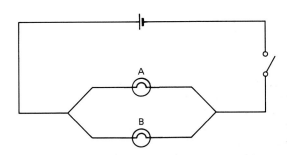

Switches

A switch is simply a device with two contacts which can either be **open** to stop the current, or **closed** to allow it to flow.

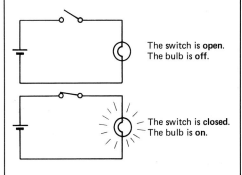

The switch is **open**.
The bulb is **off**.

The switch is **closed**.
The bulb is **on**.

Notice that this is exactly opposite to the way normally we use the words 'open' and 'closed'! If a door is open we can go through; if it is closed we can't. With a switch, 'closed' means that the circuit is closed – a closed loop – and there is a current. 'Open' means that there is a gap in the circuit, and no current.

Circuits with switches

Here is a car windscreen wiper circuit. Switch **A** is the car ignition switch, usually operated by a key. Switch **B** is the wiper switch inside the car.

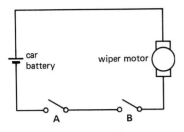

This table summarises how it works; the wipers come on only if switch **A** AND switch **B** are closed.

Switch A	Switch B	Motor
open	open	stopped
open	closed	stopped
closed	open	stopped
closed	closed	runs

This shorthand version is called a **truth table** for the combination of switches. This is the truth table for the AND combination of switches.

Switch A	Switch B	Motor
0	0	0
0	1	0
1	0	0
1	1	1

This is the circuit for the courtesy light in a car. The light comes on if either of the front doors is open. Each door has a little push button switch which is operated by the door as it opens. Switch **A** is on the driver's door; switch **B** is on the passenger's door.

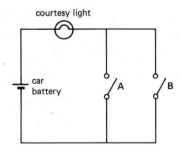

The light comes on if switch **A** OR switch **B** is closed.

Switch A	Switch B	Light
open	open	off
open	closed	on
closed	open	on
closed	closed	on

Again we can write the results in the form of a truth table. This is the truth table for the OR combination of switches.

Switch A	Switch B	Light
0	0	0
0	1	1
1	0	1
1	1	1

Two-way switches

Simple ON/OFF switches have two positions: ON (or closed) and OFF (or open). A two-way switch is ON in both its positions. The switch can be ON in either position 0 or position 1. It doesn't make much sense to use the words 'ON' and 'OFF' with this switch.

Look at the circuit diagram on the right. What will happen when the switch is in position 0? What will happen when it is in position 1? Can you think of anything this circuit might be used for?

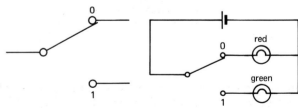

You proably have a circuit like the one below in your home. Can you work out what it does, and what it might be used for? Its truth table is also shown. Check that you agree with this truth table.

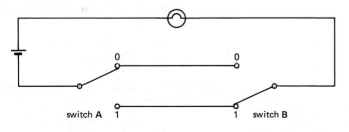

Switch A	Switch B	Light
0	0	1
0	1	0
1	0	0
1	1	1

25.4 Current

Measuring current

We can use a lamp to show if there is a current in an electric circuit, but a better way is to use an **ammeter** – an instrument for measuring electric currents.

Current is measured in amperes (or amps, for short) (A). For the moment we need not worry about exactly what 1 ampere *is*. The definition of the ampere is discussed later in Chapter 29. Nor do we need to know exactly how an ammeter works. This is also discussed in Chapter 29 in more detail. For the moment, all we need to know is that an ammeter measures the size of an electric current in units called amperes.

Current and charge

We have seen earlier that **an electric current is a flow of charge**. The size of the current at a particular point in the circuit depends on how much charge flows past that point each second.

$$\text{current} = \frac{\text{charge}}{\text{time}}$$

or, in the symbols normally used

$$I = \frac{Q}{t}$$

where Q is in coulombs, I in amperes, and t in seconds.

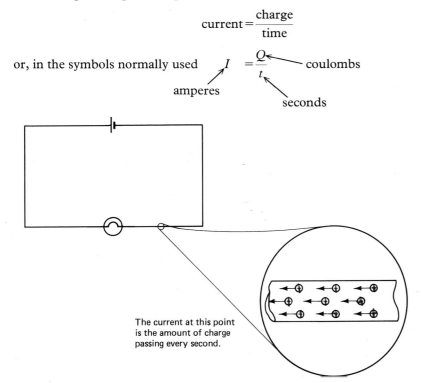

The current at this point is the amount of charge passing every second.

Currents smaller than 1 ampere are often measured in milliamperes (mA) or micro-amperes (μA).

$$1\,\text{A} = 1000\,\text{mA}$$
$$1\,\text{A} = 1\,000\,000\,\mu\text{A}$$
So $\quad 1\,\text{mA} = 1000\,\mu\text{A}$

It is useful to have some idea of how large the current might be in some common applications:

in a car headlamp bulb: 4 A
in a torch bulb: 0.2 A (200 mA)
in a transistor radio: 0.1 A (100 mA)
in a pocket calculator: 0.005 A (5 mA)
in a digital watch: 0.00005 A (50 μA)

Putting this the other way round, the amount of charge passing a point in a circuit depends on the size of the current and the time for which it flows:

$$\text{charge} = \text{current} \times \text{time}$$
$$Q = I\,t$$

So, from the unit of current (the ampere), we can define a unit of charge (the coulomb, C) in the following way:

1 coulomb is the amount of charge carried by a current of 1 ampere flowing for 1 second.

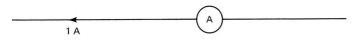

A current of 1 ampere means that 1 coulomb of charge is passing each second. 1 coulomb is equivalent to 6.24×10^{18} electrons; a current of 1 amp means that 6 240 000 000 000 000 000 electrons flow past each second!

EXAMPLE 25.1

If the current in a torch bulb is 0.2 A, how much charge flows through the bulb in 1 minute? How many electrons pass through the bulb in this time?

$$\begin{aligned}\text{charge} &= \text{current} \times \text{time}\\ &= 0.2\,\text{A} \times 60\,\text{s}\\ &= 12\,\text{C}\end{aligned}$$

1 coulomb is equivalent to 6.24×10^{18} electrons, so 12 coulombs are equivalent to

$$74.88 \times 10^{18}\,\text{electrons}$$
$$= 7.49 \times 10^{19}\,\text{electrons}$$

Current in a simple circuit

Is current different at different points in an electric circuit or is it the same? We can investigate this for the simplest electrical circuit – a single loop. Notice that the ammeter must be inserted into the circuit so that it is part of the closed loop. ▽

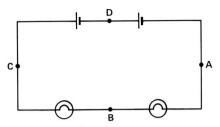

We measure the current at **A**, **B**, **C** and **D**. Here are some typical results:

At point ...	Current
A	0.25 A
B	0.25 A
C	0.25 A
D	0.25 A

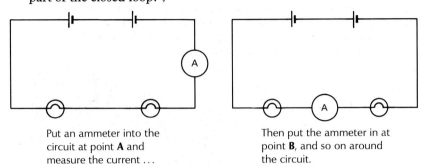

Put an ammeter into the circuit at point **A** and measure the current ...

Then put the ammeter in at point **B**, and so on around the circuit.

We get a very important result: **the current is the same at all points round a single loop circuit**.

No current is 'used up' by the lamps. Of course, some electrical energy is used up by each lamp (that is, it causes some lighting and heating), but the current is the same everywhere.

The model of electric current in wires as being like the flow of water through pipes can help to explain the result above. In a central heating system, hot water is pumped from the boiler through a number of radiators, as shown below. Since no water can escape from the closed system of pipes, the same volume of water per second must be flowing past each point round the loop. The water **loses energy** to the surroundings in each radiator, but the **amount** of water flowing stays the same.

Direction of the current

If we reverse the cells in any of the circuits above, the ammeter needle will try to move in the wrong direction. The direction of the current has reversed.

We cannot see an electric current so how can we tell which way it goes? The early electrical experimenters decided to think of an electrical current as a flow of positive charge from the positive terminal of the battery, round the circuit to the negative terminal. Now we think of electric current in wires as carried by electrons, which have a **negative** charge. The electrons flow in the opposite direction, from the negative terminal of the battery, round the circuit to the positive terminal.

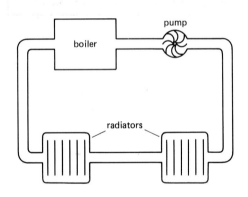

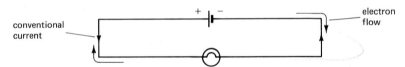

Some people still choose to use conventional current. Others prefer to think about the direction of the electron flow. It doesn't really matter as both are entirely equivalent. In this book, we will use electron flow and show the direction in which the charge carriers – the electrons – are moving.

Alternating current

Later in the book, we will come across another form of current, called **alternating current**. This is a current which changes direction at regular intervals. It is rather like having a battery which is being turned back-and-forward all the time! The electrons flow for a short time in one direction, then for a short time in the other. This is just as good as **direct current** (where the electron flow is always in one direction) for many jobs, like making a lamp light, and it is easier to produce than direct current on a large scale. The mains supplies alternating current.

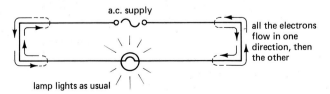

25.5 Potential difference (or voltage)

Energy changes in circuits

An electric circuit must have a cell (or some other source of electrical energy) if there is to be a current. The cell provides the electrical 'force' or 'pressure' to push the electrons; it supplies the **energy** to make the electrons move. But the 'strength' of a cell is not measured in newtons (units of force), or joules (units of energy), or even in amperes. If you look at a cell you will see that it is labelled in **volts**.

To understand what 'voltage' means, we need to think of circuits from an energy point-of-view. As each electron flows round a circuit, it is given electrical energy by the cell, and it loses this electrical energy again in the lamp.

> The word 'voltage' is really a shorthand way of saying 'the potential difference measured in volts'. So if we say that a transistor radio battery has a voltage of 9 V, we mean that the potential difference across its terminals is 9 V.

The voltage of a cell is a measure of the amount of electrical energy it gives to each electron which passes through the cell. 'Electrical energy' is rather difficult to imagine – it is like a kind of potential (or stored) energy. As the electron passes through the cell, it picks up some electrical potential energy and this is transformed into other forms of energy as the electron flows round the circuit.

Electrons are very small, so it is more convenient to think in terms of a larger unit of charge, the coulomb. **The voltage (or, more correctly, the potential difference, p.d.) of a cell is the amount of electrical potential energy which it gives to each coulomb of charge which passes through the cell.**

A cell has a p.d. of 1 volt if each coulomb of charge passing through it is given 1 joule of potential energy.

1 volt = 1 joule per coulomb

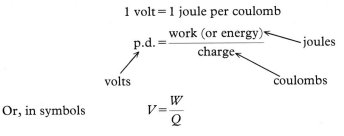

Or, in symbols $$V = \frac{W}{Q}$$

EXAMPLE 25.2

A battery is switched on for a few moments. In that time 25 C pass through the battery and they carry away 300 J of electrical energy. What is the p.d. across the battery?

$$V = \frac{W}{Q}$$

Substituting into the equation:

$$V = \frac{300 \text{ J}}{25 \text{ C}}$$
$$= 12 \text{ J/C}$$
$$= 12 \text{ V}$$

It is a 12 V battery.

Talking about circuits

Try to use the right words when talking or writing about electric circuits. We can talk about:

current **through** a conductor
current **in** a conductor or **in** a circuit
current **round** a circuit
p.d. or voltage **across** a conductor
p.d. or voltage **between** two points

It is a mistake to talk about p.d. or voltage **through** something. P.d. measures energy differences and these cannot flow **through** something! It is also wrong to talk of p.d. or voltage **in** a conductor – use the word 'across' instead.

Measuring potential difference

The potential difference (p.d.) between two points in a circuit can be measured using a voltmeter.

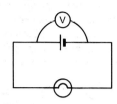
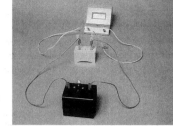

The voltmeter is connected directly to the two points concerned, without disconnecting the circuit. This is different from the way we use an ammeter, which has to be put *into* the circuit. (Look back at the diagram at the top of p. 240.)

P.d. across a series of cells

The p.d. of a cell is measured by connecting a voltmeter directly across its terminals. A group of cells connected together is called a **battery**. If we connect several cells in series, we find that the total p.d. is simply the sum of the individual p.d.s. If, however, two identical cells are connected back to back, the p.d. of the pair will be zero.

Between...	P.d.
A and B	1.5 V
B and C	1.5 V
C and D	1.5 V
A and C	3 V
B and D	3 V
A and D	4.5 V

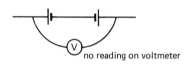
no reading on voltmeter

P.d.s round a simple circuit

Let's now look at a complete circuit. Potential difference is a measure of the electrical potential energy which charges gain as they pass through a cell. As the charges flow round a circuit, they lose this potential energy again, transforming it into other forms of energy. Taking the circuit as a whole, total electrical energy gained must be equal to the total lost.

Measuring energy gains and losses round a circuit. Each coulomb of charge gains 1.5 joules of electrical energy from each cell, making 3 joules in all. 1 joule is then transformed into light and thermal energy in each lamp, again making a total of 3 joules.

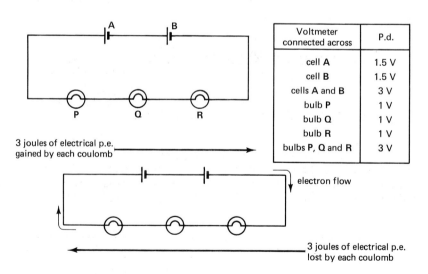

Voltmeter connected across	P.d.
cell A	1.5 V
cell B	1.5 V
cells A and B	3 V
bulb P	1 V
bulb Q	1 V
bulb R	1 V
bulbs P, Q and R	3 V

3 joules of electrical p.e. gained by each coulomb

electron flow

3 joules of electrical p.e. lost by each coulomb

The sum of the p.d.s round a circuit must be equal to the p.d. of the battery.

◁ In the 'water' model of electric current which we used earlier, you might imagine a system in which water was pumped uphill and then flowed back down in a series of stages back to the starting point.

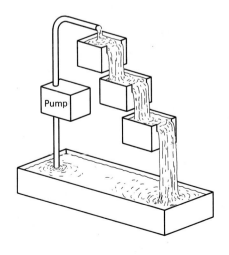

25.6 Resistance

Same battery – different currents

The other important electrical idea – along with charge, current and potential difference—is **resistance**. In a circuit, the battery supplies the energy to make charge flow round, but the circuit provides a resistance to this. We get a different current from the same battery in different situations. The size of the current depends on the **resistance** of the whole circuit.

This is why cells are labelled in 'volts' and not in 'amps'. You can buy a 1.5 V cell, but not a 1.5 A one! The current will depend on what resistance is connected across the cell.

Resistors are electrical components specially made to have a certain resistance to electrical current. The units used to measure resistance are called **ohms** (or Ω, the Greek letter 'omega', for short).

Resistance and current

If a cell is connected to a resistor, the current gets smaller as the resistance is made larger. The more resistance, the smaller the current which the cell is able to 'push' round the circuit.

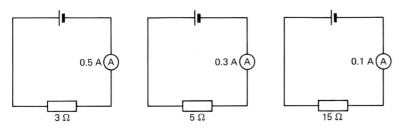

Use the same cell each time.
The larger the resistance, the smaller the current.

In terms of the 'water flow' model, a resistor is like a narrow section of pipe. The narrower it is, the more resistance it has. If a water circuit has one narrow section, then the pump will not be able to push as much water round the circuit each second.

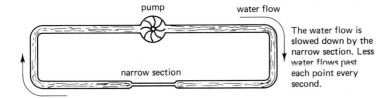

Notice too that the flow will be slowed down *before* the narrow section as well as *after* it. The same is true of the electrical circuit. The resistor limits the current all round the circuit. There is the same current before and after the resistor (at points **A** and **B** in the diagram below).

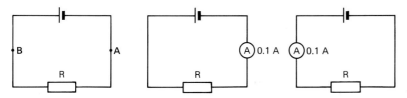

Resistors are very useful in electrical circuits because they allow us to control the size of the current. We will look in more detail at resistance in the next chapter.

25.7 Circuits with branches

Most circuits are more complicated than the simple loop. We will now take a look at what happens in circuits with branches, and see how the same basic ideas – current, voltage, resistance – enable us to understand these circuits too. For practical investigations of these circuits we can use either lamps or resistors. The lamp filament *is* simply a resistor. One advantage of using lamps is that when they light, we know that the circuit is working!

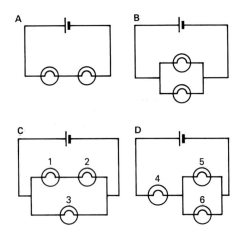

Series and parallel

There are two basic ways of joining electrical components together. In circuit **A**, the two lamps are connected in **series**. They are both in the same simple loop. In **B**, the two lamps are connected in **parallel**. Other more complicated circuits are built up of **series** and **parallel** parts. For example, in circuit **C**, lamps 1 and 2 are in series with each other; but together they are in parallel with lamp 3. In circuit **D**, lamps 5 and 6 are in parallel with each other; and this parallel combination is then in series with lamp 4.

Current in parallel branches

1. Equal branches

If there is a branch, electrons may flow in one of two directions, as shown in the diagram below.

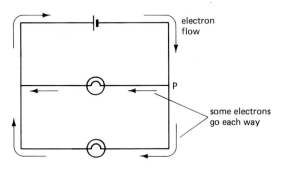

We can measure the current at different points in the circuit.

Measuring the current at **B**: Notice that the ammeter has to be inserted into the circuit at **B**. Measurements at the other points are done in the same way.

Here are some typical results:

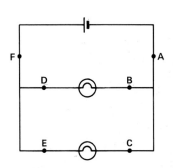

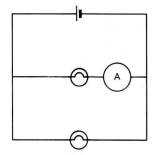

At point ...	Current
A	0.4 A
B	0.2 A
C	0.2 A
D	0.2 A
E	0.2 A
F	0.4 A

The current divides in such a way that:

    current at **A** = current at **B** + current at **C**

Again, no current is used up in the lamps:

    current at **D** = current at **B**

and   current at **E** = current at **C**

After the lamps, the branch currents join again as we would expect:

    current at **F** = current at **D** + current at **E**

2. Unequal branches

Some circuits have two parallel branches which are not identical.

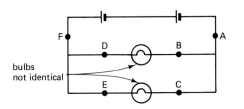

bulbs not identical

◁ We can measure the current at different points in the circuit, as before.

Here are some typical results. ▷

At point ...	Current
A	0.4 A
B	0.15 A
C	0.25 A
D	0.15 A
E	0.25 A
F	0.4 A

Even though the current does not divide into two equal parts, the sum of the two branch currents is still equal to the current from the cell.

These results support the idea of electric current as a flow of electrons.

$I_1 = I_2 + I_3$

P.d.s across parallel branches

In a simple series circuit, we have seen how p.d. is a measure of the electrical energy which each unit of charge gains from the battery and loses again in the lamps or resistors as it flows round the loop (see p. 242). What happens to p.d. in a circuit with parallel branches?

We can measure the p.d. across the battery and across each of the bulbs **A**, **B** and **C**.

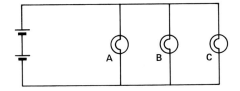

Measuring the p.d. across bulb **B**. Notice that the voltmeter is connected across **B** without disconnecting the circuit. Measurements of the other p.d.s were done in the same way.

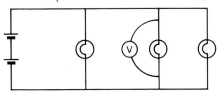

Here are some typical results:

Voltmeter connected across ...	P.d.
battery	3 V
bulb A	3 V
bulb B	3 V
bulb C	3 V

We find that the p.d. across each parallel branch is the same. The p.d. across each lamp is 3 V, equal to the p.d. across the battery.

Let us think about what is happening in the circuit in energy terms. Each coulomb of charge passing through the battery gains 3 joules of electrical potential energy.

As charge flows round the circuit, it will divide, some going through each lamp. Any particular electron can pass through one lamp only, and has to lose all its electrical energy in one go. So 1 coulomb of charge leaving the cell will divide three ways, and each portion will lose all its electrical energy in one lamp. In each lamp, the energy transferred is 1 joule per $\frac{1}{3}$ coulomb, i.e. 3 joules per coulomb. Hence the p.d. across each lamp is the full 3 V.

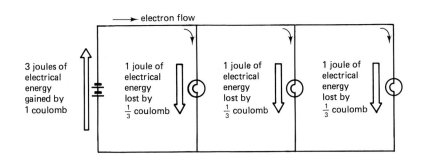

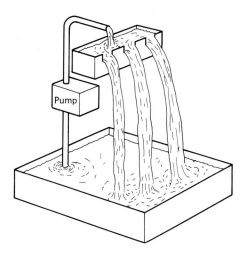

A water-flow model of a circuit with parallel branches.

Questions

1 Two pupils were asked to write down what they think electricity must be like. Which of their explanations do you think is best? List all the evidence which supports this explanation.

Peter's explanation
I think electricity is like the petrol in a car. The battery is like the petrol tank. The wire is like the tube carrying the petrol to the engine.
  The bulb uses up the electricity to make light, just like the engine which uses up the petrol to make the car go.

Susan's explanation
Electricity is like water flowing round a central heating system. The battery pumps electricity round to the bulb, like a water pump pumping hot water from the boiler to a radiator. As the water in the radiator gets cold, it flows back to the boiler to get heated up and pumped round again.
  The electricity needs another wire to let it flow back to the battery and round again.

2 Batteries and bulbs have two terminals. Which of these lamps would light?

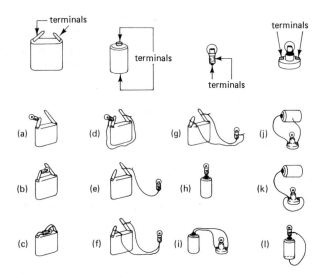

3 (a) This torch uses two 1.5 V cells. Copy the diagram and mark (with a red line) the closed loop which the current follows when the torch is switched on. Which parts of the torch must be made of metal (conductor)?

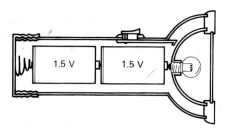

(b) When you switch on the torch, it doesn't light. List all the possible reasons. How would you test to find out what was wrong and correct the fault?

4 In these two circuits, a lamp can be switched on and off using two switches. Copy and complete the truth tables for the switches in the two circuits:

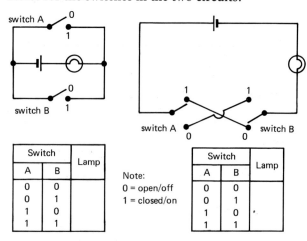

Switch		Lamp
A	B	
0	0	
0	1	
1	0	
1	1	

Note:
0 = open/off
1 = closed/on

Switch		Lamp
A	B	
0	0	
0	1	
1	0	
1	1	

5 Some motor car lighting systems use lamps with double filaments instead of using two separate lamps.

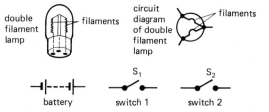

If you were given such a lamp, a battery, and two switches, draw *one circuit only*, that would allow you
(a) to switch on either filament on its own *and*
(b) to switch on both filaments at the same time.
(SEB)

6 (a) How much charge (in coulombs) passes a point in a conductor each second, when there is a current of 10 mA in the conductor?
(b) The maker of a car battery claims that it can supply a current of 3 A for 12 hours. How much charge passes through the battery in this time?
(c) A charge of 30 000 C flows through a lamp in 100 minutes. What is the current?
(d) If there is a current of 2 A in a circuit, how long will it take for 1000 C to flow past a point in the circuit?

7 In this circuit, all the lamps are identical. Is the brightness of L_1 more than, less than, or the same as the brightness of L_5? Which of the following statements are true?
(a) L_4 is brighter than L_5.
(b) L_3 is lit up.
(c) L_3 and L_5 are the same brightness.
(d) L_4 is dimmer than L_1.

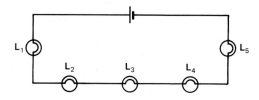

Current Electricity: Basic Ideas

8 The current measured by ammeter **1** is 0.6 A. What is the current measured by ammeter **2**?

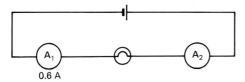

9 Most portable radios and cassette players need several 1.5 V cells to make them work. What is the total p.d. needed to run each of these, if the cells are inserted in the patterns shown below?

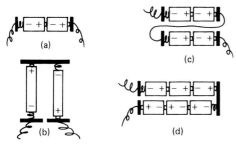

10 In these circuits, the readings on some of the ammeters are shown, but some are not. Predict what the readings on the other ammeters will be.

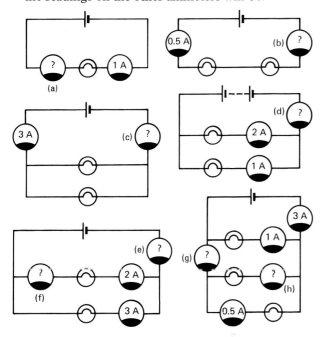

11 Redraw this circuit, showing on your diagram how you would connect a voltmeter to measure the p.d. (voltage) across lamp **P** and an ammeter to measure the current through lamp **R**.

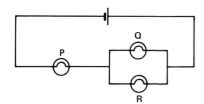

12 You can buy a 1.5 V cell or 9 V battery. Explain why you *can not* buy a 1.5 A cell or a 9 A battery.

13 The readings on some of the voltmeters in these circuits are shown. What are the readings on the other voltmeters?
(**Note:** The bulbs and cells are *not* all identical!)

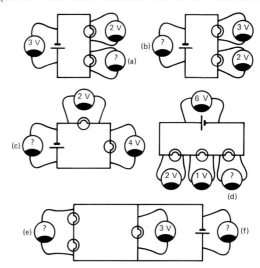

14 (a) These three bulbs are working at normal brightness from a 24 V battery. What is the correct voltage rating for each bulb?

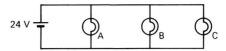

(b) Twenty Christmas tree lights work correctly from 240 V mains when they are connected like this. What is the correct voltage rating for each bulb?

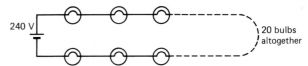

15 In which of these circuits is the current largest? In which circuit is the current smallest? What difference would you expect to **see** when you looked at the three circuits?

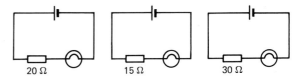

16 (a) How much work is done in transferring 2 coulombs of electric charge through a potential difference of 10 volts?
(b) What apparatus would you need to measure the number of joules of electrical energy given by a battery to every coulomb of charge passing through?

INVESTIGATION
How much current do battery-powered devices (like radios, cassette recorders and so on) require to make them work?

26: Ohm's Law and Resistance

26.1 Current and voltage: Ohm's Law

Current and voltage are two completely different things: current is a flow of charge; voltage is a measure of how much energy is gained or lost by each coulomb of charge between two points. But the two *are* related.

If we connect a length of wire to a battery, we find that the current through the wire depends on the p.d. across the wire (which is, of course, equal to the p.d. of the battery). The diagram shows one way to investigate this. We measure the current with one dry cell (1.5 V), then with 2, 3, 4, 5 and 6 dry cells in series. Typical results are shown in the table. If we draw a graph of current through the wire against p.d. across it, we get a straight line through the origin. If we double the p.d., the current doubles, and so on. The current through the wire is proportional to the p.d. across it.

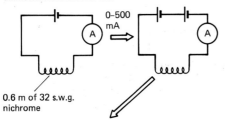

Connect a length of wire in series with a 1.5 V cell and an ammeter. Note the current.

Add a second cell. Note the current now.

0.6 m of 32 s.w.g. nichrome

Repeat with 3, 4, 5 and 6 cells to get a results table.

Number of cells	P.d. across wire V	Current mA	P.d/current V/A
1	1.5	75	20
2	3.0	150	20
3	4.5	225	20
4	6.0	300	20
5	7.5	375	20
6	9.0	450	20

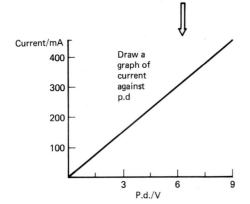

Draw a graph of current against p.d

Experiments with other lengths and types of wire give the same result. The link between p.d. and current was first discovered in 1826 by a German physics teacher, Georg Ohm, and is known as Ohm's Law:

The current flowing through a metal conductor is directly proportional to the p.d. across the conductor, provided that its temperature remains constant.

Definition of resistance

Here are some results obtained by repeating the same investigation with a different piece of wire.

P.d. across wire V	Current mA	P.d./current V/A
1.5	50	30
3.0	100	30
4.5	150	30
6.0	200	30
7.5	250	30
9.0	300	30

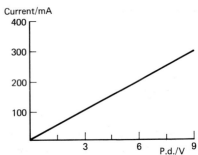

Comparing this with the first investigation, we find that current and p.d. are proportional but the current is smaller for each p.d. This second wire has a larger **resistance** than the first one. If we divide the voltage by the current, we get a number which gives us a measure of resistance. In fact, this is how resistance is defined:

$$\text{resistance of a conductor} = \frac{\text{p.d. across conductor}}{\text{current through conductor}}$$

ohms → $R = \dfrac{V \leftarrow \text{volts}}{I \leftarrow \text{amperes}}$

A conductor has a resistance of 1 Ω if a current of 1 A flows through it when the p.d. across it is 1 V.

The wire used in the first investigation has a resistance of 20 Ω and the wire used in the second investigation has a resistance of 30 Ω.

The result $R = V/I$ (or its equivalent, $V = IR$) is sometimes called Ohm's Law. However, it is really the definition of resistance. What Ohm discovered was that the resistance of a metallic conductor is constant (provided its temperature does not change). The slope of the $V-I$ graph is constant, so we will get the same value for the ratio V/I if we choose any pair of points from the graph. Not all conductors give a straight-line $V-I$ graph. The ones which do are called **ohmic conductors**.

26.2 Using the V, I, R equations

The resistance equation can be written in symbols:

$$R = \frac{V}{I}$$

where R is the resistance, V the p.d. across it and I the current through it. Two other ways of writing the same equation are:

$$V = IR \quad \text{and} \quad I = \frac{V}{R}$$

These are useful for calculating one of the three quantities, current (I), p.d. (V) or resistance (R), if the other two are known. ▷

26.3 Resistance

The resistance of a piece of wire depends on:
- its length
- its thickness
- the material it is made from.

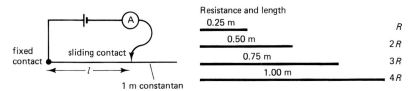

Resistance and length

The diagram below shows a method for investigating how resistance changes as the length of the wire changes. A 1 metre length of resistance wire is stretched out along the bench. By sliding the contact along, we can vary the length of wire actually in the circuit.

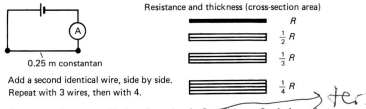

The p.d. across the resistance wire stays constant throughout. (It is equal to the p.d. of the battery.) We measure the current with different lengths of wire in the circuit and calculate the resistance each time ($R = V/I$). A graph of resistance against length is a straight-line through the origin. The resistance is proportional to the length.

Resistance and thickness

To investigate the effect of thickness on resistance, we use several equal lengths of the same resistance wire. Two pieces side by side will have twice the cross-sectional area of a single wire, three have three times the cross-sectional area, and so on.

Add a second identical wire, side by side.
Repeat with 3 wires, then with 4.

Using the same battery all the time (p.d. fixed), we find that the current increases as the thickness of wire increases. It is easier for electrons to flow through the thicker wire. The resistance gets less as the thickness increases. In fact, they are **inversely proportional** – if the cross-sectional area is doubled, the resistance is halved, and so on.

EXAMPLE 26.1

What p.d. is needed to make a current of 2 A flow through a lamp whose resistance is 6 Ω?

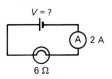

We know that: $I = 2\,\text{A}$
$R = 6\,\Omega$

We use the equation in the form:
$$V = IR$$
$$V = 2\,\text{A} \times 6\,\Omega$$
$$= 12\,\text{V}$$

A p.d. of 12 V is needed.

EXAMPLE 26.2

A coil of wire is connected across a 9 V battery. The current in the wire is 0.3 A. What is the resistance of the wire?

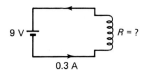

We know that: $V = 9\,\text{V}$
$I = 0.3\,\text{A}$

We use the equation in the form:
$$R = \frac{V}{I}$$
$$R = \frac{9\,\text{V}}{0.3\,\text{A}}$$
$$= 30\,\Omega$$

The wire has resistance 30 Ω.

EXAMPLE 26.3

The windscreen wiper motor of a car has a resistance of 10 Ω. When it is connected to the 12 V car battery, what current flows through it?

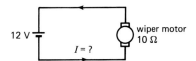

We know that: $R = 10\,\Omega$
$V = 12\,\text{V}$

We use the equation in the form:
$$I = \frac{V}{R}$$
$$I = \frac{12\,\text{V}}{10\,\Omega} = 1.2\,\text{A}$$

A current of 1.2 A flows through the motor.

Resistivity

These two results are summarised in the equation:

$$\text{resistance} \rightarrow R = \frac{\rho l}{A}$$

where ρ is resistivity, l is length, and A is cross-sectional area.

Resistivity is a basic electrical property of each material. It is measured in units ohm metres ($\Omega\,\text{m}$). If we know the resistivity of a material, we can calculate the resistance of any particular wire made from it.

Facts about resistors

Fixed resistors

A fixed resistor can be made from a length of resistance wire. Common materials to use are the alloys **nichrome** (a mixture of 60% nickel, 24% iron and 16% chromium) and **constantan** (a mixture of 55% copper and 45% nickel). Accurate laboratory standard resistors are made from a coil of resistance wire inside a protective plastic case. The resistance value is guaranteed accurate to within 0.2% of this value or better.

Some wire-wound resistors consist of a coil of wire embedded in a ceramic outer casing. These are not quite so accurate as the standard resistors but they are smaller and more robust. They are used in circuits where the currents are large and the resistor may get quite hot in use.

For many purposes, smaller and cheaper resistors are used. The resistors used in most electronic circuits do not contain any metal wire – they use carbon as the conducting material. Carbon composition resistors are made by baking carbon black with a binding material in a kiln. The finished resistor consists of a small piece of this hard carbon material with two short lengths of connecting wire attached to the ends, coated in a ceramic tube to hold it all together. Another type of small resistor is the carbon film resistor which is made by coating a small ceramic tube with carbon and then scratching a spiral groove around the tube to give the resistor a particular value. Again two wires are attached and the resistor is protected in an outer ceramic coat.

These small resistors come in a range of sizes. In use a resistor gets hotter (see Chapter 27). It must be able to lose this energy to the surroundings or it will overheat. Resistors come in different power ratings: 2 W, 1 W, 0.5 W and 0.25 W.

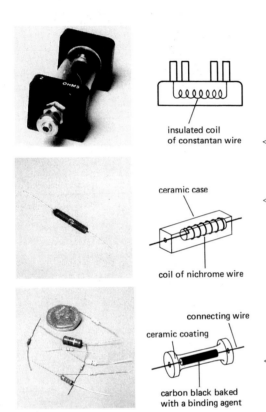

insulated coil of constantan wire

ceramic case

coil of nichrome wire

connecting wire
ceramic coating

carbon black baked with a binding agent

Resistor colour code

The resistance value of small resistors is marked on them using a special code. This uses coloured bands painted round the resistor.

How accurate it is: red = 2%
gold = 5%
silver = 10%
no band = 20%

Colour	
Black	0
Brown	1
Red	2
Orange	3
Yellow	4
Green	5
Blue	6
Violet	7
Grey	8
White	9

Examples:

	First digit	Second digit	Number of zeros	
	brown 1	black 0	red 2 zeros	$\rightarrow$ 1 000 Ω
	green 5	blue 6	orange 3 zeros	$\rightarrow$ 56 000 Ω
	grey 8	red 2	yellow 4 zeros	$\rightarrow$ 820 000 Ω
	red 2	violet 7	black 0 zeros	$\rightarrow$ 27 Ω

You will see, if you look at modern books or magazines on electronics, that a new code is now being used to label the resistor sizes on some circuit diagrams. The table below shows how it works.

Symbol	Meaning
R	ohms
K	thousand ohms (kilo-ohms)
M	million ohms (megohms)

The symbol is put where the decimal point would go; the resistor value is always *exactly* 3 characters (2 figures plus 1 symbol).

Examples: 1R0 = 1.0 Ω
39R = 39 Ω
K33 = 0.33 kΩ = 330 Ω
1K5 = 1.5 kΩ = 1500 Ω
22K = 22 kΩ = 22 000 Ω
M68 = 0.68 MΩ = 680 000 Ω
1M8 = 1.8 MΩ = 1 800 000 Ω

26.4 Variable resistors

Variable resistors enable us to alter the current in a circuit at the turn of a knob. The volume control of a radio, for example, is a variable resistor. A larger type of variable resistor used in the laboratory is the **rheostat**. This has a coil of resistance wire wound on a ceramic tube. A sliding contact can be moved to any position along the coil. The amount of coil actually in the circuit depends on where the sliding contact is placed. By moving it we change the resistance and alter the current.

Smaller variable resistors work in almost exactly the same way. Inside the outer case is a small coil of wire, curved into a circular shape. A metal slider (or wiper) moves over this coil as the knob is turned. This changes the length of wire actually in the circuit, and so changes the resistance. Cheaper versions use a curved length of carbon track rather than a coil of wire.

The variable resistor has three terminals. The resistance between terminals **A** and **C** is fixed; the resistance between **A** and **B** and between **B** and **C** changes as we turn the knob. In some applications (see section 26.7), all three terminals are used. A variable resistor used in this way is sometimes referred to as a **potentiometer** (or simply a 'pot').

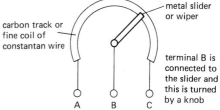

Even smaller variable resistors are used in electronic circuits. They are often used to adjust the resistance at some points in the circuit when the circuit is first being tested, and are then left at this setting. These are called **presets**. Some are open and you can see that they are just miniature versions of the larger variable resistors. They have a circular strip of carbon and a slider which can be moved using a screwdriver. Some presets are enclosed in a plastic case, but have the same construction inside. ▽

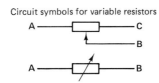

Circuit symbols for variable resistors

The resistance between A and C is fixed; the resistance between A and B can be varied.

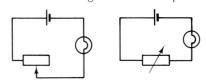

In these circuits the variable resistor controls the brightness of the lamp.

Some other useful variable resistors

Light dependent resistor, LDR

This is a device whose resistance changes depending on the brightness of light falling on it. One common type, the ORP12, uses the semiconductor material cadmium sulphide. (Semiconductors are discussed in more detail in Chapter 32.) It has two sets of metal wires, separated by a narrow strip of cadmium sulphide. This is then set in a small block of clear plastic to protect it. In the dark, the LDR has a very high resistance (over 1 MΩ) but in normal lighting this falls to around 300 Ω.

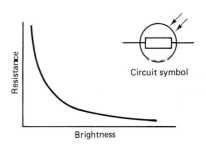

LDRs can be used to switch other circuits on when the light changes. For example, they can be used to switch on a light as darkness falls. In section 32.7, we will see how a circuit can be designed to do this. The LDR can also be used to make a simple light-meter to show how bright the light is (see also the first diagram in Chapter 22).

Thermistor

A thermistor is a resistor whose resistance changes with temperature. Thermistors are also made from semiconductor materials. They come in many different shapes and sizes. Most thermistors decrease in resistance as the temperature rises. A typical thermistor might have a resistance of 10 kΩ at room temperature (25°C), falling to 700 Ω at 100°C.

Thermistors can be used to operate thermostats – to switch other things off and on if the temperature rises too high or falls too low. They can also be used to make electrical thermometers (see p. 129).

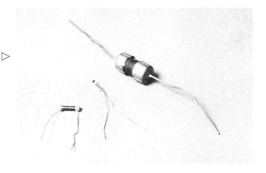

252 Ohm's Law and Resistance

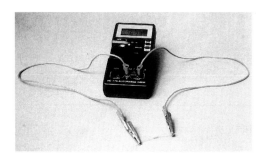

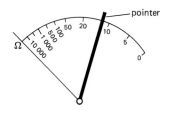

26.5 Measuring resistance

Using a multimeter

The easiest way to measure a resistance is to use a multimeter with a resistance scale. Simply connect the two ends of the resistor to the two terminals of the multimeter. It then gives a direct reading of the resistance value. If you are using a pointer-type multimeter, you will notice that the resistance scale is calibrated 'back to front' – the small values are on the right-hand side and the large values on the left. This gives a clue to how the multimeter measures resistance. There is a battery inside the meter which passes current through the resistor when it is connected. The meter actually measures this current, but the scale is marked directly in ohms. Notice that the scale divisions are not evenly spaced. As a result, measurements are not very accurate for very small (less than $0.1\,\Omega$) and very large (more than $100\,000\,\Omega$) resistors. Digital multimeters also contain a battery and work in much the same way. They are quick and convenient to use and are often accurate enough.

Substitution method

Another method is to compare the unknown resistance with resistors whose value we know. A **resistance box** makes this easier. It contains a number of standard resistors; by setting the switches you can select any resistance within the range of the box. The resistance box is adjusted until the current on the ammeter is the same for both positions of the two-way switch. Then the unknown resistor must be the same as the setting on the box. This can simply be read off.

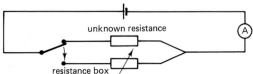

This method will work for any resistor whose value lies between the maximum and minimum values of the resistance box.

Ammeter/voltmeter method

By using the resistance equation, $R = V/I$, we can calculate an unknown resistance by measuring the current through it when a known p.d. is applied. The circuit is shown in the diagram below. Notice that the ammeter is placed in series with the resistor and the voltmeter is placed across the resistor, in parallel with it.

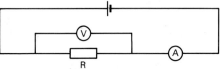

This method can be improved by varying the applied p.d. and taking a series of current and p.d. readings. There are several ways of doing this:

(i) Use different batteries to supply different p.d.s.
(ii) Use a variable power supply whose voltage can be altered by turning a knob or switch.
(iii) Add a rheostat to the circuit. By moving the slider on the rheostat, the voltage across the resistor can be changed. ▽

If we have a series of readings, we can either calculate V/I for each and take the average, or we can plot a graph of current against p.d. The resistance can be calculated from the slope of the graph (look back to p. 248).

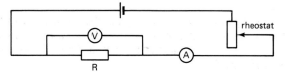

26.6 Circuit calculations

We are now going to bring together the ideas about current, p.d. and resistance which we have learned so far and see how these can be used to understand what is going on in a whole range of electrical circuits. There are so many circuits we could make that it is completely impossible to *learn* and *remember* how they all behave. Instead we have to try to *understand* what is going on. It is very useful to have in your mind a 'picture' of what current, p.d. and resistance are; try to think out how you would expect each circuit to work before you start using formulae and rules.

Let's begin by summarising the basic ideas about circuits which we have already met.

In a **series** circuit:

1. The current is the same at all points round the circuit.

2. The sum of the p.d.s across the separate resistors is the same as the p.d. across the battery.

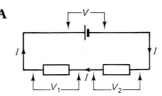

Current I is the same everywhere
$V_1 + V_2 = V$

In a circuit with **parallel** branches:

3. Parallel branches have the same p.d. across them.

4. The sum of the currents through the parallel branches is the same as the current in the main circuit before and after the branch.

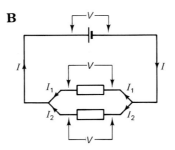

P.d. V is the same across each branch.
$I_1 + I_2 = I$

For any resistor in a circuit:

5. There is a relation between the p.d. across a resistor, the current through the resistor and its resistance:

$$\text{resistance} = \frac{\text{p.d. across resistor}}{\text{current through resistor}}$$

$$R = \frac{V}{I}$$

It is sometimes more convenient to use this in the form:

$$V = IR \quad \text{or} \quad I = \frac{V}{R}$$

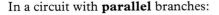

What we would like to be able to do is to calculate exactly what the currents, voltages or resistances are. For example, instead of just knowing that the two voltages in circuit **A** must add to the cell voltage, we might want to calculate exactly what each voltage is (if we know the size of the two resistors). Instead of just knowing what the sum of the two currents in the branches of circuit **B** is, we might want to know exactly what size each current is.

Instead of going straight into using equations and formulae, we will begin by trying to work out what we would *expect* to happen in these circuits, using our basic ideas about current and p.d. Then we can use the equations to confirm this.

P.d. in series circuits

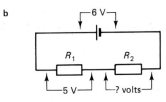

a Potential difference is a measure of the electrical energy gained or lost by each unit of charge as it passes. This battery gives 6 J to every coulomb; in the resistor, every coulomb loses 6 J again. The p.d. across the resistor is 6 V.

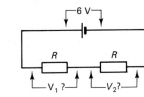

b If there are two resistors, the electrical energy is transferred in two steps. The total energy lost by each coulomb must be equal to the total energy gained. If the p.d. across R_1 happens to be 5 V, then the p.d. across R_2 must be 1 V. (**Note.** The result would be the same if there were three, or four, or any other number of resistors in series. The sum of the p.d.s would still be equal to the p.d. of the cell.)

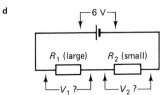

c This is a special case: the two resistors are equal. The electrical energy gained in the battery will be lost in two equal stages. The p.d. across each resistor will be 3 V.

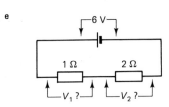

d In this circuit, R_1 is now larger than R_2. More energy will be lost in the large resistor than in the small one. The sum of the p.d.s must still be 6 V. But V_1 will be larger than 3 V and V_2 will be smaller than 3 V. The larger p.d. is across the larger resistor.

e In this circuit, we know the sizes of the two resistors. They are in the ratio 2:1. So we might predict that the two p.d.s will also be in the ratio 2:1, with the larger p.d. across the larger resistor. They must also add to 6 V. So the p.d. across the 2 Ω resistor is 4 V and across the 1 Ω resistor is 2 V.

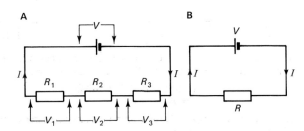

These are equivalent: same battery p.d., same current.

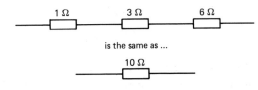

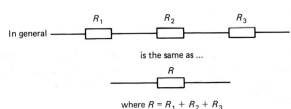

Addition of resistors in series

The circuits above have two resistors in series. Sometimes we want to find the value of the single resistor which has the same effect as several separate resistors connected in series.
◁ In circuit **A**, there is the same current I through each resistor. We can apply the resistance equation to each in turn:

$$V_1 = IR_1 \qquad V_2 = IR_2 \qquad V_3 = IR_3$$

The sum of these separate p.d.s must be equal to the p.d. across the battery:

$$\begin{aligned} V &= V_1 + V_2 + V_3 \\ &= IR_1 + IR_2 + IR_3 \\ V &= I(R_1 + R_2 + R_3) \end{aligned} \qquad (1)$$

In circuit **B**, there is the same applied p.d. and the same current:

$$V = IR \qquad (2)$$

Comparing equations (1) and (2), we can see that:

$$\boldsymbol{R = R_1 + R_2 + R_3}$$

So the effective resistance of several separate resistors in series is simply equal to their sum.

We are now in a position to confirm the result for circuit e on p. 254 by calculation. There are three steps:

Step 1
We add the two resistors to find their total resistance:
A 2 Ω resistor and a 1 Ω resistor in series are equivalent to a single 3 Ω resistor.

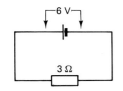

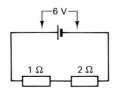

Step 2
We then use the resistance equation to calculate the current in the circuit:
$V = 6\,V, R = 3\,\Omega$

$$I = \frac{V}{R} = \frac{6\,V}{3\,\Omega} = 2\,A$$

Step 3
Finally, we apply the resistance equation to **each separate resistor** to work out the p.d.s across each:
For the 2 Ω resistor: $V = IR$
$= 2\,A \times 2\,\Omega$
$= 4\,V$
For the 1 Ω resistor: $V = IR$
$= 2\,A \times 1\,\Omega$
$= 2\,V$

This confirms our prediction. The p.d.s across the two resistors are in the same ratio as the resistances, with the larger p.d. across the larger resistance.

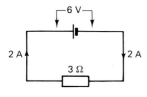
back to original circuit

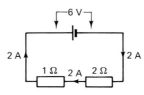
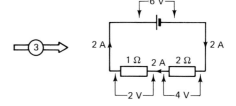

EXAMPLE 26.4

A variable resistor and a fixed resistor are connected in series. As the resistance of the variable resistor is *increased*, what will happen to the reading on a voltmeter placed across:

(i) the variable resistor;
(ii) the fixed resistor?

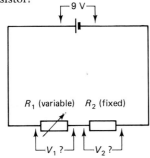

The sum of the two p.d.s must be the same as the battery p.d., in this case, 9 V. As R_1 is increased, a larger share of the total p.d. will appear across it. So V_1 will *increase*.

Because the two p.d.s must always add to 9 V, this means that V_2 will *decrease*.

(**Note.** It is important to realise that changing a resistance at one place in a circuit can affect the currents and p.d.s at other places in the circuit also. In this case, changing R_1 altered *both* p.d.s, V_1 and V_2.)

EXAMPLE 26.5

Four resistors with values 5, 6, 10 and 3 Ω are connected in series to a 12 V battery. What current will flow in this circuit? What is the p.d. across each resistor?

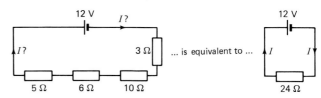

This is a series circuit so the current is the same everywhere. We begin by adding the resistors. The total resistance is:

$$R = 5\,\Omega + 6\,\Omega + 10\,\Omega + 3\,\Omega$$
$$= 24\,\Omega$$

A circuit with a single 24 Ω resistor would have the same current. Using the resistance equation:

$$V = IR$$
$$12\,V = I \times 24\,\Omega$$

$$I = \frac{12\,V}{24\,\Omega} = 0.5\,A$$

The current in the circuit is 0.5 A

The p.d.s across the separate resistors are easily calculated:
For the 5 Ω resistor: $V = IR$
$= 0.5\,A \times 5\,\Omega$
$= 2.5\,V$

In the same way, the p.d.s across the 6 Ω, 10 Ω and 3 Ω resistors are 3 V, 5 V and 1.5 V respectively. Notice that the four separate p.d.s add to 12 V and that the biggest p.d. is across the largest resistor.

Currents in circuits with parallel branches

a Current is a flow of charge. Before the branch, the current is 3 A (3 coulombs per second). If 2 A flows through R_1, then the current in R_2 must be 1 A. When the branches join up again, the current will be 3 A once more. (**Note.** The same result would apply if there were three, or four, or any other number of parallel branches. The currents would still add to equal the current before, and after, the branch.)

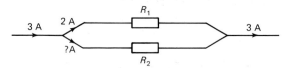

b This is a special case: the two resistors are equal. The two paths which the current can follow have equal resistance. So the two branch currents are equal, 1.5 A in each branch.

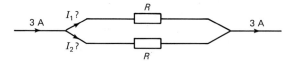

c In this circuit, R_1 is larger than R_2. The sum of the currents must still be 3 A; but the current through the smaller resistor is the larger. I_1 is less than 1.5 A and I_2 is more than 1.5 A.

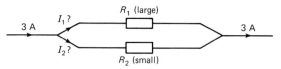

d In this circuit, we know the sizes of the resistors. They are in the ratio 2:1 (6:3). We would expect that the currents will also be in the ratio 2:1, with the larger current in the smaller resistor. They must add to 3 A. So the current in the 3 Ω resistor is 2 A and the current in the 6 Ω resistor is 1 A.

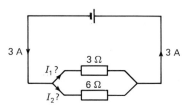

e In a circuit with parallel branches, we often know the sizes of the resistors and the p.d. across the cell. From this we can calculate the currents in the branches.

Each resistor is connected directly to the terminals of the cell. So the full cell p.d. of 6 V is applied across each resistor. Using the resistance equation:

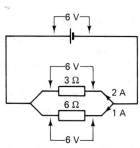

For the 3 Ω resistor: $$I = \frac{V}{R} = \frac{6\text{ V}}{3\text{ Ω}}$$
$$= 2\text{ A}$$

For the 6 Ω resistor: $$I = \frac{V}{R} = \frac{6\text{ V}}{6\text{ Ω}}$$
$$= 1\text{ A}$$

The total current from the cell is therefore 3 A (2 A + 1 A). Notice that the currents are in the ratio 2:1 and that the resistors are also in the ratio 2:1. But the current ratio is the other way round. The larger current is through the smaller resistor. The current through the 3 Ω resistor is twice the current through the 6 Ω resistor.

Addition of resistors in parallel

As with resistors in series, it is also useful to have a general method for adding resistors in parallel. Again we are looking for a single resistor which will have the same effect as the parallel combination.

In circuit **A**, the p.d. across each resistor is V. We can apply the resistance equation to each in turn:

$$I_1 = \frac{V}{R_1}, \quad I_2 = \frac{V}{R_2}, \quad I_3 = \frac{V}{R_3}$$

The sum of the branch currents must be equal to the current from the battery:

$$I = I_1 + I_2 + I_3$$

$$I = \frac{V}{R_1} + \frac{V}{R_2} + \frac{V}{R_3}$$

So $$I = V\left(\frac{1}{R_1} + \frac{1}{R_2} + \frac{1}{R_3}\right) \quad (1)$$

In circuit **B**, the same applied p.d. results in the same current:

$$I = \frac{V}{R} = V\left(\frac{1}{R}\right) \quad (2)$$

Comparing equations (1) and (2), we can see that:

$$\frac{1}{R} = \frac{1}{R_1} + \frac{1}{R_2} + \frac{1}{R_3}$$

This is the formula to use for adding resistors in parallel.

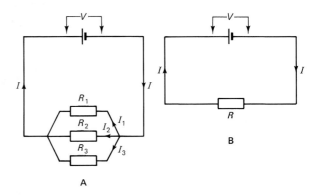

These are equivalent: same battery p.d., same current.

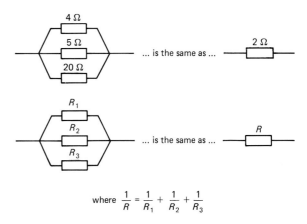

where $\frac{1}{R} = \frac{1}{R_1} + \frac{1}{R_2} + \frac{1}{R_3}$

EXAMPLE 26.6

A $12\,\Omega$ resistor and a $6\,\Omega$ resistor are connected in parallel. What is their combined resistance? If they are connected to a $12\,V$ battery, what is the current from the battery?

For a parallel combination we use:

$$\frac{1}{R} = \frac{1}{R_1} + \frac{1}{R_2}$$

$$\frac{1}{R} = \frac{1}{12\,\Omega} + \frac{1}{6\,\Omega}$$

$$= \frac{1+2}{12\,\Omega} = \frac{3}{12\,\Omega} = \frac{1}{4\,\Omega}$$

So $R = 4\,\Omega$.

A single resistor of $4\,\Omega$ is equivalent to the two parallel resistors. The current from the battery is the same as if a single $4\,\Omega$ resistor was connected across its terminals:

Using the resistance equation: $I = \dfrac{V}{R} = \dfrac{12\,V}{4\,\Omega}$
$= 3\,A$

So the battery supplies a current of $3\,A$.

Points to note about resistors in parallel

1. In Example 26.6, the sum of the two resistors is $4\,\Omega$ — less than either of the two separate resistors. This is what we always find. The resistance of a parallel combination of resistors is always smaller than the smallest resistance in the combination. If you think about it, this is not surprising. Connecting resistors in parallel provides more routes for the current. It is always easier for the current to flow through the parallel combination than through any of the individual resistors on its own — even the smallest one!

2. Equal resistors in parallel: sometimes the two resistors in a parallel combination are equal. This special case is easier than the general situation.

   Compared with a single $10\,\Omega$ resistor, two $10\,\Omega$ resistors in parallel provide two identical paths for current. The resistance is halved. The two $10\,\Omega$ resistors are equivalent to a single $5\,\Omega$ resistor.

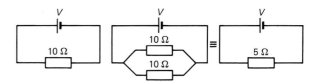

In general, two $R\,\Omega$ resistors in parallel are equivalent to a single $\tfrac{1}{2}R\,\Omega$ resistor.

Check for yourself that we get this result if we use the formula for adding resistors in parallel.

258 Ohm's Law and Resistance

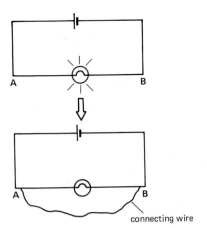

The lamp goes out.
The connecting wire short circuits the bulb.

Short circuits
When a circuit has parallel branches, the larger current is always in the branch with the lower resistance. An extreme case is when one branch is just a piece of conducting wire, with almost no resistance at all. This is called a 'short circuit', or sometimes just a 'short'. Almost all the current will go through the short circuit. In this circuit, the lamp goes OFF when a wire is connected from **A** to **B**. The current through the wire (the 'short') is so large that the battery p.d. actually drops (see section 28.6) – this will very quickly damage the battery. Most of the current goes through the wire, and very little goes through the lamp.

Series and parallel combinations
Sometimes resistors in a circuit are connected in more complicated ways. The easiest way to show how to deal with this is through some examples.

a
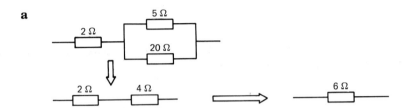

First add the $5\,\Omega$ and $20\,\Omega$ resistors in parallel.

$$\frac{1}{R} = \frac{1}{5\,\Omega} + \frac{1}{20\,\Omega}$$

$$= \frac{4+1}{20\,\Omega} = \frac{5}{20\,\Omega} = \frac{1}{4\,\Omega}$$

So $R = 4\,\Omega$

They are equivalent to a single $4\,\Omega$ resistor. This is then in series with a $2\,\Omega$ resistor. These add in series to give:

$$R = 2\,\Omega + 4\,\Omega = 6\,\Omega$$

Answer: The equivalent resistance is $6\,\Omega$.

b
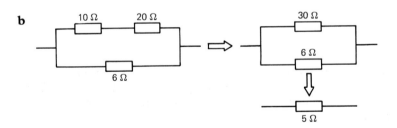

First add the two resistors in the upper branch. They are in series:

$$R = 10\,\Omega + 20\,\Omega = 30\,\Omega$$

We now have a $30\,\Omega$ and a $6\,\Omega$ resistor in parallel. Combining these:

$$\frac{1}{R} = \frac{1}{30\,\Omega} + \frac{1}{6\,\Omega}$$

$$= \frac{1+5}{30\,\Omega} = \frac{6}{30\,\Omega} = \frac{1}{5\,\Omega}$$

So $R = 5\,\Omega$

Answer: The equivalent resistance is $5\,\Omega$.

EXAMPLE 26.7
In this circuit, what is the current through the $3\,\Omega$ resistor when the switch is (a) open, (b) closed?

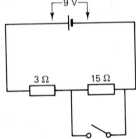

(a) Switch open
When the switch is open, the circuit consists of $3\,\Omega$ and a $15\,\Omega$ resistor in series. The total resistance in the circuit is $3\,\Omega + 15\,\Omega = 18\,\Omega$.

We can apply $V = IR$ to the whole circuit:

$$I = \frac{V}{R} = \frac{9\,\text{V}}{18\,\Omega}$$

$$= 0.5\,\text{A}$$

(b) Switch closed
When the switch is closed, the connecting wire provides a by-pass round the $15\,\Omega$ resistor – 'a short circuit'. Effectively the circuit has become just a single $3\,\Omega$ resistor. The current through this is:

$$I = \frac{V}{R} = \frac{9\,\text{V}}{3\,\Omega}$$

$$= 3\,\text{A}$$

So, with the switch open, the current through the $3\,\Omega$ resistor is $0.5\,\text{A}$. With the switch closed, it is $3\,\text{A}$. Notice how much the current through the $3\,\Omega$ resistor has risen as a result of the short circuit. This is why an accidental short circuit inside a piece of equipment can lead to problems. It can result in too large a current in some of the other resistors, which may overheat as a result.

26.7 Potential divider

The series circuit with two resistors has many important applications, particularly in electronics.

a (Recap) In a series circuit, the sum of the p.d.s across the resistors is the same as the battery p.d. If R_1 is larger than R_2, then V_1 will be larger than V_2. The ratio of the two p.d.s is the same as the ratio of the resistances.

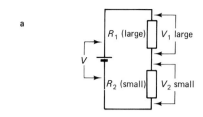

b₁ R_1 is now replaced by a light-dependent resistor (LDR). This has a very large resistance (over 1 million ohms) in the dark, but a low resistance (less than 100 ohms) in the light. R_2 has a resistance somewhere between these two extremes.

In the dark, R_1 is much larger than R_2, so V_1 is much larger than V_2. In fact, V_2 will be almost 0 V.

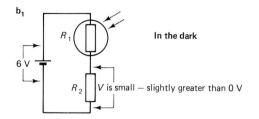

b₂ In the light, R_1 is much smaller than R_2, so V_1 is much smaller than V_2. In this case, V_2 will be almost 6 V.

The reading on the voltmeter changes from nearly 6 V to nearly 0 V as the LDR is moved from light to dark. We have made a circuit which can sense changes in the brightness of light falling on it.

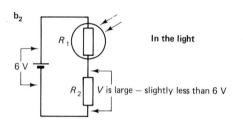

c₁ The same sort of circuit, this time using a thermistor, can be used for temperature sensing. A thermistor has a large resistance when it is cold and a small resistance when it is hot.

When the thermistor is cold, R_1 will be much larger than R_2, so V_1 will be much larger than V_2. V_2 will be close to 0 V.

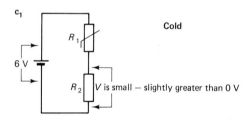

c₂ When it is hot, R_1 will be much smaller than R_2, so V_1 will be much smaller than V_2. V_2 will be close to 6 V.

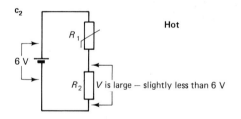

d₁ Another application of the same basic two-resistor circuit is as a variable voltage supply. By choosing the resistance values of R_1 and R_2, we can make the p.d. across R_2 take any value we want between the battery p.d. (6 V) and zero.

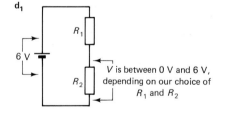

d₂ It works even better if we replace the two separate resistors with a variable resistor. R_1 and R_2 are now the resistances of the two parts of the variable resistor. The p.d. between the sliding contact and one end of the variable resistor can be varied smoothly between 6 V to 0 V. We have made a variable voltage supply. The output p.d. is controlled by moving the sliding contact.

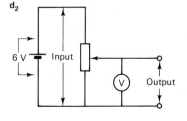

These are all examples of the **potential divider**. The input p.d. is divided into two parts.

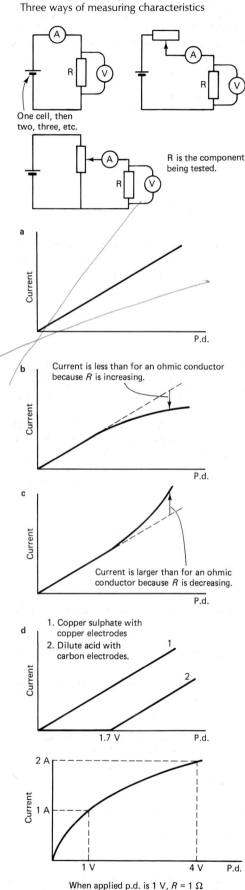

Three ways of measuring characteristics

One cell, then two, three, etc.

R is the component being tested.

a

b Current is less than for an ohmic conductor because R is increasing.

c Current is larger than for an ohmic conductor because R is decreasing.

d
1. Copper sulphate with copper electrodes
2. Dilute acid with carbon electrodes.

When applied p.d. is 1 V, $R = 1\,\Omega$
When applied p.d. is 4 V, $R = 2\,\Omega$

26.8 Characteristics

The **characteristic** of an electrical component tells us how the current through it varies as we change the p.d. applied across it. The experiment described right at the start of this chapter, where we measured the current through a length of wire as we changed the battery p.d., is an investigation of the characteristic of a length of wire. Usually the easiest way to summarise such results is in the form of a graph of current against p.d. The characteristic of a length of wire is a straight line through the origin (see the first diagram in this chapter). Some components, however, do not have straight-line characteristics.

To find the characteristic of a component, we need to vary the p.d. across it, and measure the current at various p.d.s.

Characteristics of some common components
a Resistance wire

◁ As we have seen earlier, the characteristic of a length of wire is a straight line through the origin. Current is proportional to p.d.

$$I \propto V$$

The resistance V/I is constant. Conductors with this characteristic are called **ohmic conductors**.

b Filament lamp

◁ For low currents and p.d.s the characteristic is a straight line, but at larger p.d.s, the current is less than it would be for an ohmic conductor. The filament heats up (to over 1000 °C) as the lamp gets brighter, and the resistance of the filament wire increases.

The resistance of most wires increases with temperature. In **a** we didn't apply a large enough p.d. to cause much heating of the wire and so the resistance stayed constant. A lamp filament is designed to heat up to a very high temperature even with low applied p.d.s.

c Thermistor

◁ A thermistor is made of semiconducting material. Its resistance changes as its temperature changes. For most thermistors a rise in temperature causes a fall in resistance. As larger p.d.s. are applied, the current through the thermistor rises; this makes it heat up and its resistance falls. The current is larger than it would be for an ohmic conductor.

d Ionic solutions

An ionic solution (for example, a salt or an acid dissolved in water) conducts electricity. The characteristic depends on what the solution is and which electrodes are used. With copper sulphate solution and

◁ copper electrodes, it is a straight line through the origin (ohmic conductor). For an acid with carbon electrodes, the p.d. has to reach about 1.7 V before anything happens, but then the graph is a straight line. This is because the carbon and acid behave like a cell (see section 28.1) which produces a p.d. in the opposite direction! The applied p.d. has to overcome this before any current can flow.

Non-ohmic conductors

◁ The lamp and the thermistor are **non-ohmic** conductors. Their resistance (calculated from $R = V/I$) is not constant, but changes for different applied p.d.s.

26.9 Electrical heating

In discussing characteristics, we have mentioned several times that conductors heat up when there is a current through them. In the next chapter we will look in more detail at electrical heating.

Questions

1 Find the unknown quantity (**V** or **I** or **R**) in each of these circuits.

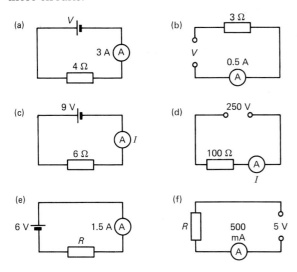

2 (a) Two pupils carry out an investigation to find how the resistance of a new material varies with length and thickness (cross-sectional area). This new material can be moulded into shapes of different length and thickness.

In their first experiment they roll the material into long cylinders of the same thickness. They insert different lengths into the circuit in order to obtain different values of resistance.

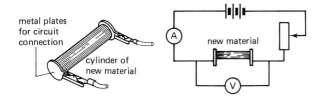

The results of their experiment are given in the table.

Length of material/cm	Voltmeter reading/V	Ammeter reading/A
24	3.2	0.1
12	3.2	0.2
8	3.4	0.3
4	2.2	0.4

(i) Calculate the resistance of each length and use the results to draw a graph of resistance against length.
(ii) What is the relationship between resistance and length for this new material?
(b) Describe clearly how the pupils could use the same apparatus to investigate how the resistance of the material varies with thickness.
(c) Name one other factor which might affect the resistance of the material. (SEB)

3 Your father opens up the speed control of the model racing car he gave you for Christmas. Inside he finds a coil of wire wound on a plastic tube. As you pull the trigger, a sliding contact moves over the coil. He asks you how this works and how it changes the car's speed. What explanation would you give him?

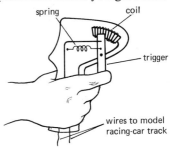

4 (a) A pupil uses the circuit shown to investigate resistors in series. He has in addition one ammeter, one voltmeter and some connecting wire.

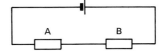

(i) Draw separate diagrams to show how these meters could be connected to this circuit to find the resistance of **A** only; the resistance of **B** only; and the combined resistance of **A** and **B** in series.
(ii) State the relationship which the pupil should find between the resistance of **A**, the resistance of **B**, and the combined resistance of **A** and **B** in series.
(b) The pupil decides to build an instrument for measuring resistance directly. He bases his design on the following circuit diagram.

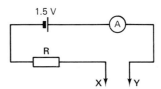

The resistance to be measured is connected between **X** and **Y**.
(i) Describe how this instrument may be calibrated so that the meter indicates resistance directly in ohms.
(ii) Give one reason why resistor **R** is included in the circuit of the instrument. (SEB)

5 In this circuit, will the brightness of the lamp increase, decrease or stay the same when:
(a) the resistance of R_1 is increased,
(b) the resistance of R_1 is decreased,
(c) the resistance of R_2 is increased,
(d) the resistance of R_2 is decreased?

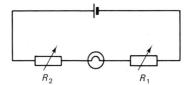

6 In both these circuits, the two resistors are equal in value. What is the reading on the voltmeter in each circuit?

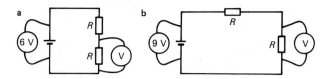

7 Find the readings on the voltmeters in each of these circuits.

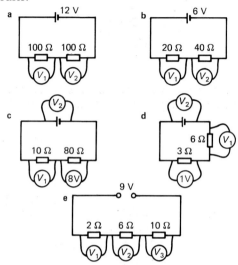

8 In all of these circuits, the voltmeter readings are either:
(a) exactly 6 V
(b) slightly below 6 V
(c) exactly 3 V
(d) slightly above 0 V
(e) exactly 0 V

For each voltmeter, pick the answer – (a) to (e) – which you think is correct.

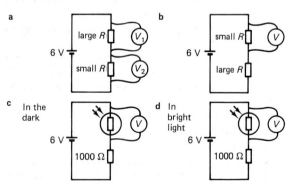

In the dark, the LDR's resistance is more than 10 kΩ; in bright light it is less than 100 Ω.

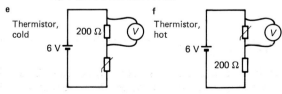

When it is cold, the resistance of the thermistor is around 1000 Ω; when it is hot, its resistance is less than 100 Ω.

9 The resistance of the variable resistor is slowly increased. Say what you think will happen to the voltmeter reading in each of the circuits. Explain your answer each time.

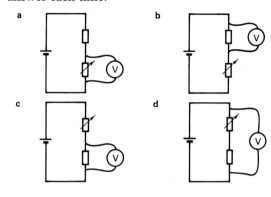

10 Find the unknown quantity (V or I or R) in each of these circuits.

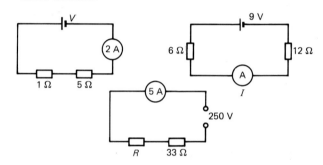

11 Which lamp in this circuit has the smallest resistance? Explain clearly how you worked this out.

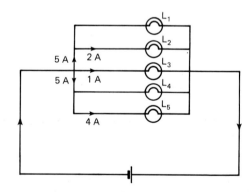

12 The ammeters A_1, A_2 and A_3 are used to measure the current at different points in this circuit. The three resistors are identical. The reading on A_1 is 0.6 A. What are the readings on A_2 and A_3?

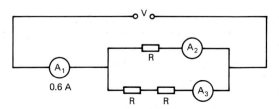

13 In this circuit:

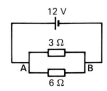

(a) what is the p.d. across **AB**?
(b) what is the current through the 3 Ω resistor?
(c) what is the current through the 6 Ω resistor?
(d) what is the current from the battery?
(e) We want to replace the 3 Ω and 6 Ω resistors by one single resistor, so that the current stays the same. What size should the single resistor be?

14 In this circuit, one ammeter is faulty. Which one is it? Explain your answer.

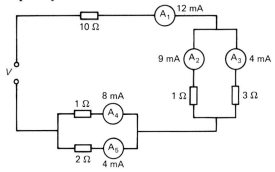

15 What is the reading on the ammeter:
(a) when the switch is open?
(b) when the switch is closed?

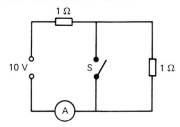

16 What is the total resistance between **P** and **Q** in each of the following?

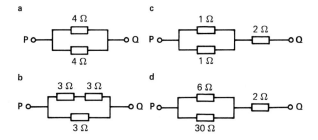

17 You are given four 2 Ω resistors. Draw diagrams to show how you could use them to make a resistance of:
(a) 8 Ω
(b) 1 Ω (Hint: use just 2 resistors)
(c) 0.5 Ω
(d) 5 Ω
(e) 2 Ω (using all 4 resistors!)

18 Find the unknown quantity (*V* or *I* or *R*) in each of these circuits.

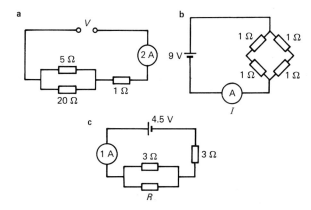

19 The diagram below shows apparatus set up by a pupil in which an ammeter is used to indicate the level of liquid in a tank.

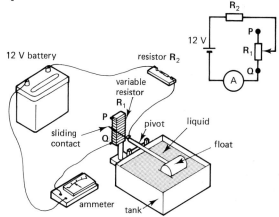

The variable resistor R_1 is made from a length of resistance wire **PQ** and a sliding contact.
When the tank is empty the sliding contact is at **P** and when it is full the sliding contact is at **Q**.
(a) Explain what happens to the ammeter reading as the tank is filled.
(b) The resistance of the variable resistor R_1 can vary from 0 to 90 Ω. The resistor R_2 has a resistance of 10 Ω. The resistance of the ammeter is so small that it can be ignored.
   Calculate the ammeter reading when:
 (i) the tank is full; and
 (ii) the tank is empty.
(c) Explain why resistor R_2 is included in the circuit.
(d) This apparatus is a model of a system used in a motor car. Which system is it? (SEB)

INVESTIGATION

The resistance of small resistors is supposed to be accurate to within a certain tolerance. Do some measurements to check whether the stated measurements seem to be correct.

27: Electrical Power and Domestic Electricity

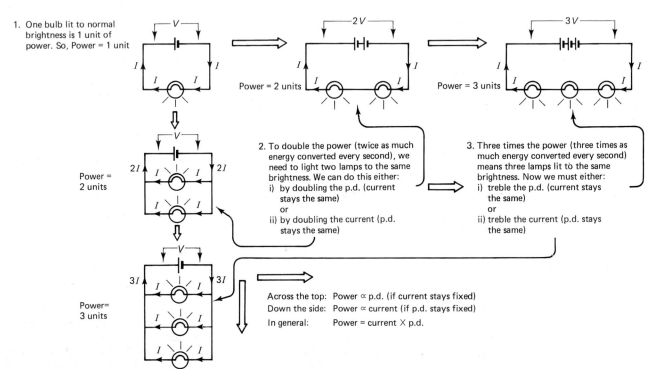

Electric circuits transfer energy. The power is the rate at which energy is transferred – the amount of energy transferred every second.

27.1 Energy converters

An electric circuit is an energy converter. The cell or battery is a store of chemical energy. It gives electrical potential energy to the electrons as they pass through it. As the electrons pass round the circuit this electrical energy is used up – to heat a wire, or to make a motor turn.

When an energy change takes place, the **power** is the rate at which the change occurs. **Power is the amount of energy transferred every second** (see section 11.8).

$$\text{power} = \frac{\text{energy change}}{\text{time taken}}$$

The electrical power of the cell or battery in a circuit is the rate at which it transforms chemical energy into electrical potential energy. The **power dissipation** in the circuit is the rate at which electrical potential energy is changed into other forms like kinetic energy in a motor, or internal energy in a resistance wire, or radiation energy when light is produced.

27.2 Current, p.d. and power

As an example of an energy conversion, let us consider a simple circuit with a cell and a lamp.

So the rate at which energy is transformed in a circuit (the power) depends on the p.d. of the cell *and* the current in the circuit.

$$P = IV$$

We can get the result $P = IV$ directly from the definition of the volt. There is a p.d. of 1 volt between two points if 1 coulomb of charge gains or loses 1 joule of electrical energy as it passes between the points.

$$1 \text{ volt} = 1 \text{ joule per coulomb}$$

$$\text{p.d. (or voltage)} = \frac{\text{energy}}{\text{charge}}$$

If we multiply both sides of this equation by **charge**, we get:

$$\text{energy} = \text{p.d.} \times \text{charge}$$

This allows us to calculate how much energy is involved when a charge moves through a known p.d. If we divide both sides by **time**, we get the power.

$$\frac{\text{energy}}{\text{time}} = \text{p.d.} \times \left(\frac{\text{charge}}{\text{time}}\right)$$

$$\text{power} = \text{p.d.} \times \text{current}$$

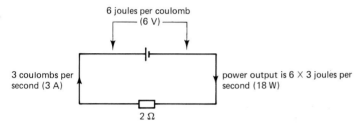

27.3 Power dissipation in a resistor

If we connect a known resistor across a battery, we can calculate the power dissipated in the resistor in two stages:

1. Use the resistance equation $V = IR$ to find the current through the resistor.
2. Use this value in the power equation $P = IV$.

By combining the two equations, we can get there in a single step:

$$P = IV$$

But current $I = \dfrac{V}{R}$, so

$$P = \left(\frac{V}{R}\right)V = \frac{V^2}{R}$$

Similarly, if we know the current through a resistor, we can calculate the power dissipated in it:

$$P = IV$$

But p.d. across the resistor $V = IR$, so

$$P = I(IR) = I^2 R$$

So three equations are available for calculating the power dissipated in a resistor.

$$P = IV, \quad P = \frac{V^2}{R}, \quad P = I^2 R$$

They are all equivalent.

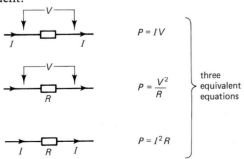

EXAMPLE 27.1

A small laboratory immersion heater uses a 12 V supply. The current through the heater is 4 A. What is the power of the heater?

We use the equation:
$P = IV$
$P = 4\,\text{A} \times 12\,\text{V}$
$= 48\,\text{W}$

The power of the heater is 48 W.

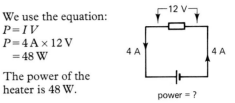

EXAMPLE 27.2

A 60 W light bulb works on mains electricity (240 V). What is the current through the bulb when it is lit?

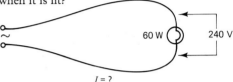

Again, we use the equation: $P = IV$
$60\,\text{W} = I \times 240\,\text{V}$

Divide both sides by 240 V:

$$\frac{60\,\text{W}}{240\,\text{V}} = I = 0.25\,\text{A}$$

The current through the light bulb is 0.25 A.

EXAMPLE 27.3

A 6 Ω resistor is connected across the terminals of a 12 V battery. What is the power dissipated in the resistor?
If a 36 Ω resistor is used instead, what is the power dissipated now?

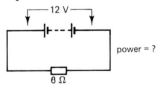

Here we know V and R. The most direct method is to use the equation:

$$P = \frac{V^2}{R}$$

When the 6 Ω resistor is used,

$$P = \frac{12^2}{6}\,\text{W} = \frac{144}{6}\,\text{W} = 24\,\text{W}$$

When the 36 Ω resistor is used,

$$P = \frac{12^2}{36}\,\text{W} = \frac{144}{36}\,\text{W} = 4\,\text{W}$$

A 6 Ω resistor connected across the battery dissipates 24 W. A 36 Ω resistor connected across the same battery dissipates 4 W.

(**Note.** The smaller the resistance connected across a given p.d., the larger the power dissipation. If, by mistake, a very low resistance is connected across a battery (a short circuit), it will heat up very rapidly.)

266 Electrical Power and Domestic Electricity

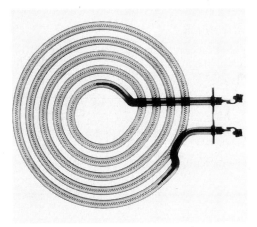

An X-ray photograph shows the coiled resistance wire inside a cooker element.

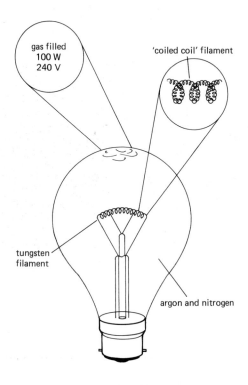

27.4 Voltage and power ratings

Heating elements

Electrical heating elements are simply made from resistance wire. The element of an electric fire – the part which glows red hot when the fire is on – is made by winding **nichrome** wire (see section 26.3) round a ceramic tube. Nichrome is chosen because it can stay red hot in air without burning or melting.

Heating elements for electric kettles and electric cookers are made in the same way. As the elements may get wet in use, the nichrome wire is coiled inside an outer metal case. This also allows more resistance wire to be fitted into a compact space.

Light bulbs

A light bulb also contains a coil of resistance wire. Inside the bulb is a filament made from a fine coil of **tungsten** wire. The metal tungsten is used because it has a very high melting point (3400 °C) and can be kept white hot without melting. At this temperature, the filament would quickly burn in air (it would react with the oxygen in the air), so the filament has to be put inside a glass bulb filled with a gas which does not react with hot metal. Argon and nitrogen are used.

Most bulb filaments are in the form of a 'coiled coil'. A very long coil of fine wire (high resistance) can be fitted into a small space. The coiled shape also reduces convection currents inside the glass bulb and keeps the wire as hot as possible – so that it gives out more light.

Power ratings

Heating elements are usually marked with a voltage and power rating. This information is printed on the end of a light bulb. A typical light bulb might be marked 240 V 100 W. This means that the bulb dissipates 100 J of electrical energy every second (100 W) *if it is connected to a 240 V supply*. It will *not* give 100 W if it is used on a different supply voltage.

From the rating, we can work out the resistance of the bulb filament, using the equation $P = \dfrac{V^2}{R}$:

Substituting: $\qquad 100\,\text{W} = \dfrac{(240\,\text{V})^2}{R}$

Multiply across by R: $\qquad R \times 100\,\text{W} = (240\,\text{V})^2$

Divide both sides by 100 W: $\qquad R = \dfrac{(240\,\text{V})^2}{100\,\text{W}}$

$$= 576\,\Omega$$

The light bulb filament has a resistance of 576 Ω.

When the bulb is made in the factory, this resistance is chosen so that the bulb *will* give 100 W when it is connected to a 240 V supply.

Less than 10% of the power output from a filament bulb is in the form of visible light. The remaining 90% is wasted in heating the bulb itself and the surrounding air. The efficiency of a light bulb is less than 10%.

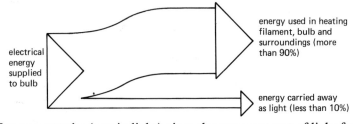

A fluorescent tube (a strip light) gives the same amount of light for only about one-third of the power input. New energy saving light bulbs also have much higher efficiency. They are really miniature fluorescent tubes coiled up inside a glass bulb, so that they can fit into a normal bulb holder.

27.5 Electrical energy

If we know the power output of a battery, we can calculate how much electrical energy it supplies in a given time.

$$\text{energy} = \text{power} \times \text{time}$$
$$E = P \times t$$
$$\text{joules} \quad \text{watts} \quad \text{seconds}$$

For example, if a battery in a particular circuit has a power output of 50 W, it will supply 500 J of electrical energy in 10 seconds.

Measuring electrical energy

Electrical energy supplied by a battery or transformed by a resistor can be measured using a voltmeter, an ammeter and a clock. We measure the p.d. by connecting the voltmeter across the battery or resistor, and the current by connecting the ammeter in series. From this we calculate **power**:

$$P = IV$$

To find the **energy**, multiply by the **time** (in seconds) for which the current is switched on.

A joulemeter measures electrical energy directly. The heating element is connected to the supply **via** the joulemeter. It then records the total amount of electrical energy supplied (see section 17.3).

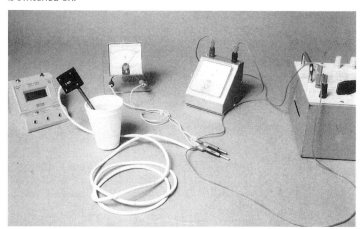
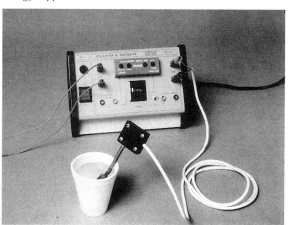

Here are some examples involving electrical power and energy. As some of them involve several steps in the calculation it is useful to work out a strategy for tackling the question before beginning.

EXAMPLE 27.4

A 5 kW (5000 W) immersion heater takes 30 minutes to heat enough water for a bath. How much energy does the heater supply to the water in this time?

The first thing to note is that the time is given in minutes. As 1 W is 1 joule per *second*, we need times in seconds.

Strategy: 1. Convert time to seconds.
2. Calculate energy supplied in 30 minutes.

We have: $P = 5000$ W
$t = 30$ minutes $= 30 \times 60$ s $= 1800$ s
$E = Pt$
$= 5000$ W $\times 1800$ s
$= 9\,000\,000$ J

The heater supplies 9 000 000 J (9 megajoules (9 MJ)) to heat the bath water. You may be surprised to know that it takes 9 million joules to heat the water for your bath!

EXAMPLE 27.5

A torch bulb is labelled 2.5 V, 0.3 A. If the bulb is connected up correctly to a 2.5 V supply, how much energy is transformed in 5 minutes?

Strategy: 1. Calculate the power of the bulb.
2. Work out the energy supplied in 5 minutes.

We have: $V = 2.5$ V; $I = 0.3$ A; $t = 5 \times 60$ s $= 300$ s

$$P = IV$$
$$= 0.3 \text{ A} \times 2.5 \text{ V} = 0.75 \text{ W}$$

In 300 s, the energy supplied is:

$$E = Pt$$
$$= 0.75 \text{ W} \times 300 \text{ s} = 225 \text{ J}$$

The torch bulb transforms 225 J of electrical energy in 5 minutes. Only a very small part of this, however, is transformed into radiation energy (light). Most of it is used in heating up the bulb and its surroundings.

EXAMPLE 27.6

An electric motor on a building site lifts a load of 60 kg of bricks to the top of a house. The house is 10 m high and it takes 15 s to lift the bricks. The power rating of the motor is 500 W. How efficient is it?

Strategy: 1. Work out how much energy is needed to lift the bricks.

2. Work out how much energy the motor actually supplies in 15 s.

3. Compare these to find the efficiency.

The energy needed to lift the bricks is equal to their gain in potential energy. We calculate this from the equation:

$$E_p = mgh$$
$$= 60 \text{ kg} \times 10 \text{ N/kg} \times 10 \text{ m}$$
$$= 6000 \text{ N m (or J)}$$

In 15 s, the energy supplied by the motor is

$$E = Pt = 500 \text{ W} \times 15 \text{ s} = 7500 \text{ J}$$

Of this 7500 J, 6000 J are used to do the job we want done – lifting the bricks. The rest is wasted in heating up parts of the motor and the surroundings, as a result of friction inside the motor, and so on.

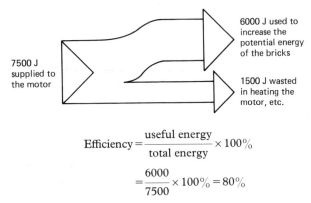

$$\text{Efficiency} = \frac{\text{useful energy}}{\text{total energy}} \times 100\%$$
$$= \frac{6000}{7500} \times 100\% = 80\%$$

This electric motor is 80% efficient when doing this job.

EXAMPLE 27.7

An electric kettle is used to boil some water to make tea. The kettle has a power rating of 3 kW, and is filled with 0.5 kg of cold tap water at 20°C. It takes a minute and a half (90 s) to boil. How efficient is this kettle? (The s.h.c. of water is 4200 J/kg K.)

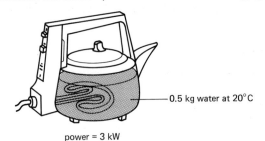

Strategy: 1. Work out how much energy is needed to heat the water to boiling point.

2. Work out how much energy this kettle supplies in 90 s.

3. Compare these to find the efficiency.

The temperature rise ΔT of the water is 80 °C (from 20 °C to 100 °C).

The energy needed to raise the temperature of 0.5 kg of water by 80°C (or 80 K) is calculated from:

$$E = cm\Delta T$$
$$= 4200 \text{ J/kg K} \times 0.5 \text{ kg} \times 80 \text{ K}$$
$$= 168\,000 \text{ J}$$

The energy supplied by this kettle in 90 s is given by:

$$E = Pt$$
$$= 3000 \text{ W} \times 90 \text{ s}$$
$$= 270\,000 \text{ J}$$

So the efficiency is $\dfrac{\text{useful energy}}{\text{total energy}} \times 100\%$

$$= \frac{168\,000}{270\,000} \times 100\% = 62\%$$

The kettle is 62% efficient. The remaining 38% of the energy is wasted in heating the kettle itself and the surroundings.

27.6 Buying electrical energy (Cost of Energy)

Every house has an electricity meter supplied by the Electricity Board. It measures the electrical energy you use for heating, lighting, cooking, and running all your mains electrical equipment. The meter does *not* measure in joules because joules are very small units, and the number of joules used every day in a typical house would be a very large number. In Example 27.4, we found that 9 million joules are needed to heat the bath water for one bath! Instead the units used are **kilowatt-hours** (kWh). Sometimes this is just referred to as a 'unit' of electricity. The electricity board charges around 6p for 1 unit.

Electrical Power and Domestic Electricity 269

1 kilowatt-hour is the energy supplied in 1 hour to an appliance whose power rating is 1 kW.

Energy in kilowatt-hours is calculated using the equation:

$$\underset{\text{kilowatt-hours}}{\text{energy}} = \underset{\text{kilowatts}}{\text{power}} \times \underset{\text{hours}}{\text{time}}$$

Electricity bills

A 3 kW immersion heater switched on for 2 hours uses 6 kWh

6 × 3,600,000 J

A 1.5 kW electric convector heater switched on for 6 hours uses 9 kWh

9 × 3,600,000 J

A 100 W (0.1 kW) light bulb switched on for 5 hours uses 0.5 kWh

0.5 × 3,600,000 J

A 5 kW electric oven switched on for 3 hours uses 15 kWh = 15 × 3,600,000 J

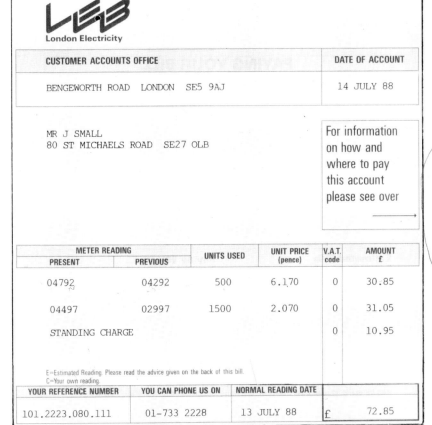

METER READING		UNITS USED	UNIT PRICE (pence)	V.A.T. code	AMOUNT £
PRESENT	PREVIOUS				
04792	04292	500	6.170	0	30.85
04497	02997	1500	2.070	0	31.05
STANDING CHARGE				0	10.95

How many joules are equal to 1 kWh?

In 1 second, an appliance with a power of 1 kW uses 1000 J of electrical energy.
So, in 1 hour, it uses 1000 × 60 × 60 J of electrical energy
= 3 600 000 J
Therefore 1 kWh = 3 600 000 J

An electricity bill shows how many 'units' you have used in the last three months. The charge is in two parts: a fixed charge which is a sort of rental for having the electricity supplied to your house; and a charge for every 'unit' used. Check that this bill has been worked out correctly.

27.7 Large and small consumers

We pay for the electrical energy we use, but some appliances use electrical energy much more quickly than others. We can measure the electrical energy used by different appliances using a kilowatt-hour meter. The appliance is connected to the mains through the kWh meter. ▷

A disc inside the meter rotates when the appliance is switched on. The number of turns indicates how much electrical energy is used. With the electric fan heater connected, the disc turns rapidly: 30 turns in 5 minutes. With a 100 W light bulb, the disc only does 2 turns in 5 minutes. When a radio is connected, the disc hardly seems to turn at all – less than 1 turn in 5 minutes.

In energy terms, heating is expensive, lighting is not so expensive and the radio is very economical. As electrical energy is what we pay for, the same thing is true in money terms! You will save more on your total energy bill if you can cut 20% off your heating costs (by better insulation) than if you reduce your lighting by 20%.

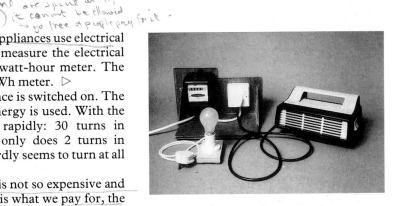

27.8 The mains electricity supply

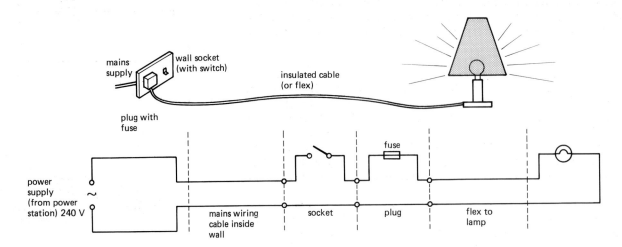

The electric sockets on the walls of your home, and the ceiling sockets which hold light bulbs, are like the terminals of a 240 V battery. The electrical energy comes from the power station. When a table lamp is plugged into a socket and switched on, a complete circuit is formed.

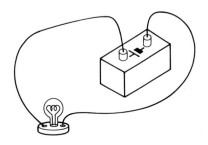
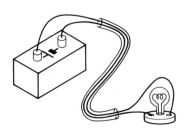

The insulated cable (or 'flex') from the mains plug to the lamp carries two wires inside it to make the circuit a closed conducting loop. The flex for many appliances also has a third wire, called the earth wire. This is a safety wire and does not usually form part of the circuit.

Alternating current (a.c.)

The mains supplies alternating current. The electrons in the wires are pushed back and forward 50 times per second. Power stations generate a.c. because it is easier to produce than direct current (d.c.) on a large scale. It is also easier to transmit it efficiently from the power station to your home, as we will see in section 30.11. Mains voltage is 240 V.

Live wire

There is usually a large potential difference between the **live wire** and earth potential (0 V). Sometimes the live wire is more than 300 V above zero; sometimes it is more than 300 V below zero. The voltage swings up and down between these levels 50 times per second.

Touching the live wire is very dangerous. Our bodies are at earth potential (0 V). If someone touches a live wire, there will be a large potential difference across their body, causing a large current through it. This is likely to be fatal.

Neutral wire

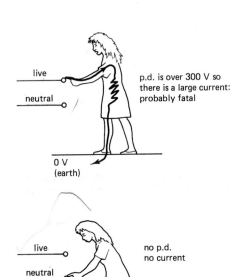

The neutral wire stays at earth potential (0 V). Although it carries current, you shouldn't get an electric shock from touching it, because it is at roughly the same potential as you. There is no p.d. across you and so no current flows. Even so, it is safer not to touch it.

Switch

The switch is always placed in the live wire. The switch would work just as well in the neutral wire. It is put in the live wire so that when the appliance is switched off, all the parts of it and all the wires in the plug are at 0 V and safe to touch.

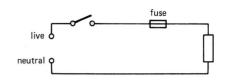

Fuse

This is a short piece of thin wire which overheats and melts if the current through it becomes too large. Fuses are made of a metal which melts easily. Fuses for mains appliances are usually in the form of a small cartridge which fits inside the plug. Like the switch, it is always placed in the live lead; if the fuse 'blows', the flex and the appliance are disconnected from the live main terminals. If a fault develops in the appliance, the fuse will melt first and switch it off, before the cable overheats and causes a fire.

fuse wire inside ceramic case

The earth wire

This is an extra safety precaution. Any metal parts of an appliance are connected directly, via the earth wire, to a large metal plate buried in the ground. This ensures that any metal part which you might touch is *always* at 0 V, even if there is a fault in the appliance. If a live wire inside the appliance happened to work loose and touch the outer metal case, a large current would flow through the circuit formed by the live and earth wires, and the fuse would blow. This immediately disconnects everything from the live terminal. Without the earth connection, someone touching the metal case of a faulty appliance could get a lethal shock.

27.9 Choosing cables and fuses

A whole range of types of cable are available for connecting electrical appliances to the mains. The cable is usually supplied with the appliance but we may need to fit it ourselves or replace old cables which are becoming broken or frayed. How do you pick the correct cable?

Cables are rated according to the current which they can carry safely without overheating. We can buy 1 A, 3 A, or 13 A flex. Some of these are twin-core (live and neutral); some are three-core (with an earth wire as well).

Do I need 2-core or 3-core?

The easiest way to find out is to look at the appliance or read its instruction book. Some appliances nowadays are **double insulated**. If they are, they should be marked with the symbol shown on the right. These appliances have a completely enclosed outer plastic body and there is no connection between any metal screws or handles on the outside and the electrical parts inside. Even if a fault occurs, there is no chance of any of the outer parts becoming live. No earth wire is needed and a 2-core flex will do. Things like electric drills, hair dryers, food mixers and table lamps are usually double insulated.

For appliances which are not double insulated, a 3-core flex is required.

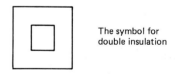

The symbol for double insulation

Electrical appliances often carry a rating plate giving electrical information, like this one for an electric cooker

What cable rating do I need?

To choose the correct current rating, you need to know what current the cable is going to have to carry in use. This can be calculated from the power rating of the appliance. Usually this is printed on a plate on the appliance itself. If we know the power and the p.d. (240 V), we can calculate the current using $P = IV$:

$$\text{power} = \text{current} \times \text{p.d.}$$

So, dividing across by p.d.

amperes ⟶ current = $\dfrac{\text{power}}{\text{p.d.}}$ ⟵ watts / volts

For example:

A 500 W hair dryer has a current of $\dfrac{500}{240}$ A = 2.1 A

A 2.4 kW electric kettle has a current of $\dfrac{2500}{240}$ A = 10.4 A

A 60 W radio has a current of $\dfrac{60}{240}$ A = 0.25 A

A 1 kW fan heater has a current of $\dfrac{1000}{240}$ A = 4.2 A

A fridge with a 1.2 kW motor has a current of $\dfrac{1200}{240}$ A = 5 A when the motor is on.

We must choose a cable whose current rating is *larger* than the normal current to the appliance. Notice from the examples above that the currents really form two groups: heaters and large motors have currents of around 5 A and above; many other appliances like lamps and radios have currents below 1 A. For mains appliances, it is now standard practice to have just two cable ratings: 3 A and 13 A. If the normal current is less than 3 A, we use a 3 A cable; if it is between 3 A and 13 A, we use a 13 A cable. Mains appliances should not have currents above 13 A because this is the largest current which the **mains plug and socket** can carry safely without overheating.

In practice there is a good safety margin and cables can usually carry up to twice their rated current before becoming dangerous.

What fuse should I fit?

We have seen already how a fuse can protect you by switching an appliance off if a fault occurs. If the live wire touches an Earth point, there is a short-circuit and the current will be very large. *Any* size of fuse would blow!

The real reason why we need to choose the *correct* fuse rating is to protect against a different sort of fault which might result in too large a current in the cable to the appliance while it *seems* to be working normally. If, for example, two parts of the heating element inside an electric toaster accidentally touched, this could short circuit part of the element so that its resistance became much lower. The current would then increase. This larger current might be more than the cable to the toaster can safely carry and it would overheat. If nothing is done about it, the cable will eventually become very hot and might cause a serious house fire. Fuses prevent this happening.

The right fuse to use is the one with the same rating as the cable. There used to be a wide range of fuse ratings available for mains appliances, but now this is being standardised on just two values: 3 A and 13 A.

To calculate the right fuse for an appliance, work out the current from the power rating just as we did above for cables. Then pick the next higher fuse rating. From our examples above, the hair dryer, radio and bedside light need 3 A fuses; the kettle, fan heater and fridge need 13A fuses.

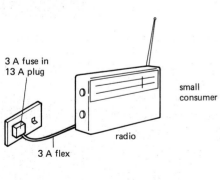

3 A fuse in 13 A plug
3 A flex
small consumer
radio

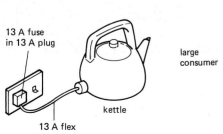

13 A fuse in 13 A plug
13 A flex
large consumer
kettle

27.10 Wiring a mains plug

Three pin plugs are a convenient method of connecting an appliance to the mains. The 13 A square pin plug is now the commonest in the UK. The diagram on the right shows where each wire should be connected inside the plug.

It is *very important* that the three wires are connected to the correct terminals. The wires are colour coded:

Live: brown
Neutral: blue
Earth: yellow and green stripes

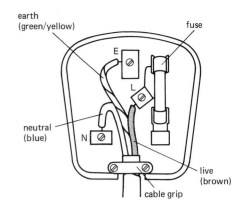

The photographs below show you how to connect the cable to the plug

1. Remove about 5 cm of the outer insulating plastic coat to get at the wires inside. Take great care not to cut the insulating plastic round each of the inner wires.

3. Remove about 1 cm of insulating plastic from each of the three wires. Twist the copper strands to tidy the ends.

5. Pick a fuse which has the right rating for the appliance. Put it into its holder.

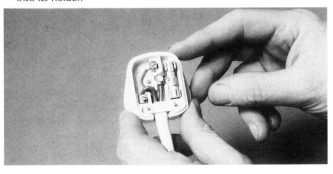

2. Place the cable over the plug and cut the wires to the correct length. The earth wire will probably need to be slightly longer than the other two. The wires are colour coded to make sure that each goes to the correct point. The outer insulation must reach beyond the cable grip. Leave about 1 cm extra on all three inner wires for making the connections.

4. Connect each wire to its own connecting point. Make sure that you connect each wire to the right pin. Tighten the screws for a good connection.

6. Check that there are no loose strands of wire anywhere, then replace the cover.

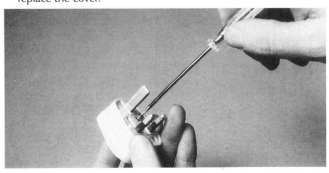

If a fuse in a plug blows, you should switch off at the mains socket and then pull out the plug. Only after doing that is it really safe to begin looking to see what is wrong. Fit a new fuse only after you have repaired the fault.

27.11 Household wiring

All the electricity you use in your house comes in through a single large electricity board cable. Inside the house there is a sealed **main fuse** and then a **meter** which records the total amount of electrical energy used in the house in kilowatt-hours. The cable comes out of the meter and into the **consumer unit** (or fuse box). Here the wires branch into several parallel circuits, some for 13 A power sockets, some for lighting, and separate ones for electric cooker and immersion heater. Before they branch, there is a double-pole switch (it breaks both live and neutral wires) for switching off all the circuits in the house.

The consumer unit also contains **fuses** or **circuit breakers** to protect each separate circuit. From the consumer unit, cables go off to a number of separate circuits round the house.

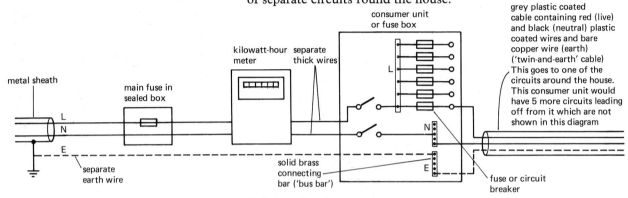

Lighting circuits

The lamps in a domestic lighting circuit are connected in parallel. This allows each lamp to be switched on and off independently. It also means that each lamp has the full mains voltage (240 V) across it, so only 240 V light bulbs need be made.

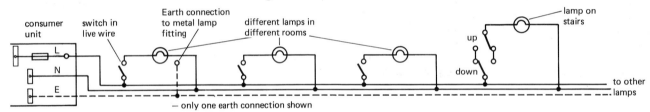

The circuit diagram shows the connections, but it is not easy to see how this relates to what you find at home. Usually the light is in the middle of the ceiling and the switch is on the wall near the door. How is the closed circuit loop formed? The diagram of a section of a room in a house shows how the cables run inside (or behind) the walls of the house. Try to follow on the diagram the closed path from the supply to the lamp and back to the supply again.

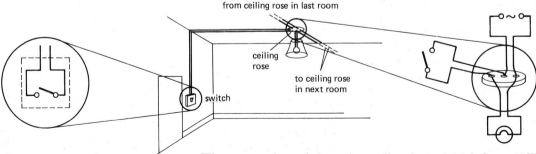

The current in each lamp is small – about 0.25 A for a 60 W lamp. The total current in the lighting circuit is the *sum* of these separate currents – currents in parallel branches add. The cable used for the lighting circuit is 5 A twin-and-earth cable which can carry 5 A without overheating. Each lighting circuit is protected by a 5 A fuse in the consumer unit.

Electrical Power and Domestic Electricity 275

Power circuits

The circuits supplying the 13 A wall sockets are different from lighting circuits. The sockets are again **in parallel** so that each one is independent; but the power circuit is in the form of a closed loop which returns to the consumer unit – a **ring main**.

Although the ring main looks like a series circuit (a closed loop), it is *not*. The sockets are in parallel. Check that you can follow on the diagram the closed conducting loop from the consumer unit to the socket, through an appliance and back to the consumer unit.

Inside the house, the cables are led from one socket to the next, usually under the floorboards or inside the wall. From the last socket in the ring, the cable returns to the consumer unit.

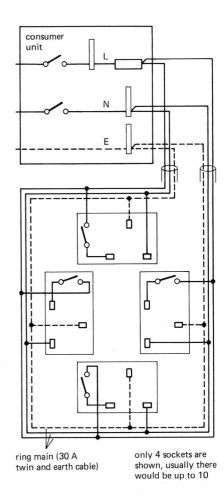

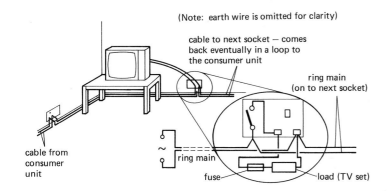

The cable is 30 A twin-and-earth. The total current to all the appliances in the ring can reach 30 A without any problems. Each socket is rated at 13 A and there may be up to 10 sockets in the ring, but it is very unlikely that all of them would be in use with a large appliance at the same time. 30 A is chosen as the likely maximum current in the ring. Each ring main is protected by a 30 A fuse in the consumer unit.

The advantage of using a ring is that there are two pathways to each socket and so the current in each is smaller. This allows thinner, lighter cable to be used.

Cookers and immersion heaters

These take such large currents that they are given separate circuits of their own.

Consumer unit fuses and circuit breakers

Each separate circuit has a fuse inside the consumer unit. The purpose of the fuse is to ensure that the lighting and ring main cables cannot overheat and cause a fire hazard. If the current becomes bigger than the safe rating of the cable, the fuse will blow before the cable overheats.

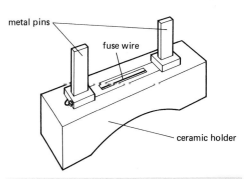

The standard type of fuse is a ceramic or plastic holder with a fuse wire connected between two terminals. If it blows, a new piece of fuse wire is fitted. Would the fuse protect you if you accidentally touched a live terminal? The answer is no. It takes a short time for the fuse to blow, and in that time you might have received a fatal shock.

This is why modern consumer units are fitted with residual circuit breakers (r.c.c.b. – sometimes also called earth leakage circuit breakers) instead of fuses. If your consumer unit still has ordinary fuses, then you are recommended to use a special adaptor plug (which contains an r.c.c.b.) when doing jobs which might be hazardous, such as cutting the grass with an electric lawnmower. An r.c.c.b. switches the current off much faster than a fuse – within about 25 milliseconds. This is fast enough to save your life. The r.c.c.b. is a fast electromagnetic switch. We will have more to say about electromagnetism in Chapter 29.

Questions

1. What is the power transferred by the following?
   (a) a 12 V car battery delivering a current of 0.5 A;
   (b) a 250 V mains supply delivering 4 A to an electric heater;
   (c) a 3 V battery delivering 0.3 A to a torch bulb.

2. What is the current in the circuit in each of the following situations?
   (a) a 500 W food mixer connected to the 250 V mains supply;
   (b) a 6 W car sidelight bulb connected to a 12 V battery;
   (c) a personal cassette player, rated at 4.5 W, running off four 1.5 V dry cells (a 9 V battery)?

3. How much power is dissipated in the resistor in each of these circuits?

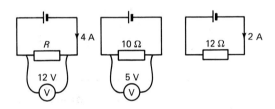

4. A light bulb is marked '240 V 60W'. What does this tell you about:
   (a) the current which it requires;
   (b) its resistance (at normal operating temperature)?

5. Which of the following requires the greatest amount of energy:
   (a) bringing water to the boil in a 1500 W electric kettle (3 minutes);
   (b) using a 500 W hair dryer to dry your hair (5 minutes);
   (c) watching television (150 W) for 2 hours;
   (d) leaving a 60 W light bulb on all night (8 hours).

6. Without doing any detailed calculation, work out which resistor will heat up quickest in each of these circuits.

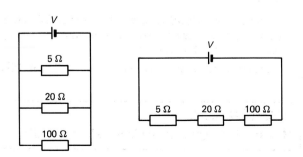

7. You are given two identical resistors and a d.c. power supply. Is more power dissipated in the resistors if you connect them to the power supply: (a) in series, or (b) in parallel? Explain how you worked it out.
   (**Hint:** You might find this easier if you begin by choosing values for the resistors and the power supply voltage.)

8. In this circuit the two lamps, **X** and **Y**, are operated at their rated voltage and power.

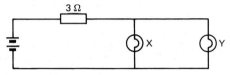

Lamp **X** is rated 6 V 12 W; lamp **Y** is rated 6 V 24 W. Calculate:
(a) the current in the 3 Ω resistor,
(b) the potential difference across the battery. (SEB)

9. Two lamps are marked 12 W 12 V and 24 W 12 V respectively.
   (a) Calculate the current drawn from a 12 V supply when the two lamps are connected in parallel across it as in diagram (A).

(A)
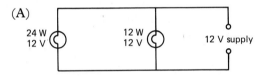

(b) Calculate the resistance of each lamp.
(c) If the lamps are connected in series across a 12 V supply as in diagram (B), estimate the current drawn from the supply.

(B)
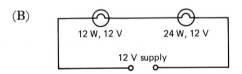

(d) What is the resistance of resistor **R** in diagram (C) which must be connected in series with the 12 W 12 V lamp to allow the lamp to operate normally using an 18 V supply?

(C)
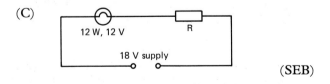

(SEB)

10. A 240 V motor runs an electric power drill. Calculate:
    (a) the power when the drill runs freely and the current is 0.2 A;
    (b) the power when the drill is drilling a hole in a piece of wood and the current is 2 A;
    (c) the efficiency when drilling wood, if the drill bit does 1800 J of work per minute.

11 A crane lifts a load of 600 N to a height of 10 m. What is the gain in potential energy of the load?

A 100 V electric motor drives the crane and the current is 2 A during the lifting time of 40 s.
(a) What is the power rating of the motor?
(b) How much electrical energy is transferred?
(c) How efficient is the crane motor?

12 In an experiment, a group of students count the turns of the rotating disc inside a kilowatt-hour meter when different appliances are connected to it. The graph summarises their findings.

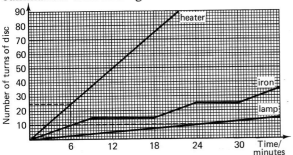

(a) Which of these appliances would be the most expensive to run?
(b) The kilowatt-hour meter is marked '250 revs/kWh'. What is the power rating of the heater used?
(c) How would you explain the flat sections of the graph obtained with the iron?

13 A bicycle lamp needs a 4.5 V battery and this supplies 0.3 A to light the bulb. The total battery life is around 20 hours.

If a battery costs 60p, how much is this per kilowatt-hour? How does it compare with the cost of mains electrical power (around 6p for 1 kilowatt-hour)?

14 Four light bulbs operate off the same domestic lighting circuit.

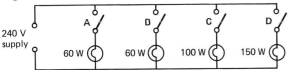

(a) What is the current through each bulb?
(b) What is the total current from the supply?
(c) What is the total power dissipated?

15 (a) A domestic ring main uses 30 A cable. Would it be safe to run six 1 kW electric fires off different sockets in the ring at the same time? (Take the mains voltage to be 250 V.)
(b) Each individual socket in the ring is rated at 13 A maximum. A four-way extension lead is connected to one socket. Would it be safe to run the following appliances off the extension lead at the same time?
– a 1 kW electric heater;   – a 750 W iron;
– a 1500 W electric kettle; – a 250 W television set.

16 This extract, from an Electricity Council information leaflet, tells customers how to choose the correct cable and fuse ratings for their appliances. Read it carefully and then explain clearly why the changeover from 3 A to 13 A cable and fuses comes at 720 W.

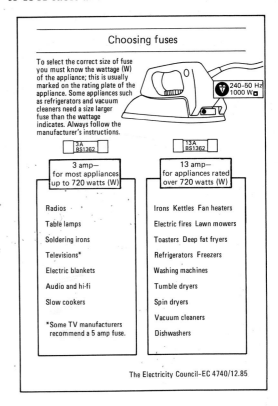

17 Imagine that you are an electrician and that the following problems and questions are brought to you by some of your customers. Remember, your business depends on satisfied customers.

In each of the first four cases, explain a possible cause for your customer's trouble and how it might be solved.

Customer 1: 'A lamp in my house has gone out.'
Customer 2: 'The lights downstairs have gone out.'
Customer 3: 'All the lights in my house have gone out.'
Customer 4: 'All the lights in the houses in my street have gone out.'

The next customer would like a clear wiring diagram.

Customer 5: 'I would like two lamps in the dining room which both worked from the same switch.'

Draw a suitable wiring diagram.

The last customer does not even know there is a problem!

Customer 6: 'I should like you to come and fit a mains socket in the bathroom so that I can have the electric fire on when I have a bath.'

Explain what the problem is.   (Sp.NEA)

INVESTIGATION

How much energy is stored in a 1.5 V torch cell?

28: Sources of Electric Current

Cells and batteries

We can buy cells and batteries in a whole range of shapes and sizes. They all give electrical potential energy to electrons passing through them, as a result of chemical reactions going on inside the cell.

28.1 Cells and batteries

The first cell was the **Voltaic pile**, invented by Alessandro Volta just before 1800 (see section 25.1). The cells we use today have developed from Volta's original design. Inside an electric cell, chemical reactions cause a potential difference of a volt or two between the two terminals. If the terminals are linked by a conducting path, this potential difference will make charge flow. The current may last for several hours before the chemical reactions in the cell have used up all the chemical energy available. Then the current stops – the cell is 'flat'.

A **battery** is really the name given to a collection of several cells (though we also use the word 'battery' in everyday language to mean just one cell). By connecting cells in series we can produce a larger potential difference between the two terminals.

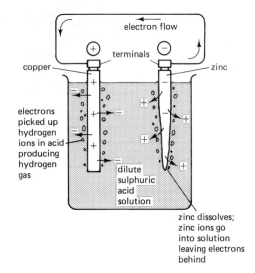

The simple cell

A zinc rod and a copper rod placed in dilute sulphuric acid produce a p.d. of about 1 V between them. The zinc rod slowly dissolves in the acid and zinc **ions** go into the solution. These are positively charged and so they leave electrons behind on the zinc rod. The electrons move through the connecting wire to the copper rod. Here they can be picked up by some of the hydrogen ions in the acid solution, changing them to hydrogen atoms. As the cell runs, the zinc slowly dissolves and bubbles of hydrogen collect on the copper rod.

This simple cell has two defects:

Polarisation: the cell works for only a short time because of the build-up of hydrogen bubbles round the copper rod. The bubbles prevent the copper rod coming into good contact with the acid and stop the reaction proceeding.

Local action: tiny impurities in the zinc, such as traces of iron, form tiny cells on the zinc plate itself. Hydrogen is given off even when the cell is not in use, and the zinc rod dissolves away too quickly.

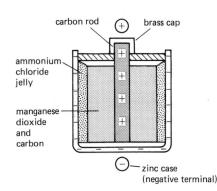

28.2 The dry cell

A dry cell contains no liquid, so it can be used in any position. The positive terminal is a carbon rod, surrounded by a mixture of manganese dioxide (a depolariser) and powdered carbon (to make the depolariser conduct better). The outer case, which is also the negative terminal, is made of zinc. Between the two is a paste or jelly of ammonium chloride.

The p.d. between the terminals of a zinc–carbon dry cell is 1.5 V. Most batteries have p.d.s which are a multiple of this. They are made of several identical zinc–carbon cells in series. For example, a 4.5 V cycle lamp battery has three 1.5 V cells inside; a small 9 V transistor radio battery contains six 1.5 V cells.

278

28.3 Primary and secondary cells

The dry cell is a **primary** cell. The chemical reactions cannot be reversed and the cell cannot be recharged once it has run down. A **secondary** cell is one which must first be charged by passing current through it. Once it is flat, it can be recharged again. It stores energy in chemical form, so that it can easily be released again as electrical energy.

Lead–acid cell

The lead–acid **accumulator** (the car battery) is a secondary cell. It holds much more charge than a dry cell and can give a very large current – over 100 A – for short periods. This is needed to operate the car's starter motor. The cell consists of plates of lead and lead oxide, dipping into dilute sulphuric acid. Its p.d. is 2 V. A 12 V car battery has six cells in series. It supplies the starter motor, the headlamps and rear lights, windscreen wiper motor, and so on. While the car is running, the battery is recharged by a **generator** (an alternator or dynamo – see section 30.4) driven by the engine.

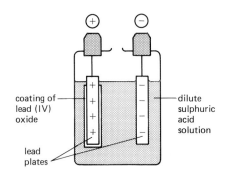

Nicad cells

These are small rechargeable dry cells. Instead of zinc and carbon, they use nickel oxide for the positive terminal and cadmium for the negative outer case – hence the name **Nicad** (from **ni**ckel and **cad**mium). Between the two electrodes there is a paste of potassium hydroxide. The p.d. of a Nicad cell is around 1.2 V.

Recharging a Nicad cell. Although these are more expensive than ordinary dry cells they can be recharged more than 100 times. The cost of recharging is very small.

28.4 Other types of cell

The silicon solar cell

A solar cell has no internal store of chemical energy. When light shines on the cell, a p.d. appears across its terminals. The p.d. depends on the brightness of the light – around 0.5 V in bright sunlight, with a current of about 25 mA for every square centimetre of cell surface. By connecting several in series, we can get a higher p.d.

Photovoltaic cells are used in solar powered calculators and in camera light meters. They are also used on satellites to provide the current to recharge the satellite's batteries. In countries where there is a lot of sunlight, they can be used to generate electricity on a small scale.

Solar cells used for pumping water in Mali, West Africa

28.5 P.d. and e.m.f.

A 'perfect' cell or battery would have a constant p.d. Unfortunately cells are not perfect! If we measure the p.d. across the terminals of a cell, we find that it gets smaller when the cell is delivering a current.

1. First measure the p.d. when the cell is supplying no current. It is said to be on 'open circuit'.

   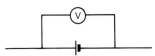

2. Varying the rheostat changes the current. Measure the p.d. across the cell for different currents.

   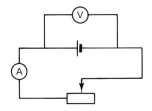

3. As the current gets bigger, the p.d. across the cell falls. Some voltage appears to be 'lost'.

Current/A	P.d. across terminals/V	'Lost voltage' /V
0	1.5	0
0.5	1.25	0.25
1.0	1.0	0.5
1.5	0.75	0.75
2.0	0.5	1.0
2.5	0.25	1.25

4. A graph of p.d. against current shows that the p.d. across a cell falls steadily as the current flowing from it increases.

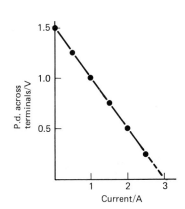

The p.d. across a cell when it is on 'open circuit' and supplying no current is called the **e.m.f.** of the cell. This is the maximum p.d. which the cell can produce. (**Note.** The initials e.m.f. stand for **electromagnetic force**, an old-fashioned term from the early days of electrical investigations.)

The zinc–carbon dry cell used in the experiment above has an e.m.f. of 1.5 V. A quick way to measure the e.m.f. of a cell is to connect a voltmeter across its terminals when it is supplying no current. This method gives only an approximate value, since the voltmeter itself takes a small current from the cell. For accurate measurements, a very high resistance voltmeter, such as a good digital voltmeter, is needed.

28.6 Internal resistance

The reason why the p.d. across a cell drops when it is supplying current is that the cell itself has a resistance. This is called **internal resistance** – inside the cell itself. The chemical reactions inside the cell produce a p.d. which makes electrons flow through the cell, but as they flow through they have to pass through the material of the cell itself. This has some resistance. The internal resistance is a result of the way a cell is made; the cell cannot simply be opened up and have its internal resistance removed!

In the results table on p. 279, the p.d. across the dry cell falls to 1.0 V when the current is 1 A. The 'lost voltage' is 0.5 V (1.5 V–1.0 V). This is the p.d. needed to push electrons through the cell itself. If we imagine the internal resistance as a separate resistor inside the cell, then the p.d. across it is 0.5 V. We can calculate its resistance using $R = V/I$:

$$\text{internal resistance}, R = \frac{0.5\,\text{V}}{1\,\text{A}} = 0.5\,\Omega$$

If we use any of the other results from the table, we get the same result.

Maximum current

The graph of p.d. against current indicates that there is a maximum current which a cell can provide. If the current was 3.0 A, the p.d. across the cell would be zero. This is the current you would get if you connected a very small resistance across the terminals – a short thick length of copper wire. The lower the internal resistance of a cell, the larger the maximum current it can supply. It is not a good idea to short-circuit a cell in this way. The cell can be damaged by making it provide too large a current.

Questions

1 The p.d.s across the two parallel branches in circuit (A) are equal. But if we close the switch in circuit (B), the bulb immediately goes out – we have a 'short circuit'. Explain why the p.d. across the bulb has suddenly dropped.

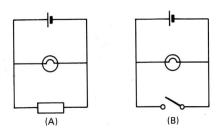

2 When an ordinary 1.5 V torch cell is connected directly across the terminals of an ammeter (assume it has zero resistance), the current from the cell is about 1 A. Calculate the internal resistance of the torch cell.

3 The graph shows how the output p.d. of a small solar cell (photovoltaic cell) drops when the cell supplies current.

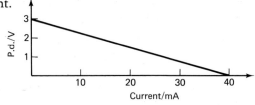

(a) Explain why the p.d. drops in this way.
(b) To run a small radio, a p.d. of 1.5 V supplying a current of 50 mA is needed. Several of these solar cells could be joined together in parallel to provide this. How many cells would you need? Explain your answer.

> **INVESTIGATION**
>
> A potato battery consists of two pieces of different metals stuck into a potato. Find out as much as you can about the behaviour of a potato battery: its p.d., the current you could get from it, its internal resistance.

29: Magnetism and Electromagnetism

The ancient Chinese knew of the magnetic properties of lodestone – a rock containing iron – and used pieces of it as simple compasses for navigating. The planet we live on is, itself, a huge magnet. Many of the machines which make our daily lives easier use the magnetic effects of electric currents and magnetic materials.

29.1 Magnets and magnetic materials

Magnets are very common and perhaps we take their strange properties for granted because we are so used to them! Permanent magnets can attract small pieces of iron, such as paper clips or nails, and will stick to the steel door of a fridge or cooker. Only a few materials are attracted by magnets. Anything made of **iron** or **steel** (which is mainly iron) is attracted: safety pins, ball bearings, pieces of iron ore. Only two other metals, **cobalt** and **nickel**, behave the same way. The same materials are also the ones from which permanent magnets can be made.

29.2 Properties of magnets

Poles

The places on a magnet where the magnetic forces are strongest are called the **poles**.

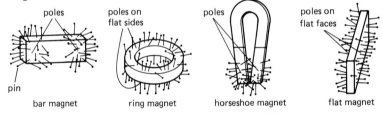

Iron filings or small pins show where the poles of the magnets are.

Magnets point north

If a bar-shaped magnet is hung from a thread so that it is free to turn, and is kept well away from any iron pipes or girders which might themselves be magnetic, it always ends up pointing in the same direction. One end points nearly (but not exactly) to the north, the other to the south. This is why magnets can be used as compasses for finding one's direction.

It is also used to name the poles of a magnet:
The pole at the end pointing north is called the north-seeking pole, or N pole.
The pole at the end pointing south is called the south-seeking pole, or S pole.

Forces between magnets

If two magnets are brought close together, they exert forces on each other. If the N pole of a bar magnet is brought towards the N pole of another suspended bar magnet, the second magnet moves away. There is a repulsion between the two poles. Two S poles behave the same way. But a N pole attracts another S pole. The results can be summarised as follows:

Like poles repel each other.
Unlike poles attract each other.

This result is very similar to the one we found for charged rods in Chapter 24. However, magnetic forces are usually stronger than electrostatic forces and this makes the experiment easier to do.

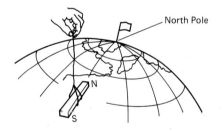

The compasses used by sailors or hill-walkers contain a small permanent bar magnet which points (approximately) north. At sea, or in mist, it is important to know the direction one is travelling in.

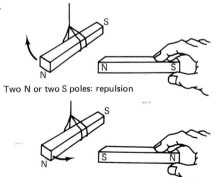

Two N or two S poles: repulsion

One N and one S pole: attraction

Testing for magnetic poles

The repulsion of like poles is the best method for testing a material to find out if it is a magnet. We use one known magnet to see if it can repel the material we are testing. If we can make the two repel, then the test material is also a magnet.

Attracting the test material is *not* a good way to test if it is a magnet. This only shows that it is a magnetic material (such as iron), not that it is a magnet.

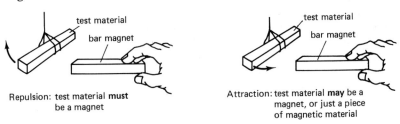

Repulsion: test material **must** be a magnet

Attraction: test material **may** be a magnet, or just a piece of magnetic material

29.3 Induced magnetism

When a small steel object (a paper clip) is attracted to a magnet, it becomes a magnet itself. It can attract other paper clips. Once it is removed from the magnet, it loses its magnetism again. While it was in contact with the magnet, the paper clip was an **induced magnet**.

The paper clip does not have to be actually touching the magnet for this to happen. Having the magnet nearby is enough to induce magnetism. If we use the N pole of the magnet, the end of the paper clip closest to it will become an induced magnetic S pole. The two unlike poles then attract and the clip is pulled towards the magnet. Induced magnetism always results in a force of attraction.

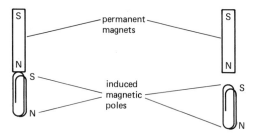

29.4 Making magnets

By induction: Some materials can hold on to their induced magnetism longer than others. If we attract an iron nail to a magnet, it is an induced magnet while it is close to the magnet, but loses its magnetism as soon as it is removed. Steel objects retain a little of their magnetism, and may still be able to attract other objects. They will be rather weak magnets, however, and this is not really a satisfactory way to make a magnet.

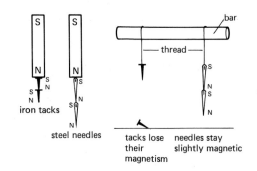

Stroking with a magnet: A piece of iron or steel can be permanently magnetised by stroking it with one pole of a bar magnet. The pole is moved over the bar several times in the same direction. The pole at the end of the stroke is always the opposite of the one used to stroke. If we stroke with a N pole, we will produce a S pole at the end of the bar where the stroke ends, and vice versa. The magnet produced is stronger than by the simple induction method, but is still rather weak.

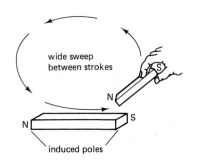

Electrical method: Stronger permanent magnets are made by an electrical method. We will discuss this later in section 29.9.

Losing magnetism: Permanent magnets gradually lose their magnetism. If a permanent magnet is hammered or shaken, this can reduce its strength as a magnet. Another way to demagnetise a magnet is by heating it strongly.

29.5 Theory of magnetism

If a magnetised steel rod is cut or broken into pieces, each separate piece is a magnet. If we cut these up, we get still more small magnets. We cannot separate the N pole and the S pole – each small piece has its own N pole and S pole. It is a **magnetic dipole**. One model which could explain this is to imagine that a magnet is made up of very small magnets all lined up in the same direction. Indeed the dipoles may even be the molecules of the material itself. Wherever we break the magnet, we will be left with new magnets. An unmagnetised bar is one which has its little dipoles completely disorganised and pointing in different directions.

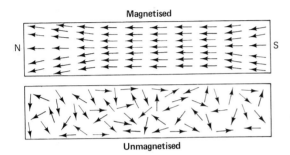

This model explains how magnetism can be induced. Near to a permanent magnet, the little dipoles will turn (like small compasses) to point in the same direction. When the permanent magnet is removed, they will become disordered as they were before. In some materials (hard magnetic materials), the dipoles may not go back completely to the disordered state and so some magnetism is left. Stroking a bar makes it become a magnet by aligning the dipoles. If a bar is stroked with a N pole, then the S poles of all the dipoles will be attracted to it and will turn to point towards the same end of the bar.

It is also easy to see how demagnetising methods work. Shaking or striking a magnet may disturb the neat alignment of the dipoles. Heating makes the molecules of the material vibrate more strongly and this too can destroy any neat lined-up pattern of dipoles.

This model of tiny dipoles is a very useful one, but there is one fact which it cannot explain. Why are the magnetic properties of iron, cobalt and nickel so different from other materials? The simple dipole theory would seem to apply equally well to *all* materials. It seems that in ferromagnetic materials like iron, the tiny molecular magnets are already lined up within small areas, called **domains**, inside the material. The domains are still tiny (less than 0.1 mm across), but each one might contain over 10^{17} molecules! In an unmagnetised piece of iron, the tiny domain magnets are all pointing in different directions, tending to form closed loops and cancel each other out. The bar is not a magnet. Magnetising it involves getting the domains to line up. In fact what seems to happen as the bar is magnetised is that the domains which are pointing in the right direction grow and others pointing in the wrong direction shrink. Once all the domains have lined up, the material is said to be **magnetically saturated**. It cannot be made into a stronger magnet than this.

Magnetic materials

Materials containing iron, cobalt and nickel can be magnetised strongly and are strongly attracted to magnets. Steel, an alloy of iron (containing mainly iron with some carbon added), is an example. Materials which can be magnetised and which are strongly attracted to magnets are known as **ferromagnetic** materials. They are classified as **hard** if they hold on to their magnetism well when magnetised, or **soft** if they lose it again easily.

Hard magnetic materials are more difficult to magnetise, but once they have been magnetised they hold on to their magnetism well. Alloys used for making permanent magnets are made from hard magnetic materials. Some very hard magnetic materials are nowadays also made by a process similar to making pottery. Powdered metal oxides (including iron oxide) are baked under high pressure to form a hard brittle material which can be made strongly magnetic. One such material is 'magnadur'. Cassette tapes are coated with iron oxide particles – another hard magnetic material. They record sounds in the form of variations in the strength of magnetism along the tape.

Soft magnetic materials also have their uses. They are easily magnetised and easily lose their magnetism. They are used in making electromagnets – magnets which can be switched on and off by an electric current (see section 29.8). Iron and mumetal (an alloy of nickel (74%) and iron (20%), with a little copper (5%) and manganese (1%)) are soft magnetic materials.

Other metals and non-metals like aluminium, copper, glass, plastic are usually described as non-magnetic. They are, however, influenced slightly by very strong magnets, though the effect is **very** weak compared with ferromagnetic materials.

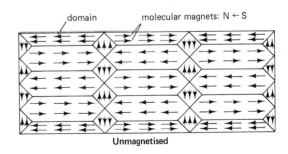

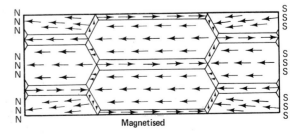

284 Magnetism and Electromagnetism

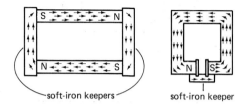

Repulsion makes the ends of the pins splay out.

The dipoles inside the material are also splayed out at the ends.

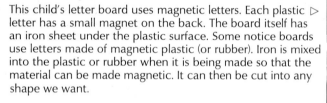

soft-iron keepers   soft-iron keeper

Storing magnets

If we use a bar magnet to pick up a number of pins at one end, the free ends of the pins will 'splay out'. The pins are induced magnets and the free ends all have the same polarity. They repel and this makes them move apart.

Of course, the same thing will be going on, though we cannot see it, inside the magnetic material itself. This breaks down the neat parallel arrangement of the dipoles and weakens the magnet. This is why magnets gradually get weaker just sitting in a drawer; it is why soft magnetic materials demagnetise easily.

This **self-demagnetisation** problem can be prevented by storing magnets with soft-iron keepers joining their poles. The dipoles can then form closed loops, with no free ends to disturb the neat pattern.

Using permanent magnets

Magnets are in use every day all round us. Fridge and freezer doors are held closed by magnetic strips around the door frame. The magnet attracts the metal body of the fridge, and this holds the doors shut. ▷

Small magnetic catches are used for cupboard doors. The magnet is fastened to the cupboard and a small steel plate to the door. This provides a neat and effective catch. ▷ ▷

This child's letter board uses magnetic letters. Each plastic ▷ letter has a small magnet on the back. The board itself has an iron sheet under the plastic surface. Some notice boards use letters made of magnetic plastic (or rubber). Iron is mixed into the plastic or rubber when it is being made so that the material can be made magnetic. It can then be cut into any shape we want.

Magnetic soap holders attract a small steel disc which is embedded in the surface of the bar of soap. ▷ ▷

The magnetic blade of a screwdriver attracts a steel screw-nail. This can be a help when working in awkward corners, where it may be difficult to hold the screw by hand. ▷

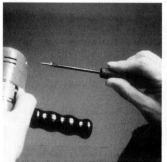

The brown strip on a Cash Card is a magnet. It carries a code which can be read by the cash dispenser machine, so the bank knows whose account to take the money from. ▷ ▷

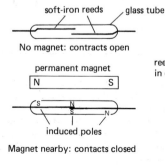

One simple type of burglar alarm for doors and windows uses a reed switch. This is fixed to the frame and a magnet is fixed to the point on the door directly opposite. The reed switch is a little glass capsule with two iron strips (or reeds) inside. Near the magnet (when the door is closed) these become induced magnets and their ends attract. The switch is closed (ON). When the door is opened, the reeds lose their induced magnetism and the switch opens. This can be used to set off a bell or a buzzer warning that someone has opened the door. This type of reed switch is called 'normally open' – a nearby magnet closes the contacts. Reed switches are also made 'normally closed' – a nearby magnet opens the contacts (see also page 328).

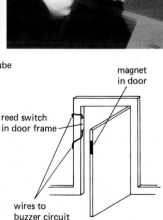

29.6 Magnetic fields

Two magnets attract or repel each other without touching – the force between them acts at a distance. As with charged rods (section 24.8), it is useful to think about this in terms of **fields**. In the region around a magnet there is a **magnetic field**. Any other magnet coming into this field will experience a force.

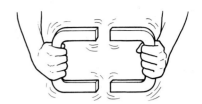

If you want to get a sense of how 'real' a field is, take two flat ceramic magnets and try to press the same poles together against the repulsion force. It really feels as if there is something spongy in the space between. If you are fortunate enough to have two large Eclipse magnets, you can ▷ get an even better 'feel' for the field by trying to put their like poles together! There really *does* seem to be something in the space between the poles.

Mapping the shape of the field

Using iron filings: A quick way to get an idea of the shape of a magnetic field is by using iron filings. These are scattered on to a thin card placed over the magnet. The iron filings become small induced magnets and are attracted and repelled by the poles of the magnet. If the card is tapped gently, the filings form patterns which show the shape of the field. ▷

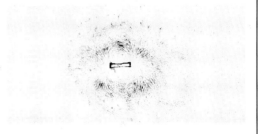

Using a plotting compass: A complete magnetic field map will show the **direction** of the forces at different points in the field. We define the direction of the magnetic field at any point as the direction of the force on another N pole placed at that point. So if we place a small compass in a field, the needle will turn so that its N pole points in the direction of the field at that point. The compass box can be used to plot the entire field:

1. Place it near one pole of the magnet. Mark a dot at the position of the N pole of the compass needle.

2. Now move the compass box so that the S pole of the needle is over this dot. Mark the new position of the needle's N pole.

3. Go on moving the compass box in this way until the line of dots comes back to the magnet. Then join the dots by a smooth curve; this is one **field line**.

4. Start again from a slightly different point on the magnet, and draw a second field line in the same way.

etc.

5. Go on doing this until you have a set of field lines mapping out the field all around the magnet.

Using a floating magnet: A rather nice way to see the shape of field lines is to use a magnetised steel needle, which is inserted in a cork so that it floats upright in water with its N pole upwards. It behaves as if it was a 'free' N pole, though of course there is a S pole under water at the other end of the needle. A fixed magnet is placed at the side of the water dish with the floating magnet nearby. The floating N pole feels a force in the direction of the field lines and moves slowly along the curved field line from one pole to another.

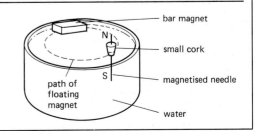

286 Magnetism and Electromagnetism

Fields of some common arrangements of magnets

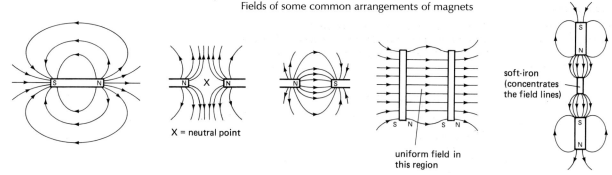

29.7 Currents and fields

Until about 170 years ago no one had discovered any connection between electricity and magnetism, though some experimenters thought they were related. For 14 years, the Danish physicist Hans-Christian Oersted searched for a link. Then one day in 1819, during a lecture, he found it by accident! He noticed that the needle of a compass box moved when he passed an electric current through a wire parallel to the needle. The current was producing a magnetic field.

Field of a straight wire

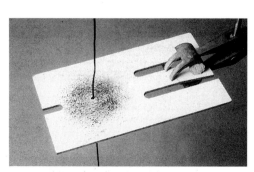

We can investigate the field round a wire using iron filings. The wire carrying the current passes through the centre of the card. Filings sprinkled on the card form a pattern of circles centred on the wire. A compass box indicates the direction of the field at different points.

Changing the direction of the current changes the direction of the field lines but not the general shape of the field. Notice the symbols used in the diagram below to show the current direction. They are very useful in 3-dimensional diagrams. A point indicates the current coming towards you; a cross, the current going away from you. Think of the point of a dart coming towards you, and the tail of a dart going away from you.

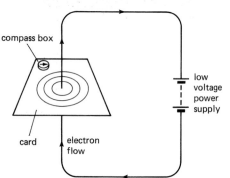

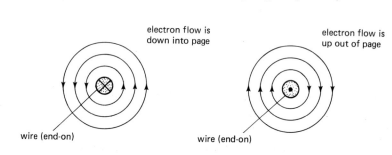

Ampère's grip rule

It is difficult to remember the field direction for each current direction, so a memory aid (a mnemonic) is useful. Imagine grasping the wire with your left hand, with your thumb pointing in the direction of the electron flow. Your fingers now point round the wire in the direction of the magnetic field. This is called **Ampère's left hand grip rule**. It does *not* explain why the field is in this direction – it is just a convenient way of remembering it.

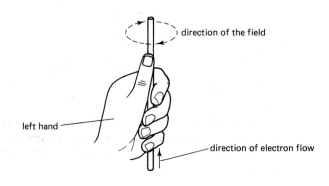

Field of a single coil

The field of a single coil of wire can be investigated in the same way. We can think of the coil as two lengths of wire, one carrying current upwards and the other downwards. In the centre of the coil, both sets of field lines point in the same direction. The fields add, so that there is a stronger field here.

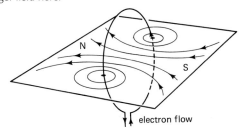

Field of a solenoid

A **solenoid** is a special term for a long coil made up of several turns of wire of the same diameter. It is like having a row of single turn coils placed side by side. The magnetic fields of each coil add to give quite a strong field along the centre line (the axis) of the solenoid. In fact the magnetic field of a solenoid is very similar to the field around a bar magnet.

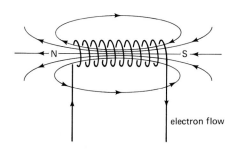

The coil behaves as if it has a N pole at one end and a S pole at the other. Changing the direction of the current through the coil changes the poles over. Again it is useful to have a memory aid for working out which end is which. ▷

The **left hand grip rule for solenoids**:

1. Grip the coil in your left hand with your fingers pointing in the direction of the electron current.
2. Your thumb is now pointing towards the N pole end.

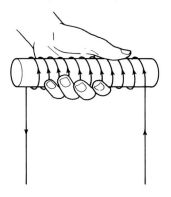

29.8 Electromagnets

The strength of a solenoid can be increased by:
- increasing the current through it
- increasing the number of turns.

But there is another way to make a much bigger increase in strength. It is to put an **iron core,** a soft magnetic material, inside the solenoid. When there is a current in the coil, this magnetises the core and the magnetic field produced is over 1000 times stronger than for the coil alone! When the current is switched off, the soft magnetic material in the core loses its magnetism almost immediately. This is an **electromagnet** – a magnet which can be switched on and off.

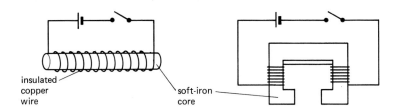

A C-core with 20 turns of wire is an electromagnet. If the object being lifted bridges the gap between the poles, the electromagnet becomes even stronger. This is because the 'magnetic circuit' is now complete and there are no demagnetising effects (section 29.5). ▷

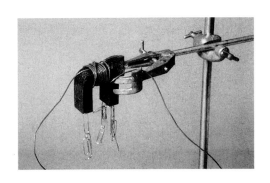

Uses of electromagnets

Electromagnets have medical uses. An electromagnet can be used to remove steel splinters which have accidentally entered a patient's eye.

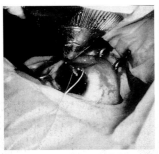

Electromagnets are very useful such as for lifting iron and steel loads, scrap metal or girders.

The repulsion between permanent magnets on the single rail track and powerful electromagnets underneath the vehicle keeps this monorail train hovering about 1 cm above the track. It can then be propelled along with very little friction to slow it down.

A telephone earpiece contains an electromagnet. When you speak into the mouthpiece of a telephone, the sound vibrations in the air make a thin metal plate (diaphragm) move inwards and outwards. Each time it moves in, it squashes the carbon granules behind it closer together. They then have a slightly smaller electrical resistance and so the current in the wires increases slightly. The sound vibrations have been changed into electrical pulses.

In the earpiece, these are changed back into sound. The pull on the iron diaphragm increases every time there is a current pulse and so it vibrates along with the changes in current. It makes a sound which is a copy of the original sound which produced the electrical pulses.

An electric bell contains an electromagnet which can switch itself on and off rapidly. Current in the electromagnet attracts a soft iron plate, but as this moves, it breaks the circuit. The electromagnet loses its magnetism and the hammer springs back. Then the whole cycle happens again... and again...

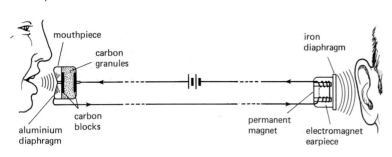

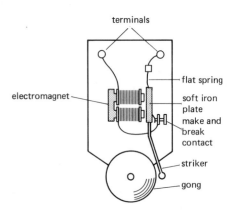

Relays

A relay is a switch operated by an electromagnet. When there is a current in the relay coil, a soft iron plate (or armature) is attracted. This operates a lever which closes the contacts. Other designs can use the current to open the contacts, or to switch over from one pair of contacts to another.

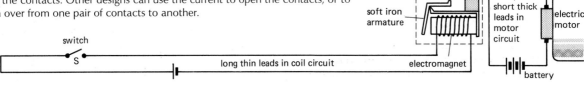

Relays are used to make a small current to switch on a larger one. Notice that the two circuits in the diagram above are completely separate. A small current in the coil circuit switches on a large current in the motor circuit. This means that the switch circuit can use lightweight switches and cables, with heavy duty cables in the motor circuit only.

For example, the switch in the diagram above might be under the rubber mat at a supermarket entrance. As you step on it, the switch closes. Current flows in the relay coil and this switches on the motor which operates the automatic doors. Long thin wires can run from the mat to the relay, with short thick cables from the relay to the motor.

Relays are also used for computer control. Computers run on very small currents and a computer cannot supply enough current to run a motor directly. However, the output from the computer can be used to operate a relay which switches the motor on and off.

Large robotic systems use relays to switch motors on and off, controlling the position and operation of the robotic arms. ▷

29.9 Making magnets electrically

Permanent magnets can be made electrically. If the specimen to be magnetised is bar-shaped, it is placed inside a solenoid; if it is a more unusual shape the coil can be specially wound round it. A large current is used to produce a strong field which lines up the magnetic domains inside the material. When the current is switched off, the material has been permanently magnetised.

A solenoid can also be used to *de*magnetise a specimen. An alternating current is passed through the solenoid, and the size of the current is gradually reduced. The alternating current magnetises the bar first in one direction and then the other. As the current drops, the magnetism gradually gets weaker and weaker.

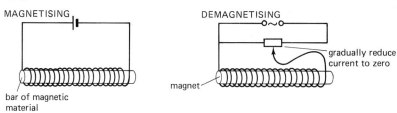

Tapes and tape recording

Magnetising and demagnetising are important aspects of how tape recorders work. A cassette recorder has two small horseshoe shaped electromagnets (called 'heads') which rub against the tape as it passes. The tape itself is made of strong plastic, with a coating of hard magnetic material (usually containing iron oxide).

One head is used for recording and playback. When used for recording, varying electric currents from a microphone make the field of the electromagnet increase and decrease. This results in variations in the strength of magnetisation of the tape.

The same head can be used for playback. This time the variations in magnetisation along the tape cause changing currents in the electromagnet's coils. This process is **electromagnetic induction**, and will be discussed in Chapter 30.

The second head is called the erase head. If you make a recording on a tape that has been used before, the original recording has to be erased first. The erase head is exactly the same as the record/playback head, but it is supplied by an alternating current whose frequency is well above our upper limit of hearing. It records a pattern on the tape which we are unable to hear.

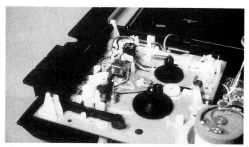

The heads in a cassette recorder are electromagnets. △ ▽

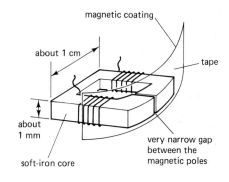

29.10 The Earth's magnetic field

The Earth itself has a magnetic field. No one is sure exactly what causes this, though it may be due to currents in the molten iron in the Earth's core. In any case, the Earth behaves as if it has a huge bar magnet in the centre!

A compass needle turns to point along the Earth's field lines. The N pole of the compass needle points in a direction close to north. As unlike poles attract, this means that there is a magnetic S pole lying somewhere close to the geographical North Pole! Just to be even more confusing, it is often referred to as magnetic north.

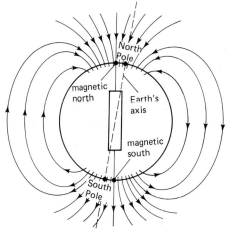

Declination

At most places on the Earth's surface, a compass needle doesn't point towards true north. The magnetic poles are not exactly at the geographical poles. The angle between the compass needle direction and true north is called the **angle of declination**. In Britain, magnetic north is about 5° west of true north (in 1989), and the angle is getting smaller by about $\frac{1}{2}$° each year – the magnetic poles move slightly! If you are navigating by map and compass, you must allow for this difference of 5° when you go from a map bearing to a compass bearing or vice versa.

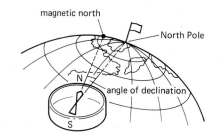

29.11 Forces, currents and fields

When a magnet is placed in a magnetic field it experiences a force. We have also seen that when there is a current in a wire, it sets up a magnetic field. So if we have a current-carrying wire in a magnetic field, there will be a force between them.

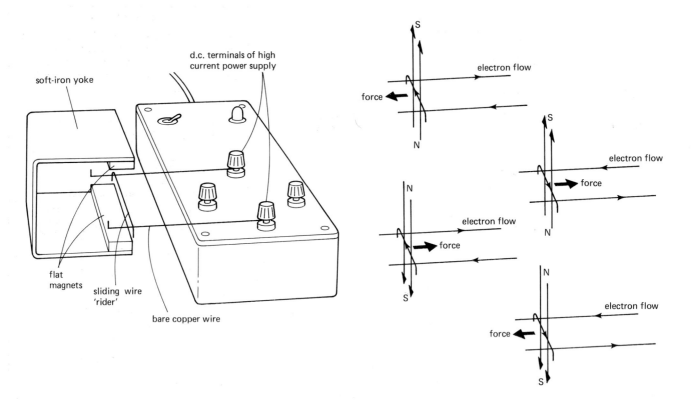

This diagram shows an experiment to investigate this force. The electrical circuit consists of three straight lengths of wire, one of which can move without breaking the circuit. The movable wire is in the strong uniform magnetic field between two flat ceramic magnets. When the current is switched on, the movable link slides along the parallel wires. The direction of the force on the wire can be reversed by changing either the current direction or the field direction. If we change both we are back to where we started.

Notice that the force is at right angles to both the current and the field. The wire is not attracted to either pole of the magnets. To help us remember the direction of the force, we can use the **right-hand motor rule**. (We will see later why it is called the 'motor' rule.)

The right-hand rule gives the direction of the force if the current and field are at right angles. If current and field are parallel, there is no force between them. At other angles, there *is* a force but its direction is more difficult to predict.

Right hand motor rule
1. The magnetic field direction is from the N pole to the S pole.
2. The current direction is the direction of electron flow (from the −ve terminal of the supply to the +ve).

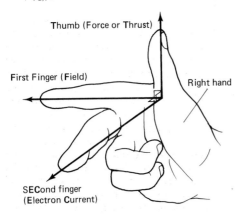

Size of the force

The force on a current-carrying conductor in a magnetic field depends on:

- the strength of the magnetic field
- the size of the current
- the length of the conductor.

Increasing any one of these makes the force larger.

29.12 Forces between parallel currents

Two current-carrying conductors both produce magnetic fields and so there may be a force between them. The easiest situation to understand is when the conductors are parallel. If the currents are in the same direction, the two conductors attract each other. The force is weak.

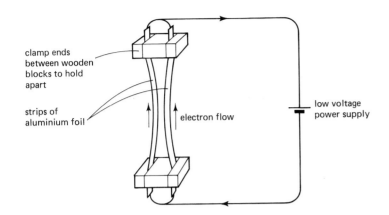

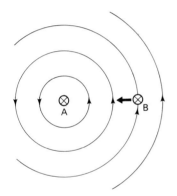

1. There is a magnetic field round strip A. Its direction is given by the left-hand grip rule.
2. The direction of the force on strip B is then given by the right-hand motor rule.
3. If we start with the field of strip B, we find that the force on strip A is also inwards.
4. If the currents are in opposite directions, the conductors repel each other.

Defining the ampere

The force between two currents is actually used to define the unit of current, the ampere:

1 ampere is the current which, in two very long straight parallel wires placed 1 metre apart in a vacuum, produces a force of 2×10^{-7} newtons on each metre length of wire.

It is not important to remember this definition but it is worth knowing how the ampere is defined. In theory, any current could be measured by passing it through two parallel wires and measuring the force between them! But the force is so small that this is not practical. An accurate instrument called a **current balance** is used to measure the force between two parallel coils. This is then used to **calibrate** (mark the scale on) other more convenient ammeters.

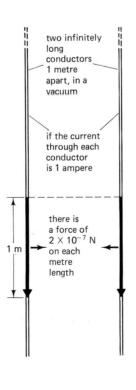

Loudspeakers

Every radio, cassette player or hi-fi system contains at least one loudspeaker. Most loudspeakers are of the moving-coil type, and make use of the force on a current-carrying conductor in a magnetic field.

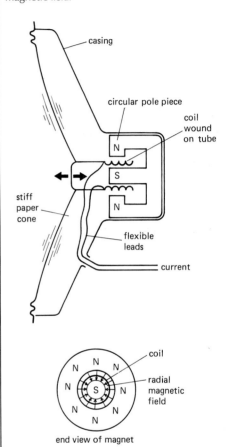

1. A specially shaped strong permanent magnet produces a radial magnetic field between its poles (the field lines are like the spokes of a wheel).
2. The wire of the coil is at right angles to the magnetic field. When there is a current, the force moves the coil inwards or outwards. The coil is wound on a tube and can move small distances back and forth in the magnetic field.
3. The paper cone is connected to the coil. As the coil moves, it makes the cone move in and out.

If the current is a.c., the cone will move inwards and outwards in time with the current. The cone vibrates and sound is produced. The frequency of the sound is the same as the frequency of the vibrations of the cone. If the current comes from a microphone and amplifier, the vibrations of the loudspeaker cone will match the original sound vibrations picked up by the microphone.

29.13 Turning force on a coil in a magnetic field

Imagine a flat square coil in a magnetic field. When there is a current, the two sides of the coil will experience forces. By the right-hand rule, side 1 will be pulled upwards; side 2 will be pulled downwards. The coil will turn into the vertical position.

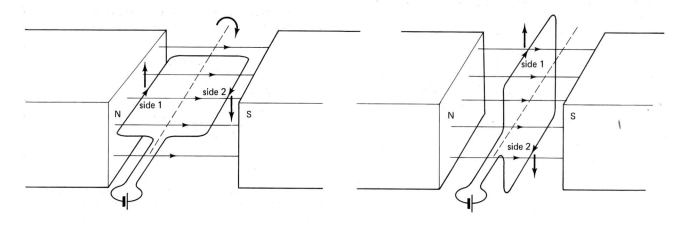

The turning effect (the couple) on the coil is biggest when it lies flat in the field, because the two parallel forces are now furthest apart. By the time the coil is vertical, the forces are in line and there is no couple.

The size of the turning force can be increased by:
– increasing the current
– increasing the strength of the magnetic field
– increasing the number of turns in the coil (each turn is like a separate coil and the forces add)
– increasing the coil area: making it longer increases the force on each side; making it wider increases the turning effect of the forces.

The d.c. motor

A flat coil in a magnetic field moves when a current is supplied. To make a **motor**, we need to improve the design so that the coil turns continuously. When the coil above has reached the vertical position, we want to reverse the direction of the forces, so that side 1 is now pulled downwards and side 2 upwards. The easiest way to do this is to reverse the current direction. We use an arrangement of sliding contacts called **brushes**.

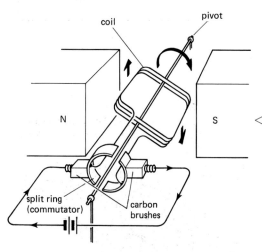

The two ends of the wire of the coil are connected to the two halves of a **split-ring commutator**. The brushes are held against the split-ring by springs. Notice where the join in the ring is. When the coil is in the vertical position, the brushes are touching the join between the two halves of the ring.

If the coil is lying flat when a current is switched on, it begins to turn. These forces turn it until it is vertical, but by this point is has picked up quite a bit of momentum and overshoots the vertical position. The commutator now changes the direction of the current (each brush is touching the *other* half of the split-ring) and the forces change direction. The coil goes on turning through another complete half-turn. Then the current changes direction again... and so on. The coil rotates continuously – a motor.

Making a d.c. motor

1. Start with two flat ceramic magnets joined together. Pull them apart and 'stick' them to a metal yoke. There is now a strong uniform magnetic field in the gap.

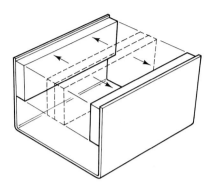

2. Take a wooden armature. Wrap a short piece of sticky tape round one end of the metal tube which runs through it.

3. Wind a coil of about 10 turns of insulated wire on to the wooden armature. Make sure that the two bare ends of the wire lie on either side of the metal tube. Use two small elastic bands to hold them in place.

4. Put a wire spindle through the centre of the armature and pass its ends through two split-pins. Check that the armature is free to turn.

 Make connections from a low voltage power supply to the two ends of the coil using two more pieces of wire. Bare about 1 cm of their ends and use the rivets on the baseboard to hold them so that they press against the two ends of the coil.

 Put the whole assembly into the field between the two magnets.

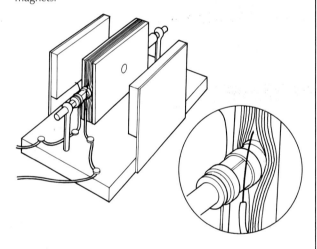

When the current is switched on, the coil should turn. It is best to start the coil from a horizontal position. It may need a small 'flick' to get it going, but then it should continue to spin on its own.

Practical motors

Real motors work on the same principles as the simple d.c. motor described above. However, there are some further improvements we can make.

1. The simple motor has a rather jerky turning action. The turning forces on the coil are largest when it is horizontal and so it gets two large 'kicks' every turn. If we use several coils set at different angles on the same armature, each with its own pair of commutator contacts, the turning force is much more even. The commutator now has many segments, switching on the current in each coil in turn as it rotates past the brushes.
2. The coils are wound on a soft-iron armature. This becomes magnetised itself and this greatly increases the strength of the magnetic field.
3. The pole pieces are curved. This combines with the effect of the soft-iron armature to produce an almost radial magnetic field. Each coil is parallel to the field for most of its rotation, so the force on it is more constant.

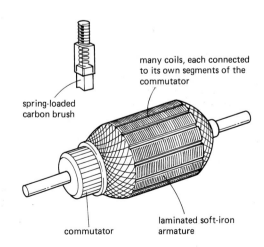

294 Magnetism and Electromagnetism

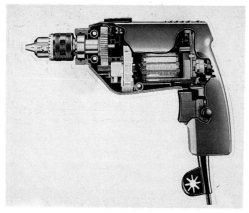

A power drill contains a powerful a.c. motor.

In some motors, electromagnets rather than permanent magnets are used to provide the field. This has the advantage that the motor can run from an a.c. supply. Although the current in the coil is constantly changing (because of the a.c.), the current in the electromagnet's coils is changing at exactly the same time. So the two stay in step and the turning effect is always in the same direction! Electromagnets are smaller than permanent magnets of the same strength, so the motor can be made smaller.

29.14 The moving-coil galvanometer

The turning force (the couple) on a coil in a magnetic field depends on the current in the coil, so it can be used to *measure* the size of the current. A **galvanometer** is the name for any instrument that measures small currents.

This is how a **moving-coil galvanometer** works. When there is a current through the coil, it turns in the magnetic field and moves a pointer across a scale. As it does so, the spring tightens and exerts a turning force (a couple) back in the other direction. This restoring couple gets bigger the more the coil turns. At some point the two couples balance and the coil stops. The bigger the current, the further it turns before this happens.

Making a model galvanometer

If we remove the brushes from our model motor, and then wind the ends of the coil into two loose spiral springs, we have a model meter. When there is a current, the coil begins to turn. The straw pointer indicates how large the current is.

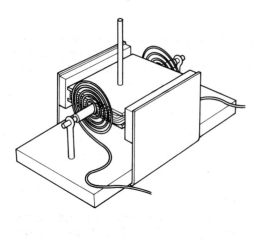

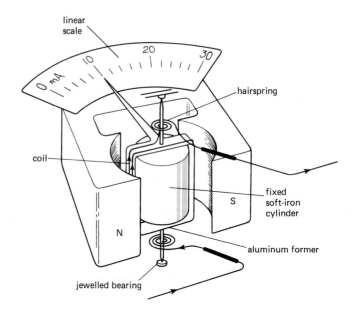

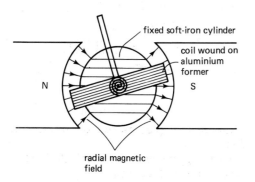

The diagrams above and on the left show some design features of a good moving-coil galvanometer.

1. The coil rotates around a soft-iron cylinder. Together with the curved pole pieces, this produces a radial magnetic field. As the coil turns, it is always parallel to the field lines and so the couple is constant. As a result the scale is **linear** – equal divisions all the way along.

2. The coil is wound on to an aluminium former which turns along with it. Because of an effect which we will discuss in section 30.6, the aluminium resists being moved through a magnetic field and this **damps** the movement of the coil. The coil moves smoothly to its final position; the pointer doesn't oscillate around but comes quickly to rest.

Questions

1. You are given a painted metal rod and a bar magnet. Explain clearly how you could show if the painted rod were:
   (a) non-magnetic (e.g. aluminium);
   (b) magnetic but not magnetised (e.g. iron);
   (c) a magnet.

2. Copy the diagrams and add the compass needles, showing clearly which direction they will point.

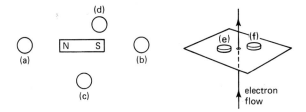

3. Diagram (A) shows the magnetic field near two bar magnets.

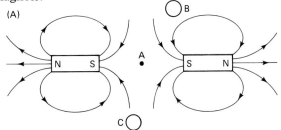

   (a) Describe the field at **A**.
   (b) Draw a sketch diagram to show the direction the two compass needles at **B** and **C** will point.
   (c) One of the magnets is now turned round as in diagram (B). Draw a sketch showing what the field lines inside the dotted area would look like.

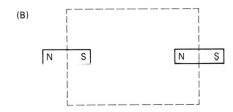

4. The diagram shows an electromagnetic relay. Explain why the motor starts when switch **S** is closed.
   This seems a rather complicated way to switch on a motor! Explain why a relay might be used instead of switching the motor on directly.

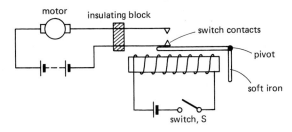

5. An experiment is carried out to investigate how the strength of an electromagnet depends on the current in the coil. The electromagnet is used to pick up small steel pins, and the mass of pins collected is measured for a number of different currents. The mass lifted is a measure of the magnet's strength.

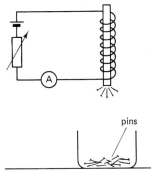

Current/A	Mass of pins/g
0	0
0.2	1
0.4	4
0.6	8
0.8	15
1.0	27
1.2	50
1.4	68
1.6	73
1.8	75
2.0	75

   (a) Draw a graph showing how the mass of pins (*y*-axis) varies with current (*x*-axis).
   (b) The graph seems to level off at the top. Explain what this means about the electromagnet's strength. Suggest an explanation for this.
   (c) The coil used in the experiment had 50 turns. What would you expect the results to look like if a 100-turn coil was used instead?

6. Two metal rods are placed in a long coil as shown. When a direct current flows through the coil, the rods move apart. When the current is switched off, the rods return to their original positions.

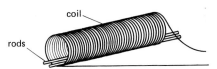

   (a) Why did the rods move apart?
   (b) From what metal are the rods likely to be made? Give a reason for your answer.
   (c) If alternating current from a mains transformer is passed through the coil, what effect, if any, will this have on the rods? Explain your answer.
   (Sp.O&C,MEG)

7. A very simple loudspeaker can be made by sticking a small magnet to the bottom of a polystyrene cup and clamping an electromagnet nearby. When there is an alternating current in the electromagnet coil, the base of the cup vibrates and sound is produced.

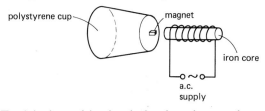

   (a) Explain how this simple loudspeaker works and why it produces sound.
   (b) Show, using a diagram, how this simple design has been improved in a 'real' loudspeaker.

8 **A** is an iron strip clamped at one end. The strip passes through coil **C**.

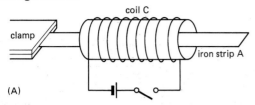

(A)

(a) When the switch is closed, what will happen to the iron strip?
(b) What difference would there be if the current in the coil were reversed?
(c) What happens if an alternating current is used (e.g. mains, which changes direction at a frequency of 50 Hz)?
(d) If the free end of the strip is now placed between the poles of a horseshoe magnet, what will happen in situations (a), (b) and (c) above?

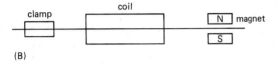

(B)

(e) Which common laboratory device is made in this way?

9 Make a list of all the electric motors you can find in your home. Try to find out the power and the current rating of each of them by reading the rating plate on the appliance.

10 Imagine that you are a journalist for your local newspaper back in the 19th century! You have just been lucky enough to see a demonstration of the newly invented electric motor.
 Write an article for your newspaper explaining to your readers what this wonderful new invention is and how it works.

11 (a) What are the functions of parts **A** and **B** of this simple motor?

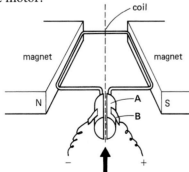

(b) In which direction, looking along the arrow, will the coil rotate?
(c) What would be the effect of reversing: (i) the magnets; (ii) the current; (iii) both the magnets and the current?

12 The diagram shows a model of an electric motor.

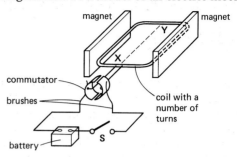

When the switch **S** is closed, the coil rotates about the axis **XY**.
(a) Explain what the commutator is for and why it is needed.
(b) State two ways in which the direction of rotation of the coil could be reversed.
(c) State two differences between this model motor and the kind of motor you would find in an electrical appliance.

13 A pupil constructed a model ammeter as shown below.

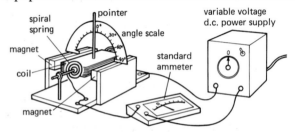

The pupil calibrated the model ammeter using the standard ammeter in the circuit shown. He recorded the angle of deflection of the pointer for known values of current. The pupil's readings were as follows:

Standard ammeter reading in amperes	0.00	0.10	0.30	0.50	0.70
Deflection of pointer in degrees	0.00	9.0	20	29	36

(a) On graph paper, plot a graph to show how the deflection of the pointer varied with current.
(b) From your graph find the angle through which the pointer is deflected when the current is increased:
(i) from 0.0 A to 0.2 A; (ii) from 0.4 A to 0.6 A.
 Give a reason why the answers to (i) and (ii) are different even though the change in current is the same.
(c) (i) What would have been the effect on the deflection of the pointer for the same current if the spiral springs had been stiffer? Explain your answer.
(ii) What would have been the effect on the deflection of the pointer for the same current if the magnets had been stronger? Again explain your answer.
(SEB)

INVESTIGATION

How long does an electromagnet stay electromagnetic after the current is switched off?

30: Electromagnetic Induction

We can produce motion by passing current through a conductor in a magnetic field. But is it possible to reverse this process – to move a conductor in a magnetic field and produce a current? In the 1830s, Michael Faraday carried out a famous series of experiments which showed that this could be done – he made the first dynamo, or electrical generator. The methods we use today to generate electricity in power stations come directly from Faraday's work.

30.1 Moving a wire in a magnetic field: induced e.m.f. and current

If we connect a loop of wire to a sensitive galvanometer and then move the wire quickly down through the field between two flat magnets, the meter pointer shows a small momentary deflection while the wire is moving. As we quickly pull it up again, we get a small 'kick' in the other direction. The effect is called **electromagnetic induction**. A small current has been **induced** in the circuit while the wire is moving through the magnetic field. When the wire is stationary in the field, there is no induced current. Moving the wire parallel to the field lines also produces no effect – the wire has to cut across the field lines to produce an induced current.

Induced e.m.f.

What would happen if we removed the meter, so that the circuit loop was broken, and then moved the wire through the field? The two ends of the wire would become like a small cell (or battery) while the wire was moving. There would be a small **induced e.m.f.** (see section 28.5) across the two ends of the wire. When we complete the circuit and close the loop, this induced e.m.f. makes an **induced current** flow in the wire. ▽

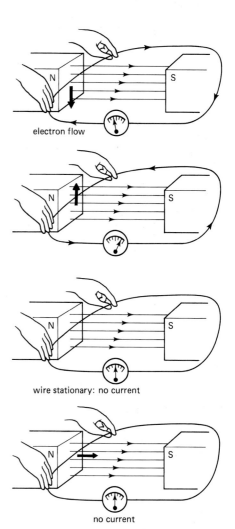

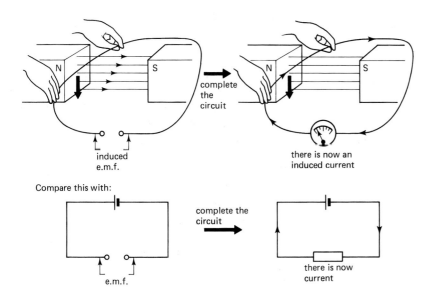

The size of the induced e.m.f. (and hence of the induced current) can be increased by:
- moving the wire faster
- using stronger magnets to produce a stronger field
- increasing the length of wire moving through the field, for example by making a coil of several loops and moving all of these through the field together.

298 Electromagnetic Induction

30.2 Faraday's coil and magnet experiments

Another way to induce an e.m.f. and a current is by moving a magnet in and out of a coil.

1. As the N pole of a bar magnet moves into the coil, a small e.m.f. is induced across the ends of the coil. If the coil is part of a circuit, a small induced current flows while the magnet is moving.

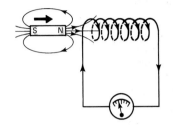

2. As the magnet is removed there is again a small 'kick' on the meter, but this time in the other direction. There is a small induced current in the opposite direction.

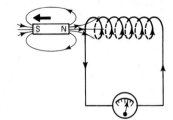

3. If we use the S pole of the magnet, we reverse the direction of the induced e.m.f. and current each time.

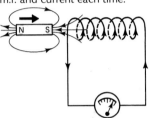 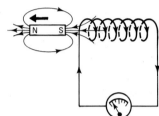

4. There is no induced e.m.f. or current while the magnet is stationary inside the coil.

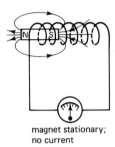

magnet stationary; no current

The size of the induced e.m.f. (and hence of the induced current) can be increased by:

– moving the magnet at a higher speed
– using a stronger magnet
– increasing the number of turns in the coil. Each turn is like a separate coil and they are all connected in series, so the total e.m.f. is the sum.

We get the same results if we hold the magnet steady and move the coil over it and back again. It is the **relative** motion of the coil and magnet that matters.

Cutting field lines

The two methods of inducing an e.m.f. and current (moving a wire through a uniform field, and moving a magnet in and out of a coil) have one thing in common. In both cases, we get an induced e.m.f. and current while the conductor is cutting through the lines of magnetic field. Being *in* the field is not enough – the conductor has to *cut through* the magnetic field lines if we are to get any induced effects. It doesn't matter whether we cut the field lines by moving the conductor or the magnet. From experiments like these, Faraday worked out his first law of electromagnetic induction:

The e.m.f. induced in a conductor is proportional to the rate at which the conductor cuts through the magnetic field lines.

Can you see how this agrees with what we have said above about the size of the induced e.m.f.? Remember that the stronger a magnet, the closer together are the field lines around its poles.

30.3 Direction of the induced current

Straight wire in a uniform field

When a straight wire is moved through a uniform magnetic field, the induced current direction is at right-angles both to the direction of motion and to the magnetic field direction. The **left-hand generator rule** is a method of remembering the induced current direction.

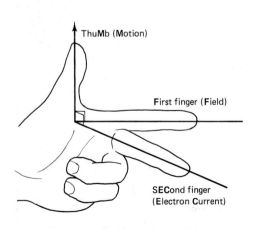

It is important to try not to confuse this with the right-hand motor rule on p. 290. The motor rule applies when we are using a current to produce motion; the generator rule applies when we are using motion to produce a current.

Lenz's Law

In 1834, Heinrich Lenz noticed something about the direction of induced currents:

An induced current always flows in such a direction that it will oppose the change which is causing it.

This is known as **Lenz's Law**. To see how it works, look at the diagrams of the coil and magnet experiment. As the N pole approaches the coil, the induced current flows to make this end of the coil a N pole – repelling the N pole of the magnet. (Remember the left-hand grip rule (p. 287) for working out which end of a solenoid is N.) As the N pole is removed again, the induced current is in the opposite direction, making the end of the coil a S pole – attracting the magnet. Each time the current tries to *oppose* the movement of the magnet. Check that this also works when the magnet's S pole is used.

The induced currents have to oppose the magnet's motion if energy is to be conserved. When we induce an e.m.f. and current, we have produced a small amount of electrical energy. Where has this energy come from? It has come from the chemical energy stored in our muscles as we move the magnet against the small opposing forces caused by the induced current!

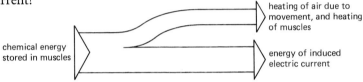

Lenz's law applies not only to the coil and magnet experiment, but to any situation where current is induced.

Bicycle dynamo

A bicycle dynamo improves on the simple coil and magnet experiment by having a magnet rotate near a stationary coil. The magnet is rotated by a small bevelled wheel which grips the side of the tyre when the dynamo is on. As the magnet rotates, the induced magnetic field in the soft-iron core is constantly changing in direction and so the coil is constantly cutting through magnetic field lines. The induced e.m.f. is large enough to provide the current to light the cycle lamps.

Since the coil is stationary, there is no problem about making the electrical connections to it. This is a big advantage of this design. The current produced is a.c. – constantly changing in direction – but this is perfectly satisfactory for lighting lamps.

Moving coil microphone

A moving-coil microphone is like a loudspeaker working in reverse. Sound waves – compressions and rarefactions in the air – set the diaphragm of the microphone vibrating. Connected to the diaphragm is a small coil which moves backwards and forwards through the magnetic field of a cylindrical-shaped permanent magnet. Small currents are induced in the coil as it moves. These are electrical 'copies' of the sound vibrations striking the diaphragm. The electrical signal can then be amplified (made bigger) and used to make a recording on magnetic tape, or to drive a loudspeaker.

Playing back a tape recording

We saw in Chapter 29 how the heads of a tape recorder work as small electromagnets and put patterns of magnetisation on to the magnetic surface of the tape. When we play a tape back, the head now works as a detector. The magnetic patterns on the tape induce small currents in the coils around the head. The pattern of the varying current is the same as the magnetic pattern on the tape. The electrical signals can then be amplified and used to drive a loudspeaker.

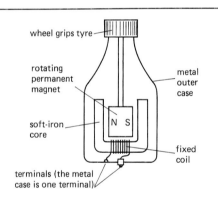

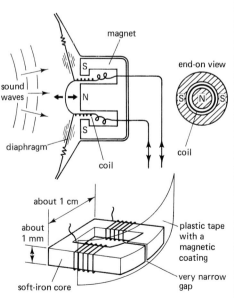

30.4 Induced e.m.f. and current in a rotating coil

A bicycle dynamo generates current by rotating a magnet near a coil. It is also possible to rotate the coil and keep the magnets fixed.

As the coil rotates, one side moves upwards through the magnetic field and the other side moves down. An e.m.f. is induced across the ends of the coil as a result. If the coil is part of a circuit, there will be a current. The direction of the induced current is given by the left-hand generator rule.

A quarter turn later, the coil is in the vertical position. The sides of the coil are now moving parallel to the field lines and so the induced e.m.f. and current are zero.

Another quarter turn later, the coil is back in the horizontal position. Again there is an induced e.m.f. and current but now it is in the opposite direction.

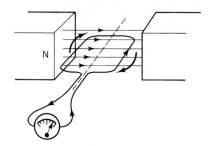
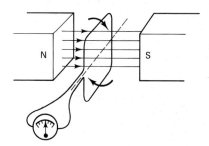
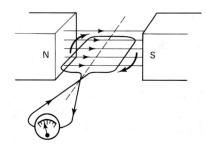

If we could keep rotating the coil like this we would generate an alternating current in the circuit. But the diagrams show clearly one problem with this — the leads from the coil will twist and stop it turning!

A simple a.c. generator

The a.c. generator (or **alternator**) overcomes the problem of twisting wires by connecting the two ends of the coil to two **slip rings**. These rub against two **brushes** which carry the current to the rest of the circuit. The slip rings turn with the coil; the brushes are fixed and stationary.

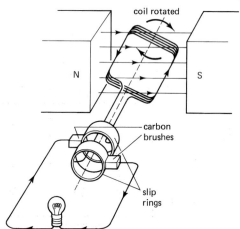
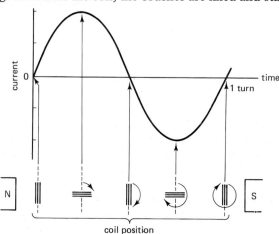

The current produced as the coil rotates is alternating current, since the induced e.m.f. in each side of the coil is in one direction while it is moving up, and in the other direction when it is moving down. The diagram above shows how the changes in current correspond to the different positions of the coil. The current is largest when the coil is horizontal and is zero when it is vertical.

We can increase the size of the alternating current by:

- using stronger magnets
- winding the coil on a soft-iron core to increase the magnetic field strength
- using a coil with more turns
- rotating the coil at a higher speed.

Increasing the speed of rotation of the coil also increases the **frequency** of the alternating current.

Practical generators

Most practical generators for producing a.c. work differently from the simple alternator described on p. 300. Instead of having stationary magnets and a moving coil, they have a rotating electromagnet (the **rotor**) which turns inside a fixed coil (the **stator**). The current for the rotor has to be supplied via slip rings since this coil is rotating, but the output from the fixed stator does not need brushes and rings. This is a big advantage: the current needed for the electromagnet is much smaller than the current output from the stator coils, so the brushes and slip rings can be much smaller.

▽ Generators in power stations are of this design. The rotor is driven by a turbine and a large alternating current is delivered from the stator coils.

Cars also have alternators, driven by the engine as it runs, which generate enough electrical energy to run the car lights, windscreen wiper motor, radio and so on. The alternating current from the alternator is also converted into direct current (we will see in section 32.4 how this can be done). This is then used to keep the car battery charged.

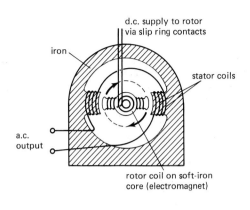

30.5 Generating direct current (d.c.)

If we change the slip rings of an alternator to a single split-ring, the output of the generator is a kind of d.c. The split-ring commutator reverses the current direction once every half turn.

The output is called **full-wave rectified a.c.** It is not smooth d.c. but it is always in the same direction and the average value of the current is no longer zero.

In practical d.c. generators it is common to use electromagnets rather than permanent magnets to produce the magnetic field.

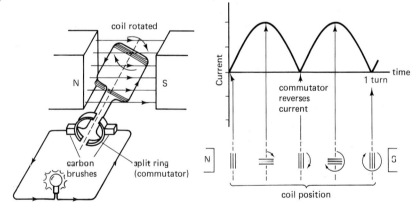

Making a model generator

The model motor described in section 29.13 can be run as an electrical generator. No power supply is used. Instead, connect the two wires from the brushes to a sensitive milliammeter. Spin the coil. A good way to do this is to wind a short piece of thread round the spindle; pull the thread and the coil will turn. The milliammeter will show a small direct current as it turns.

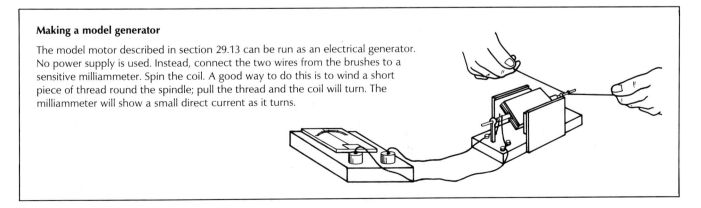

Electromagnetic Induction

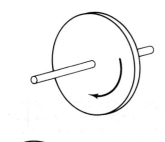

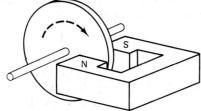

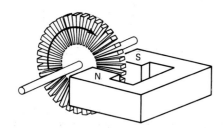

30.6 Eddy currents

◁ If a copper or aluminium disc is set spinning, it will go on turning for quite a long time. If, however, the same disc is spun in a strong magnetic field, it slows down and stops almost immediately. The movement is said to be **damped**. Aluminium and copper are not magnetic materials but they *are* good conductors of electricity. As the disc moves in the magnetic field, currents are induced inside the material of the disc itself. These currents flow in small closed loops within the aluminium or copper disc. They are known as **eddy currents** – they are like eddies (or whirlpools) in water when it is stirred. Each eddy current produces its own small magnetic field. The effect of this is always to produce forces which slow the rotating disc and oppose its motion – another example of Lenz's Law.

◁ If we use a disc with many slits cut in it and spin this in a magnetic field, it continues to turn for longer than a solid disc. The slits cut down the number of possible paths for eddy currents and so there is less damping.

The damping effect of eddy currents is used in the design of the moving-coil galvanometer (p. 294). The galvanometer coil is wound on an aluminium former. As the coil turns in the magnetic field, eddy currents are generated in the former and this damps the motion of the coil – the pointer moves smoothly to its final position rather than oscillating around the correct reading and taking a long time to settle.

Induction motors

Eddy currents can be used not only to stop motion but also to *cause* motion. In fact most of the electric motors used today are **induction motors** which make use of eddy currents. The details of these motors are complicated but the basic principle is simple enough.

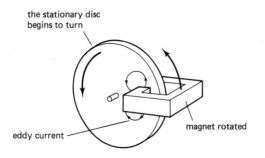

If an aluminium disc is set spinning in a strong magnetic field, it soon stops moving.

If the field is rotated by making the magnet move round the disc, the same eddy currents are induced and try to stop the *relative* motion of the disc and magnet. The aluminium disc begins to rotate.

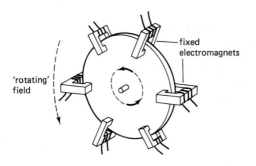

In an induction motor, the rotating field is produced, not by moving permanent magnets around, but by several fixed electromagnets placed round the disc. Each has an a.c. supply. The motor is designed so that the maximum current occurs in each electromagnet coil slightly later than in the one before. So the fields produced by each electromagnet reach maximum at slightly different times and this has the effect of a rotating magnetic field.

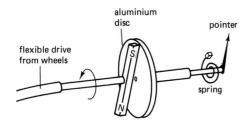

◁ This same idea is used in car speedometers. A flexible drive cable (rather like a bicycle brake cable, with an outer sheath and an inner wire) comes from the wheels of the car and has a bar magnet attached to its end. This is near an aluminium disc which is connected directly to the speedometer pointer. The faster the car is moving, the faster the magnet rotates. This then induces eddy currents in the aluminium disc which try to make it rotate. As it turns, a coil spring opposes its motion. The position where it stops depends on how big the induced eddy currents are – and this depends on how fast the magnet is turning.

30.7 Mutual induction

Moving a magnet in and out of a coil induces an e.m.f. and current in the coil. The same thing happens if we place an electromagnet near the coil and switch it on and off.

1. As the magnet moves, field lines are cut by the coil. An e.m.f. is induced in the coil.

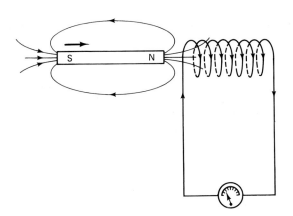

2. When the electromagnet is switched on, the magnetic field increases rapidly. You can imagine the number of field lines increasing as they 'sweep out' from the electromagnet. As they do so, the field lines are cut by the coil. An e.m.f. is induced in the coil and there is a short pulse of induced current.

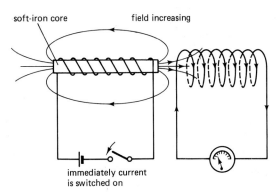

3. While the magnetic field is steady there is no induced e.m.f. The pattern of field lines is steady and none is being cut by the coil.

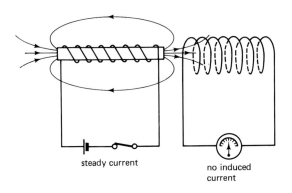

4. When the current is switched off, the magnetic field falls rapidly to zero. It is as though the field lines are now rushing back out of the coil again! They are again cut by the coil and an e.m.f. is induced. There is another short pulse of current – this time in the opposite direction.

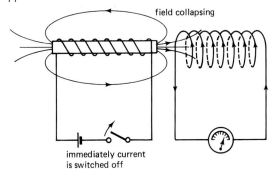

Another way to set this experiment up is to wind both coils on the same iron core. If the core is a closed loop so that there is a complete magnetic circuit, the effects are stronger. The **primary** coil is an electromagnet. The magnetic field produced is concentrated in the iron core so that it has to pass through the **secondary** coil. When the current *changes* in the primary coil, an e.m.f. and a short pulse of current is generated in the secondary coil.

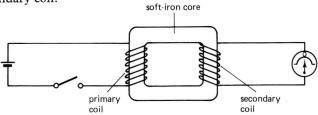

The experiment will work just as well in the other direction. Changing the current in the secondary coil induces an e.m.f. in the primary also. Changing the current in one coil causes an induced e.m.f. in the other. The effect is called **mutual induction** – both coils influence each other in exactly the same way.

30.8 Transformer

In the arrangement shown on p. 303, changing the current in the primary coil induces an e.m.f. and current in the secondary coil. We can do this by continually switching the current on and off in the primary, but there is an easier way! If a.c. is used in the primary, the current is changing all the time. This induces an alternating e.m.f. and current (of the same frequency) in the secondary coil.

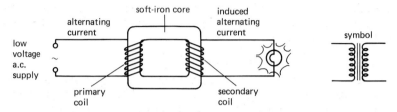

The primary and secondary coils and the core make a **transformer**. Notice that there is no wire connecting the primary and secondary circuits. They are linked only by the changing magnetic field in the core.

Step-up and step-down transformers

By changing the number of turns in the primary and secondary coils, we can vary the alternating e.m.f. produced across the secondary. We can make the secondary voltage different from the primary. A **transformer** is so called because it can *transform* an alternating voltage from one value to another.

If we measure the primary and secondary voltages, we find that they are linked by a simple rule. If there are enough turns on both coils, we find that:

$$\frac{\text{secondary voltage}}{\text{primary voltage}} = \frac{\text{number of turns on secondary}}{\text{number of turns on primary}}$$

In symbols, $\dfrac{V_s}{V_p} = \dfrac{N_s}{N_p}$

If there are only a few turns, or if either coil has an appreciable resistance, then this equation will give only a rough approximation. In most cases, it works well enough.

N_s/N_p is called the **turns ratio** of the transformer. The **volts ratio** V_s/V_p is equal to the turns ratio.

If the secondary or output voltage is bigger than the primary or input voltage, the transformer is called a **step-up** transformer. The transformer in example 30.1 is a step-up type. It has more turns in the secondary than in the primary. On the other hand, if the secondary voltage is lower than the primary we have a **step-down** transformer (example 30.2). This has more turns on the primary than the secondary. Many items of domestic electrical equipment contain step-down transformers to change the mains voltage (240 V) to the lower and safer voltage needed for most purposes.

EXAMPLE 30.1

The primary coil of a transformer has 50 turns and the secondary has 200 turns. A 5 V alternating supply is connected to the primary. What will be the reading on an a.c. voltmeter connected across the secondary?

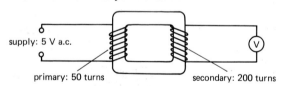

We use the equation: $\dfrac{V_s}{V_p} = \dfrac{N_s}{N_p}$

Here we have: $V_p = 5\text{ V}$
$N_s = 200$ turns
$N_p = 50$ turns

Substituting: $\dfrac{V_s}{5\text{ V}} = \dfrac{200}{50} = 4$

Multiplying both sides by 5 V:
$V_s = 5\text{ V} \times 4 = 20\text{ V}$

A voltmeter across the secondary coil will read 20 V.

EXAMPLE 30.2

You have a 12 V a.c. supply and a 2 V lamp. The lamp runs at full brightness using a transformer with 240 turns on the primary coil. How many turns must the secondary have?

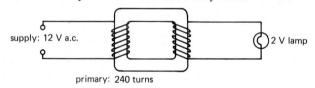

We use: $\dfrac{V_s}{V_p} = \dfrac{N_s}{N_p}$

In this case: $V_s = 2\text{ V}$
$V_p = 12\text{ V}$
$N_p = 240$ turns

Substituting into the equation: $\dfrac{2\text{ V}}{12\text{ V}} = \dfrac{N_s}{240}$

Multiply both sides by 240: $\dfrac{2\text{ V}}{12\text{ V}} \times 240 = N_s$

$N_s = 40$ turns

The secondary coil should have 40 turns.

Cardiac pacemaker

Electromagnetic induction helps many people suffering from heart conditions to live normal lives. The regular beating of the heart is triggered by electrical impulses generated by a special group of cells inside the heart itself. If this fails to function correctly, an operation can be done to implant a small coil inside the chest wall with two electrodes wired to the heart. A small pacemaker unit is then strapped to the *outside* of the patient's chest, over the implanted coil. The pacemaker contains a pulse generating electrical circuit and another coil. The pulsing magnetic field of this primary coil is picked up by the secondary coil inside the chest which stimulates the heart to beat correctly.

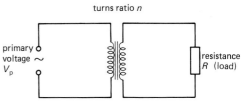

Power and current in transformers

In a step-up transformer, the output voltage is bigger than the input voltage. Are we getting something for nothing? No—the output power (energy supplied per second) is always less than the input power.

Using $P = IV$: Power supplied to primary coil $= I_p \times V_p$
Power delivered by secondary coil $= I_s \times V_s$

If the transformer is perfectly efficient and wastes no energy, these are equal:

$$I_p \times V_p = I_s \times V_s \quad \text{or} \quad \frac{I_s}{I_p} = \frac{V_p}{V_s}$$

This means that currents are transformed in the opposite way to voltages. If V_s is bigger than V_p, then I_s is smaller than I_p, by the same ratio. For example, if V_s is twice V_p, then I_s will be half I_p and so on.

The current ratio is the opposite of the volts ratio:

$$\frac{I_s}{I_p} = \frac{V_p}{V_s} = \frac{N_p}{N_s}$$

We need to be clear about what 'transforming current' means. It doesn't mean that we can control the output current from a transformer just by changing the turns ratio. The secondary current always depends on the resistance (the load) connected to it. The flow diagram on the right shows which quantities are given and how the others depend on them. ▷

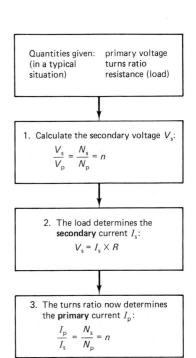

Transformer design

Up to now we have assumed that transformers are 100% efficient. In fact, transformers *are* very efficient devices, but they are never quite 100% efficient. A good transformer is designed so that virtually all the magnetic field lines produced by the primary are 'trapped' in the core and pass through the secondary. This is called having a good **flux linkage**.

All transformers have some energy losses, and heat up slightly in use. The most important ways in which energy can get wasted are:

1. Coil resistance: both primary and secondary coils have some resistance and heat up as current passes through them.
2. Magnetising and remagnetising the core: on each cycle of a.c., the core has to be magnetised in one direction and then in the other. This is going on all the time! It takes energy to turn the domains round and this heats the core slightly. If we choose a soft magnetic material for the core, these losses are kept to a minimum, because it magnetises and demagnetises easily.
3. Eddy currents: the changing magnetic field around it induces eddy currents in the transformer core, and these also heat it slightly. To reduce eddy currents, transformer cores are not made from solid pieces of soft iron, but from thin strips glued together. Each strip is electrically insulated from its neighbours. This is called a **laminated core** and cuts down the number of possible paths for eddy currents.

Well designed transformers are over 99% efficient.

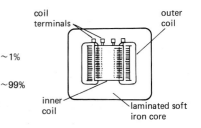

306 Electromagnetic Induction

30.9 The National Grid

The National Grid is the name given to the network of cables and transformers which link all our homes and schools and offices and factories to the power stations which generate electrical energy. The grid connects every power socket in your house back to the power station! It also connects all the power stations together, so that the demand for electrical energy can be spread out among all the available stations.

Large transformer substations transform the 'supergrid' to lower grid voltages.

A local transformer sub-station finally reduces the voltage to 240 V for use in houses.

The voltage used for long distance transmission is very high.

Pylon lines distribute the main cables from power stations to all parts of the country.

Why is grid voltage so high?

The advantage of high voltage transmission is that energy losses are much lower. At high voltage, the current in the cables is small and they heat up less. To see how this works, imagine we are supplying electrical energy to a village. The houses have electric lights and other appliances, all connected in parallel, working at 240 V. The total power required comes to 24 kW.

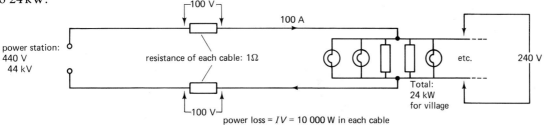

So (using $P = IV$), the total current to the village is 100 A. This is the current in the two cables from the power station some distance away. Let us suppose the total resistance of each cable is just 1 Ω.

The p.d. across each cable (using $V = IR$) is 100 A × 1Ω = 100 V. So the p.d. at the power station will have to be 440 V (240 V + (2 × 100) V).

The power loss in each wire (using $P = IV$) is 100 A × 100 V = 10 000 W. The total power loss in the two cables is 20 kW. So the power station will have to generate 44 kW if the village is to receive 24 kW.

Improved method

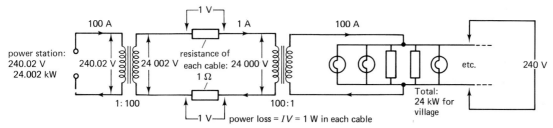

Now imagine that we use transformers to step the voltage up 100 times at the power station end, and to step it down again at the village end. The voltage across the primary of the village transformer will be 24 000 V.

The current in the grid cables will be stepped down by 100 times, to 1 A. The p.d. across each wire is now just 1 V (using $V = IR$). The power loss in each wire is just 1 W (using $P = IV$).

So the voltage across the secondary of the power station transformer is 24 002 V. Its primary voltage must be 240.02 V. The power station will have to generate 24 002 W for the village to receive 24 000 W.

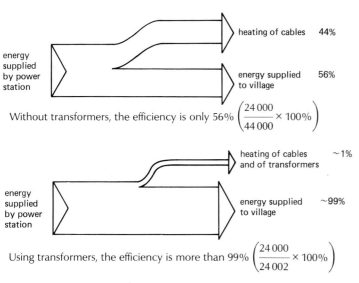

Experimental demonstration

The effect of using step-up and step-down transformers for power transmission can be demonstrated in the laboratory. A 12 V power supply is used to light a 12 V lamp, 5 m away. The 'grid' cables are two 5 m lengths of resistance wire. Without transformers the lamp is very dim, or completely out.

The transformers are two small coils, one of 120 turns, the other 2400 turns, mounted on the same C-core. Using the transformers, the lamp lights brightly. Take care not to touch the transmission wires during the experiment – they are at 240 V!

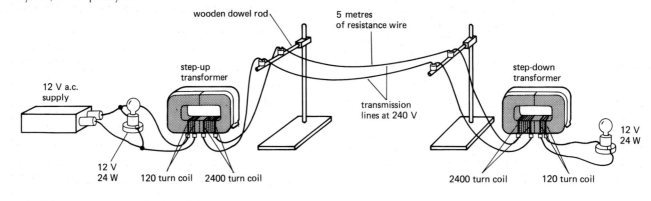

Questions

1. Draw an energy arrow diagram for the energy transfers shown in the drawings of the moving wire in the field experiment, and the coil and magnet experiment on pages 297 and 298.

2. A bar magnet is attached to one end of a long wooden rod **AB** which is hung from a pivot at **X** so that the magnet can swing in and out of the coil **PQ**.

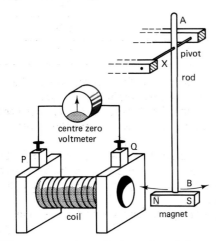

The magnet is pulled to the right and released. It makes ten swings before coming to rest. Describe the changes which take place in the voltmeter reading during this time.
   What difference will there be in the voltmeter readings if the experiment is repeated but with:
(a) the coil **PQ** replaced by one of more turns of the same wire, (b) the magnet reversed?   (SEB)

3. A man makes an anemometer, a device for measuring wind speed, out of the apparatus shown and positions it on the roof of his garage.

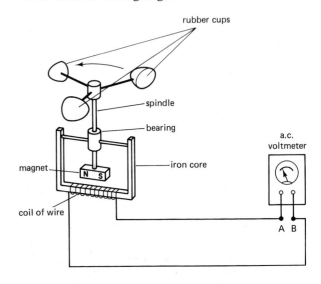

When the wind blows, the spindle rotates and the a.c. voltmeter pointer is deflected to give a reading.
(a) Explain how the rotating spindle causes the voltmeter pointer to be deflected.
(b) When the wind speed increases, what difference, if any, would you expect in the voltmeter reading? Why?
(c) Give two ways in which the apparatus could be modified so as to obtain a bigger reading on the same voltmeter.   (SEB)

4 A public address system consists of a microphone, an amplifier and several loudspeakers. Here is an energy arrow diagram for the system.

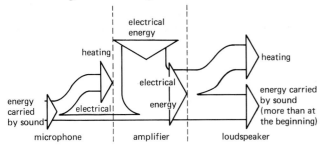

(a) Explain in your own words the energy changes shown in the arrow diagram.
(b) Read the sections on the loudspeaker and the moving-coil microphone (pages 291 and 299) and explain how these devices produce the energy transfers shown in the diagram.

5 Copy and complete this diagram to show how to make a simple generator, which will produce electricity continually when the coil is rotated.

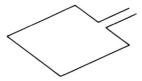

6 What is the output p.d. of these transformers (assuming they are 100% efficient)?

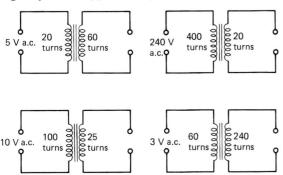

7 How many turns are needed on the windings of these transformers (assuming they are 100% efficient)?

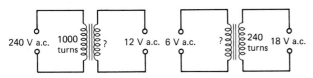

8 (a) It has been said that the transformer is one of the most efficient of all machines. Explain what it means to say that a transformer is 'efficient'.
(b) No real transformer is 100% efficient. Make a list of the main causes of energy losses in transformers.

9 (a) A pupil is given three transformers labelled **A**, **B** and **C**. She is asked to investigate the relationship between the primary (input) and secondary (output) voltages of each transformer. She applies different voltages to the primary coil of each transformer and measures the voltage produced across the secondary coil.
 The following graphs were plotted from her results.

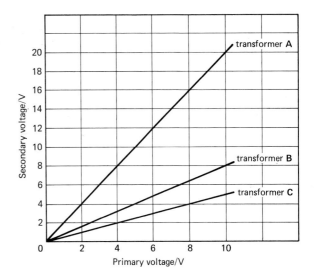

 (i) Which transformer is operating as a step-up transformer?
 (ii) The primary coil of each transformer has 250 turns of wire. Copy and complete the following table, using information from the graph where required.

Trans-former	Primary voltage V	Secondary voltage V	Primary turns	Secondary turns
A	7		250	
B	6		250	
C		4	250	

 (iii) Draw a diagram of the circuit which the pupil could have used to obtain her results.
(b) In the National Grid transformers are used so that electrical energy can be transmitted over long distances.
 (i) Draw a diagram to show how this use of transformers can be demonstrated in the laboratory. Indicate on your diagram where there is a high voltage.
 (ii) Why are high voltages used in the National Grid?
(SEB)

INVESTIGATION
How efficient is a transformer made from two coils of wire wound on a pair of C-cores?

31: Electron beams and the CRO

An 'electronic revolution' is going on around us. Computers, electronic keyboards, digital watches and calculators are just some of the many applications of electronics in everyday and working life. All depend on the properties of **electrons**. Early electronics experiments with beams of free electrons tell us much about electrons and how they behave. They also led directly to the development of television.

31.1 A sensitive detector for charge

In Chapter 24, we looked at the effects of electric charge. Two charged rods exert small forces on each other and attract (if they are unlike charges) or repel (if they are like charges). However the forces are very small and this is not a very sensitive way to test for the presence of small amounts of charge. A more sensitive detector is the **gold leaf electroscope**. It works in a very simple way. When its cap is charged, the charge spreads to the stem and leaf; the leaf then rises because it has the same charge as the stem and they repel each other.

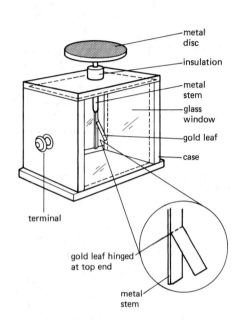

The electroscope itself can be charged up, either by rubbing its cap with a plastic rod or by connecting the cap briefly to a high voltage power supply.

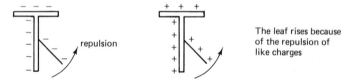

The leaf rises because of the repulsion of like charges

31.2 Thermionic emission

A small coil of resistance wire glows red hot if a large enough current passes through it. If the coil is held near a positively charged gold-leaf electroscope, it quickly discharges the electroscope. A negatively charged electroscope is not discharged in the same way. Something seems to be coming from the hot wire – something which can discharge a positive electroscope.

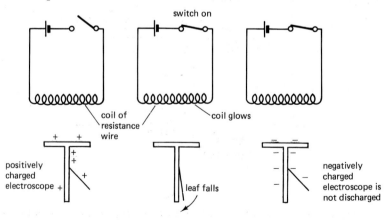

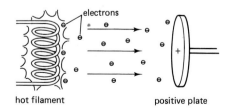

One possible explanation is that negatively charged particles are being given off from the wire when it is heated. We now believe that the electrons in the hot wire are being given enough energy to enable some of them to escape from the surface of the wire and become free electrons. This process is called **thermionic emission**. It is a bit like electrons being evaporated off from the hot wire!

310

The diode

We can investigate thermionic emission further using a resistance wire sealed inside a glass bulb from which most of the air has been pumped out. This allows any free electrons to move around freely without colliding with air molecules. A second metal plate is also sealed into the bulb. This device – a sealed bulb with two **electrodes** – is called a **diode**.

When the filament is heated by passing a current through it, we find that we can make a small electric current flow through the diode in one direction but not the other. ▷

What seems to be happening is that electrons are being emitted from the hot cathode. These form a 'cloud' around it. If the cathode is negative and the anode is positive, electrons will be pushed away from the cathode and attracted to the anode. They cross the gap between the two electrodes and a current flows in the circuit. If the cathode is positive and the anode is negative, the thermionic electrons are pushed back towards the cathode and there is no current.

The **thermionic diode** allows current to flow in one direction only. This is a useful property. Nowadays, however, thermionic diodes have been almost completely replaced by semiconductor diodes. So we will put off looking at the usefulness of the diode as a 'one-way street' for electrons until Chapter 32.

If the hot filament (called the **cathode**) is connected to the negative terminal of a 500 V d.c. supply, and the metal plate (or **anode**) is connected to the positive terminal, there will be a small current of a few milliamperes.

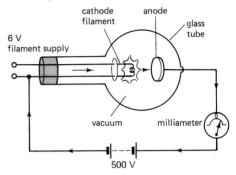

If the high p.d. is connected the other way round, there is no current.

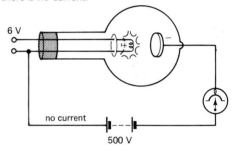

31.3 Cathode rays

Despite the development of semiconductor devices (see Chapter 32), thermionic emission is still very important as a method of producing beams of electrons. It was also very important historically in helping scientists to find out more about the electron and its behaviour.

The 'electron gun'

If the anode of a thermionic diode is made with a hole in it, a beam of electrons can be produced inside the vacuum tube. Electrons are emitted from the hot cathode. They are accelerated towards the anode, and some will pass on through the hole as a beam. With a small hole, we produce a fine, narrow beam; with a larger hole, a broad beam.

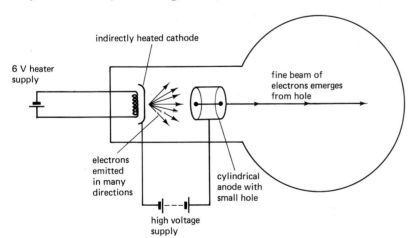

Some facts about thermionic emission

All metals emit thermionic electrons, but some do it at lower temperatures than others. A tungsten filament at 2500 K is a good thermionic emitter, but other oxides of metals emit well at lower temperatures. A nickel cathode coated with barium oxide or strontium oxide (or a mixture of the two) will release lots of thermionic electrons at 1200 K. Oxides are non-conductors so the cathode is shaped like a cap which sits over the wire filament and is heated indirectly.

The early researchers, in the nineteenth century, produced beams like this and called them 'cathode rays', because they came from the cathode of the vacuum tube. There was, however, a fierce argument about what these 'rays' were. Were they a new form of radiation or a stream of particles? The experiments which they carried out over a period of more than 20 years led eventually to the discovery of the electron, as it became clear that the 'rays' were best explained as streams of particles.

312 Electron Beams and the CRO

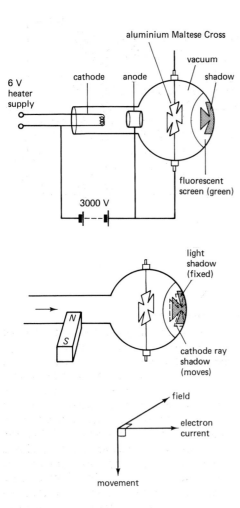

Maltese Cross tube

This tube has a cylindrical anode, which produces a broad beam of electrons. Beyond the anode is a metal obstacle, shaped like a Maltese Cross. The end of the vacuum tube is coated with a fluorescent material. When the cathode heater is switched on, and the anode is at a high positive potential compared with the cathode, the fluorescent screen glows with a green light. A Maltese Cross-shaped shadow appears on the screen.

With this tube we can discover several useful facts about electron beams:

1. The bright green light is produced where the electrons strike the fluorescent coating. This tells us that the electrons carry energy. The kinetic energy of the electrons is carried away as light energy from the screen.

2. There is a shadow because some electrons are stopped by the metal Maltese Cross inside the tube. The electron beam cannot pass through a metal obstacle, but is stopped by it.

3. The electrons travel in straight lines. If we turn off the high anode–cathode p.d., the glowing filament (cathode) casts a light shadow on the screen. When the p.d. is switched back on again, the electron beam shadow region (the dark part within the green glow) is in exactly the same place. So the electrons must travel by the same path as light – in straight lines.

If we now bring a magnet close to the Maltese Cross tube, the electron beam shadow on the screen begins to distort and change shape. The light shadow (with the p.d. switched off) does not, of course. The direction in which the shadow region moves agrees with what we would expect (using the right hand motor rule – see section 29.11) if the 'cathode rays' are a stream of negatively charged particles.

The deflection tube

This tube gives us further evidence that 'cathode rays' are a stream of negatively charged particles. Its anode has a slit, so that a narrow beam of electrons emerges. A fluorescent screen shows the path of the electron beam across the tube. At the top and bottom of the screen are two metal plates with connecting leads to outside the tube.

If we apply a high p.d. across these plates, the beam is deflected by the electric field. When the upper plate is at a more positive potential, the beam bends upwards; when the lower plate is at a more positive potential, the beam bends downwards. Again this is what we would expect if the beam is a stream of negatively charged particles.

The deflection tube can also be used to investigate the behaviour of the beam in a magnetic field. Two large coils placed on either side of the tube produce a fairly uniform magnetic field in the centre. If we consider the moving electrons as an electron current, we can use the right hand motor rule (section 29.11) to predict the direction of the force on this current in a magnetic field. The direction of bending which we observe turns out to agree with this prediction.

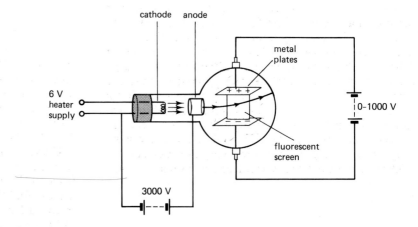

The Perrin tube

The final evidence that 'cathode rays' are negatively charged particles comes from the Perrin tube. This is based on a very simple idea – if the beam is a stream of particles, we should be able to collect them and test their charge. The French physicist, Jean Perrin, designed a tube with a metal can to collect the electrons. A magnet is used to deflect the fine beam of electrons into the can. The fluorescent coating on the end of the tube makes it easier to direct the beam into the collecting can.

When the beam enters the can, the gold leaf electroscope shows a deflection. The beam carries charge. If we bring a negative rod (a rubbed polythene rod) close to the electroscope, the leaf rises further. This shows that the charge on the electroscope is the *same* as that on the rod – negative.

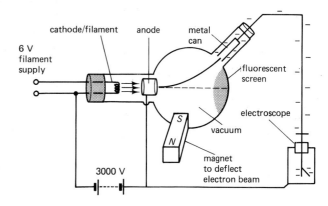

31.4 Measuring the electron

Eventually the scientific community agreed that 'cathode rays' were beams of electrons. Work could then begin on finding out more about the electron. Physicists were keen to know what the mass of an electron was, and how much charge it carried.

Thomson's experiment

In a famous experiment in 1897, J.J. Thomson succeeded in measuring the ratio charge/mass (or e/m) for the electron. He used a deflection tube. We have seen that an electron beam can be deflected either upwards or downwards by an electric or a magnetic field. Thomson used a deflection tube with *both*. He adjusted the current in the magnetic field coils so that the two effects opposed each other, and the beam passed straight through. This enabled him to calculate the speed of the electrons in the beam.

He then switched off the magnetic field and measured the deflection of the beam in the electric field alone. Because he now knew the speed of the electrons, he was able to work out the ratio of their charge to their mass. The result he obtained was:

e/m for electrons $= 1.76 \times 10^{11}$ C/kg

This was a surprisingly large number. The largest charge-to-mass ratio known for any particle at that time was 9.6×10^7 C/kg for hydrogen ions. Thomson's result for the electron was about 2000 times bigger than this. It might mean that the electron had 2000 times more charge than a hydrogen ion and the same mass; or that the electron had the same charge as the hydrogen ion but a mass 2000 times smaller; or some combination of the two. Thomson's hunch, which turned out to be supported by later evidence, was that the electron had much the same charge as the hydrogen ion but was very tiny indeed, only 1/2000th of the mass of the smallest ion known!

Millikan's oil-drop experiment

If we just know e/m, we can only guess at what the charge and mass might be. The first person to measure the charge on a single electron was the American physicist, Robert Millikan. Millikan watched small droplets of oil from a spray falling in the electric field between two plates. As the oil drops come out of the spray nozzle, they rub against the sides and become charged. In a region with no field, the drops soon reach a terminal velocity (when the air resistance force at the speed they are moving with is equal to their weight). By measuring this velocity, it is possible to work out the mass of the drop. The electric field is then switched on. Since the drop is charged, this changes the forces on it. Soon it settles down again to move at a different terminal velocity. This new terminal velocity depends on the charge it carries.

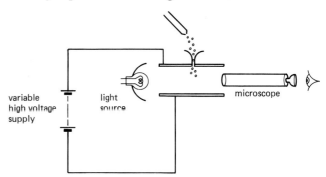

By doing hundreds of measurements on drops during the years 1909–1913, Millikan was able to show that the charge was always a multiple of one basic charge. The drops were so small that the charge on a drop was always just 1, or 2, or 3, or some small number of electrons. The basic 'unit' of charge – the electronic charge – was 1.6×10^{-19} C.

Using Thomson's result for the charge-to-mass ratio, this means that the mass of an electron is just 9.1×10^{-31} kg. That means that you would need about 1 100 000 000 000 000 000 000 000 000 000 electrons to make 1 kilogram of electrons! More than 1 million million million million million. That's how small an electron is.

Electron Beams and the CRO

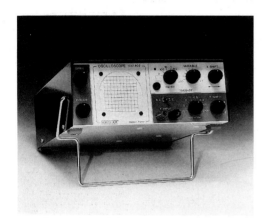

31.5 The cathode ray oscilloscope (CRO)

The cathode ray oscilloscope, or CRO for short, is one of the most important electronic test and measurement instruments. The screen of the CRO is the flat end of a cathode ray tube. This is coated with a fluorescent material which glows (usually green) when a fine beam of electrons strikes it. The controls on the front panel of the CRO are used to make adjustments to the intensity and movement of the electron beam inside the tube.

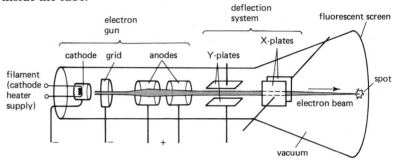

The cathode ray tube has two main parts:

1. The electron gun

This produces a fine beam of electrons. The **on/off control** switches on the filament current and the anode–cathode p.d. so that electrons are emitted from the cathode. To make the beam sharp, the anode is in the form of two cylinders at a positive potential. These accelerate the electrons from the cathode and also push the beam back towards the centre line of the tube. The **focus control** makes small changes to the p.d. between the two anode cylinders to adjust this focusing effect.

Between the cathode and anode is an extra electrode called the **grid**. If the grid is more negative than the cathode, it repels some electrons back towards the cathode and weakens the beam. So the spot becomes less bright. When you turn the **brightness control**, you are altering the potential of the grid.

2. The deflection system

The electron beam passes between two sets of parallel plates on its way to the screen. The two plates above and below the beam are called the Y-plates. If there is a p.d. between these plates, the electric field will bend the electron beam upwards or downwards – in the y-direction. The X-plates are on either side of the beam. If there is a p.d. between them, this electric field will bend the beam to one side or the other – in the x-direction.

Vertical deflection

1. When the p.d. across both the Y-plates and the X-plates is zero, the electron beam is in the centre of the screen.

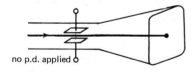

2. If a steady d.c. voltage is now applied to the Y-plates, so that the upper plate is more positive, the electron beam will bend upwards. The top plate attracts the negatively charged electrons.

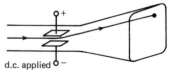

3. If the lower Y-plate is more positive, the beam will bend downwards.

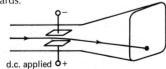

4. If a 50 Hz alternating p.d. is applied to the Y-plates, the beam will be attracted upwards and downwards 50 times a second. The spot moves up and down the screen so rapidly that it looks like a continuous vertical line.

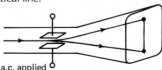

We do not connect an input *directly* to the Y-plates. We can apply a p.d. to the two **input terminals** on the front of the CRO. This p.d. is then amplified, by an amplifier inside the CRO, before reaching the Y-plates. The **gain** of the amplifier (the number of times it magnifies the input p.d.) can be altered using a control called **Y-gain**.

If we connect a 1.5 V cell to the input terminals, with the Y-gain set at 1 V/cm, the spot will be deflected 1.5 cm up the screen. On this setting the amplifier produces just enough gain for a 1.5 V input to give a 1.5 cm deflection. ▷

If we now change the Y-gain setting to 0.5 V/cm, the spot will be deflected 3 cm up the screen. The input has stayed the same, but the gain of the amplifier is now twice what it was before. ▷

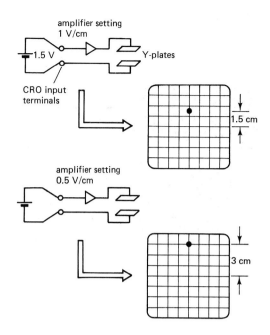

The amplifier also has a second use. Sometimes we want to be able to move the spot away from the centre of the screen even when there is no p.d. applied to the input. This can be done with the **Y-shift** control. As it is turned, the amplifier applies a p.d. to the Y-plates to move the spot to a new starting position.

Horizontal deflection

When we apply an alternating p.d. to the Y-plates, the spot moves up and down the screen. But we cannot see *how* it is moving as it only produces a straight line on the screen. Imagine what we would see if we could make the spot move sideways at a steady speed at the same time as it is moving up and down. Think of what would happen if you were moving a pencil ▷ backwards and forwards across a long roll of paper – and someone then began slowly to pull the paper sideways under your pencil. The result would be a wavy trace on the paper which would show exactly where the pencil had been at every moment.

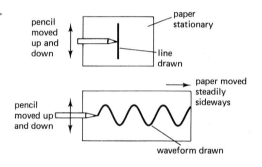

If we can make the electron beam move steadily across the screen, it will produce a **waveform** of the alternating p.d. The trace on the screen is a graph showing how the alternating p.d. varies with time. The **time base** circuit inside the CRO moves the spot sideways at a steady speed. It applies a p.d. to the X-plates which steadily increases and then drops suddenly – and keeps on repeating this.

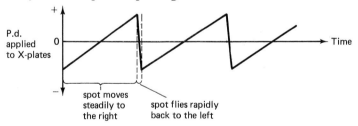

The spot moves steadily across the screen from left to right as the p.d. rises, and then flies rapidly back to the start when the p.d. suddenly falls again. The cycle repeats, so that the spot is continually sweeping across the screen and then rapidly back to the start again. The **time base control** allows you to adjust the speed of the spot across the screen.

To get a steady trace on the screen, you have to adjust the time base control so that each new waveform is drawn exactly on top of the last one. The time taken for the spot to cross the screen must be just right to give a whole number of waves of the input p.d. on the screen. Then the spot will be ready to begin the next trace at exactly the same position when it jumps back to the start.

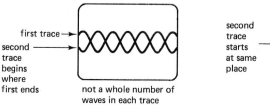
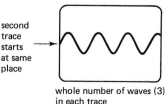

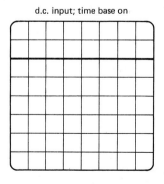

Of course, the time base can also be switched on when a steady p.d. is applied to the input terminals. The horizontal line moves up or down to a new position. The displacement of the line upwards or downwards is measure of how big the p.d. is.

31.6 Using the CRO

1. Displaying waveforms

The CRO is very useful for displaying waveforms. Signal generators, and some power supplies, can produce a variety of output waveforms. These can be displayed on the CRO screen.

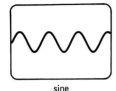
sine

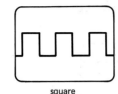
square

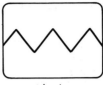
triangle

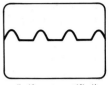
'half-wave rectified'

If a microphone is connected to the input terminals, sound waveforms can be studied (see section 23.4). The pressure wave in the air is converted by the microphone into an electrical waveform and this is displayed on the CRO screen. Remember that the sound wave is a longitudinal wave even though the wave on the CRO screen is transverse. The CRO trace is just a graph showing how the air pressure in front of the microphone varies with time.

2. Measuring p.d.

The vertical deflection of the CRO spot depends on the input p.d., so a CRO can be used to as a voltmeter. It has a very high resistance so it is a very good voltmeter.

The time base is switched off. The Y-gain is set at 1 V/cm. With two torch cells connected, the spot is deflected by 3 cm. So the p.d. of the two cells is 3 V, i.e. they are 1.5 V cells.

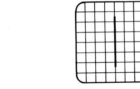

With the time base switched off, the alternating p.d. from a low voltage power supply produces a vertical line on the screen. The Y-gain is now set at 0.5 V/cm and the line on the screen is 5 cm long. This corresponds to a p.d. of 0.5 × 5 V, or 2.5 V. The peak-to-peak p.d. of the power supply is 2.5 V.

The time base is now switched on. The Y-gain is set at 2 V/cm. This square wave rises and falls 3 cm above and below the centre line. The peak value of the p.d. is therefore 6 V and the minimum value −6 V.

Sometimes the Y-gain settings are not accurate enough for precise measurements. If you are in any doubt, the best solution is to calibrate the CRO on a particular setting by using a known input p.d. (say, a 1.5 V cell). Note the deflection and *work out* what the Y-gain (in volts/cm) really is. Then, *using the same Y-gain setting*, measure the deflection produced by the unknown p.d. You can then calculate the unknown value.

Although the CRO is basically a voltmeter, it can be used to measure current and to study changes in current in circuits. To do this we use the CRO to look at the p.d. across a known resistor. The current through the resistor is then given by $I = V/R$.

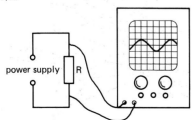

EXAMPLE 31.1

A test engineer uses a CRO to measure the p.d. at a particular point inside a computer. This is what she sees on her CRO screen. ▷

The Y-gain setting is 0.5 V/cm. What is the peak value of the p.d.?

The waveform has an amplitude on the screen of 1.5 cm.

So, peak voltage = 1.5 cm × 0.5 V/cm
   = 0.75 V

The peak value of the test p.d. is 0.75 V.

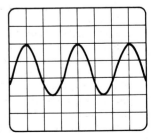

3. Measuring frequency

We can use the CRO to measure the frequency of an alternating p.d. by making horizontal measurements on the screen. To do this, we need to know the time base setting in seconds/centimetre (or milliseconds/centimetre). On many CROs, there are two time base control knobs: one with a range of fixed settings and the other continuously variable for fine adjustment. For frequency measurements, the fine control must be set at the position marked 'CAL' or '× 1'.

The time base is set at 5 ms/cm. This means that it takes the spot 5 milliseconds to move 1 cm across the screen from left to right. On the screen, the distance from one peak to the next is 4 cm. The spot takes 4 × 5 ms, or 20 ms, to move this distance. So one complete wave takes 20 ms. In one second (1000 ms), there will be $\frac{1000}{20}$ complete waves, i.e. 50 complete waves. The frequency of the input p.d. is 50 Hz.

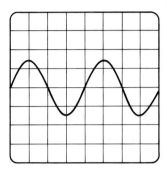

Television

The screen of a television set is really the end of a cathode ray tube. A black and white TV has an electron gun and a screen coated with a material which fluoresces, that is it emits white light when the electron beam strikes it. The electron beam is deflected by magnetic fields produced by coils around the cathode ray tube.

A television has two separate time base circuits, which are used to make the beam cross the screen in a particular way. The horizontal time base moves the spot across the screen rapidly; the vertical time base moves it down the screen at a rather slower speed. The spot takes $\frac{1}{50}$ second to make a complete sweep down the screen. In this time it makes 312.5 complete sweeps across the screen. The spot is really drawing a series of parallel lines across the screen very close together.

In the next $\frac{1}{50}$ second, the spot moves down the screen again, scanning another 312.5 lines, this time mid-way between the first set! So after $\frac{1}{25}$ second (two $\frac{1}{50}$ second sweeps), the spot has ruled a set of 625 lines on the screen. The reason for doing it in two sets of 312.5 is to reduce the amount of 'flicker' of the TV picture to a level where it is not noticeable.

The picture itself is produced by varying the intensity of the electron beam as it scans. This is done by altering the potential of a grid electrode in the electron gun. The brightness of the spot on the screen changes with the intensity of the electron beam, and this produces a picture. An image on the retina of the eye persists for about $\frac{1}{19}$th second, so the television picture appears steady. Although we are really seeing 25 still pictures each second, motion on the screen appears smooth.

A colour TV is more complicated. The screen is covered with tiny strips of three different fluorescent materials (or **phosphors**), one which emits red light, one green and one blue (the three primary colours – see section 19.5). These are arranged in groups of three. There are also three electron guns. A screen with accurately drilled holes, called a **shadow mask**, makes sure that electrons from one gun can strike only the strips of one phosphor. So one electron gun causes the red phosphor strips all over the screen to glow; another makes the green strips glow; and the third affects the blue. The three electron beams scan the screen just as the single beam does in a black and white television. As in the black and white TV, the brightness of the light from each strip can be controlled by the grid voltage of the electron gun. In the colour TV we have separate control of the red, green and blue light coming from each tiny group of three strips. By mixing the three primary colours of light in different proportions, we can make any colour we want and the eye sees a full colour picture on the screen.

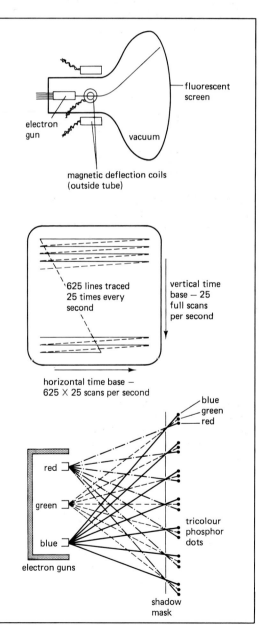

Questions

1. In the 19th century, there was a fierce argument amongst scientists about whether cathode rays were waves or streams of particles.
   Make a list of all the pieces of evidence you can think of which suggest that cathode rays are negatively-charged particles.

2. Copy this diagram of a cathode ray oscilloscope (CRO) and use the information in section 31.5 on p. 314 to write brief captions explaining what each part does.

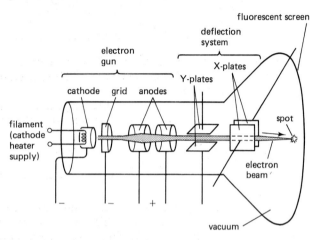

3. Connecting a 1.5 V cell to the Y-input of a CRO made the spot move up by $1\frac{1}{2}$ squares on the screen.

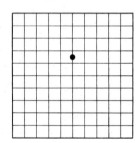

(a) Where would the spot go if the cell was connected the other way round?
(b) What voltage would move the spot up by 5 squares?
(c) How far would a 4.5 V battery make the spot move?
   Draw what you would expect to see if you used the shift controls to put the spot in the middle of the screen and then connected the following to the Y-input:
(d) a 2 V d.c. supply;
(e) a 12 V d.c. car battery;
(f) an 8 V peak-to-peak a.c. supply.

4. Explain what the 'time base' on an oscilloscope does.

5. The diagram shows some traces obtained on a CRO with different settings on the Y-gain and time base controls. For each trace, work out the following:
(a) the number of squares from peak to centre line;
(b) the peak voltage of the input signal;
(c) the number of squares for one complete cycle of the input waveform;
(d) the frequency of the input signal.

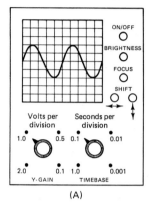
(A)
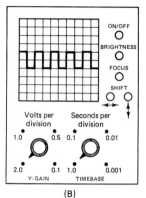
(B)

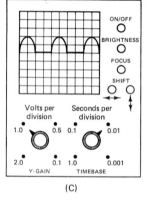
(C)
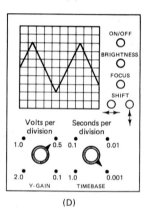
(D)

6. An oscilloscope is used to study the output p.d. from a bicycle dynamo. The diagram shows the trace observed at one particular cycling speed.

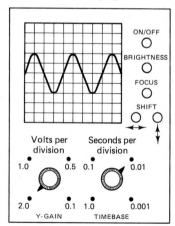

(a) What is the frequency of rotation of the dynamo spindle?
(b) Copy the trace and mark on your diagram the new trace you would expect to see if the bicycle was moving faster.

7 The diagram shows the screen, Y-gain and time-base controls from a typical oscilloscope displaying a waveform.

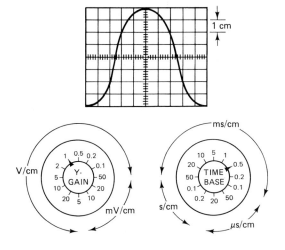

(a) What is the setting of the Y-gain control?
(b) What is the peak voltage of the waveform?
(c) What is the time base setting?
(d) What is the period of the trace?
(e) What is the frequency of the waveform?
(f) Copy the graticule shown below and on it draw the resultant trace if the time-base setting is altered to 1 ms/cm and the Y-gain to 2 V/cm. (Sp.NEA)

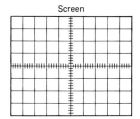

8 The diagram shows some traces obtained on a CRO screen with a 50 Hz a.c. input. In each case, what was the time taken for the spot to cross the screen from left to right?

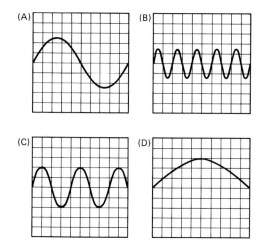

9 A bar magnet is suspended by a spring so that it can oscillate freely in and out of a coil as shown. The coil is connected to an oscilloscope which has its time base switched off. The oscilloscope is adjusted so that the spot is in the middle of the screen.

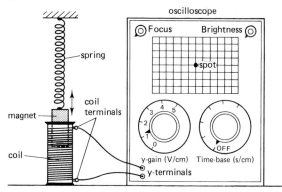

(a) The magnet is set oscillating. Describe with the aid of a sketch the resulting movement of the spot on the screen of the oscilloscope.
(b) If you wished to increase the amplitude of the movement of this spot, explain how you would do this by:
  (i) making a change which would alter the input signal to the oscilloscope;
  (ii) making a change to the controls of the oscilloscope.
(c) With the magnet still in motion, the time base of the oscilloscope is switched on. The controls are suitably adjusted until the following trace is seen on the oscilloscope screen.

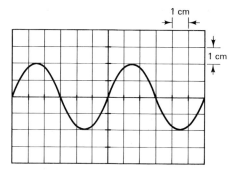

Redraw the trace and mark on it a position of the spot which corresponds to the position of the magnet when it is
  (i) moving at its greatest speed;
  (ii) at the highest point of its oscillation.
(d) If the trace shown above is obtained when the time base control is set to 0.05 s/cm, estimate the frequency of the oscillating magnet.

INVESTIGATION
Does a telephone transmit all sound frequencies equally well? (You will need a CRO and microphone, and a signal generator and loudspeaker for this investigation, as well as two telephones!)

32: Electronics

Electronic devices made from semiconductor materials have taken over from thermionic valves. They are much smaller and cheaper and they last a long time as there is no filament to burn out. They are at the heart of the 'microelectronics revolution'.

32.1 Semiconductors

The most important part of a modern electronic component is the small piece of **semiconductor** material inside the outer packaging. Semiconductors are materials which conduct electricity better than insulators, but not so well as ordinary conductors.

Semiconductor devices: inside each is a small specially treated piece of silicon crystal (a 'chip'). The outer plastic or metal case is for protection. The legs are connected by fine gold wires to the silicon chip itself.

Resistivities of some materials (the resistance of a 1 metre cube of the material). The semiconductors lie in the middle of the range.

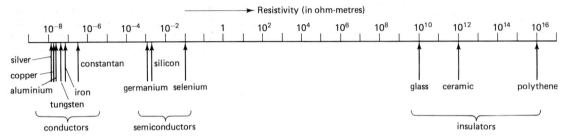

Semiconductors have several very important properties. We can make a big change in their electrical resistance by adding very small amounts of other substances. Their resistance also changes with temperature and with the brightness of the light shining on them. This is how thermistors and light dependent resistors work.

n- and p-type semiconductors

A 2D picture of a silicon crystal. The real crystal is three-dimensional of course!

A crystal of pure silicon has a very regular structure of atoms. Every atom has four neighbours. Each atom shares its four outer electrons with its neighbours, and this holds the crystal together. At room temperature, these electrons can move only with difficulty through the material, jumping from one atom to the next, so the silicon crystal conducts electricity badly. At higher temperatures, the electrons have more energy and it conducts better.

◁ **n-type semiconductor** is made by adding a small amount of impurity to molten silicon and allowing it to cool and crystallise. Phosphorus has five outer electrons and so one is 'spare'. This electron is very free to move through the material and so the semiconductor now conducts much better. It is called n-type because the *free* charge is a *n*egative electron. n-type material is *not* negatively charged.

If the impurity added is boron, this produces **p-type semiconductor**. Boron has three outer electrons, so there is a 'missing' electron at each boron atom. An electron from nearby may move into this gap but it will leave another gap behind. This is rather like what happens when a bubble rises in a tube of liquid. It looks as if the bubble is moving up, but really it is the liquid which is moving down! The 'bubble' in the semiconductor is called a **positive hole**. It behaves just as if a real positive charge was moving, hence the name p-type.

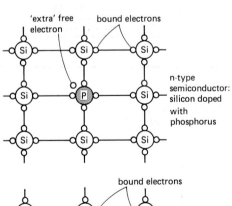

n-type semiconductor: silicon doped with phosphorus

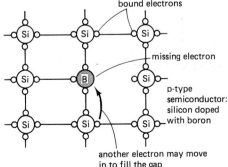

p-type semiconductor: silicon doped with boron

another electron may move in to fill the gap

32.2 p–n junction diode

A **diode** is a device which allows charge to flow through it in one direction only. A semiconductor diode can be made by joining pieces of n-type and p-type semiconductor.

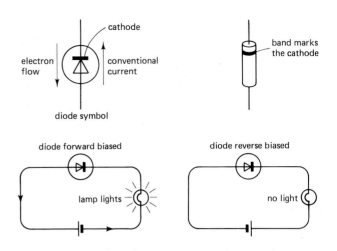

diode symbol

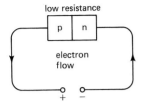

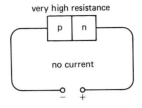

1. If a p.d. is applied across the diode in this direction, there is a current. The diode has low resistance. It is said to be **forward biased**. Note that the p-type material is at the more positive potential and the n-type more negative.

2. If the p.d. is applied in the opposite direction, there is no current. The diode has a very high resistance. It is said to be **reverse biased**.

3. The diode symbol is arrow shaped, pointing in the direction of the 'conventional current'. Electrons can flow through the diode in the opposite direction to the arrow. The vertical line represents the cathode. This should be connected to the negative terminal of the supply for the diode to conduct. On most diodes, the cathode is marked by a band.

How does the p–n junction diode work?

1. When two pieces of p- and n-type semiconductor are joined, some free electrons from the n-type cross the junction into the p-type and fill the holes there. So there are fewer free charges, both electrons and holes, in the region near the junction. This is known as the **depletion layer**. Since it has no free carriers, it is a very effective insulator – a barrier to the current.

2. When the diode is forward biased, the applied p.d. tends to push electrons and holes **into** this depletion layer. This increases the number of free carriers in the region and effectively reduces the depletion layer. The diode conducts.

3. When it is reverse biased, the applied p.d. tends to move electrons and holes away from the depletion layer, making it wider. It is now an even bigger barrier to the current, and the diode does not conduct.

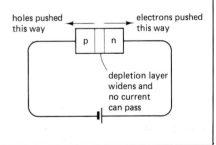

32.3 Diode characteristics

An 'ideal' diode would have no resistance in one direction and an infinite resistance in the other. But real diodes are not quite as good as this!

Diode characteristics: when the diode is reverse biased, the current is very small – almost zero. When it is forward biased, a silicon diode needs a p.d. of around 0.6 V to make it conduct, but above this the current rises very quickly with the p.d. The diode has a very small resistance once the applied p.d. is above 0.6 V. For a germanium diode, the 'switch-on' p.d. is smaller – around 0.2 V.

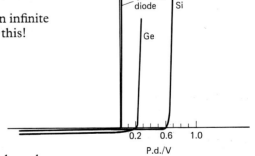

Silicon diodes are the commonest type. Two simple rules of thumb can be used in circuit calculations where silicon diodes are involved:

– when a silicon diode is reverse biased, its resistance is very large.
– when it is forward biased, there is a p.d. of 0.6 V across the diode, no matter what the current is.

32.4 Rectification and power supplies

We have seen in Chapter 30 that it is much easier to generate a.c. on a large scale than to make d.c. For many applications, however, we need direct current. D.c. power supplies, running off the mains, do the job of converting a.c. into d.c. The process is called **rectification**. Diodes play an important role in it.

1. Imagine what will happen in this circuit. An alternating p.d. is applied. For half of the cycle, **A** is more positive than **B**; for the other half cycle **A** is more negative than **B**.

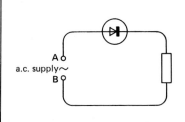

2. During the positive half-cycle, the diode conducts; for the negative half-cycle, no current can flow.

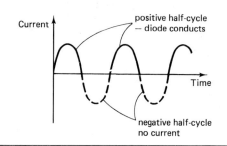

3. We could inspect the current by using a CRO to look at the p.d. across the resistor. This is called **half-wave rectified** a.c. The average value of the current is no longer zero.

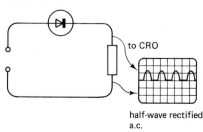

Full-wave rectification

Half-wave rectification wastes all the electrical energy available during the negative half cycle. We can improve on this by using four diodes arranged as a **diode bridge**.

1. The diagrams below show the path taken by electrons flowing round the circuit on both half cycles of the a.c. supply. Remember that electrons can only flow 'against' the arrow direction of the diode symbol. This means that the electrons have to follow the paths shown. Notice that the current through the load resistor is in the *same* direction during both half cycles.

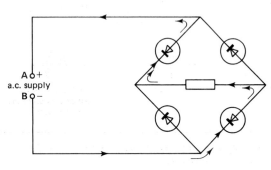

2. The upper CRO trace shows the supply p.d. The lower one shows the p.d. across the load resistor. This is called **full-wave rectified a.c.**

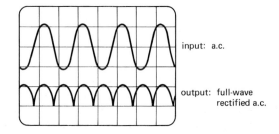

3. Diode bridges can be bought in the form of a single package containing all four diodes. The a.c. supply is connected to the two terminals marked ∼. The full-wave rectified output comes from the other two terminals marked + and −.

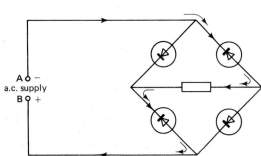

Both half- and full-wave rectified a.c. provide current in one direction only. They could scarcely be called 'direct current' though they *are* good enough for some applications. To produce genuine d.c., we need to 'smooth' out the variations in the rectified output. This can be done using a **capacitor**.

Capacitors

A capacitor is a device which can store electric charge. In fact, any conductor is able to store some charge. But as the amount of charge increases, the potential of the conductor also increases. If the potential becomes too big, a spark will jump from the charged conductor to earth, and the charge will escape.

If we put 1 µC of charge on this van de Graaff generator dome, its potential will rise to 10 000 V. With 2 µC, the potential is 20 000 V, and so on...

If we put 1 µC on this smaller dome, its potential rises to 20 000 V. With 2 µC, the potential is 40 000 V, and so on...

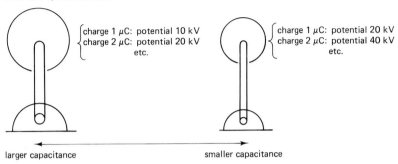

We say that the larger dome has twice as much **capacitance**. It can hold twice as much charge *for the same rise in potential*. Capacitance is defined in the following way:

$$\text{capacitance} = \frac{\text{charge on conductor}}{\text{potential of conductor}}$$

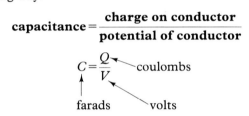

$$C = \frac{Q}{V}$$

where C is in farads, Q in coulombs, V in volts.

If a charge is measured in coulombs and potential in volts, capacitance is in farads. The farad (F) is a very large unit. Most practical capacitors have capacitances of several microfarads (10^{-6} F), nanofarads (10^{-9} F) or even picofarads (10^{-12} F).

Practical capacitors are made from two conducting plates separated by a layer of insulator. To keep the size down, this is sometimes done by sandwiching a layer of insulating material between two thin metal sheets and rolling it into a cylinder. This is then coated with a protective layer of plastic.

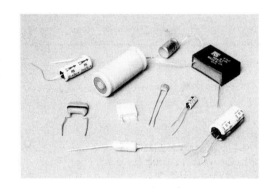

capacitor   electrolytic capacitor (must be connected the right way round)

1. One way to think about capacitance is to imagine a beaker being filled with water. The more water the beaker contains, the higher the water level rises.

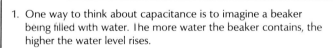

2. In the same way, the more charge a capacitor contains, the higher the p.d. across its plates.

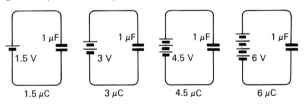

3. If we have a series of beakers of different cross-section, the wider beakers can hold more water if filled to the same level.

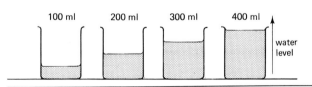

4. Large capacitors can store larger amounts of charge for the same p.d. across the plates.

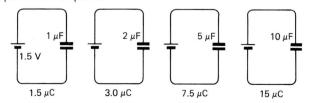

Charging and discharging a capacitor

When the switch is moved to position **A**, we have a series circuit consisting of capacitor, resistor and cell. Electrons flow round and the capacitor charges up.

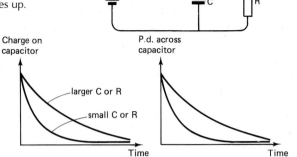

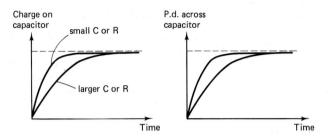

The amount of charge on the capacitor's plates rises quickly at first, then more slowly as the capacitor reaches the charge it can hold at this applied p.d. A digital voltmeter placed across the capacitor shows that the p.d. across the capacitor also rises quickly at first and then more gradually until it is equal to the cell p.d. The bigger the capacitor or the resistor, the more slowly it charges up.

When the switch is then moved to position **B**, the capacitor discharges through the resistor. Both charge and p.d. fall back to zero, falling quickly at first and then more slowly as they approach zero. The bigger the capacitor or the resistor, the longer it takes to discharge.

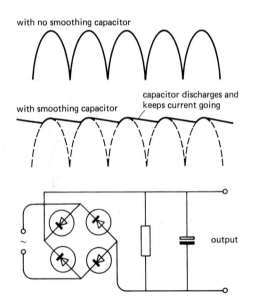

Smoothing

If a large capacitor (usually 1000 μF or more) is placed across the output terminals of a full-wave rectified a.c. supply, the waveform is partly **smoothed** out. The capacitor charges up when the current from the diode bridge is large. Then when the full-wave rectified output begins to drop towards zero, the capacitor takes over and begins to discharge through the load resistor. The current through the load is almost constant – d.c. with just a slight ripple. With a large capacitor and a large resistance load (so that the output current is small), the ripple is very small.

Power supplies

A simple mains d.c. power supply needs a step-down transformer, a diode bridge and a large smoothing capacitor. Notice that the switch and fuse have been placed in the live wire supplying the transformer primary coil.

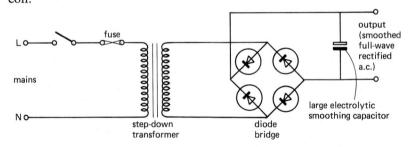

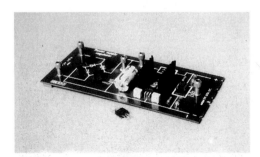

This power supply board takes an a.c. input and produces a smooth 5 V d.c. output. It uses a voltage regulator similar to the one shown beside the board. On the board, the regulator is fitted with cooling fins to stop it overheating in use.

The output p.d. depends on the transformer turns ratio. It is equal to the peak p.d. across the transformer secondary, since the capacitor smooths the output at the level of the peaks of the full-wave rectified a.c.

A simple power supply like this, however, behaves as if it has quite a large internal resistance – the output p.d. falls as we make it supply more current. The basic reason is that the smoothing capacitor holds only a limited amount of charge. If it has to supply a large current, it discharges too quickly and the ripple becomes more noticeable. This means that the *average* output p.d. gets less. Nowadays it is easy to improve on the simple power supply by using a voltage regulator chip. This contains a more complicated circuit which gives a very smooth and accurately known output p.d.

Other semiconductor devices

1. Light-emitting diode (LED)

An LED is a diode made of semiconductor material (usually gallium arsenide or gallium phosphide) which glows when there is a current through it. As with all diodes, current can flow through an LED in one direction only. If it is reverse biased, the LED has a very large resistance and there is no current.

LEDs have the advantage over filament lamps that they need only a very small current to make them light and they do not heat up in use. There is no filament to burn out so they have a long lifetime. An LED will be damaged, however, if it is made to carry too large a current. ▷

To use the LED described on the right, we must put a resistor in series to limit the current. For example, if we want to light the LED using a 6 V supply, we know that there will be a p.d. of 4 V across the resistor, as there is always 2.0 V across the LED itself when it is conducting. The current should be 10 mA – through both the LED and the resistor. We can use the resistance equation $R = V/I$ to calculate the value of R:

$$R = \frac{V}{I} = \frac{4\,V}{10\,mA} = \frac{4\,V}{0.01\,A} = 400\,\Omega$$

The protecting resistor should be around 400 Ω.

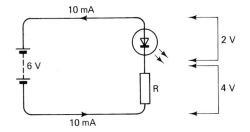

Technical specification (quoted at 25°C)		
I_F max.	30	mA
I_F typ.	10	mA
V_R max.	3	V
V_F at I_F typ.	2	V
Peak wavelength	697	nm

These manufacturers' specifications for a red LED tell us that the typical forward current to light the LED is 10 mA, and the maximum current which the LED can take is 30 mA. With a typical current of 10 mA, the forward p.d. across the LED (V_F) is 2.0 V. The maximum p.d. which can safely be applied across the diode in the reverse direction (V_R) is 3 V.

2. Seven-segment displays

Many digital displays, like those in clocks and calculators, use **seven-segment displays**. Any of the digits from 0 to 9 can be displayed by lighting different combinations of the seven segments. Each segment is a separate LED and needs its own protecting resistor and a switch to turn it on and off.

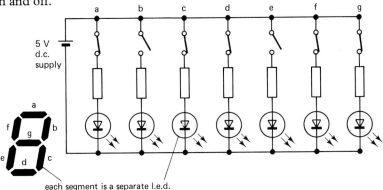

each segment is a separate l.e.d.

The diagram above shows how the seven LEDs are wired up. This seven-segment display is 'common cathode' type – all the LED cathodes joined together. All seven-segment displays are either common cathode or common anode. What number is showing on the display when the switches above are closed.

In a clock or a calculator, the segments are switched on and off electronically rather than by separate switches.

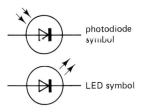

3. Photodiode

One important property of semiconductors is that light can make them conduct better. The energy of the light hitting the semiconductor is enough to free electrons or positive holes inside the material, and this improves its conduction. This property is used in **photodiodes**. Most photodiodes are used with reverse bias. In the dark, there is no current. But when light falls on them, their resistance drops and they begin to conduct. The light switches on the current. ▷

The light-activated switches (photodetectors) used for measuring the speeds of trolleys and air-track gliders (see sections 5.3 and 5.9) are often made using photodiodes.

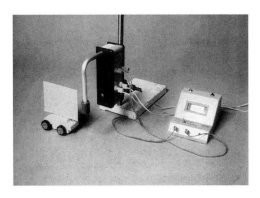

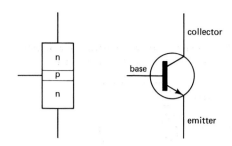

32.5 The transistor

The invention of the **transistor** in 1948 by three American physicists, William Shockley, John Bardeen and Walter Brattain, changed the face of electronics. Eight years later they were awarded the Nobel prize for their work. An **n-p-n transistor** is a sandwich of a thin layer of p-type semiconductor (usually silicon) between two layers of n-type; a p-n-p transistor has an n-type slice between two p-type layers. In this book, we will deal with n-p-n transistors only.

◁ A transistor has three leads, one connected to each of the three layers. These are called the **emitter**, the **base** and the **collector**. This means that a transistor is like two junction diodes connected back-to-back. No matter which way we connect a p.d. across the collector and emitter terminals, one of the two diodes is reverse biased and no current can flow.

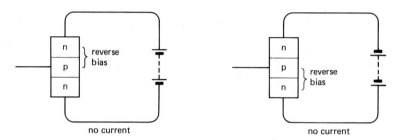

The surprising thing about the transistor is what happens if we connect a *second* p.d. to forward bias the base–emitter junction. (The p.d. must be greater than 0.6 V – the p.d. needed to make a silicon junction diode conduct.) Electrons can now flow from the emitter into the base layer. The base is very thin and most of these electrons are attracted across into the collector because it is at a more positive potential. By making the bottom junction conduct, we have allowed a current to pass from the emitter to the collector – despite the reverse bias!

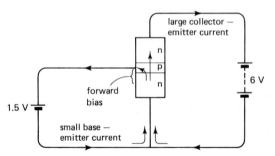

The basic behaviour of a transistor can be summarised as follows:

If there is a small current in the base–emitter circuit, this allows a bigger current to flow in the collector–emitter circuit.

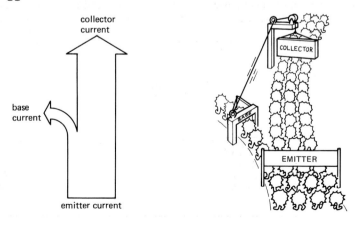

Electronics

32.6 The transistor as a switch

A transistor works as an electronic switch.

If there is a base–emitter current, then the transistor switches ON, and there is a collector–emitter current.

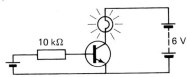

With no base–emitter current, the transistor is switched OFF, and there is no collector–emitter current.

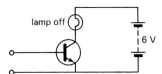

A small base–emitter current can switch on a much bigger collector–emitter current. However, we don't need two separate power supplies for this.

If the base lead is connected (through a resistor to limit the current) to the positive terminal of the supply, a small current will flow through the base–emitter junction. This switches the transistor ON and a larger current flows through the collector–emitter circuit. The lamp lights.

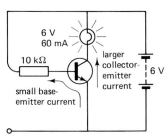

If the base lead is connected to the negative terminal of the supply, there is no p.d. across the base–emitter junction and so there is no base–emitter current. The transistor is switched OFF and the lamp is out.

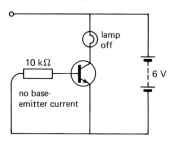

Often we refer to the positive terminal of the supply as HIGH potential and the negative or earth terminal as LOW potential. Looking at it in this way, the rule becomes:

If the input to the base is at HIGH potential, the transistor switches ON and the lamp lights.

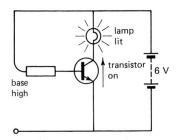

If the input to the base is at LOW potential, the transistor switches OFF and the lamp is out.

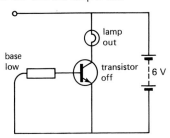

Is your body a conductor when the applied p.d. is low? If we connect a 6 V battery and lamp and hold the two leads **A** and **B** in each hand, the lamp is unlit. It appears that there is no current.

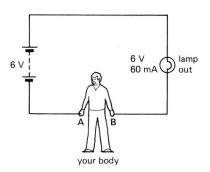

But if we make this circuit and then take hold of **A** and **B** in each hand, the same bulb will light. The current through the body is small – not enough to light the lamp directly. But it *is* enough to provide a base–emitter current to switch the transistor ON. The larger collector–emitter current lights the lamp.

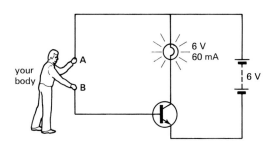

32.7 Electronic systems

The transistor and lamp on p. 327 are a simple **electronic system**. The transistor and base resistor is the *PROCESSOR* and the lamp is the *OUTPUT* stage. The system works according to the following 'rule':

input HIGH, lamp ON
input LOW, lamp OFF

Many useful electronic circuits are built out of these building blocks: *INPUT*, *PROCESSOR* and *OUTPUT*.

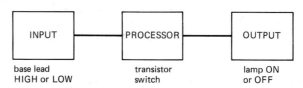

1. Simple intruder alarm

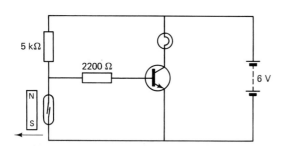

◁ The *INPUT* is a reed switch, operated by a magnet. The switch might be set in a door frame and the magnet fixed to the door. When the door is closed, the magnet closes the reed switch, making the base LOW. When the door is opened, the magnet moves away and the reed switch opens. The base terminal now goes HIGH and the light comes on.

Of course this alarm has the disadvantage that when the intruder closes the door again, the warning lamp goes out.

2. Light-operated switch

The light-dependent resistor has a very high resistance in the dark (over 1 million ohms) and a low resistance in bright light (less than 100 Ω). **R** is a fixed resistor of several thousand ohms. Together they form a **potential divider** (see section 26.7). The supply p.d. is
◁ shared across the two resistors.

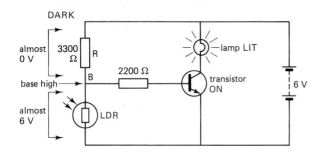

DARK

◁ In the dark, the LDR has a much larger resistance than **R**. Most of the 6 V p.d. is across the LDR, so the point **B** is at a HIGH potential. The transistor is switched ON and the lamp is lit.

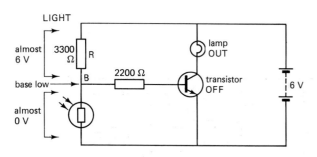

LIGHT

◁ In the light, the LDR has a smaller resistance than **R**, so that most of the 6 V p.d. is now across **R**. The point **B** is at a LOW potential. The transistor switches OFF and the lamp goes out. This circuit could be used to switch on a light after dark, using the LDR as the sensor.

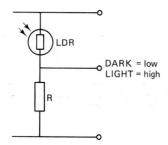

◁ The LDR and the fixed resistor in the input circuit could be swapped over. This input circuit provides a LOW input in the dark and a HIGH input in the light. This could be used in an intruder alarm for a bedroom. If someone switches on a light, the transistor will switch ON an alarm.

3. Temperature controlled switch

This works in much the same way as the light-operated switch. A thermistor has a high resistance when cold, and a low resistance when hot. The fixed resistor is between these.

When it is cold, the thermistor has a larger resistance than **R**, so most of the 6 V is across the thermistor. The base is at LOW potential and the transistor is OFF.

When the temperature rises, the resistance of the thermistor falls. Most of the 6 V p.d. is now across the fixed resistor. The base of the transistor is at a HIGH potential. The transistor switches ON and the lamp is lit. ▷

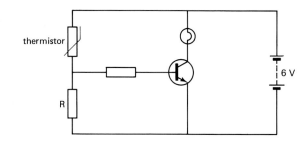

This circuit could be used as a high temperature warning light. An improvement on the basic circuit is to use a variable resistor instead of a fixed resistor for **R**. By adjusting this variable resistor, we can adjust the temperature at which the light comes on.

Again the thermistor and **R** could be swapped over, making an input circuit which goes HIGH when the thermistor is cold. This could be used in a thermostat, to switch on a heater when the temperature falls below a certain point.

4. Using a transistor to switch a relay

Sometimes we want a transistor switch to do more than just switch a lamp on and off. One possibility is to use an electromagnetic relay (see section 29.8) as the OUTPUT stage. When the base is at a HIGH potential, the transistor switches the relay on. The emitter–collector current is less than 1 A but this is enough to operate the relay; much bigger currents can be switched on and off by the relay itself. ▷

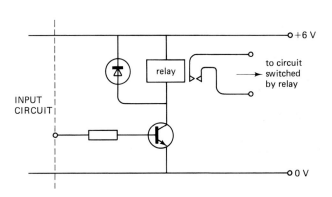

When relays are used in this way, a diode is always connected across the relay coil. There is a good reason for this. When the relay switches off, the magnetic field of the relay coil suddenly falls to zero. This changing magnetic field induces a large e.m.f. across the relay coil itself, in the opposite direction to the original applied p.d. (Lenz's Law) (see section 30.3). If there was no diode, this large e.m.f. would push a large current through the transistor from collector to emitter which might damage it. The diode prevents this. Normally it is reverse biased and there is no current through it. If, however, a large e.m.f. is induced across the relay coil when it switches off, the large induced current will flow harmlessly through the diode instead of through the transistor.

5. Feedback

A temperature controlled switch with relay OUTPUT can be used to make a simple thermostat for keeping a room at a steady temperature. If the temperature is too low, the thermistor switches the transistor ON and this closes the relay contacts. This can be used to switch a heater on. Once the temperature has risen sufficiently, the thermistor will detect this and will switch the transistor OFF again. This is a simple example of **feedback** — one of the most important ideas in electronics. The output — the heater — affects the input, which in turn alters or adjusts the output.

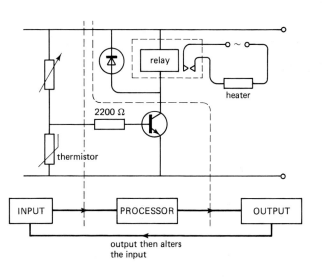

32.8 Digital systems: the inverter

We have seen in section 32.6 how a transistor works as an electronic switch. The transistor switch is sometimes called an **inverter**.

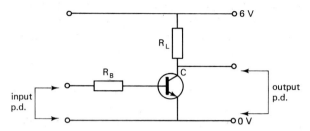

When the input is HIGH, the transistor switches ON. The p.d. across R_L is 6 V. **C** is at a LOW potential. When the input is LOW, the transistor is switched OFF. The 6 V p.d. is across the transistor itself. **C** is at a HIGH potential. We can summarise this as follows:

Input	Output
LOW	HIGH
HIGH	LOW

or

Input	Output
0	1
1	0

This table is called the **truth table** for the inverter. A truth table lists the output for all the possible inputs. Another name for the inverter is a NOT gate. The output is NOT the input – it is always the opposite.

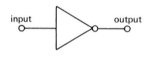

◁ The symbol for an inverter. The temperature-controlled relay switch on p. 329 could be shown like this. ▷

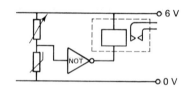

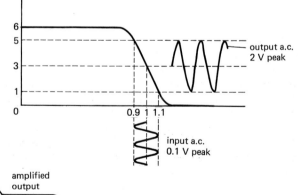

Amplifying with a transistor

When its input changes from 0 V to 6 V, a transistor switches ON. In fact, the output p.d. changes rapidly over a very small range of input p.d.s.
◁ For input p.d.s between about 0.6 V and 1.4 V, the graph is a steep straight line. In this region, small changes in the input voltage produce big changes in the output voltage. For example, if we can hold the base at an average voltage of 1 V (called the bias voltage), then small changes of around 0.1 V cause much larger changes in the output voltage. The output voltage is an exact replica of the input, but **amplified**.

▽ A simple transistor amplifier. The two resistors hold the base at an average potential of 1 V. The crystal microphone causes small changes in this input voltage and these are amplified by the transistor. The earphone changes the amplified voltage back into sound. (Note: The capacitor stops current from the battery passing straight through the 50 kΩ resistor and the microphone. A capacitor has two plates separated by an insulator and so d.c. cannot pass through it.)

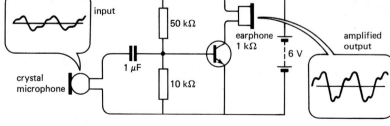

32.9 Digital systems

The single transistor inverter circuit is an example of a **digital system**. A digital system is one where the input and output cannot take any value but must always be in one of two states: HIGH (1) or LOW (0). Information is transferred or stored in the form of patterns of 1s and 0s. Digital systems have become very important and are used in computers, calculators, digital clocks and watches, and so on.

Logic gates

A very important group of digital systems is the **logic gates**. They are called 'gates' because they 'open' and give a 1 on the output only when a particular combination of 1s and 0s is present on the inputs. This combination is the 'key' to open the gate! A logic gate is really a combination of transistor switches. The five basic gates are called NOT, OR, NOR, AND and NAND. The easiest way to summarise how they work is by writing their truth tables.

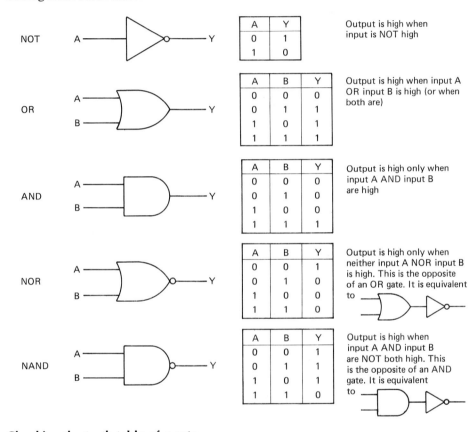

Checking the truth table of a gate

As an example, let us see how we could check the truth table of an OR gate. This has two inputs and one output. We need an indicator to show whether the output is HIGH or LOW. An easy way to do this is to use a LED in series with a 400 Ω resistor, connected between the output and the 0 V line. When the output is LOW, there is no p.d. across the LED and resistor so there is no current and the LED is off. When the output is HIGH, the diode will conduct and light up. The protecting resistor keeps the current down to a safe value, around 10 mA. (Remember that there will be a p.d. of 2 V across the LED.)

With both inputs LOW, the LED is off. The output is LOW. When input 1 is connected to the HIGH potential, the LED comes on; the same happens if input 2 is made HIGH. With both inputs HIGH, the LED is also on. So the output is HIGH if *either* input 1 OR input 2 is HIGH (or if both are HIGH). Hence the name OR gate.

The other gates can be tested in similar ways.

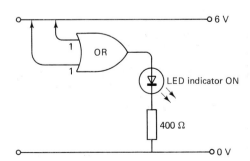

Logic gates in action

Here are some possible input circuits:

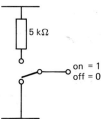
on = 1
off = 0

Could be manually operated, or by a pressure pad (someone standing on a mat or sitting on a seat), or by a magnet (reed switch), or by liquid level in a tank (float switch).

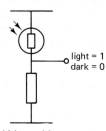
light = 1
dark = 0

Light-sensitive switch

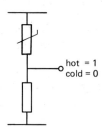
hot = 1
cold = 0

Temperature-sensitive switch

Burglar alarm (door and window)

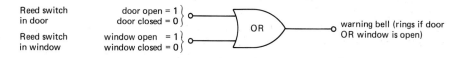

Reed switch in door: door open = 1, door closed = 0
Reed switch in window: window open = 1, window closed = 0
OR → warning bell (rings if door OR window is open)

Supermarket automatic door

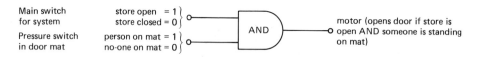

Main switch for system: store open = 1, store closed = 0
Pressure switch in door mat: person on mat = 1, no-one on mat = 0
AND → motor (opens door if store is open AND someone is standing on mat)

Automatic greenhouse heater: a gardener wants his greenhouse heater to switch on at night if there is a frost. During the day he will control the heating manually.

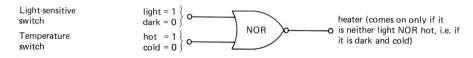

Light-sensitive switch: light = 1, dark = 0
Temperature switch: hot = 1, cold = 0
NOR → heater (comes on only if it is neither light NOR hot, i.e. if it is dark and cold)

Car door warning light: a warning light indicates if either door is not properly closed. There are push button switches in each door.

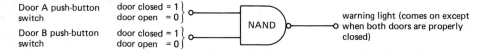

Door A push-button switch: door closed = 1, door open = 0
Door B push-button switch: door closed = 1, door open = 0
NAND → warning light (comes on except when both doors are properly closed)

Car passenger seat-belt warning light: a warning light comes on if there is a passenger in the front seat who is not wearing a seat belt. There is a pressure switch in the car seat and a switch in the seat-belt catch.

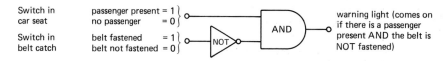

Switch in car seat: passenger present = 1, no passenger = 0
Switch in belt catch: belt fastened = 1, belt not fastened = 0
NOT, AND → warning light (comes on if there is a passenger present AND the belt is NOT fastened)

32.10 Counting and measuring with digital systems

Inputs and outputs of digital electronic systems can only take the values 0 or 1. How can complex circuits like computers or digital measuring instruments work when only two states are possible? The answer is that several parallel digital lines are used to allow us to represent larger numbers. ▷

For example, with four lines we can represent numbers up to 15. Each line can be either HIGH (1) or LOW (0). The LEDs indicate the pattern of 1s and 0s. The smallest number is 0000 and the largest is 1111. Since the only digits possible are 0 and 1, we are working in the **binary** counting system (base 2). The number 1111_2 is the same as 15_{10}. By using parallel lines like this, we can represent numbers as large as we want.

A set of parallel lines connecting two parts of a digital system is called a **bus**. ▷

Sending information in digital form, as a binary number, has certain advantages. Any electrical circuit is 'noisy' – the p.d. at any point can vary slightly because of pick-up from other electrical equipment nearby. If a single lead is used to carry an electrical waveform, this noise may change it slightly as it travels across. After many transfers inside a large circuit, it could have become very different. Noise affects digital signals too, but these signals are always either 1 (5 V) or 0 (0 V). The noise is small by comparison and is never enough to confuse the receiving system. So the digital signal is transmitted accurately without any loss of clarity. ▷

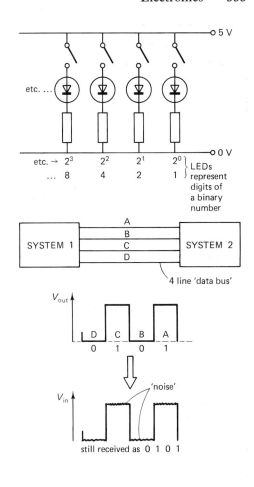

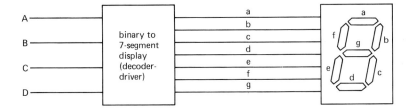

Numbers in base 2 are rather difficult to read, so when a display is needed, special decoding chips are used to convert binary data into a form where it can operate a 7-segment display. △

Digital measurement

Digital systems can be used for making measurements in the laboratory. Most of the quantities we want to measure are not digital, however, but can take any value. Things like p.d., temperature, pressure and so on vary smoothly, not in digital 'steps'. All digital measuring devices use an **analogue-to-digital converter** (A/D converter) to change the varying input into a digital number. Most common systems use 8 parallel digital lines and can handle numbers up to 11111111_2, or 255_{10}. An input of 0 V produces a digital output of 0. The maximum input (say, 5 V) gives an output of 255 on the digital lines. Other p.d.s between 0 V and 5 V give digital outputs between 0 and 255. For example, a 2.5 V input, which is half-way through the range, gives a digital output of 128, and so on.

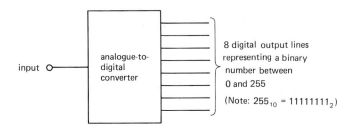

Questions

1. Draw a circuit showing a lamp, a semiconductor diode, a cell and a switch connected in series so that the lamp will light when the switch is closed.

2. Sketch diagrams to show the trace you would expect to see on the CRO screen in each of the following situations.

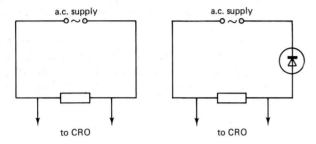

3. (a) What is meant by the word 'rectification'?
   (b) (i) Diagram (A) shows an incomplete bridge rectifier. Copy and complete the circuit by drawing the four diodes.

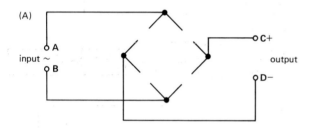

   (ii) A CRO is connected first across **AB** and then across **CD**. Graph (B) shows the p.d. across **AB** for two cycles. On graph paper draw the same axes and time intervals. Label the y-axis 'p.d. **CD**' and sketch the p.d. across **CD** for the same time.

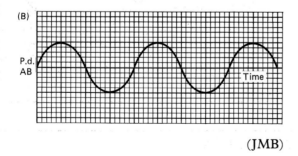

(JMB)

4. To produce a smoother d.c. output from the power supply in question 3, one extra component is added.
   (a) Name this extra component and draw a diagram to show where it should go.
   (b) Explain briefly why it smoothes the output.

5. How much charge is stored in a 1000 μF capacitor charged to a potential of 12 V?

6. Copy the diagram, adding the names of the three terminals of the transistor.

7. In this circuit, when you take the two leads **A** and **B** in each hand, lamp L_1 comes on but lamp L_2 does not. Explain why this happens. What does it tell you about the action of a transistor?

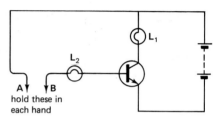

8. The circuit diagram below shows how a transistor may be used to make a switch which makes a light come on when it gets dark.

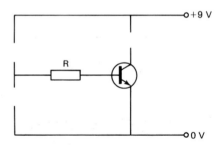

(a) Copy and complete the circuit by adding in the correct spaces and labelling the following:

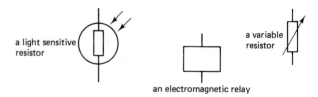

(b) Explain how the circuit will operate the relay and hence switch a bulb on as it gets dark.
(c) Why is a variable resistor preferred to a fixed resistor?
(d) Why is a relay preferred to inserting a bulb directly in the circuit?
(e) How could the circuit be adapted to switch on a warning light when an old person's bedroom becomes too cold?

(Sp.LEAG)

9 Diagram (A) shows the truth table for a single transistor switch. Work out the truth table for the arrangement shown in diagram (B), which has two separate inputs to the same transistor. What sort of gate is this?

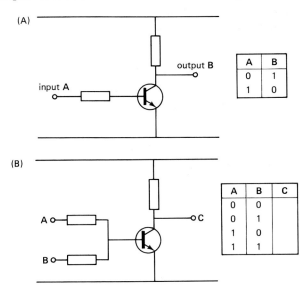

10 Work out the truth tables for each of the following combinations of gates.

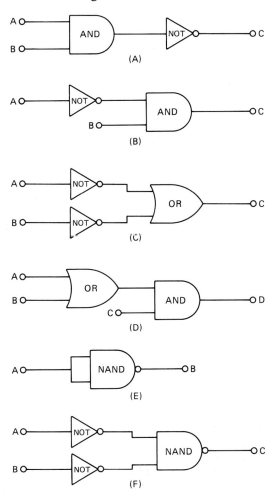

11 The circuit in diagram (A) switches on a lamp if it is both hot and light. It can also be represented by a 'block diagram' (diagram (B)).

(a) Draw similar block diagrams for circuits which will switch on the lamp:
– if it is hot or if it is light;
– except when it is both hot and light;
– if it is cold and light;
– if it is cold and dark.

(b) Sometimes it is useful to be able to test a circuit to see if it is working. Diagram (C) shows one way to add a test button to the 'hot and light' detector. Draw the block diagram for this arrangement and explain clearly how it works.

(c) A final refinement is to add an enable/disable switch, so that the automatic 'hot and light' detector only operates when you switch it on. Diagram (D) shows a circuit to do this. Draw its block diagram and explain how it works.

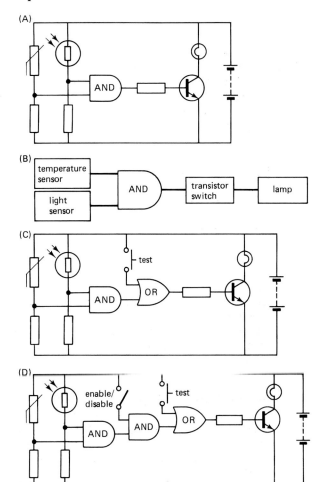

INVESTIGATION

The output and inputs of a logic gate can be either high (1) or low (0). For the TTL series, 5 V is high and 0 V is low. But what is the output of the gate if the input is somewhere between 0 V and 5 V? (Hint: it is easiest to start by investigating a NOT gate, but you could go on to look at other gates too.)

33: Radioactivity

Reports and articles about radioactivity and nuclear power are often in the newspapers. Technical words like 'radiation', 'isotope', 'half-life' and so on are used. To understand the nuclear power issue and take part in the debate, you need to know what these words mean – you have to understand some of the physics which lies behind them. This is one reason for learning about radioactivity. Another is that the study of radioactive materials has led physicists to a much greater understanding of the atom and of the basic building blocks of which the atom itself is made.

33.1 Radioactivity: discovered by accident

In 1896, the French physicist Henri Becquerel discovered, rather by chance, that substances containing uranium emitted radiation which could pass through paper and blacken a photographic plate. Nothing was needed to start the radiation off; it came from the uranium substances all the time. ▽

By 1898, Marie Curie had managed to extract two other substances from an ore called **pitchblende** which emitted radiation much more strongly than uranium. They were the elements polonium and radium. She named the effect **radioactivity** and the substances which produced it **radioactive materials**. ▽

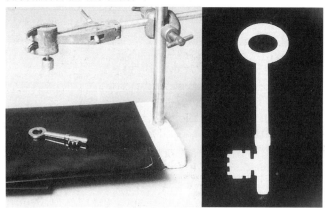

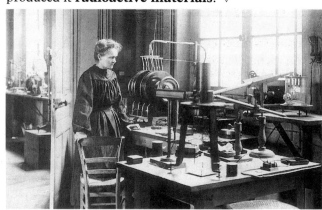

Three types of radiation

In the years around 1900, many experiments were done to find out more about the newly discovered radiation from radioactive materials. These involved finding out how penetrating the radiation was (what thicknesses of different materials it could pass through) and whether it was deflected by electric and magnetic fields. It was quickly realised that there were three different types of radiation involved. These were given the names **alpha** (α), **beta** (β) and **gamma** (γ).

Alpha radiation is stopped by a thin sheet of paper. Beta radiation can pass through paper but is stopped by a few millimetres of aluminium. Gamma radiation is very penetrating, needing several centimetres of lead to reduce the intensity to almost zero.

Alpha radiation is slightly deflected by both electric and magnetic fields. The direction of bending suggests that it carries a positive (+) charge. Beta radiation is deflected very strongly in the other direction – it seems to carry a negative (−) charge. Gamma radiation is not deflected by electric or magnetic fields.

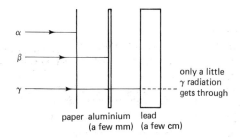

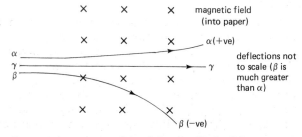

From these experiments, it appears that alpha 'radiation' is a stream of positively charged **alpha particles**; beta 'radiation' consists of negatively charged **beta particles**; whilst **gamma radiation** behaves like very short wavelength X rays.

33.2 Detecting radioactive emissions

IONISING RADIATION

Most of the methods of detecting radiation from radioactive substances depend on the fact that the radiation causes **ionisation** in any material it passes through. An alpha particle, beta particle or gamma ray can knock an electron off an atom, leaving a positive ion. The free electron attaches itself to another nearby atom, making a negative ion.

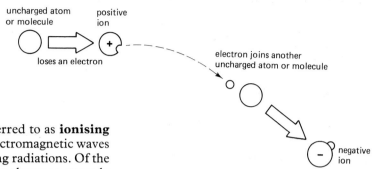

For this reason alphas, betas and gammas are referred to as **ionising radiation**. X rays also cause ionisation, but other electromagnetic waves such as ultraviolet, visible and infrared are *not* ionising radiations. Of the three types of radioactive emission, alpha particles are the most strongly ionising, beta particles produce less ionisation and gamma radiation least of all.

METHODS OF DETECTION

1. Photographic film

Photographic film is sensitive to ionising radiation and can be used to detect all three types of radiation from radioactive sources.

2. Gold-leaf electroscope

A radioactive source held near the cap of a charged electroscope produces positive and negative ions in the air around the cap. Ions with a charge opposite to that on the electroscope are attracted and this gradually discharges the electroscope. The leaf falls.
Works: with alpha particles; beta particles and gamma radiation do not produce enough ionisation in the air.

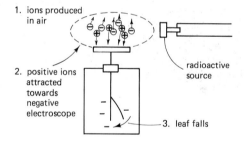

3. Ionisation chamber

An ionisation chamber 'collects' the ions produced by ionising radiation. A battery maintains a small p.d. between the outer metal can and the central electrode. Ions produced in the air inside the chamber are attracted to the electrode of opposite sign. This movement of ions is equivalent to a small electric current in the circuit, and produces a p.d. across the load resistor. This p.d. is then amplified by a sensitive amplifier.
Works: best with alpha particles, but can be used to detect any type of ionising radiation.

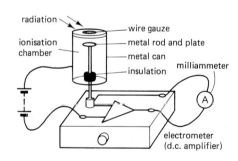

4. Spark counter

In a spark counter, a large p.d. is connected across the narrow gap between a metal wire and a metal gauze. The p.d. is turned up to the point where sparks jump across the gap, and then reduced until the sparking just stops. If an alpha particle source is now brought close, sparks will begin to jump across the gap. What happens is that the alphas produce ions in the air in the gap, and these are attracted towards the electrode of opposite sign. The p.d. is very large so the ions move very fast and as they go they collide with more air molecules causing more ionisation – a so-called **avalanche effect**. The sudden rush of ions is the spark. Each alpha particle produces one spark, so we can actually count the number of alphas detected.
Works: with alpha particles but not with beta particles or gamma rays.

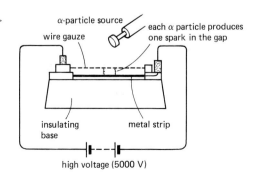

5. Geiger-Müller tube

This is similar in design to an ionisation chamber, but works like a special kind of spark counter! Inside the tube is argon gas at low pressure. The thin window on the end allows ionising radiation to enter. A p.d. of about 400 V is applied between the central wire electrode and the outer tube. When an alpha or beta particle or a gamma ray enters, it causes some ions to form in the argon. The p.d. is enough to cause an avalanche effect but not enough to make a spark. The avalanche of ions is like a brief pulse of current between the electrodes. There is one pulse for every alpha, beta or gamma detected. The Geiger-Müller tube is connected to a **scalar** ◁ (which counts the pulses) or a **ratemeter** (which indicates the number of pulses per second). The complete apparatus is called a **Geiger counter**. Works: with alpha and beta particles and gamma radiation.

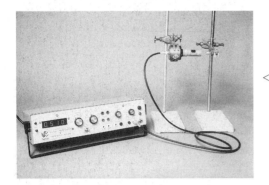

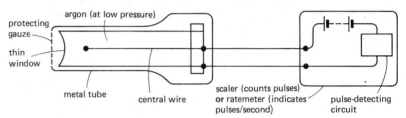

6. Cloud chamber

◁ A cloud chamber actually allows us to see the tracks of charged particles. A felt ring round the top of the chamber is soaked in alcohol. Under the base of the chamber is 'dry ice' (solid carbon dioxide). Alcohol vapour spreads downwards and is cooled below the point where it would normally condense. Alpha particles from the source produce a trail of ions along their path and the alcohol condenses more readily around these ions. A narrow 'cloud' forms along the track of the alpha particle, just like the vapour trail of a high-flying aircraft.

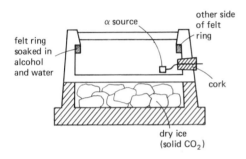

a. Alpha particle tracks are straight and thick, because alphas are strongly ionising.

b. Fast-moving beta particles produce fainter straight tracks; slow-moving betas produce short, twisted tracks, because of the frequent collisions between the beta particle and electrons in the air. This photograph was taken in 1923 by Charles Wilson, who invented the cloud chamber.

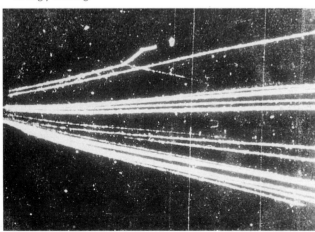

c. Gamma rays do not leave a track in a cloud chamber but the electrons they knock out of atoms leave short tracks which can be seen.

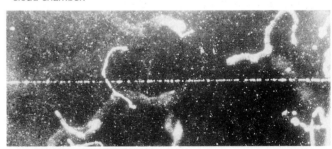

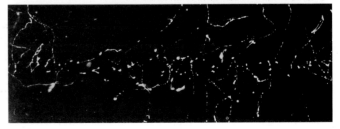

7. Bubble chambers

An improvement on the cloud chamber is the bubble chamber which is used for research work on high-speed particles. The ionising particles pass through a tank of liquid hydrogen at very low temperature (a few K). The hydrogen is normally kept under pressure but, when this is suddenly released, streams of bubbles form, just like the bubbles which appear in a bottle of lemonade as the cap is unscrewed. The bubbles form more readily along the tracks of charged particles, and so their paths can be seen and photographed. ▷

33.3 More about alphas, betas and gammas

More detailed measurements of the deflection of alpha and beta particles in electric and magnetic fields were carried out in the years just after 1900. These showed that the charge-to-mass ratio of beta particles was the same as that of the electron (measured by J.J. Thomson in 1897 – see section 31.4). As a result, it was concluded that **beta particles are fast moving electrons**.

Alpha particles were found to have a much smaller charge-to-mass ratio – about half that of a hydrogen ion. By 1908 Geiger and Rutherford had found that the alpha particle had a charge twice as large as an electron but opposite in sign. All of this suggested to them that **an alpha particle is the same as the nucleus of an atom of helium**. To check this, Rutherford and Royds devised an experiment to 'collect' alpha particles. They then tested these trapped alphas and showed that they were helium nuclei.

The Big European Bubble Chamber at CERN near Geneva. The workers standing in front of it give you an idea of how big it is.

The Rutherford–Royds experiment: Radioactive radon gas emits alpha particles which get through the thin inner glass tube but cannot penetrate the thicker outer glass tube. After some time, mercury is pumped up to compress the trapped alphas at the top of the tube. An electric discharge through the gas produces light which contains the characteristic line spectrum of helium. The alphas had picked up two electrons from the glass and changed into helium gas!

The properties of alpha particles, beta particles and gamma radiation are summarised in this table:

Radioactive emission	What it is	Mass (proton = 1 unit)	Charge (proton = 1 unit)	Speed	Absorbed by	In an electric or magnetic field
Alpha particle (α)	Helium nucleus 2 protons + 2 neutrons	4	2	Up to $\frac{1}{10}$ of the speed of light	Sheet of paper or a few centimetres of air	Deflected a little
Beta particle (β)	Fast moving electron	$\frac{1}{1840}$ (almost zero)	-1	Up to $\frac{9}{10}$ of the speed of light	A few millimetres of aluminium	Strongly deflected
Gamma radiation (γ)	Short bursts of high frequency (short wavelength) electromagnetic radiation	0	0	Speed of light	Several centimetres of lead or concrete	Not deflected

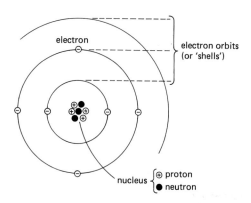

33.4 Atoms and radioactivity

Radioactive materials are constantly emitting alpha and beta particles and gamma rays. But where do these alphas, betas and gammas come from and what provides the energy to kick them out at high speed? To understand the answers to these questions, we need to know something about the structure of atoms.

The atom

◁ At the centre of each atom is a **nucleus**. Surrounding this nucleus are the very much lighter **electrons**. We can think of the electrons as orbiting around the central nucleus like the planets going round the Sun. The nucleus itself is made up of even smaller particles called **protons** and **neutrons**.

Electrons and protons carry electric charge:
  electrons have a negative (−ve) charge
  protons have a positive (+ve) charge.

The *size* of the charge on the electron and proton is the same. Every atom has the same number of electrons and protons, so that their charges neutralise each other and the atom as a whole is electrically neutral. Neutrons carry no electrical charge.

Electrons have very little mass compared to protons and neutrons. The mass of a proton is almost exactly equal to the mass of a neutron – the mass of an electron is 1840 times smaller.

This is summarised in the table on the left.

Particle	Proton (p)	Neutron (n)	Electron (e)
Mass (proton = 1)	1	1 (approx)	$\frac{1}{1840}$ (almost zero)
Charge (proton = +1)	+1	0	−1
Where it is in the atom	Nucleus	Nucleus	Around the nucleus

Atoms and elements

All the materials we see around us are made up from about 90 different substances called **elements**. The elements are like the building blocks out of which everything is made. The smallest 'piece' of an element which can exist on its own is an atom. Each element has its own distinctive atom, and the atoms of different elements are different.

Let us see what this means with a few examples. ▽

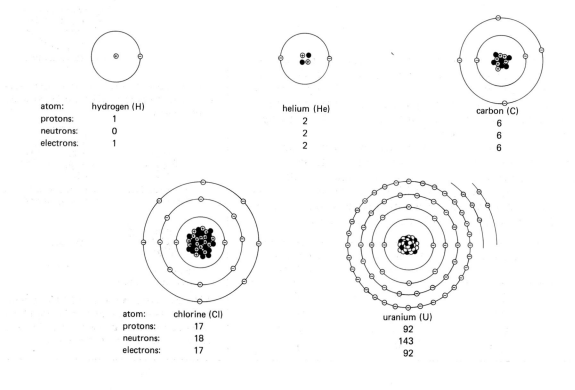

atom: hydrogen (H)
protons: 1
neutrons: 0
electrons: 1

helium (He)
2
2
2

carbon (C)
6
6
6

atom: chlorine (Cl)
protons: 17
neutrons: 18
electrons: 17

uranium (U)
92
143
92

The chemical behaviour of an element depends on the number and arrangement of the electrons round the atom, because this determines the way it can react with other atoms and form chemical bonds with them. It is having six electrons which makes carbon behave like carbon! The number of electrons, however, is always equal to the number of protons. The number of protons in the nucleus of an atom is called its **atomic number** (or sometimes, proton number). Every element has its own atomic number.

The mass of an atom is almost entirely due to the protons and neutrons in its nucleus. The total number of protons and neutrons in the nucleus of an atom is called its **mass number** (or sometimes, nucleon number – a 'nucleon' is the name for a proton or neutron).

Using the symbols for the elements, the information about the atoms on p. 340 can be summarised as follows:

$$^{1}_{1}H \qquad ^{4}_{2}He \qquad ^{12}_{6}C \qquad ^{35}_{17}Cl \qquad ^{235}_{92}U$$

mass number (number of protons and neutrons) ← symbol of element ← atomic number (number of protons)

Another way of writing this is to use the names: hydrogen-1, helium-4, carbon-12, chlorine-35 and uranium-235.

Isotopes and nuclides

The number of electrons (or protons) – the atomic number – must be the same for all atoms of any particular element. But the number of neutrons does not necessarily have to stay the same. Some elements occur in more than one form, with different numbers of neutrons in their nuclei. For example, there is a second form of chlorine, with 20 neutrons. This still has 17 protons and 17 electrons, so chemically it behaves exactly like chlorine-35.

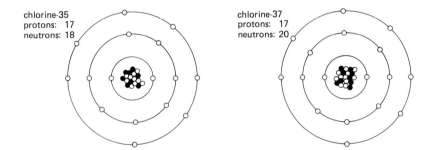

Atoms with the same number of protons but different numbers of neutrons are known as **isotopes**. Chlorine-35 and chlorine-37 are two isotopes of the element chlorine. Many elements have isotopes. In the naturally-occurring element these usually occur in a fixed proportion. For example, naturally-occurring chlorine is about three-quarters chlorine-35 with one-quarter chlorine-37.

For talking about one specific isotope of an element, the word **nuclide** is sometimes used. The element chlorine has two main isotopes. One of them, chlorine-35, is a nuclide which has 17 protons and 18 neutrons in its nucleus. Frequently, however, the word 'isotope' is simply used for both purposes. Data on some isotopes of uranium are shown in the table below. Note that the number of protons is the same each time.

Isotope (or nuclide)	Symbol	Number of protons	Number of neutrons
Uranium-233	$^{233}_{92}U$	92	141
Uranium-234	$^{234}_{92}U$	92	142
Uranium-235	$^{235}_{92}U$	92	143
Uranium-238	$^{238}_{92}U$	92	146

Radioactive decay

The reason why some materials are radioactive is that their nuclei are **unstable**. Imagine the atoms sitting on a little ledge above the bottom of a cavity. They are not entirely at rest but 'jiggling around' a little. From time to time, one atom will fall off the ledge into the bottom of the cavity where it has less energy and is more stable.

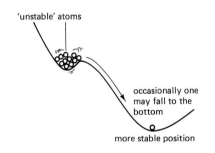

An unstable nucleus can become more stable by releasing a particle which carries off some energy with it.

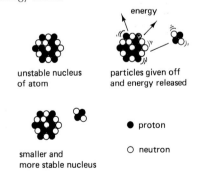

All the isotopes of elements with atomic numbers over 84 are radioactive and many of the elements with lower atomic number also have some isotopes which are radioactive.

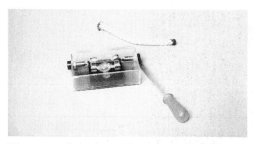

We can see Brownian motion in a smoke cell.

Ernest Rutherford (1871–1937). Born in New Zealand, Rutherford worked first at Manchester, then at the Cavendish Laboratory in Cambridge.

Models of the atom

A lot of evidence suggests that matter is made of tiny atoms. The regular shape of crystals, Brownian motion and diffusion all provide clues (see section 1.12). Just over 200 years ago, the Swiss physicist Daniel Bernoulli suggested that a gas is made up of millions of tiny particles in rapid random motion. This model – the kinetic model of gases – enables us to explain the Gas Law results (see section 15.5).

To understand electrical phenomena, we need a model of the atom itself. We can explain charging by rubbing by imagining charges – electrons – being transferred from one object to another. The electrons are on the *outside* of the atoms where they can be easily rubbed off (section 24.3). Thermionic emission of electrons from a hot wire also suggests that the electrons are near the outside of the atom and can escape (section 31.2).

Early this century (around 1902), J.J. Thomson proposed the 'plum-pudding' model of the atom: a ball of positive charge, with electrons dotted over the surface like the currants in a pudding!

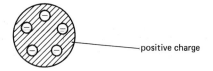

By 1911, Rutherford and his co-worker Geiger had begun to suspect that all the positive charge in an atom was concentrated into a very tiny region at the centre. Rutherford had worked out a mathematical theory to predict how many alpha particles would be deflected at each angle as they passed through a thin metal foil.

Geiger and Marsden planned and carried out an experiment to test Rutherford's mathematical predictions. They fired a stream of alpha particles at a very thin gold foil and counted how many alpha particles were scattered at a number of different angles. The results agreed well with the theory.

angle of deflection	number of alphas counted
15°	132 000
30°	7 800
60°	477
105°	69
150°	33

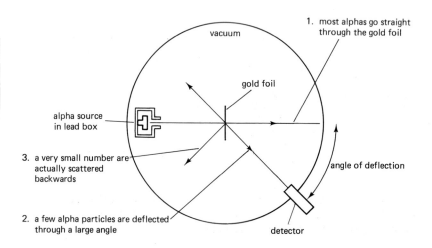

Most of the alpha particles passed straight through the foil. Even though it appears very thin, gold foil is still at least 100 atoms thick. So most of the space taken up by the atom must be completely empty! A few alphas, however, are deflected through very large angles. Rutherford is supposed to have said that this was about as strange as firing a shell at a piece of tissue paper and having it bounce back at you! His explanation was that the positive charge inside the nucleus must be very small and concentrated. If a positively charged alpha particle happens to come near it, it is repelled with a very large force.

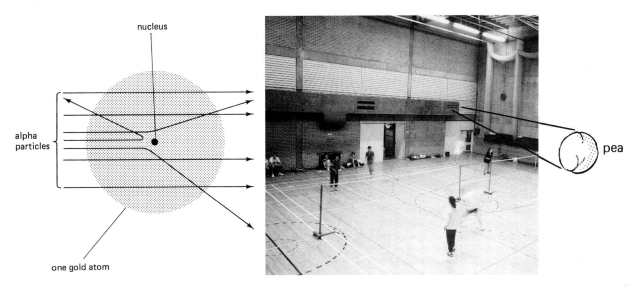

The nuclear model of the atom. Most of the atom is empty space. From Geiger and Marsden's results, Rutherford was able to calculate how large the nucleus is compared to the whole atom. An atom the size of a sports hall would have a nucleus no bigger than a pea placed in the middle.

More clues from spectra

Rutherford's nuclear model still left many questions unanswered. For example, why are the electrons not attracted by the nucleus and fall into it? The next improvement came when the Danish physicist, Niels Bohr, used the nuclear model to explain why the light emitted from each element had a characteristic line spectrum. Bohr suggested that electrons can only be in certain 'allowed' orbits. Different orbits have different energy levels. If an electron stays in one orbit, it is stable and does not emit any radiation. But if an electron moves from one orbit to another, the energy difference is emitted as a **photon**, or **quantum**, a short burst of electromagnetic radiation.

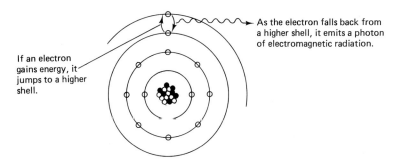

Bohr was able to calculate theoretically what the wavelengths in the spectrum of the simplest atom, hydrogen, should be – and his calculations agreed exactly with what was observed!

A mathematical model

Bohr's model simply *assumed* that certain allowed orbits are stable but didn't explain why. Eventually a new mathematical theory emerged called **quantum mechanics**. In this theory, the atom is described by mathematical equations which are difficult to put into words. The best we can do is to imagine the electrons as 'clouds' or 'orbitals' around the nucleus. The theory will tell us what the **probability** is of finding the electron in a particular place, but cannot tell us exactly where it is. Despite this, the predictions made by quantum mechanics turn out to be very exact and reliable – it is the most accurate and reliable theory in the whole of science!

33.5 Radioactive processes

Alpha decay

When an unstable nucleus emits an alpha particle, it loses two protons and two neutrons. The alpha particle also carries away some energy. The atomic number has dropped by two, so the atom is now an atom of a different element:

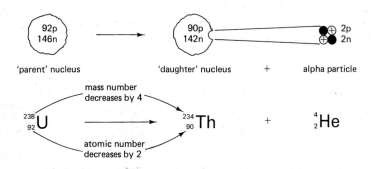

Notice that the top number must balance on both sides of the equation. This is because the total number of nucleons (the total mass) must stay the same. The bottom numbers must also balance. This is because the total amount of positive charge must also stay the same.

It might seem that when an atom loses an alpha particle, it has two 'spare' electrons. This is true, and for a short time at least the new atom will really be an ion, with two extra electrons. However, electrons are easily lost and picked up from the other atoms around and the newly formed ion will soon lose its spare charge.

Beta decay

In beta decay, an unstable nucleus emits a beta particle – an electron. This may seem strange since our model of the atom does not have any electrons in the nucleus. What appears to happen is that a neutron changes into a proton and an electron, and the electron is emitted as a beta particle.

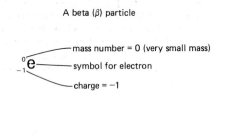

A beta (β) particle
- mass number = 0 (very small mass)
- symbol for electron
- charge = -1

$_{-1}^{0}e$

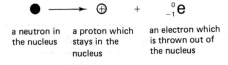

a neutron in the nucleus → a proton which stays in the nucleus + an electron which is thrown out of the nucleus

An example of a beta-emitter is the isotope of iodine, iodine-131. As a result of the beta decay, the nucleus has an extra proton. The atomic number has increased by one and the new atom is an atom of a different element.

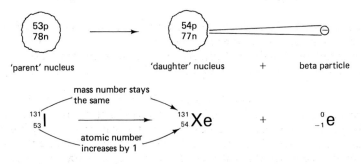

Another tiny particle, the neutrino, is also emitted along with the electron. It has been mentioned in section 10.8. For most purposes, we can simply think of beta decay as the emission of an electron from the nucleus.

Gamma emission

After emitting an alpha or beta particle, a nucleus is sometimes still left in an 'excited' or 'too energetic' state. As the protons and neutrons inside the nucleus rearrange themselves to become more stable, more energy may be emitted in the form of a burst of gamma radiation. Emitting a gamma does not change the atomic number of the atom; it also has very little effect on the mass.

33.6 Rate of decay and half-life

During radioactive decay, unstable nuclei disintegrate by emitting alpha or beta particles and gamma radiation. The rate at which the nuclei decay depends only on which radioactive isotope is involved – it cannot be changed by heating or cooling it, by making it react chemically, or by any other means. Radioactive decay is also a completely **random** process – we cannot predict when any particular nucleus will decay. Despite this, we can still make some predictions about what will happen to a large collection of nuclei.

Interval	1	2	3	4	5
Counts in 1 minute	256	243	271	262	249

Radioactive decay is random. The number of counts recorded by the Geiger counter in one minute is not the same for successive one-minute intervals. This is not because of errors in the counting but because the number of radioactive decays in each interval varies slightly.

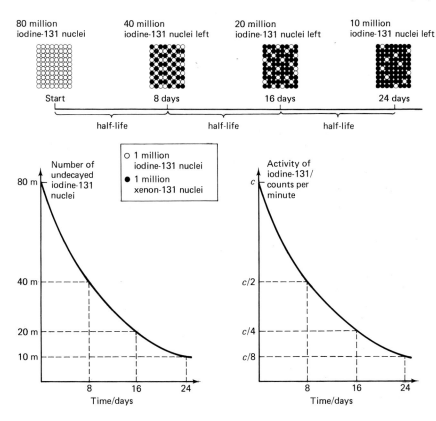

Iodine-131 is a beta emitter. A sample of iodine-131 will contain millions of atoms. Although we cannot predict when any particular nucleus will decay, we find that the rate at which beta particles are emitted from the sample (the number of betas per minute) drops to half of its original value after 8 days. Another 8 days later, it is down to half of that again – that is, to one-quarter of what it was at the beginning. And the rate goes on halving like this in 8-day intervals.

The rate at which betas are emitted drops because the number of iodine-131 nuclei is also dropping. Imagine that there are 80 million iodine-131 nuclei at the beginning. After 8 days, 40 million have decayed by changing into nuclei of xenon-131 and a beta particle. Only 40 million iodine-131 nuclei remain. The *chances* of each of these now decaying is still the same – but there are only half as many of them. So in the next 8-day period, 20 million decay, leaving 20 million after 16 days. And so it goes on.

The **half-life** of a radioactive nuclide is the time taken for half of the nuclei present in any sample of it to decay.

It also follows that the half-life of a radioactive nuclide is the time taken for the count-rate of radioactive emissions from it to fall to half its original value.

The half-life of iodine-131 is 8 days. There is no period of time after which we can be sure *all* the iodine nuclei will have decayed. All we can say is that half the nuclei in any sample will decay in an 8-day interval. Other nuclides have different half-lives as the table on the right shows.

It might seem strange that there are any isotopes with very short half lives around. We might expect them to have decayed long ago. But they are the 'daughter' products of other radioactive decays and some come from 'parents' which have very long half-lives. Some other short half-life nuclides have been artificially produced in nuclear reactors.

Isotope	Symbol	Half-life
Polonium-212	$^{212}_{84}$Po	3×10^{-7} s
Radon-220	$^{220}_{86}$Rn	52 s
Sodium-24	$^{24}_{11}$Na	15 hours
Iodine-131	$^{131}_{53}$I	8 days
Carbon-14	$^{14}_{6}$C	5730 years
Plutonium-242	$^{242}_{94}$Pu	400 000 years
Potassium-40	$^{40}_{19}$K	1300 million years
Uranium-238	$^{238}_{92}$U	4500 million years

Activity

The **activity** of a radioactive source is measured in units called **becquerels**. This tells us how many nuclei in the radioactive sample decay in one second (which is the same as the number of alpha or beta particles emitted in one second). **A sample has an activity of 1 becquerel if, on average, one nucleus in the sample decays each second.**

RADIATION AND RISK

Radiation around us

Becquerel discovered radioactivity in 1896 – but ionising radiation has always been part of our environment. This is known as **natural background radiation**.

Our food contains small amounts of naturally occurring radioactive substances, so our bodies are slightly radioactive.

Some radiation reaches the Earth from outer space and from the Sun. This is called **cosmic radiation**. Most of it is absorbed by the atmosphere, but some penetrates to ground level.

There are naturally occurring radioactive materials in the soil and in rocks, in water, plants and animals.

Many rocks contain small amounts of uranium, as do building materials made from them, such as bricks and concrete blocks. Uranium decay produces a radioactive gas, **radon**. In well insulated houses and buildings, radon can accumulate and become a health hazard. This problem is now being studied in the UK and abroad.

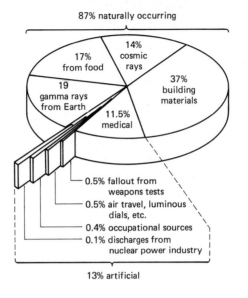

87% naturally occurring
14% cosmic rays
17% from food
19 gamma rays from Earth
37% building materials
11.5% medical
0.5% fallout from weapons tests
0.5% air travel, luminous dials, etc.
0.4% occupational sources
0.1% discharges from nuclear power industry
13% artificial

Radiation and cells

Although we are surrounded by natural sources of ionising radiation, we must treat radiation with caution. All ionising radiation damages living cells. The energy carried by the radiation can break up the chemicals inside the cell. The natural repair mechanisms of the cell may restore it. Otherwise this damage results in the death of that particular cell. Cells are always dying and being replaced with new ones, so this usually causes no ill effects. Sometimes, however, the effects may be more serious.

1. If a person is exposed to very intense radiation the cell damage may not be repaired in time. Some important cells (like the cells in bone marrow which make new blood cells) may not recover in time. Large amounts of radiation can kill.

2. Sometimes the chemical DNA which carries the instruction 'code' inside each cell can be *slightly* damaged by ionising radiation – not enough to kill the cell but enough to change its 'code'. The cell may then begin to divide and multiply in the wrong way – out of control. It produces a mass of faulty cells – a tumour or cancer.

3. If the male or female sex cells are slightly damaged by ionising radiation, a baby which grows from one of them may be abnormal in some way. For this reason, it is particularly important to protect the ovaries and testes from ionising radiation.

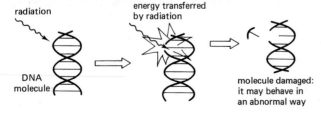

Low-level radiation

The ionising radiation we are normally exposed to is **low-level radiation**. It is much too weak to produce any detectable effects. But does that mean it does us no harm? Doctors and other radiation experts now believe that any exposure to radiation – even at very low levels – involves some risk. It is estimated that about 1% of all cancers and genetic abnormalities are caused by low-level radiation. This is just an estimate, and it is impossible to be sure that any individual case is due to radiation effects. All we can do is to estimate the risks.

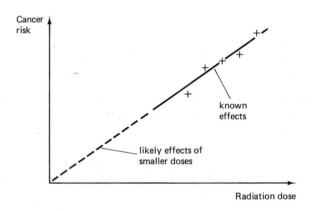

Surveys of people who have been exposed to larger amounts of ionising radiation (sometimes as a result of accidents) show that the larger the **dose** of radiation received, the greater the risk of developing cancer later. At low doses, the risk is very low and impossible to measure accurately. Even so, the safest estimate is that the graph is a straight line and that there is some small risk with even very low doses of radiation.

Radiation dose

The graph on the previous page introduces the idea of **radiation dose**. If we want to be able to avoid running too great a risk, we need a way of measuring the amount of radiation we are exposed to. Radiation dose is measured in **sieverts**.

> The basic unit of **absorbed dose of radiation** is the **gray**. It measures the energy transferred by the radiation to 1 kg of body tissue. 1 gray (Gy) is equivalent to 1 joule per kilogram. However, alpha particles do about 20 times more cell damage than the same absorbed dose of betas, gammas and X rays. **Radiation dose equivalent** in sievert (Sv) takes account of this by multiplying the absorbed dose by a 'quality factor' (Q):
>
> dose equivalent (in sievert) = absorbed dose (in gray) × Q
>
> (For alpha particles, $Q = 20$. For betas, gammas and X rays, $Q = 1$.)

So the sievert is the unit which really measures the damage done by the radiation to your body. The sievert is a very large unit. In one year, each person in the UK receives a dose equivalent of about 2 millisieverts on average.

The table on the right shows how this average dose is divided between the different sources of radiation. The amount from artificial sources seems relatively small, and the extra amount caused by discharges from nuclear power stations looks very small indeed. So it is important to remember that these figures are *averages* – some people, because of where they work or live, or just by chance, may receive much more than the average dose.

Average radiation dose per year for a person in the UK (in millisieverts) (courtesy CEGB, *Radiation: Its Origin and Effects*)

Naturally occurring:	
Gammas from Earth	0.4
Cosmic radiation	0.3
Food	0.37
Radon from building materials	0.8
Total natural	1.87
Man-made sources:	
Medical	0.25
Weapons fallout	0.01
Occupational	0.01
Miscellaneous	0.01
Discharges from nuclear power stations and disposal of radioactive waste	0.002
Total man-made	0.282

External and internal radiation sources

We are at much greater risk from radioactive sources *inside* our bodies than outside. An alpha emitter outside the body is hardly dangerous at all, since alphas are stopped by a few centimetres of air. But an alpha emitter in the form of dust, which can be breathed in, is very hazardous.

A radioactive isotope can sometimes become concentrated in the food we eat. For example, radioactive iodine-131 in rainwater reappears in milk once the cows have eaten the grass. Inside the human body, iodine concentrates in the thyroid gland in the neck. The radioactive iodine has got caught up in the **food chain**. This happened in May 1986, after the Chernobyl accident. Iodine-131 has a half-life of 8 days, so the risk becomes less and less as the weeks go by.

Radiation protection

Precautions are taken wherever radioactive materials and ionising radiation are involved. Radiographers giving X rays are shielded by screens. Other parts of the patient's body, especially the sex organs, are shielded using lead aprons. Workers handling radioactive materials use protective clothing and sometimes handle particularly dangerous materials using remote control equipment through a lead-glass screen.

All radiation workers are also checked regularly to make sure that the dose they receive is within safety limits. One method is to use a film badge. Every few weeks the film is developed; the amount of darkening of the film measures the dose received by the wearer. At present the limit for radiation workers in the UK is 50 millisieverts per year. Some scientists think this limit is much too high and, indeed, it is lower in some other countries. Most radiation workers in fact receive much smaller doses, around 1.5 millisieverts extra due to their work.

Radiation and you

So is radiation safe and are our safety limits for the public and for radiation workers good enough? There is no single answer to this. It is not something we could all agree on even if we had all the facts. Each of us must decide what is an acceptable dose for us.

USING RADIOACTIVITY

Radioactive isotopes have many practical uses, in medicine, agriculture and industry. Most of these use man-made artificial radioactive isotopes. If samples of non-radioactive materials are placed inside a nuclear reactor (see section 33.8), they will be bombarded with neutrons and this can result in new isotopes being formed. The different isotopes are then separated and purified by chemical methods.

Radioactive emissions carry energy

The energy of radioactive emissions can be used deliberately to kill or change living cells. It also causes heating; this can be used to generate small amounts of electricity (e.g. by using thermocouples—see section 14.3).

1. Radiotherapy is used to treat some types of cancer. The radiation kills the cancer cells; of course it will also damage any other cells it passes through on the way. During treatment, a strong beam of gamma rays from the isotope cobalt-60 is directed towards the tumour for a short time. By repeating this from different angles the radiotherapist can give a large radiation dose to the tumour without doing too much damage to other body tissue in between. The directions and doses are calculated by computer.

2. Medical products, such as dressings, syringes and needles must be sterilised before use. This can be done by pre-packing them in sealed bags and then passing the pack through a powerful beam of gamma radiation. Cobalt-60 is often used as the gamma source. The radiation passes through the packaging and kills any microorganisms on the object inside. Plastic objects which would be damaged by strong heating or by chemicals can be sterilised in this way – and it is usually cheaper.

3. Radiation can be used for food preservation. Some foodstuffs, like grain and dried beans, are often attacked by pests while in store. Radiation treatment kills these pests and reduces the losses. Potatoes treated with low doses of radiation can be stopped from sprouting.

These strawberries have been stored for 15 days at 4°C, and the irradiated ones are still fresh. Irradiation of food to increase its shelf life is approved in many countries, though not yet in Britain. The food itself does not become radioactive to any measurable extent, of course, though there are worries that it could be altered by radiation in other ways which might be harmful.

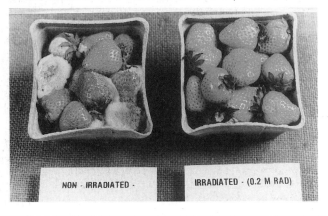

4. The so-called Sterile Insect Technique (SIT) is helping in many parts of the world to control insect pests which damage crops. Male insects are bred in the laboratory and then irradiated. This does not kill them but damages their sex cells, making them sterile – unable to produce offspring. These males are then released in great numbers in affected areas. They breed as usual with normal females, but no new generation of the insects is hatched. So the pest is controlled.

Radioactive isotopes can be detected in very small quantities

Modern detectors of ionising radiation give us more information than Geiger counters. They don't just count the numbers of alphas, betas or gammas but measure the energy of each one. This can tell us which isotope they have come from. Each isotope emits radiations of a characteristic energy – its 'fingerprint'. So very small amounts of the isotope can be detected.

5. Possibly the most important of all uses of radioactivity is the process called **radio-immunoassay**. It is used for medical tests on samples of a patient's blood. Radioactive isotopes are added to the sample after it has been taken. It is difficult to explain exactly how the tests work – but the method depends on the fact that very accurate tests can be done on small samples.

6. Because small amounts can be detected, we can follow the route taken by a radioactive isotope during a process. For example, the way the tides will wash the waste entering the sea from a pipe can be studied by adding a small amount of radioactive isotope to the water in the pipe. Samples can then be taken from different points in the sea over the following days and weeks to find where the material has spread to. The isotope is used as a **tracer**.

Tracers are also used in biology. For example, to study how much of the nitrogen in a fertiliser is taken up by plants, some fertiliser containing the radioactive isotope nitrogen-15 can be used. Tests will later show how much nitrogen-15 is present in the plants.

7. **Nuclear medicine** is the name given to the branch of medicine which uses radioactive isotopes as tracers. Many isotopes are used for different purposes. In studying lung function, a patient can be asked to breathe air containing the gas xenon-133 (xenon is an inert gas present in all air). The gamma rays emitted are studied using a **gamma camera** – an array of detectors linked by computer, which produces a picture of the patient's lungs from the gamma radiation each one receives. The patient soon breathes out all the xenon and receives only a tiny dose of radiation.

A gamma camera

Another very important isotope in medicine is technetium-99m (the m stands for 'metastable'). This is a special isotope because it is a gamma-only emitter and produces no harmful alphas or betas inside the body. The technetium is combined into samples of the protein albumin, and this is injected into the patient. Its movement round the body can be followed by a gamma camera, picking up the gammas emitted from the technetium-99m. For example, it can be used to study blood flow in the heart (shown below) and lungs. As technetium-99m has a half-life of just 6 days, its effects soon disappear. The hospital, however, must continually renew its supplies!

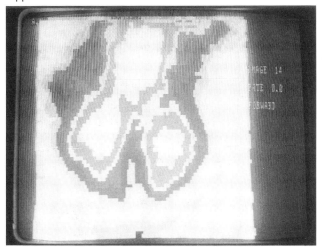

Radiation is absorbed by materials it passes through

The amount of radiation passing through a specimen gives us information about the composition and thickness of the material between the source and detector.

8. In a smoke detector isotopes measure the change in the composition of the air. The photograph below shows the inside of a detector. The black object on the right contains a chamber which has a p.d. across it. It also contains a tiny amount of the radioactive isotope americium-241, which emits beta radiation. The radiation ionises the air in the chamber, which produces a current. When smoke enters the chamber the smoke particles get in the way of the beta radiation, reducing the ionisation and the current across the chamber. This drop in current sets off the alarm.

9. The thickness of materials like paper and polythene, which are produced in continuous rolls, can be monitored using radioactive isotopes. Beta emitters are best for this (alphas are absorbed too easily and gammas not enough). In the process for making plastic sheeting, the beta particles from a source below the sheet are counted by a detector above it. When the plastic is coming out at the correct thickness, the detector will give a certain count rate. If the count rate becomes too low, the plastic must be too thick. If there are too many counts per minute, the plastic is too thin. The signal from the counter can be used to control the rollers automatically. The whole system is a feedback loop (section 32.7).

10. Gamma rays can be used to look for small flaws and cracks in aircraft engine parts, or in pipes. The result is a photograph showing up any defects. This method has the advantage that the source can be used in remote and inaccessible positions.

The rate of radioactive emissions from an isotope decays with time

If we know the half-life of an isotope, we may be able to calculate how old a specimen is by measuring its activity.

11. The method of **carbon dating** to measure the age of an archaeological specimen depends on the half-life of the rare isotope carbon-14. Ordinary carbon contains a very small proportion of carbon-14, produced when cosmic rays from space collide with nitrogen-14 in the atmosphere. Living plants take up the carbon-14 in the carbon dioxide they use for photosynthesis, as do animals when they eat the plants for food. While the plant or animal is alive, the proportion of carbon-14 to ordinary carbon-12 in their tissues stays constant; but once they die, the carbon-14 begins to decay – with a half-life of 5730 years!

To date an archaeological specimen, a small sample of carbon is extracted from it. Detecting and counting the radioactive emissions from the carbon-14 tells us how much of it is still present. As we know its half-life, this allows us to estimate the time since the specimen was alive. Carbon dating can be used on samples of wood, bone, textiles, paper and ivory. It works well for objects between 2000 and 20 000 years old.

Lise Meitner and Otto Hahn

33.7 Fission

In addition to alpha, beta and gamma emission, there is another way in which unstable nuclei may decay. There are a few isotopes, all of elements with very large atomic number, which split into two roughly equal halves, instead of just throwing out a small alpha or beta particle. This splitting is called **nuclear fission**. It was first discovered by two German physicists, Otto Hahn and Lise Meitner, in 1938. They found the element barium (atomic number 56) among the daughter products from a uranium 'parent' (the atomic number of uranium is 92). At first they were very puzzled by this; then it occurred to them that the uranium nucleus might have split in half! To test they looked for traces of the element krypton (atomic number 36). Can you see why they decided to search for krypton? They found it! The uranium nucleus had split into two parts.

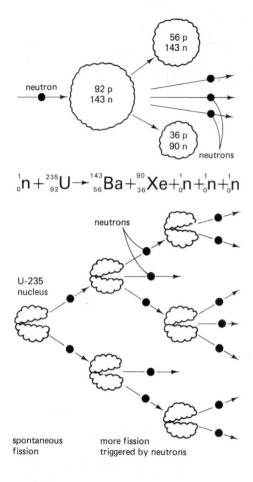

$${}^{1}_{0}n + {}^{235}_{92}U \rightarrow {}^{143}_{56}Ba + {}^{90}_{36}Xe + {}^{1}_{0}n + {}^{1}_{0}n + {}^{1}_{0}n$$

Triggering fission

For some nuclei, fission can be triggered by bombarding the nucleus with neutrons. The neutron is first absorbed; this makes the nucleus very unstable and it decays by fission. Two important isotopes which behave this way are uranium-235 and plutonium-239. Uranium-235 occurs naturally. About 1% of natural uranium is uranium-235; the remaining 99% is uranium-238. Plutonium-239 is an artificial isotope, produced when uranium-238 is bombarded with neutrons. If a neutron is absorbed, the isotope uranium-239 results. This quickly decays by beta emission to plutonium-239.

$${}^{238}_{92}U + {}^{1}_{0}n \rightarrow {}^{239}_{92}U \rightarrow {}^{239}_{93}Pu + {}^{0}_{-1}e$$

When fission occurs, the two parts of the nucleus fly apart at great speed, carrying large amounts of kinetic energy. They gradually lose this kinetic energy in collisions with other atoms, and the absorbing material heats up.

Chain reaction

There is something else special about the way that nuclei of uranium-235 and plutonium-239 split. In addition to the two halves of the old nucleus, a few stray neutrons are released.

The first fission may happen spontaneously, or be triggered by a neutron. The free neutrons produced can then trigger more nuclei to split – this produces more free neutrons which go off and split other uranium atoms – and so on. We have a **chain reaction**. The number of nuclei undergoing fission increases rapidly. Each fission releases energy; a very large amount of energy is released in a short time.

An atom bomb is made by allowing a nuclear chain reaction to get out of control. On the other hand, in a nuclear power station, the chain reaction is kept under control, so that the energy is released slowly and steadily. The power station is constructed in such a way that it is impossible for it ever to explode like an atom bomb.

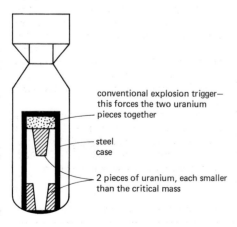

Critical mass

In a small lump of uranium-235, a chain reaction cannot occur because too many neutrons escape from the sides before they can cause further fission. The piece of uranium must be larger than a certain **critical mass**. The first atom bomb consisted of two hemispheres of uranium. On its own, each was too small for a chain reaction; together they were bigger than the critical mass. In the bomb, a small conventional detonator forced the two half spheres together, triggering the explosion.

33.8 Nuclear reactors

Nuclear reactors are designed to allow a nuclear chain reaction to take place in a steady and controlled way.

The advanced gas-cooled reactor (AGR) is widely used in Britain for generating electricity. The energy released by the fission of uranium heats high pressure carbon dioxide gas inside the reactor. This is used to boil water, which then drives turbines as in a conventional coal-fired power station (see section 34.4).

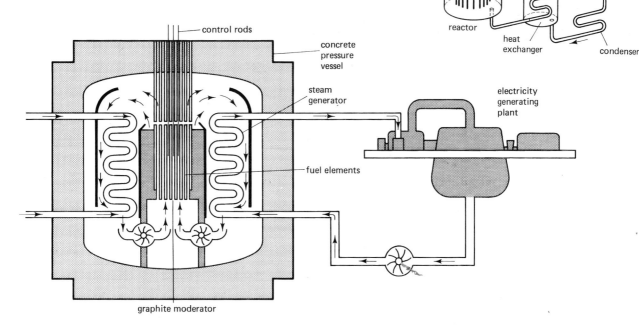

Fuel elements: The fuel for an AGR consists of pellets of uranium dioxide. The natural uranium is first **enriched**, to increase the amount of uranium-235 it contains. The pellets are about the size of a sugar cube. A number of pellets are put together into a sealed stainless steel can to form a fuel pin. A group of these pins forms a fuel assembly. The total amount of uranium present is more than the critical mass, but it is arranged so that the rate of the chain reaction can be kept under control. 1 kg of enriched uranium fuel produces as much electricity as 55 tonnes of coal. On the other hand, it takes several tonnes of uranium ore from a mine to produce 1 kg of enriched fuel!

Moderator: Nuclei of uranium-235 absorb slow neutrons best. The neutrons produced by fission are very fast neutrons. The chain reaction runs better if these fast neutrons can first be slowed down. This is the job of the **moderator** – blocks of graphite (carbon) in this case. Inside the moderator, fast neutrons collide with carbon nuclei and this gradually slows them down.

Control rods: In a reactor, we need to keep a chain reaction going, but it must never be allowed to get out of control. The rate of the chain reaction is controlled by raising or lowering the boron–steel **control rods**. Boron is good at absorbing neutrons.

Safety features: The entire reactor is contained within a thick steel pressure vessel and this is surrounded by a concrete outer pressure vessel. The concrete absorbs the radioactive emissions from the core of the reactor. If the reactor temperature begins to rise, the control rods are automatically lowered into the core, stopping the chain reaction. The carbon dioxide gas will become radioactive, so care must be taken to see that none can escape from the pressure vessel. This is particularly important when fuel assemblies are being changed. The water for the boiler is in a closed circuit. Even so, it may become slightly radioactive and cannot simply be released into the environment.

Where does the nuclear energy come from?

Nuclear energy is just potential energy stored within the nucleus. Nuclear reactions release much more energy than ordinary chemical reactions. This is because the potential energy stored in unstable nuclei is very large. In fact, it is so large that the mass of an unstable nucleus is greater than the sum of the masses of the separate protons and neutrons it is made from. The extra mass is the mass of the potential energy!

The idea that energy has mass seems a strange one. In 1905, Albert Einstein published his Special Theory of Relativity. One conclusion of the theory is that energy (E) and mass (m) are related by the famous equation:

$$E = mc^2$$

where c is the speed of light (3×10^8 m/s).

Nuclear fuel cycle

The nuclear fuel cycle is the sequence of stages involved in making the fuel for nuclear power stations and dealing with the waste that they produce.

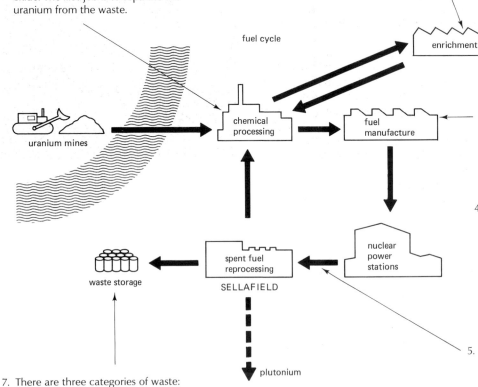

1. Uranium ore leaving the mine may contain less than 1% of uranium oxide. The first job is to separate the uranium from the waste.

2. Some of the uranium-238 is removed, leaving the remaining uranium slightly richer in the isotope uranium-235. The left-over uranium-238 can be used in so-called **fast reactors**.

3. Pellets of enriched uranium oxide are made into fuel assemblies.

4. Spent fuel contains a mixture of strongly radioactive isotopes from the fission process. It is stored under water for several months in storage ponds, to allow all the isotopes with short half-lives to decay. The fuel rods cool down as their activity drops.

5. Spent fuel is transported to a **reprocessing plant** in specially-constructed steel flasks. These have been thoroughly tested to make as sure as possible that in an accident no radioactivity from the spent fuel could escape.

6. At the reprocessing plant, the fuel is treated chemically to separate the usable plutonium and uranium from the unusable waste isotopes.

7. There are three categories of waste:

   Most waste is **low-level**. This includes disposable protective clothing, slightly radioactive water, and so on. Some is buried in waste disposal sites; the water is treated and some is released. Discharges are kept within agreed safety limits, though some people now believe that these are not strict enough for adequate public safety.

   Intermediate-level waste includes the steel cladding from fuel elements and other components which have come into close contact with the fuel. This is mixed with concrete and stored in steel drums.

   Only a little **high-level** waste is left but it is extremely dangerous. This is the strongly radioactive material left over after reprocessing. Some of the isotopes have half-lives of thousands, or even millions of years. There is no real solution at present to its disposal, though the idea of embedding it in glass (vitrification) before deep burial is being considered. The problem is that it must remain isolated from all water supplies and from living things in general for such a long period of time.

Power stations of the future?

Nuclear power stations of the sort we have now are *not* a long term solution to our electricity needs. Reserves of uranium are limited and will eventually be used up, just like any other fuel. There are two developments which might change the future of nuclear power:

1. **Fast reactors**

   The fast reactor is sometimes called a **breeder reactor**. Around its core of enriched uranium and plutonium fuel is a 'blanket' of uranium-238 – the isotope of uranium which does not undergo fission. Fast neutrons released by the fission of the fuel are absorbed by the uranium-238, changing it to plutonium-239. The reactor actually produces *more* fuel as it runs, turning the unusable uranium-238 into a valuable fuel.

   This looks a very useful process at first sight but there are many technical difficulties. Plutonium is a very dangerous substance. It is highly poisonous as well as being radioactive with a long half-life. It is also the material most widely used for making nuclear weapons. In short, it is one of the most dangerous substances known; and fast reactors use and make a lot of it! Also, the coolant used in fast reactors is molten sodium, a highly reactive substance which would explode if it came in contact with water. The technical problems are expensive to solve and so fast reactors may not be economical to build and operate. We also need to consider whether it is wise to produce such large amounts of plutonium, which has to be stored and protected from possible theft by terrorist groups.

2. **Nuclear fusion**

   When very large heavy nuclei split, energy is released. If two very light nuclei can be made to stick together, a lot of energy is also released.

   $$^2_1H + ^3_1H \rightarrow ^4_2He + ^1_0n$$

   We can predict that this process will release energy because measurements show that the mass of a helium nucleus plus a neutron is smaller than the mass of the hydrogen-2 (deuterium) and hydrogen-3 (tritium) nuclei we started with. The difference is the mass of the energy released. Using $E = mc^2$, we can calculate how much energy this will be. The process is called **fusion**.

   To make it happen, very high temperatures are needed – as much as 100 million K. Making it happen in a controlled way is very difficult and no one has succeeded to date. Fusion reactions take place in the Sun, releasing the vast amounts of energy which are continually radiated away from it. On Earth, fusion has only been used destructively – in the hydrogen bomb. This uses the enormous temperatures produced by an atom bomb (fission) to trigger a second fusion reaction. The total effect is the most destructive weapon ever produced.

   Controlling fusion involves containing hydrogen at very high temperatures and extracting the energy steadily if fusion does occur. Research into fusion is going ahead through projects such as JET – the Joint European Torus. A 'torus' is a doughnut-shaped ring, referring to the magnetic field shape used to try to hold the hot, ionised hydrogen in place. If fusion can be made to work, the rewards would be enormous. The fuel is hydrogen, which is present in water. Also the waste product, helium, is not radioactive and is quite inert.

   In April 1989 two physicists startled the scientific world with the surprise announcement that they had produced nuclear fusion at room temperature. They passed an electric current through 'heavy water' (water containing the isotope ^2_1H), using a platinum anode and a cathode of another metal palladium. Other physicists have tried to repeat this experiment and some have confirmed that larger amounts of energy than expected *are* produced. But the question remains whether or not the process really involves nuclear fusion, or if it could be developed as a large-scale energy source.

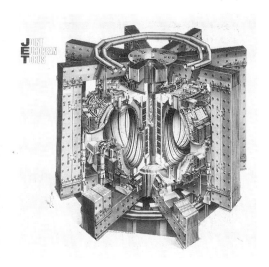

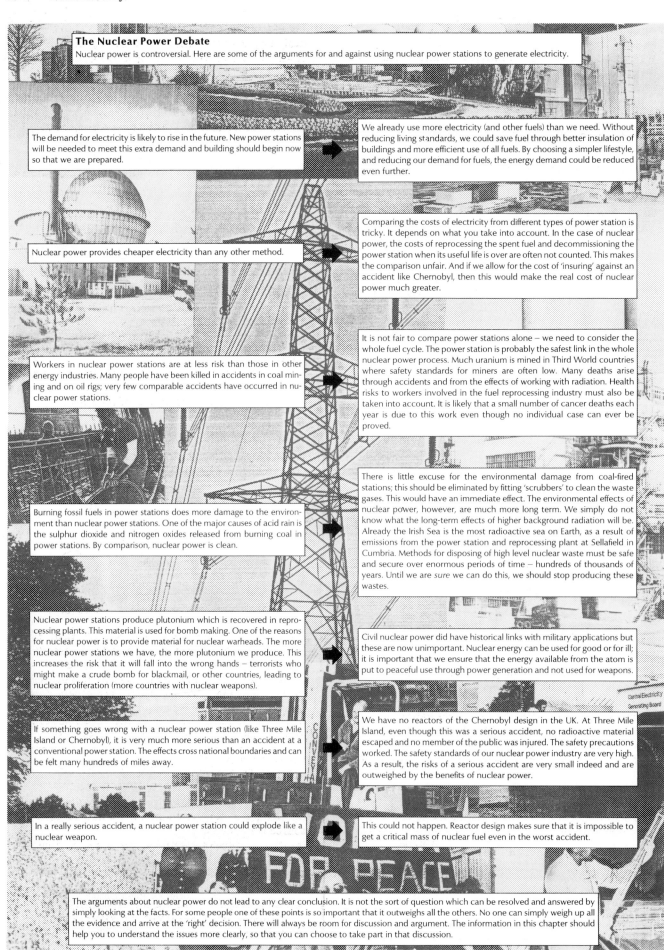

The Nuclear Power Debate

Nuclear power is controversial. Here are some of the arguments for and against using nuclear power stations to generate electricity.

The demand for electricity is likely to rise in the future. New power stations will be needed to meet this extra demand and building should begin now so that we are prepared.

→ We already use more electricity (and other fuels) than we need. Without reducing living standards, we could save fuel through better insulation of buildings and more efficient use of all fuels. By choosing a simpler lifestyle, and reducing our demand for fuels, the energy demand could be reduced even further.

Nuclear power provides cheaper electricity than any other method.

→ Comparing the costs of electricity from different types of power station is tricky. It depends on what you take into account. In the case of nuclear power, the costs of reprocessing the spent fuel and decommissioning the power station when its useful life is over are often not counted. This makes the comparison unfair. And if we allow for the cost of 'insuring' against an accident like Chernobyl, then this would make the real cost of nuclear power much greater.

Workers in nuclear power stations are at less risk than those in other energy industries. Many people have been killed in accidents in coal mining and on oil rigs; very few comparable accidents have occurred in nuclear power stations.

→ It is not fair to compare power stations alone — we need to consider the whole fuel cycle. The power station is probably the safest link in the whole nuclear power process. Much uranium is mined in Third World countries where safety standards for miners are often low. Many deaths arise through accidents and from the effects of working with radiation. Health risks to workers involved in the fuel reprocessing industry must also be taken into account. It is likely that a small number of cancer deaths each year is due to this work even though no individual case can ever be proved.

Burning fossil fuels in power stations does more damage to the environment than nuclear power stations. One of the major causes of acid rain is the sulphur dioxide and nitrogen oxides released from burning coal in power stations. By comparison, nuclear power is clean.

→ There is little excuse for the environmental damage from coal-fired stations; this should be eliminated by fitting 'scrubbers' to clean the waste gases. This would have an immediate effect. The environmental effects of nuclear power, however, are much more long term. We simply do not know what the long-term effects of higher background radiation will be. Already the Irish Sea is the most radioactive sea on Earth, as a result of emissions from the power station and reprocessing plant at Sellafield in Cumbria. Methods for disposing of high level nuclear waste must be safe and secure over enormous periods of time — hundreds of thousands of years. Until we are *sure* we can do this, we should stop producing these wastes.

Nuclear power stations produce plutonium which is recovered in reprocessing plants. This material is used for bomb making. One of the reasons for nuclear power is to provide material for nuclear warheads. The more nuclear power stations we have, the more plutonium we produce. This increases the risk that it will fall into the wrong hands — terrorists who might make a crude bomb for blackmail, or other countries, leading to nuclear proliferation (more countries with nuclear weapons).

→ Civil nuclear power did have historical links with military applications but these are now unimportant. Nuclear energy can be used for good or for ill; it is important that we ensure that the energy available from the atom is put to peaceful use through power generation and not used for weapons.

If something goes wrong with a nuclear power station (like Three Mile Island or Chernobyl), it is very much more serious than an accident at a conventional power station. The effects cross national boundaries and can be felt many hundreds of miles away.

→ We have no reactors of the Chernobyl design in the UK. At Three Mile Island, even though this was a serious accident, no radioactive material escaped and no member of the public was injured. The safety precautions worked. The safety standards of our nuclear power industry are very high. As a result, the risks of a serious accident are very small indeed and are outweighed by the benefits of nuclear power.

In a really serious accident, a nuclear power station could explode like a nuclear weapon.

→ This could not happen. Reactor design makes sure that it is impossible to get a critical mass of nuclear fuel even in the worst accident.

The arguments about nuclear power do not lead to any clear conclusion. It is not the sort of question which can be resolved and answered by simply looking at the facts. For some people one of these points is so important that it outweighs all the others. No one can simply weigh up all the evidence and arrive at the 'right' decision. There will always be room for discussion and argument. The information in this chapter should help you to understand the issues more clearly, so that you can choose to take part in that discussion.

Questions

1 When you hold a luminous watch near a Geiger counter, there is a noticeable increase in count-rate. Explain in detail how you would go about investigating which type (or types) of radiation or particles the radioactive material in the watch is emitting.

2 Workers in a Nuclear Power Station wear small badges. The badge is used to find out if the worker has received a dangerous dose of nuclear radiation.

The badge contains a photographic film in a light-proof packet. After being worn for several weeks the film is taken from the badge and developed. The parts of the film which have received radiation will then show up as dark areas.

The diagram below shows how the badge is made up.

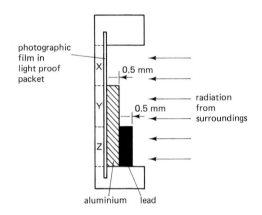

(a) Why would it be wrong to keep the badge in a pocket?
(b) How would the developed film show that the worker had received a large dose of alpha (α) radiation?
(c) How would the developed film show that the worker had not been exposed to gamma (γ) radiation?
(d) Why is the film in a light-proof packet? (Sp.NEA)

3 Copy and complete the following table:

	Mass	Charge
proton	1 unit	+1 unit
neutron		
electron		
alpha particle		
beta particle		
gamma ray		

4 Use the information on pages 342 and 343 to help you design a poster showing how scientists' models of the atom have changed over the past two hundred years.

5 The commonest isotope of uranium is $^{238}_{92}U$. What do the numbers 238 and 92 tell you? How many protons, neutrons and electrons are there in every atom of $^{238}_{92}U$?

6 Describe what happens in the following nuclear reactions.

$$^{239}_{94}Pu \rightarrow ^{235}_{92}U + ^{4}_{2}He$$

$$^{90}_{38}Sr \rightarrow ^{90}_{39}Y + ^{0}_{-1}e$$

7 The isotope of uranium, $^{238}_{92}U$, decays in a series of stages. Here is the first part of the decay series:

$$^{238}_{92}U \rightarrow ^{234}_{90}Th \rightarrow ^{234}_{91}Pa \rightarrow ^{234}_{92}U \rightarrow ^{230}_{90}Th \rightarrow ^{226}_{88}Ra$$
$$\quad (1) \quad\quad (2) \quad\quad (3) \quad\quad (4) \quad\quad (5)$$

(a) Which particles are emitted at each of the five stages?
(b) The information above does not give a complete picture of what happens at each stage. What else might be emitted during some (or all) of the stages?

8 A Geiger–Müller tube with an automatic counter is used by a pupil to examine a radioactive source which emits one type of radiation only.

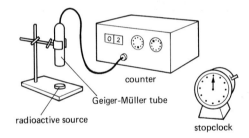

When the Geiger-Müller tube is more than 10 cm from the source, an average count rate of 5 counts per minute is obtained. The Geiger–Muller tube is now moved until it is 2 cm from the source and the results in the following table are taken over a period of 80 minutes. The source is then removed.

Time (minutes)	0	20	40	60	80
Count rate (number of counts in one minute)	101	65	43	29	21

(a) What type of radiation is emitted by the source?
(b) Suggest what the count rate would be after the source is removed.
(c) Plot a graph of count rate for the source against time, and use it to estimate the half-life of the source.
(d) During the experiment, the pupil reported that 'the counter might be sticking since it appeared to be counting irregularly during the experiment'.
Comment on the pupil's interpretation of the irregular counting. (SEB)

9 (a) The activity of radioactive material is measured in **becquerels**. What is 1 becquerel?
   (b) During the monitoring of foodstuffs after the Chernobyl accident, a sample of milk was found to have an activity of 1600 becquerels per litre.
   The activity of the milk sample was measured again at intervals of seven days:

Time from start/days	Activity/becquerels per litre
0	1600
7	875
14	470
21	260
28	140
35	77

   (i) Draw a graph to show how the activity varies with time.
   (ii) From your graph, estimate the half-life of the isotope causing the radioactivity of the milk sample.
   (c) The maximum permitted level of radioactivity in milk in the UK is 500 becquerels per litre. Explain why this milk would not be safe to drink as fresh milk, but could be used to make cheese or milk powder.

10 The following information appeared in a magazine article shortly after the Chernobyl nuclear power station accident:

   'The isotopes in the Chernobyl fallout which are now causing most concern are iodine-131 and caesium-137. Both are beta and gamma emitters. Iodine-131 in rainfall finds its way into milk. As it has a half-life of 8 days, the activity of the iodine-131 will have dropped to insignificant levels within a few weeks. Caesium-137, with a half-life of 30 years, may cause more health problems in the longer run.'

   (a) Explain clearly the meaning of the terms underlined in the passage.
   (b) What is the meaning of the numbers 131 and 137 in the names 'iodine-131' and 'caesium-137'?
   (c) Explain how iodine-131 in rainfall 'finds its way into milk'.

11 A nuclear power station supplies an average power of 150 MW to the National Grid. The power station's reactor consumes 0.3 kg of uranium fuel per hour.
   (a) (i) Calculate the energy supplied to the National Grid in 1 hour.
   (ii) If the energy extracted from 1 kg of uranium fuel in the nuclear reactor is 5.4×10^{12} J, calculate the efficiency of the power station.
   (b) (i) Part of the radioactive waste from the reactor has a half-life of 30 years. A drum of this waste material is encased in a concrete block before storage. The measured radiation level outside the concrete block is 512 units.
       If the maximum safe allowable level is 1 unit, how long must the concrete block be kept in secure storage before it is safe to approach it?
   (ii) How does the concrete casing help to make the radioactive waste safe during storage? (SEB)

12 Isotopes are used in industry and medicine to solve problems. Some of the isotopes commonly used are:

Isotope	Solid, liquid or gas	Radiation emitted	Half-life
strontium-90	solid	beta	28 years
cobalt-60	solid	gamma	5 years
iridium-192	solid	gamma	74 days
xenon-133	gas	gamma	5 days
water containing hydrogen-3	liquid	beta	12 years
polonium-210	solid	alpha	140 days

   Each of the following problems can be solved by using one of the radioactive isotopes from this list. For each problem:
   (a) explain briefly how the isotope would be used;
   (b) state which isotope would be the best one to use, and explain why.

   Problem 1: A doctor suspects that some air passages in a patient's lungs are blocked and wants to check this.

   Problem 2: A company makes rolls of paper. An automatic method is needed for checking that the thickness of the paper is constant.

   Problem 3: In a dry region of Africa, local water supply experts want to drill deep wells to obtain drinking water from underground reserves. It is important to know which streams feed these underground reservoirs, so that pollution can be avoided. They need to be able to check whether the water from each stream gets into the supply from the boreholes.

   Problem 4: It is important that the pipes which carry water from a power station's boilers are free from any small cracks or faults. The firm which makes the pipes needs to check this thoroughly before the pipes are installed.

   Problem 5: Disposable plastic syringes, needles and dressings must be completely sterile before being sealed into bubble packs. The plastic might be damaged by using boiling water to sterilise. Another method is required.

   Problem 6: To separate an ore from waste rock and sand, a mining company washes the ore and plans to get rid of the unwanted slurry by piping it several hundred metres out to sea. In order to decide where to site the pipe outlet, it is important to know exactly where the waste silt and sand will be carried by the tides after leaving the pipe. The local authority asks the company to provide this information before they will grant planning permission.

INVESTIGATION

Is the dust in your house radioactive? Collect a sample by fastening a gauze over the inlet tube of a vacuum cleaner and switching it on for a time. If it is radioactive, what kind(s) of particle does it emit?

34: Energy and Energy Supply

The first section of the first chapter of this book began with **energy**. We looked at events which happen around us from an energy point of view. We can think of every event as a transfer of energy from one place to another, and summarise it by an arrow diagram. We found two important energy laws: the law of conservation of energy (which says that the amount of energy is always the same after an event as before it); and the law of spreading of energy (which says that in every energy transfer energy spreads out from concentrated sources and ends up dispersed in many places).

In later chapters, we found out how to calculate amounts of energy in different situations, gravitational potential energy ($E_P = mgh$), kinetic energy ($E_k = \frac{1}{2}mv^2$), internal energy ($E = cm\Delta T$) and electrical energy ($E = IVt$). We also developed the important idea of efficiency – the fraction of the energy input which becomes useful energy output.

Now in this final chapter, we are going to use all these ideas to look at the way we use energy in our everyday lives, at home and at work. We will see how the ideas we have learnt about energy help us to understand the important issues of energy supply and energy use.

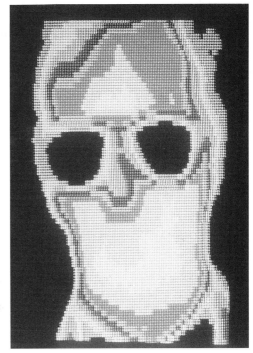

An infra-red photograph shows up energy losses by radiation.

34.1 Energy and you

A healthy adult eats enough food every day to provide about 10 000 kilojoules of energy. A growing teenager may need as much as 15 000 kilojoules. Some of our food (the protein) is used to make new body tissue and to replace old tissue – it provides the raw materials for making new cells. The carbohydrate foods (plus any spare protein) are the energy stores which we use to keep our bodies warm (at a steady temperature of 37°C) and for moving around and keeping active. Different foods contain very different amounts of energy.

Energy in various foods

Cup of tea with 1 sugar	200 kJ
Boiled egg	400 kJ
Apple	400 kJ
Slice of bread and butter	550 kJ
Bowl of porridge	650 kJ
100 g cheese	1500 kJ
Bag of chips (200 g)	2000 kJ
Steak and kidney pie with potatoes (1 helping)	3500 kJ

Note: these are all approximate and depend on the size of the helping.

Even when we are resting we are transferring about 100 J every second from our energy store. Most of it goes into keeping us warm. The biggest energy loss from a resting person is the energy carried away by infra-red radiation! (See the photograph above.) Other more 'energetic' activities involve transferring much more energy every second. ▷

Our energy intake has to be balanced against our energy needs. If we take in more than we transfer during our daily activities, the extra energy is stored as fat. Fat is a chemical energy store.

Activity	Energy transferred in one second J/s or W
Sleeping	70
Sitting reading or watching TV	100
Playing the piano, typing, working at a computer	130
Walking slowly	200
Running, digging the garden, playing a game	400–700
Swimming	500
Climbing the stairs	700

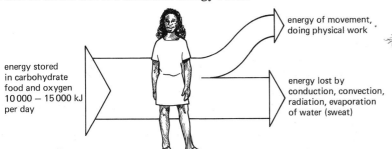

34.2 Energy at home

Take a look around your home. It is worth listing all the things you have which need a fuel – a concentrated energy source – to make them work. Until about 200 years ago, almost all jobs had to be done by hand. For very heavy tasks, like ploughing and pulling heavy loads, animals were used to help. In the industrial revolution of the 18th century, the invention of the steam engine meant that the energy stored in fuels like coal could be harnessed and used. This transformed the way we live and work. In the 1850s, the internal combustion engine (the petrol engine) was invented. Faraday's experiments led to electricity generation on a large scale. Nowadays our homes, offices and factories are full of machines and gadgets which transfer energy in various ways and do useful jobs for us.

We often talk of 'using' energy in the home, and we think of our heaters and boilers, lights and all the other electrical appliances as energy 'users' or 'consumers'. But, of course, energy is **conserved** in all processes and events. The various devices in the home are not really **consumers** of energy but **transformers** of energy. They need a fuel – a concentrated form of energy – to make them work and they use up the fuel as they do their job. In the process, the concentrated stored energy in the fuel is transferred into other places and other forms, and ends up widely spread and dispersed.

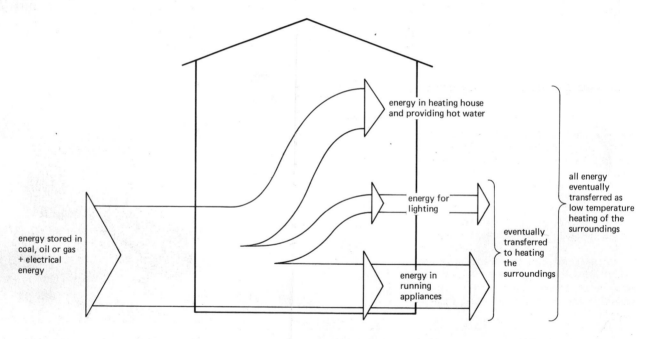

We can think of the whole house as an energy transferring unit, with energy entering in the form of fuels like coal, oil and electricity and being transferred into many other places and forms by the equipment in the house. Eventually all the energy will be in the form of low temperature heating, making the material of the house, the air inside it and the surroundings a little warmer.

Over the past 200 years, the amount of fuel which each of us uses has grown enormously. The fuel used each year by an average family living in a centrally-heated house today amounts to about 10^{11} J of stored energy. What is this all used for?

The pie-chart shows that by far the biggest share is used for keeping the house warm and for heating water. Cooking counts for about 5%, lighting for another 4% and all the other electrical appliances together add up to about 6%. Of course, these are just average figures.

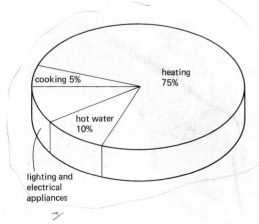

This means that the biggest savings could be made by cutting down on the amount of fuel used for heating. A reduction of 25% (one-quarter) in the heating bill means saving one-quarter of the biggest 'slice' in the pie-chart. Cutting the lighting bill by one-quarter would save much less fuel (and therefore much less money)!

There are many ways of reducing the amount of fuel needed for heating in most homes:
– insulating the cavity between the two layers of the walls with rockwool or some similar material;
– insulating the loft using fibreglass wool or insulating granules;
– installing good draughtproofing on all doors and windows;
– fitting double glazed windows;
– lagging the hot water tank with a thick insulating jacket;
– taking a shower (which uses much less water) rather than a bath;
– turning the room thermostats of the central heating system down a few degrees.

34.3 Energy supply to the home

Homes in the United Kingdom use four main fuels: natural gas, oil, solid fuel (mainly coal) and electricity. The top pie chart shows how much of each is used.

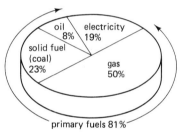

81% is in the form of **primary fuels**: coal, oil and gas. Electricity, which makes up the remaining 19%, is not a primary fuel because it has to be generated from another fuel in a power station. The lower pie chart shows the amounts of different fuels used in power stations in the UK. By far the largest share is taken by coal.

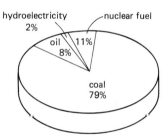

Electricity is a very useful fuel. Many of the appliances we have at home and in offices and factories can only work with electricity as their fuel. Things like washing machines, food mixers, televisions, computers, industrial machinery and lights can only run on electricity. Others, like room heaters, water heaters and fridges can run on other fuels if we choose. ▷

Electricity is also a convenient fuel because it is available where we want it. Electric circuits inside buildings are designed to provide sockets where we need them. We can easily add extra sockets if we want to. Electricity is also a 'clean' fuel when we use it, compared with solid fuel which leaves ash and produces waste gases. This impression is slightly misleading because if we also consider the production of electricity, we see that the whole process is not so 'clean'! Power stations produce ash, waste gases and waste heating water in large quantities. It is easy to forget about these when we use the 'clean' fuel, electricity, which reaches our homes.

These can only run using electricity as the fuel.

These use electricity but the job they do could be done using other fuels.

34.4 Generating electricity in power stations

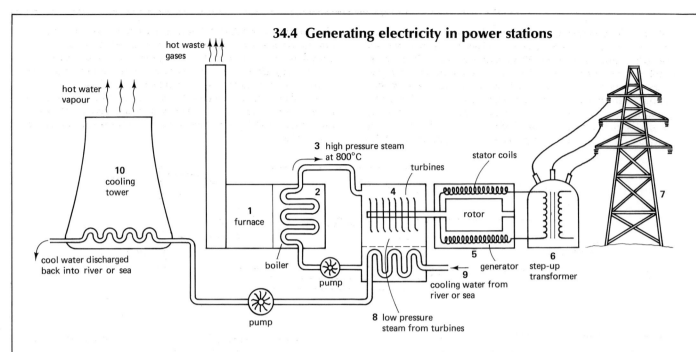

In a coal-fired power station, the primary fuel is burnt in large **furnaces** (1) to heat water in the **boilers** (2). The water boils to produce steam. By keeping the steam in a closed system of pipes so that the pressure can build up, the steam is heated to high temperatures, around 800 °C (3). This high pressure steam drives **turbines** (4) and the spinning turbine shaft turns the rotor coil of a **generator** (5). By means of gears and other controls, this is made to turn at a steady rate of 50 revolutions per second, generating alternating current with a frequency of 50 hertz. A **step-up transformer** (6) transforms this to very high voltage (225 kV) so that it can be transferred over long distances to the consumer through the **National Grid** (7).

The steam emerging from the turbines has to be cooled to reduce its pressure (8). This is essential because it is the *pressure difference* across the turbine which makes it turn. Large quantities of water from a nearby river or lake or from the sea are used (9) to cool the steam, which then recirculates back into the boilers. In the process, the cooling water gets heated up by the hot steam. It is often necessary to cool the water in **cooling towers** (10) before releasing it back into the river, lake or sea that it has come from.

A power station can be thought of as a large device for transferring energy from one place to another.

The overall efficiency of the whole process is about 35%. That means that only 35% (just a little over one-third) of the energy stored in the fuel is transformed into electrical energy. For every 100 units of stored energy in the coal, we get only 35 energy units of electrical energy. Where does the rest go? Most of it is wasted in heating things we don't really want to heat. About 15 units are wasted in heating the flue gases from the furnace and boiler. The hot gases escape up the power station chimney and heat the surrounding air. Even more – as much as 45 units – is wasted in heating up the cooling water. The other 5 units are wasted in heating up various parts of the turbine and generator as a result of friction and in electrical heating in the transformer and wires.

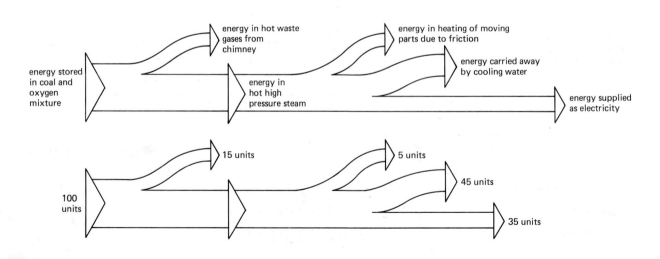

We can also think of the flow of matter through a power station. A large coal fired power station uses over 12 000 tonnes of coal every day. To burn this, we need over 30 000 tonnes of oxygen from the air! All this matter must come out of the power station again. Some goes up the chimney in the form of carbon dioxide and water vapour. There are also impurities in the coal which lead to gases like sulphur dioxide and various oxides of nitrogen in the waste gases. These are the gases which contribute to **acid rain**. They *could* be reduced by modifying the boilers and adding a **flue gas desulphurisation (FGD) plant** to the power station. This would add to the cost of building and running the power station and so would increase the cost of electricity by about 15%. We must decide whether we would prefer to pay a little more for electrical energy in order to have cleaner air and less damage to buildings and the environment from acid rain.

The other waste product is ash. As much as 3000 tonnes of ash have to be removed from a large power station daily. Much of it is used in road building and some is made into breeze blocks for the building industry.

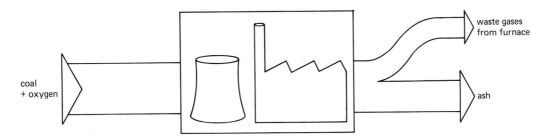

Similarities between power stations

An oil-fired power station is almost exactly the same as the coal-fired station described above – just using a different fuel. Many of the parts of a nuclear power station are also very similar to the coal-fired power station. Instead of using a furnace and boilers to heat the water and make steam, the energy comes from a controlled nuclear reaction (see section 33.8). The steam drives a turbine which turns a generator, just as in a coal or oil-fired power station.

In hydroelectric generating stations there is no need to heat steam at all. Water is collected in a high reservoir. As it falls, it is used directly to drive large water turbines, which then turn the generators.

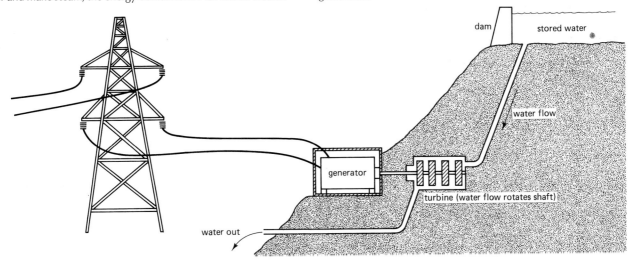

In all these power stations, the electricity is produced by turning the rotor coil of a generator. The differences lie in the methods used to make the rotor turn.

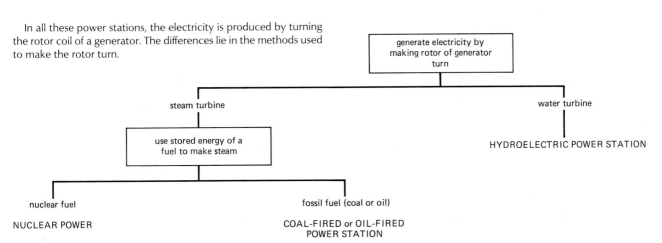

34.5 Why are thermal power stations so inefficient?

A **thermal power station** (one which heats steam to turn the turbines) is never better than about 37 to 38% efficient. Why is this figure so low? Can it not be improved by better design? The answer is *no*. There is a theoretical limit to the best efficiency we can get from such a power station. The problem lies in the second energy law: energy tends to spread out from concentrated stores and be dispersed.

Energy losses in the power station are a result of heating the waste gases and cooling water. We cannot avoid these losses. We simply cannot get *all* the energy which is released from the fuel back into a useful concentrated form – electricity.

When a fuel is used to heat something, the chemical energy stored in the fuel is transferred to internal energy in the hot object. The molecules of the hot object speed up and have a faster random motion. The problem is to convert this random motion back into an orderly motion – to transfer the internal energy back into kinetic energy of a moving body.

We *can* transfer some of the energy of a hot object in this way. For example, if the gas inside a cylinder is heated, it will push the piston up.

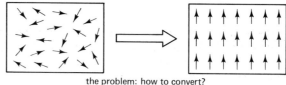

the problem: how to convert?

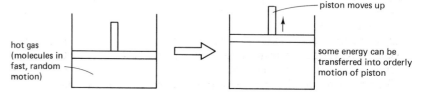

The gas cools slightly as it does this. Some of the energy in the random motion of the gas molecules is transferred into the orderly motion of the molecules in the piston. But if the cylinder is part of an engine, the piston must move back down again to repeat the cycle. This means either removing the hot gas and getting rid of it (which wastes the energy it still has left), or cooling it down again (which will heat the cooling water and waste energy in this way). It is impossible to extract *all* the energy available in the random motion of the gas molecules.

Theoretical calculations, which go well beyond the level of this book, show that the maximum possible efficiency of any machine which transfers random internal energy into orderly kinetic energy is:

$$\% \text{ efficiency} = \left(1 - \frac{T_2}{T_1}\right) \times 100$$

where T_1 is the temperature (in K) of the hot steam which supplies the energy, and T_2 is the final temperature (in K) of the steam afterwards.

So if a power station boiler supplies steam at 800 °C (1073 K) to the turbine, and the steam leaving the turbine is cooled to 400 °C (673 K), the maximum possible efficiency is:

$$\% \text{ efficiency} = \left(1 - \frac{673}{1073}\right) \times 100$$
$$= (1 - 0.63) \times 100\%$$
$$= 37\%$$

So although electricity is a very useful fuel when it reaches our homes and factories, the process of making electricity is not efficient. The fuels which we can use to make electricity are very valuable because they are the only stores of concentrated energy which we have. We need to look after them carefully!

34.6 Fuel supplies

Every fuel is an energy store. The table on the right shows how much energy is available in some of the fuels we use.

The total UK consumption in one year is over 8×10^{18} J. This is a very large amount of fuel indeed. The pie-charts show how much is provided by the different fuels.

There are other small contributions from burning wood and peat, solar energy and wind energy, but these are too small to be shown clearly on the diagrams.

All the nuclear fuel, all the hydroelectric 'fuel' and most of the coal is used to generate electricity. So the shares of the different fuels actually supplied to the consumer are different – electricity now plays a major part.

Notice that the lower pie chart is not the same as the one on p. 359. This is because the two diagrams show different things. One just shows the share of different fuels supplied to our *homes*; the other is for *all* uses – in homes, factories, offices, for transport, and so on.

Fuel reserves

The world faces a fuel crisis. 99.5% of all the primary fuels we use are things which will eventually run out – coal, oil, gas, uranium. Of the fuels we have discussed so far, only the hydroelectric supplies are constantly being renewed. Coal, oil and gas are **fossil fuels**, formed over the last 600 million years. Coal is the fossilised remains of huge forests which covered many parts of the world in prehistoric times. As the trees and plants died, they were covered in layers of silt and sand and gradually became formed into coal deposits.

In a similar way, oil comes from the remains of tiny creatures which lived in the prehistoric seas. Along with the oil deposits, gas frequently forms and collects in porous (spongy) rock.

The processes which form fossil fuels are extremely slow – slower than we can imagine. Yet we are using up our stocks of coal, oil and gas very quickly. The table below shows how much longer the known stocks of each fossil fuel will last if we go on using them at the same rate as we are doing now. If we make an optimistic guess about how much more fossil fuel may yet be discovered in new oil and gas fields and coal deposits, the lifetime is longer – but not much. Unless something changes, all the oil and gas will probably be finished within your lifetime; and coal stocks will last just a few hundred years more.

Fuel	Lifetime of known reserves at current rate of use	Lifetime of predicted reserves at current rate of use
Coal	250 years	600 years
Oil	30 years	400 years
Natural gas	17 years	200 years

Fuel	Energy stored
Natural gas	38 MJ/m³
Oil	44 GJ/tonne
Coal	27 GJ/tonne
Wood	14 GJ/tonne
Water (in a hydroelectric scheme)	1 MJ/tonne for every 100 m drop
Uranium	75 000 GJ/kg

1 MJ = 1 000 000 J
1 gigajoule (GJ) = 1 000 000 000 J (10^9 J)

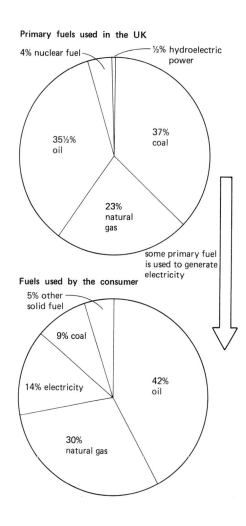

The future for nuclear power is not much better. There is a very limited amount of uranium ore for making fuel for the types of nuclear power station we have today. At best, it can last only another hundred years or so. With breeder reactors (see section 33.8), the fuel will last much longer, but with all the problems involved in handling the plutonium produced.

It is also important to remember that oil, gas and coal are not just useful as fuels. They are also valuable raw materials for the chemical industry – used for making plastics, synthetic fibres, detergents, drugs and many specialised chemicals. At some stage we may have to decide that they are simply too valuable to burn and must be kept for these other uses. ▷

34.7 Energy in the future

Fuel stocks are limited. So how can we plan for our future fuel needs? Two decisions we need to take quickly are to:

1. Develop alternatives to fossil fuels and nuclear fuels. In particular, we need to move towards greater use of renewable energy stores – ones which will never run out.
2. Use all our fuels and energy supplies more efficiently.

Let us look at these in more detail.

ALTERNATIVES TO FOSSIL FUELS: RENEWABLE ENERGY

Renewable energy sources are those which come to us continually from the Sun or the Earth and will last as long as the Solar system itself.

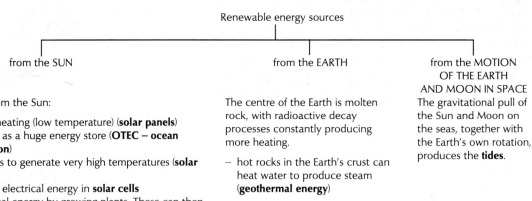

Radiation reaching Earth from the Sun:
- can be used directly for heating (low temperature) (**solar panels**)
- heats the seas, which act as a huge energy store (**OTEC – ocean thermal energy conversion**)
- can be focused by mirrors to generate very high temperatures (**solar furnace**)
- is transformed directly to electrical energy in **solar cells**
- is transferred into chemical energy by growing plants. These can then be used for fuel in different ways (**biomass**)
- drives the Earth's weather, producing **wind** and **waves** whose energy can be harnessed. Rain collecting in high lakes and reservoirs is used in **hydroelectric** power stations.

The centre of the Earth is molten rock, with radioactive decay processes constantly producing more heating.
- hot rocks in the Earth's crust can heat water to produce steam (**geothermal energy**)

The gravitational pull of the Sun and Moon on the seas, together with the Earth's own rotation, produces the **tides**.

1. Solar thermal: low temperature

Infra-red radiation reaching the Earth from the sun can be used directly for heating. A solar panel on a south-facing roof absorbs this radiation and heats water. The pipes carrying the water are painted black for good absorption and a glass cover reduces the energy losses (the 'greenhouse effect' – see section 22.4).

As much as 85% of the fuel used in homes in the UK is for heating, so it is worthwhile trying to make some savings here. Even in the UK's climate, it is estimated that 4 or 5 m² of solar panels on the roof could provide half of the energy needed to hot water. This cuts the bill for heating fuel by half. A disadvantage, of course, is that you get most hot water in the summer when you need it least.

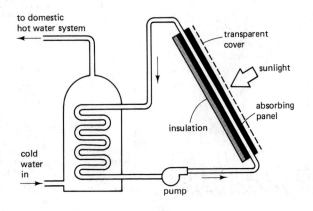

2. Ocean thermal energy conversion – OTEC

Solar radiation also heats the seas. Sea water acts like an enormous energy store. The water near the surface is warmer than the deeper water. This warmer water is used to boil a liquid with a low boiling point, such as ammonia or propane. As the vapour expands, it drives a specially designed turbine. It is then cooled again by colder water pumped up from the deep sea to make it condense, allowing more vapour to pass through the turbine. Experimental OTEC devices have shown that the idea works – at least in regions where the surface sea temperatures are high.

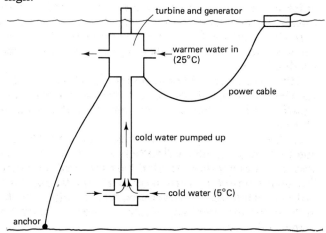

3. Solar thermal: high temperature

Concave mirrors can be used to focus the Sun's radiation on to a small area, producing very high temperatures there. The French solar furnace at Odeillo in the Pyrenees is the largest in the world, reaching temperatures over 3000 °C. Sunlight is reflected onto its large parabolic mirror by over 60 smaller mirrors, which are all controlled electronically to track the sun and reflect radiation precisely on to the furnace at the focus (see section 18.8).

4. Solar electric

Solar cells are made of semiconductor materials which develop an e.m.f. when exposed to sunlight. The cell transforms the energy carried by the radiation directly into electrical energy. The efficiency of conversion is quite low (about 5%) and the cells are expensive to produce, so the technology of solar cell production will have to improve greatly before they could be used on a large scale. For special applications where only small amounts of electrical energy are needed, solar cells are ideal as they provide a permanent energy supply.

Solar cells provide an ideal power supply for pocket calculators. This is possible only because the calculator's circuitry, including the liquid crystal display, needs very little electrical energy.

5. Biomass

The most efficient way of trapping the Sun's energy is by growing plants! Trees can be used for fuel. Straw can be produced even more quickly and burnt for heating. New plant matter is growing six times as fast as we are using up the fossil fuels!

Some countries have schemes to produce liquid fuels from biomass. In Brazil, sugar from sugar cane is fermented to make alcohol. Car engines can be easily modified to run on a mixture of petrol with 20% alcohol – or 'gasohol'.

Another method of using biomass is to generate methane gas in large **digesters** from agricultural wastes and dung – rather like a large controlled compost heap!

6. Wind energy

Windmills used to be used for grinding corn and pumping water. Now they are making a comeback, with new and different designs. A 3 MW wind turbine on the Orkney islands has two blades with a span of about 60 m and is part of the National Grid. The drawing on the right compares its size with an electricity pylon and a traditional windmill.

Other more novel designs are the Savonius rotor and the Darrieus rotor. These have a vertical axis and work equally well no matter which direction the wind is blowing.

It would take about 600 wind turbines like the Orkney one to generate as much electricity as a 900 MW power station (assuming that the wind blows for half the year). We might object to such large numbers of big wind turbines on hilltops. However, smaller machines could be very useful for providing electrical energy locally for homes and farms.

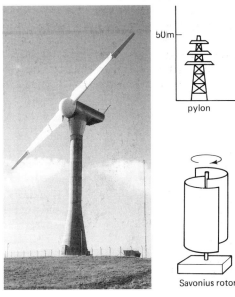
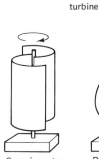

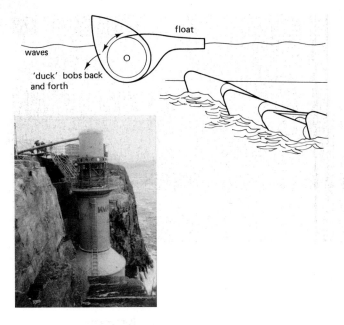

7. Wave energy

The waves around our shores carry an enormous amount of energy. If we could transform just a small fraction of this into electrical energy, it would make a great saving on our need for other fuels. The 'Salter duck', devised in the early 1970s by Stephen Salter at Edinburgh University, is a specially shaped float which rocks back and forth as the waves strike it. This rocking motion can be used to drive a turbine and generate electricity. The whole wave energy generator has a row of ducks lined up facing the oncoming waves.

◁ Another approach is to build a breakwater on the sea bed, specially shaped to trap a water column inside its hollow centre. As waves pass, the water column moves up and down, compressing and expanding the air above it. Again this can be used to drive turbines.

One disadvantage of wave generators is the very rough seas they would sometimes have to withstand. However, peak output would occur in the winter months when there is the greatest demand.

8. Geothermal energy

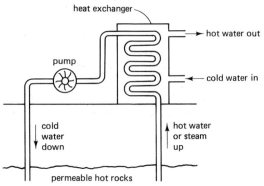

The Earth's core is molten rock. In some areas, hot rocks can be found quite near the surface. Here there is the possibility of drilling bore holes down to the hot rocks, passing cold water down and getting warm water or steam back up in return.

The hot water can be used directly for heating buildings in the district nearby, in a 'group' central heating system. It can also be used to run turbines and generate electricity, but this is not very efficient.

9. Tidal energy

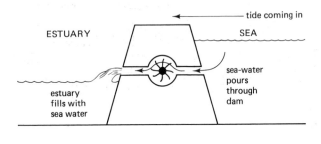

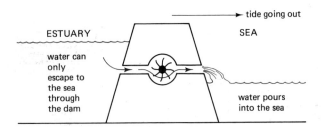

Britain is particularly well sited for making use of tidal power. The idea is to build a barrier across an estuary. Sluice gates in the barrier can be opened to let the incoming tide through. At high tide, this water is then trapped inside the barrier by closing the gates while the level outside falls. The water collected inside the barrier then flows out through water turbines built into the tidal barrier itself. This generates electrical energy. Another method is to make the *incoming* tide pass through the turbines or ◁ even to use *both* rising and ebbing tides.

Since 1966 a tidal power station has been operating at La Rance in Brittany. Some people think that the Bristol Channel is the best site for tidal energy in the world. It has been calculated that a tidal power station there could generate nearly one-tenth of Britain's entire electricity supply. It would, however, be expensive to build and would have a major effect on wildlife in the whole area involved.

Renewable sources are 'spread out' sources

Renewable sources seem an obvious and sensible alternative to fossil fuels and nuclear fuel. But there are snags. The renewable sources are all rather 'spread out' energy stores. They are not concentrated energy stores, like fuels. The energy carried by wind, waves and tides is dispersed and difficult to collect and use. Solar and geothermal sources produce low temperature heating but are not very suitable for generating electricity. Renewable energy sources are hard to exploit!

USING FUELS EFFICIENTLY

1. Better insulation

Heating accounts for a very large part of our total fuel consumption. The most obvious way to save on fuel for heating is by better insulation of buildings. The main methods have been listed in section 34.2 (see also 16.5 and 16.6).

Insulation cuts down the energy losses from the warm air inside the house to the colder air outside. So less fuel is needed to keep the inside temperature at a comfortable level. It has been calculated that a programme to insulate every house in the UK would cost less than building extra power stations to heat the poorly insulated homes which many of us have now!

2. Combined heat and power (CHP)

Generating electricity from fossil or nuclear fuel is less than 40% efficient. Most of the waste energy goes into the waste gases and cooling water from the power station. We cannot improve this efficiency figure because there is a theoretical limit to how good it can be. But we *can* use the energy in the waste gases and cooling water rather than just throw it away. The energy wasted in the cooling water from a large power station is enough to meet the heating needs of one million people.

A combined heat and power scheme uses the energy in the cooling water to provide piped central heating for houses near the power station. The efficiency of electricity generation is a little less, but the overall efficiency of the whole system (electricity *and* useful heating of houses) is much better. CHP schemes are widely used in European cities but there are only a few examples at present in the UK. Ideally the district heating scheme needs to be planned when the power station is being built. Future power stations may be designed to include some CHP.

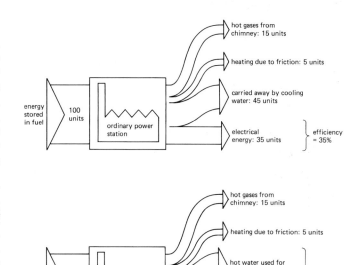

3. Think twice before using electricity for heating!

When fossil fuel is burnt in a power station, only 35–38% of the energy released is transformed into electrical energy. So it does not make good sense then to use this electrical energy for heating. We could have got about 3 times as much heating from the original fossil fuel! It makes good fuel sense to keep electrical energy for the jobs which can only be done by electricity – running machinery, computers, domestic appliances, televisions, radios and so on. It is more efficient to use fossil fuels directly for heating.

In the same way, it makes good energy sense to use renewable sources like solar radiation and geothermal energy directly for heating. Both are examples of low temperature, 'spread out' energy. If we can use them for heating purposes, we save on concentrated fuels.

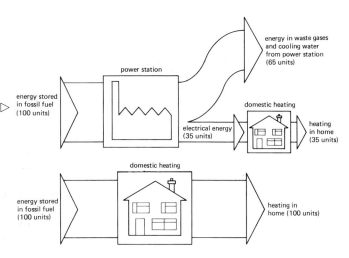

In 1980, the amount of fuel used, on average, by each person throughout the world was equivalent to about 2.5 tonnes of coal per year. This is how it was shared:

People living in North America used 14 tonnes per person.
People living in Western Europe used 4 tonnes per person.
People living in Eastern Europe used 3 tonnes per person.
People living in the Far East and Africa used 0.3 tonnes per person.

34.8 Energy and lifestyle

We have all got used to a way of living which uses a lot of fuel. In many other countries, people use very much less fuel on average. It is true that in hotter climates less fuel is needed for heating buildings, which is the largest item in our fuel bill in Britain. But is the way we share out the world's fuel really fair?

Perhaps part of the solution to the fuel supply problem is for people in countries like Britain to make changes in the way we live and the amount of fuel we need. Let's look at just one aspect of this, the way we use fuel for transport.

Energy and transport

All forms of transport use fuel. Aircraft use kerosene, trains and buses use diesel, cars and motorcycles normally use petrol. They can travel very different distances on one gallon:

- a passenger airliner travels about 0.3 miles on one gallon
- a train travels about 1 mile
- a bus travels 5 miles
- a car travels 40 miles
- a small motorcycle travels 100 miles

The energy stored in one gallon of all these fuels is about the same. But the comparison is not a fair one because a motorcycle will usually carry just one passenger whilst a train may carry 400 passengers. For a fair comparison, we can use the unit of **passenger-miles** – the number of passengers multiplied by the number of miles travelled for each gallon. This is how the comparison now comes out:

These vehicles are carrying ...

69 people who could all ...

be on this one bus.

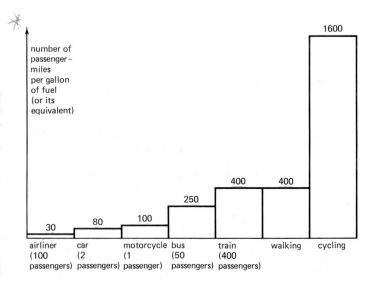

We have added two extra forms of transport: the bicycle and walking! The fuel for these is food. The figures in the diagram are based on the amount of carbohydrate food which stores the same amount of energy as one gallon of petrol. In terms of fuel efficiency, the bicycle is one of the most successful inventions ever!

The choices we make about transport are just one example of how the way we decide to do things affects the amount of fuel we need.

Questions

1. The table at the bottom of page 357 shows the energy transferred every second by a person in different situations. Present the same information in the form of a bar graph.

2. Design a questionnaire that you could use with the pupils in your class to find out what fuels are used to heat peoples' homes and what forms of insulation they use.

3. (a) Explain what is meant by the term **renewable energy source**.
   (b) A magazine article about energy supply contains the following passage:

   'One of the problems with the renewable energy sources is that their energy is very "spread out". So it is very difficult to use them to produce a "concentrated" form of energy like electricity.'

   Explain what you think this means and give some examples to show how several of the renewable sources are 'spread out'.

4. The angle of tilt of a solar panel greatly affects the amount of energy it receives at different times of the year. The diagram on the right shows what is meant by the angle of tilt. The table of data below shows the effect of different angles of tilt for the summer months.

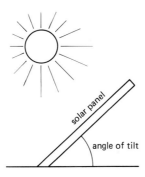

Maximum daily input of energy in megajoules to a 1 m² panel

Month	Angle of tilt of panel to the horizontal					
	20°	30°	40°	50°	60°	70°
Apr	23.8	24.9	24.8	24.1	22.7	20.5
May	28.4	28.8	27.4	25.2	23.0	19.8
Jun	29.2	29.2	27.4	25.2	22.3	19.1
Jul	28.8	29.2	27.4	25.6	23.0	20.2
Aug	25.6	25.9	26.3	24.8	22.7	20.5
Sept	20.5	21.6	22.3	22.7	21.6	20.5

Use the table above to answer the following questions.
(a) What angle of tilt would be ideal for a solar panel in April?
(b) Is it better to have the solar panel tilted at an angle of 40° or at an angle of 50° for all the months shown in the table? Give reasons for your answer.
(c) What is the maximum amount of energy that a 4 m² panel could receive during a day in July?

(Sp.LEAG)

5. (a) The following table shows five ways that are available for producing electricity, and their costs. 1 unit is 1 kilowatt-hour.

Oil	1.09p	per unit
Coal	0.97p	per unit
Nuclear	0.67p	per unit
Hydro	0.18p	per unit
Wind	1.4p–3.2p	per unit

(i) Explain why oil and coal are known as fossil fuels.
(ii) What was the original source of energy stored in fossil fuels?
(iii) Which of the above methods of generating electricity is the cheapest?
(iv) Give **one** reason why the method given in (iii) is cheap compared to other methods listed.
(v) Suggest **one** reason why it costs more to get electricity from wind power.

(b) The graph shows actual and estimated figures for the production of coal and oil for the past, present and future.

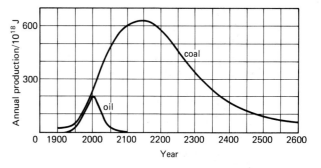

(i) Using the graph above, estimate the date for **maximum** oil production; and estimate the date for **maximum** coal production.
(ii) Why will the production of oil and coal reach a maximum and then decline?
(iii) Suggest **two** reasons why the estimated production of oil and of coal for the future are almost certain to be inaccurate.

(c) It is expected that people's energy needs will soon be so great that energy production will have to be increased to meet demand. It has been suggested that one answer to this problem could be a greater use of nuclear energy.
  (i) State **one** advantage and **one** disadvantage of using nuclear energy.

  Another answer to the problem might be to develop alternative energy sources such as wind energy.
  (ii) State **two** alternative energy sources not already mentioned.

(Sp.LEAG)

6. Use the information about renewable energy sources on pages 364 to 366 to draw up a table, showing the advantages and disadvantages of using each source in the UK.

7 Domestic hot water can be provided by using solar panels. These are metal boxes placed on the roof facing the sun. Water is circulated through the boxes by a pump. The boxes absorb some of the sun's radiation and heat the water.

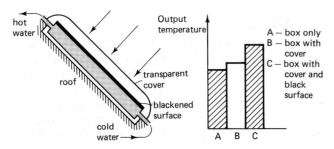

(a) The histograms show the effect on the output temperature of adding a cover over the box and of blackening the upper surface of the box.
 (i) Suggest why the output temperature is changed by each of these alterations.
 (ii) What happens to radiation which is not absorbed?
(b) In UK in the middle of a summer day, the power of sunlight falling on a 1 m² solar panel is about 900 W. About 50% of the energy falling on the panel goes to heat the water. A householder wants to install a panel which will have a useful power of 2 kW.
 (i) How big should the panel be?
 (ii) One kilowatt hour of electrical energy costs about 5p. Estimate the cost saving in water heating which might be made in a summer month. (Make clear any assumptions you make in your calculation.)
 (iii) The householder borrows money to pay for the panel. His repayments of capital and interest come to £200 a year. Discuss whether or not it is worth his while to install the panels.
(Sp.O&C, MEG)

8 A remote farmhouse uses a wind-generator to produce electricity. This is stored by charging lead-acid batteries, so that electricity is available even when no wind is blowing. Draw an energy arrow diagram to show all the energy transfers involved when the farmer uses one of the batteries to run a food mixer. Include in the diagram all the places where useful energy is lost.

9 A hydroelectric power station takes its water from a reservoir whose water level is 100 m above the turbines.
 (a) How much gravitational potential energy is lost by 1 kg of water falling from the level of the reservoir down to the power station?
 (b) If the station is 40% efficient, how much electrical energy is produced from the energy carried by this 1 kg of water?
 (c) What mass of water must pass through the turbines every second to produce a power output of 1 MW?

10 (a) A large coal-fired power station uses the energy stored in coal to produce electricity. The table shows the total energy input and output (in megajoules per second):

Input energy	Output energy	
From burning coal 5400	Electrical energy to Grid	2000
	Heating the cooling water	2740
	Hot gases up the chimney	460
	Electrical energy used in running the power station	110
	Total	5400

 (i) Draw a pie-chart to show how the total output energy is made up.
 (ii) Which is the largest single share of the output energy?
 (iii) What is the percentage efficiency of the power station?
(b) A combined heat and power (CHP) station makes use of the hot cooling water. It is pumped through large pipes to heat homes and factories near the power station. This table shows the total energy input and output (again in megajoules per second) of a CHP scheme:

Input energy	Output energy	
From burning coal 400	Electrical energy to Grid	120
	District heating using the cooling water	210
	Hot gases up the chimney	33
	Heating losses in the generators	7
	Electrical energy used in running the power station	30
	Total	400

 (i) What is the percentage efficiency of this power station if we count **only** the electricity as useful output?
 (ii) What is the percentage efficiency if we count **both** the electrical energy **and** the district heating as useful output?
 (iii) The CHP station has a much lower total energy output than the conventional one. Why is this usually the case?

INVESTIGATION

What is the efficiency of a small generator in transferring mechanical work into electrical energy? You might use a bicycle dynamo, or a small motor working 'backwards', as a generator.

Appendix 1: Mathematical Skills

This is a summary of the mathematical skills which are needed in some parts of the book. You should turn to this section at any time when you are unsure about the mathematics involved.

EQUATIONS

1. Many physics results (or laws) can be summarised as simple equations of the form $x = a/b$. If we know a and b, we can calculate x. Sometimes we must rearrange the equation, to work out the answer to a calculation.

   When we know x and b, and want to calculate a:

   Multiply both sides by b: $\quad x \times b = \dfrac{a}{b} \times b \quad x \times b = a$

   When we know x and a, and want to calculate b:

   Multiply both sides by b: $\quad x \times b = a$

   Divide both sides by x: $\quad \dfrac{\cancel{x} \times b}{\cancel{x}} = \dfrac{a}{x} \quad b = \dfrac{a}{x}$

   Examples of equations of this type are:
   average speed = distance/time ($v = s/t$)
   acceleration = change of speed/time ($a = \Delta v/t$)
   density = mass/volume ($\rho = m/V$)
   pressure = force/area ($p = F/A$)
   resistance = p.d./current ($R = V/I$)

2. Some equations are usually remembered in the form: $x = ab$. These are even easier to rearrange:

   When we know x and b, and want to calculate a:

   Divide both sides by b: $\quad \dfrac{x}{b} = \dfrac{a\cancel{b}}{\cancel{b}} = a$

   When we know x and a, and want to calculate b:

   Divide both sides by a: $\quad \dfrac{x}{a} = \dfrac{\cancel{a}b}{\cancel{a}} = b$

   Equations of this type include:
   $F = ma$; $V = IR$; $P = IV$; $v = f\lambda$.

3. Some equations have more than three terms, or include a squared term. Examples are: $E = cm\Delta T$; $E_p = mgh$; $E_k = \frac{1}{2}mv^2$
   These are rearranged in much the same way as before.
   For example:

   (i) Calculating s.h.c. (c):
   Divide both sides by $m\Delta T$: $\quad c = \dfrac{E}{m\Delta T}$

   (ii) Calculating h (from the potential energy formula):
   Divide both sides by mg: $\quad h = \dfrac{E_P}{mg}$

   (iii) Calculating v from the kinetic energy equation:
   First multiply both sides by 2: $\quad 2 \times E_k = mv^2$
   Then divide both sides by m: $\quad v^2 = \dfrac{2E_k}{m}$
   To find v, take the square root of both sides.

4. A few common equations involve addition or subtraction. They can also be rearranged. An example is $v = u + at$. If we want to find a:

   Subtract u from both sides: $\quad v - u = (\cancel{u} + at) - \cancel{u} = at$

   Divide both sides by t: $\quad \dfrac{v-u}{t} = \dfrac{a\cancel{t}}{\cancel{t}} = a$

PROPORTION

In physics we often want to find the relationship between two things we are measuring (two **variables**). How does one change as the other changes? For example: how does the acceleration of an object change as the force changes? How does the volume of a sample of gas change as the pressure changes? How does the current through a resistor change as the p.d. across it changes? And so on.

Direct proportion

The commonest relationship between two variables is **direct proportion**. In the table below, y is proportional to x. When x doubles, y doubles too; when x increases to three times its original value, so does y; and so on.

x	1	2	3	4	5
y	5	10	15	20	25

We say that y is **directly proportional** to x. This is written as:
$$y \propto x$$

It also follows that the **ratio** y/x is the same number for each pair of values in the table of results. The ratio has a **constant** value. In the table above:
$$\dfrac{y}{x} = \text{constant} = 5$$

The constant is called the **constant of proportionality** and is often given the symbol k. In general, if x and y are directly proportional, then:
$$\dfrac{y}{x} = k \quad \text{or} \quad y = kx$$

Inverse proportion

When two variables are in direct proportion, they both increase and decrease together. Sometimes, however, one variable **decreases** as the other **increases**. An example is pressure and volume of a gas sample – as pressure goes up, the volume goes down.

The values of p and V in the table below are in **inverse proportion**. When p doubles, V halves; when p increases to three times its original value, V is one-third what it was originally; and so on.

p	1	2	3	4	5
V	60	30	20	15	12

We say that V is **inversely proportional** to p. This can be written as:
$$V \propto \dfrac{1}{p}$$

It also follows that the **product** $p \times V$ is the same number for each pair of values in the table of results ($= 60$).
In general, if p and V are inversely proportional, then:
$$pV = \text{constant} = k$$

GRAPHS

A graph is a convenient way to summarise the results of an experiment. It shows how one variable changes when we alter another. As an example, think about the relationship between the **load** on a spring and its **extension**.

Plotting a graph

Step 1: Which way round? Usually we plot the variable we are able to control on the horizontal axis (the *x* axis). In this case, we would have *load* across the bottom and *extension* up the side.

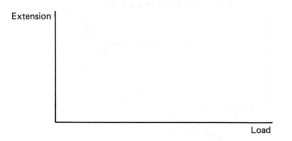

Step 2: Labelling axes. Unless there is a special reason not to, begin each axis from zero. Look at the results you have to plot and see what the largest number will be on each axis. Choose a scale that lets you fit this in. Don't choose a scale which will cramp the graph in one corner of the page. Remember to write the name of the variables (*load* and *extension*) on the axes and also the units they are measured in.

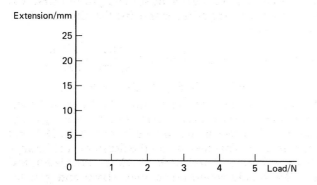

Step 3: Plot the points. For each *load* value, go across the *x*-axis to the correct point. Then go up to the corresponding *extension* value. Plot a point here. Do this for every result.

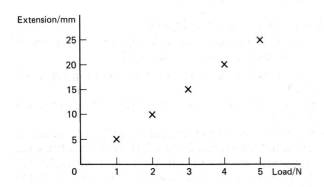

Step 4: Making the graph. In almost every experiment, the graph of the results will show a smooth relationship between the variables – either a straight line or a smooth curve. **Never join the dots!** Always decide whether it looks like a straight line or a curve. Then either *rule* the best line, or draw a smooth curve free-hand. It might not go exactly through the points (because of experimental errors), but it should go close.

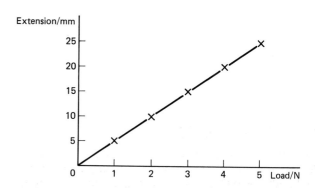

Some experimental results which lead to a curved graph. The curve goes close to all the points.

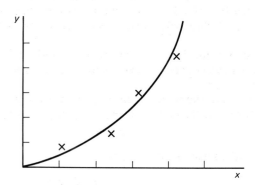

Straight-line graphs

If two variables are in direct proportion, their graph will be a straight line through the **origin** (the zero point of both scales). When we do an experiment, getting a straight line graph shows that there *is* direct proportionality between the two variables, i.e. $y \propto x$.

To find the constant of proportionality, we have to measure the **slope** or **gradient** of the graph.

Slope of a graph

The slope of a graph tells you *how quickly* one variable changes as the other changes. To measure the slope of a straight-line graph, pick any two points on the graph. The slope is the change along the *y*-axis divided by the change along the *x*-axis.

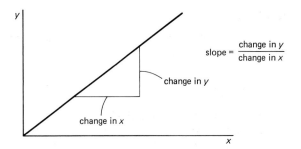

If the graph is a curved line, its slope changes from place to place. Its slope at any point is the same as the slope of a straight line drawn to touch the curve there.

The distance–time graph of a body with steady acceleration is a curve. Its slope increases as the body speeds up.

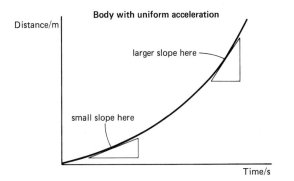

POWERS OF TEN NOTATION

Some of the numbers we meet as we study the natural world are very large or very small. For example, the speed of light is 300 000 000 m/s. The wavelength of yellow light is about 0.000 000 6 m. The mass of an electron is 0.000... (30 zeros altogether)...091 kg! Very big and small numbers like these are awkward to deal with, and so we use powers of ten to make it easier:

speed of light $= 3 \times 10^8$ m/s
wavelength of yellow light $= 6 \times 10^{-7}$ m
mass of the electron $= 9.1 \times 10^{-31}$ kg

The value is written as a number between 1 and 10, multiplied by a certain power of 10. This is called **standard form**. $\times 10^8$ means multiplied by $10 \times 10 \times 10 \times 10 \times 10 \times 10 \times 10 \times 10$ (multiplied by 10 eight times). Similarly, $\times 10^{-7}$ means divided by 10 seven times.

When we multiply two numbers in standard form, we **add** the powers of ten. When we divide, we **subtract** the powers of ten.

Here is an example. Using $v = f\lambda$ and the values given above, we can calculate the frequency of yellow light:

$$v = f\lambda$$

So, dividing across by λ: $\quad \dfrac{v}{\lambda} = f$

Substituting: $\quad f = \dfrac{3 \times 10^8 \text{ m/s}}{6 \times 10^{-7} \text{ m}}$

$$= \dfrac{3}{6} \times \dfrac{10^8}{10^{-7}} \text{ Hz}$$

$$= 0.5 \times 10^{15} \text{ Hz}$$

(**Note.** Subtract the powers: $8 - (-7) = 15$)

$$= 5 \times 10^{14} \text{ Hz}$$

SIGNIFICANT FIGURES

Calculators work out answers to six or seven figures. Often, however, it is silly to give all these figures in your answer. To see why, think about this example. You are measuring the density of a steel bolt. By weighing it, you find its mass is 150 g. To find its volume, you immerse it in water in a measuring cylinder and find that the water level rises by 19 cm³.

Calculate the density: $\quad \text{density} = \dfrac{\text{mass}}{\text{volume}} = \dfrac{150 \text{ g}}{19 \text{ cm}^3}$

$$= 7.8947368 \text{ g/cm}^3 \text{ (by calculator)}$$

But this is much too accurate. After all, the value 19 cm³ for volume is just a measurement. If we are reading to the nearest cm³, it means the volume is more than 18.5 and less than 19.5. That means that the density could be anywhere between 8.11 and 7.69! (check the calculation and see). We should round our answer above and give far fewer figures. In fact, an answer of 8 g/cm³ is the best we can do. A good rule of thumb is to give no more **significant figures** than you had in the numbers you started with. 150 and 19 both have 2 significant figures; so the final result cannot have more than 2 significant figures: 7.9 g/cm³ in this case.

How to count significant figures

Usually this just means count the digits in the number. If there are any zeros, special rules apply:

- if the zero is at the start, you never count it, e.g. 0.056 has 2 s.f.
- if the zero is in the middle, you always count it, e.g. 207 has 3 s.f.
- if the zero is at the end, you can count it if you want, e.g. 250 could have either 2 or 3 s.f. If we claim it has 2 s.f., this means we know the value is between 245 and 255; if we say it has 3 s.f., this means it is between 249.5 and 250.5.

Appendix 2: Microcomputer Methods in Mechanics

In Chapter 5, microcomputer methods for timing and for measuring speed and acceleration were described. These are easy to implement on the BBC microcomputer. The BBC microcomputer has a **user port** on the underside, which can receive 'messages' from the outside world or send messages out from the computer. The 'language' it uses for this communication is a pattern of HIGH (5 V) or LOW (0 V) voltages on eight parallel lines. To use it for timing, we arrange for a sensor to switch one of these lines from HIGH to LOW (and vice versa). The computer's internal clock is switched ON and OFF when these signals arrive.

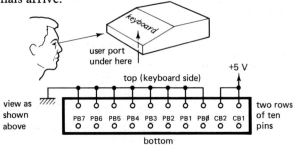

The diagram below shows two suitable sensor circuits. When a light beam falling on the sensor is interrupted, the sensor's output switches (OFF→ON or vice versa). This starts and stops the timing.

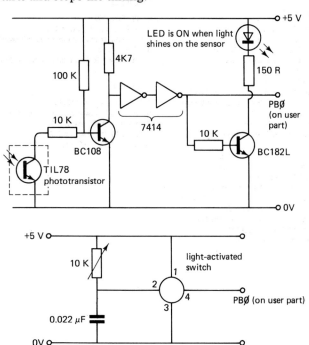

Simple programs to measure speed and acceleration are listed below. In both cases, the user has to tell the computer the length of the card which will cut through the light beam. These simple programs use the BBC micro's BASIC centisecond timer. This is not quite fast enough to measure the acceleration due to gravity, but works satisfactorily for trolleys on sloping ramps where the acceleration is smaller. A versatile suite of more powerful, but very user-friendly, programmes is the *Leicester Physics Interfacing Programes* package (available from the School of Education, University of Leicester, 21 University Road, Leicester LE1 7RF).

VELA can be used to do the same sort of measurement, provided it has been fitted with the PHYSICS EPROM (information from ISL, 7 Gledhow Wood Avenue, Leeds, LS8 1NY). The same sensor circuits will work. VELA stores all the times and (assuming a standard card length) will calculate speeds and accelerations.

The advantage of microcomputer methods is that accelerations can be measured quickly, and so new investigations become possible. For example, we could measure the acceleration of an air track glider with different shaped cards mounted transversely on it like sails, to investigate the effect of air resistance; or we could see how the acceleration of a trolley on a sloping board becomes less, due to air resistance, as it speeds up.

```
10 REM *** VELOCITY CALCULATION ***
20:
30 MODE 7:PROCINIT
40:
50 PROCTIME
60 PROCCALC
70 END
80:
90 DEFPROCINIT
100 PRINTTAB(0,10);"How long is the card (in metres) ";:INPUT L
110 PRINTTAB(0,24);"PRESS ANY KEY TO START";
111 A=GET:CLS:ENDPROC
120 DEFPROCTIME
130 FLAG=FALSE
140 REPEAT:UNTIL?&FE60 MOD 2 =0:TIME=0
150 REPEAT:UNTIL?&FE60 MOD 2 =1:T=TIME
180 ENDPROC
190:
200 DEFPROCCALC
210 SEC=T/100:VEL=L/SEC
230 PRINT"VELOCITY = ";VEL;" m/s"
240 ENDPROC
```

```
10 REM *** ACCELERATION CALCULATION ***
20:
30 MODE 7:PROCINIT
40:
50 PROCTIME
60 PROCCALC
70 END
80:
90 DEFPROCINIT
100 PRINTTAB(0,10);"How long is L1 (in metres) ";:INPUT L
110 PRINTTAB(0,11);"How long is L2 (in metres) ";:INPUT L2
120 PRINTTAB(0,24);"PRESS ANY KEY TO START";
130 A=GET:CLS:ENDPROC
140 DEFPROCTIME
150 REPEAT:UNTIL?&FE60 MOD 2 =0:TIME=0
160 REPEAT:UNTIL?&FE60 MOD 2 =1:T1=TIME:TIME=0
170 REPEAT:UNTIL?&FE60 MOD 2 =0:T2=TIME:TIME=0
180 REPEAT:UNTIL?&FE60 MOD 2 =1:T3=TIME
190 ENDPROC
200:
210 DEFPROCCALC
220 VEL1=L/(T1/100)
230 VEL2=L2/(T3/100)
240 PRINT"ACC =";(VEL2-VEL1)/((T1/2)+T2+(T3/2));" m/s^2"
250 ENDPROC
```

Answers to Questions

Chapter 1
9 1.25×10^{17}

Chapter 2
3 95 m; 72 m **4** 14 m
5 (a) 0.1 mg (b) 0.5 mg
6 0.7 g/cm³; 24 cm³; 156 g: 70 cm³; 11.4 g/cm³
7 2700 kg/m³; 19300 kg/m³; 250 kg/m³
8 A silver; B, G, H bronze; C gold; D ceramic; E, F iron
9 (i) 22 cm³; 70 cm³; 48 cm³ (ii) 384 g (iii) 8 g/cm³ **10** (a) 75 m³ (b) 97.5 kg

Chapter 3
2 (a) (in cm) 2.4, 4.8, 7.2, 4.8, 0, 7.2, 10.4, 16.0, 9.0 (b) 3.6 cm; 13.6 cm (c) 7 N **3** 3.5 N
4 half as much; twice as much
6 tension: a, c, e, f, g, h, k; Compression: b, d, i, j **7** 17 N; 7 N; 13 N at 23° to 12 N force **8** 26 N

Chapter 4
1 (a) 50 N (b) 40 N (c) 2800 N
3 1 L; 2 R; 3 R; 4 R; 5 L
4 (b) A **5** (a) 50 000 N
7 C; A

Chapter 5
1 1 m/s **2** 4800 m **3** 200 s
4 (in m/s) 10.1, 10.1, 9.2, 7.9, 7.6, 7.2; 9.5, 9.4, 8.4, 7.1, 6.6, 6.5
6 (a) 200 m (b) 11 s (c) 80 m (d) 15 s
(e) 8 m/s (f) staying the same (g) 0 m/s
7 40 cm/s; 90 cm/s; 55 cm/s
8 (b) 150 km/hour **9** 80 cm/s
10 0.5 m/s↓ **11** 7.5 m/s
12 0.5 m/s **13** (a) 25 m/s (b) 2.5 m/s
15 (b) 2.5 m/s²; 20 m; 2.8 m/s **16** 180 m
17 dead heat **18** 40 m/s; 80 m
19 1.8 m/s²
20 (a) 21 m (b) 0.7 s (d) 4.7 m/s

Chapter 6
6 (b)(i) 0.07 N; 0.175 N; 0.28 N; 0.35 N

Chapter 7
7 A, A, A, A **8** 0 N; 60 N
9 downwards in all three cases

Chapter 8
1 (a) 10 N (b) 16 N (c) 1600 N
2 (a) 0.25 m/s² (b) 8 m/s² **4** 0.8 kg
5 (a) 0.5 m/s² (b) 400 N (c) 800 N
6 (a) 120 N (b) 2 m/s² (c) 10 m/s; 25 m
7 (a)(i) 16 m/s (ii) −5 m/s² (iii) 5000 N (iv) 25.6 m **8** 14 N
9 (a) 6 m/s (b) 0.1 s (c) 120 m/s² (d) 8400 N—12 times his weight
10 (b) 200 N
12 Earth: 30 kg, 300 N, 2 m/s²; Moon: 30 kg, 50 N, 2 m/s²; Jupiter: 30 kg, 750 N, 2 m/s²
13 (a)(i) A: 0.8 m/s² B: 0.6 m/s² C: 0.4 m/s² D: 0.2 m/s² (b)(i) 51 N

Chapter 10
1 (a) 1.5 kg m/s (b) 225 kg m/s (c) 150 000 kg m/s (d) 200 000 000 kg m/s
2 (a) 12 m/s (b) 4 m/s **3** 800 kg
4 (a) 50 kg m/s (b) 100 kg m/s (c) 200 kg m/s
5 (a) 2 kg m/s (b) 0.017 s **6** (a) 25 N
7 (a) 12 kg m/s (b) 12 kg m/s (c) 2 m/s
8 (a) 0.2 m/s (b) 0.125 m/s **9** 1.3 m/s
10 0.2 m/s
11 (a)(i) BC (ii) CD (iii) 0 m/s
(b)(i) 3150 m/s² (c)(i) −1.2 m/s²

Chapter 11
1 (a) 8000 J (b) 1500 J (c) 3 J (d) 6750 J
2 2000 J **3** 1600 J
4 (a) 100 J; none (b) 137.5 J; none
5 (a) 3000 J (b) 600 J (c) 24000 J
6 1333 m **7** (a) 100 J (b) 6.4 J (c)1.5 J
8 (a) 30 m (b) 40 m/s (c) 0.1 kg
9 (a) 900 J (b) 900 J (c) 30 m/s
10 same **11** 10 m/s
12 (a) 80 J (b) 80 J (c) 8 m/s (d) same
13 (a) 312 500 J (b) 25 kW; 39 kW
14 (a) 50 W (b) 100 W (c) 6 W (d) 50 W
(e) 200 W (f) 250 W (g) 1000 W (h) 400 W
15 (a)(iii) 12 m/s (b) 72 J (c) 2400 W
16 (a) 5000 N (ii) 7500 J (b)(i) 1000 J (c) 11.3%
17 (a)(i) 5 (ii) 1 (iii) 1.5 m (b)(i) 1500 J (ii) 200 N **18** 25 N

Chapter 12
1 (a) 50 N/cm² (b) 20 N/cm² (c) 50 N/cm²
3 (a) 128 N (b) 3200 N **3** 0.01 m²
4 (a) 2 N/cm² (b) 20 000 Pa **4** upright
6 (a) 15 N/cm² (b) 1500 N/cm²
7 (a)(i) 12.5 N (ii) 20 cm²
(iii) 0.625 N/cm² (iv) 0.31 N/cm²
(b)(i) 160 cm³ (ii)7.8 g/cm³
8 50 000 Pa; 4.8 m
11 (b)(i) 50 000 Pa; 50 kPa (ii) 5 N

Chapter 14
3 (a) 16.5 mm (b) 0.53 m
7 (c)(i) 20°C (ii) 120°C
8 (a) 300 s (b) 600 s (c) 50%

Chapter 15
1 313°C; −127°C
3 (a) 100°C (b) −50°C (c) 5727°C (d) 273°C (e) −173°C (f) −269°C
4 (a) 2 × 10⁴ Pa (b) 8 × 10⁴ Pa (c) 16 cm
5 (b)(ii) 100 cm³ (d) 150 kPa
6 (b)(i) 2.73 × 10⁶ Pa

Chapter 16
6 (a) windows: 750 J; walls: 1200 J; roof: 500 J; floor: 120 J (b) windows: 29% walls: 47% roof: 19%; floor: 5%
7 (a) 200 J (b) 17.3 MJ (c) 2500 MJ
8 (a) 1500 MJ (b) 1000 MJ (c) £5

Chapter 17
1 11 000 J **2** 40°C **3** 0.5 kg
4 250 W **5** 460 J/kg K
6 (a) gas (b) 200°C; 100°C (c) freezing (d) all (e) s.l.h. of vaporisation
7 (a) 300 s (b) 150 000 J/kg
8 1130 s; 0.68 kg
9 (i) 20 kg (ii) 6720 000 J (6.7 × 10⁶ J) (iii) 2688 s (b) 2.26 × 10⁷ J
10 (a) 2 × 10⁵ J/kg
12 (a)(i) 3024 s (b)(i) 320 min; 120 min (ii) 60 MJ **13** 39°C

Chapter 18
6 (a) B, D (c) 0.8 m

Chapter 19
3 air, water, glass, diamond

Chapter 21
1 (a) 13 mm (b) 4.5 mm **4** 48 cm/s
5 4 cm **6** 4 Hz **7** (a) 40 cm/s (b) 10 Hz

Chapter 22
2 A: infra-red; B: X rays
3 (a) 0.06 s (b) 1.1 s (c) 500 s (d) 9000 S (e) 3.8 × 10¹³ km **5** (b)(i) 200 m

Chapter 23
2 (ii) 1320 m **4** 22 m; 16.5 mm
5 (a) 70 m **6** (a) 800 m (b) 200 m
9 (d) 0.33 m

Chapter 25
6 (a) 10 mC (b) 129 600 C (c) 5A (d) 500 s
7 (b) and (c) are true **8** 0.6 A
9 (a) 3 V (b) 3 V (c) 6 V (d) 9 V
10 (a) 1 A (b) 0.5 A (c) 3 A (d) 3 A
(e) 5 A (f) 2 A (g) 2 A (h) 1.5 A
13 (a) 1 V (b) 5 V (c) 6 V (d) 3 V (e) 3 V
(f) 3 V **14** (a) 24 V (b) 12 V
16 (a) 20 J (b) voltmeter

Chapter 26
1 (a) 12 V (b) 1.5 V (c) 1.5 A
(d) 2.5 A (e) 4 Ω (f) 10 Ω
2 (a)(i) 32 Ω, 16 Ω, 11.3 Ω, 5.5 Ω
5 (a) decrease (b) increase (c) decrease (d) increase
6 (a) 3 V (b) 4.5 V
7 (a) $V_1 = V_2 = 6V$ (b) $V_1 = 2V$; $V_2 = 4V$ (c) $V_1 = 1V$; $V_2 = 9V$ (d) $V_1 = 2V$; $V_2 = 3V$
(e) $V_1 = 1V$; $V_2 = 3V$ (f) $V_3 = 5V$
8 (a) $V_1 − $ (b); $V_2 − $ (d)
(b) $V − $ (d) (c) $V − $ (b) (d) $V − $ (d)
(e) $V − $ (d) (f) $V − $ (d)
9 (a) increase (b) decrease (c) decrease (d) stay the same
10 12 V; 0.5 A; 17 Ω
11 L_5 **12** $A_2 = 0.4A$; $A_3 = 0.2A$
13 (a) 12 V (b) 2 A (c) 2 A (d) 6 A (e) 2 Ω
14 A_3 **15** (a) 5 A (b) 10 A
16 (a) 2 Ω (b) 2 Ω (c) 2.5 Ω (d) 7 Ω
18 (a) 10 V (b) 9 A (c) 3 Ω
19 (b)(i) 1.2 A (ii) 0.12 A

Chapter 27
1 (a) 6 W (b) 100 W (c) 0.9 W
2 (a) 2 A (b) 0.5 A (c) 0.5 A
3 (a) 48 W (b) 2.5 W (c) 48 W
4 (a) 0.25 A (b) 960 Ω **5** light bulb
6 parallel, 5 Ω; series, 100 Ω **7** parallel
8 (a) 6 A (b) 24 V
9 (a) 3 A (b) 6 Ω, 12 Ω (c) 0.67 A (d) 6 Ω
10 (a) 48 W (b) 480 W (c) 6%
11 6000 J (a) 200 W (b) 8000 J (c) 75%
12 (a) heater (b) 1 kW **13** £22
14 (a) 0.25 A, 0.25 A, 0.42 A, 0.62 A
(b) 1.54 A (c) 370 W
15 (a) yes (b) no.

Chapter 28
2 1.5 Ω **3** (b) 3

Chapter 29
11 (b) anticlockwise
13 (b)(i) 15° (ii) 8°

Chapter 30
6 15 V a.c.; 12 V a.c.; 2.5 V a.c.; 12 V a.c.
7 50 turns; 80 turns
9 (a)(i) A (ii) 14 V; 500 turns; 4.8 V; 200 turns; 8 V 125 turns

Chapter 31
3 (b) 5 V (c) 4½ squares
5 A: 2; 2V; 5; 2 Hz
B: 1; 2 V; 2; 0.5 Hz
C: 2; 2 V; 4; 25 Hz
D: 3; 1.5 V; 6; 167 Hz
6 25 Hz
7 (a) 1 V/cm (b) 4 V (c) 0.5 ms/cm (d) 5 ms (c) 200 Hz
8 (A) 0.02 s (B) 0.1 s (C) 0.05 s (D) 0.01 s
9 (d) 3.3 Hz

Chapter 32
5 0.012 C **7** (a) AND or OR **9** NAND

Chapter 33
5 92p; 92e; 146n;
7 (a) 1, α; 2, β; 3, β; 4, α; 5, α
8 (b) 5 counts per minute (c) 30 minutes
9 (b)(ii) 8 days
11 (a)(i) 5.4 × 10¹¹ J (ii) 33% (b)(ii) 270 years

Chapter 34
4 (a)30° (b) 40° (c) 117 MJ
5 (a)(iv) hydroelectric (b)(i) 2000, 2150
7 (b)(i) 4.4 m² (ii) about £25
9 (a)1000 J (b) 400 J (c) 2500 kg
10 (a)(iii) 37% (b)(i) 30% (ii) 82%

Index

absolute zero 137, 141
acceleration
    and force 62, 68–9
    definition of 44
    due to gravity 50
    measurement 48–9
accommodation 189
accumulators 279
acoustics 219
action and reaction 66
action at a distance 12, 231
activity, of radioactive source 345
addition of resistors
    in parallel 257
    in series 255
air resistance 60, 63–5
aircraft pressurisation 119
alpha decay 344
alpha particles 336–9
    scattering experiment 342
alternating current 240, 270
    generation 300–1
    transmission 307–8
alternative energy sources 364–6
alternators 300
altimeter 119
ammeters
    moving coil 294
    use in circuits 240
ampere 239
    definition of 291
Ampére's grip rule 286
amplifier, transistor 330
amplitude
    of a wave 197
    of a sound wave 218
analogue to digital conversion 333
AND combination 238
AND gate 331
antinodes 222–3
aperture, camera lens 190
Archimedes' principle 122
area 17
astigmatism 189
atmospheric pressure 115–18
atom 7
    charges in 227, 340
    models of 227, 340–3
    structure of 227, 340–3
atomic energy (see nuclear energy)
atomic number 341

background radiation 346
barometers
    aneroid 118
    mercury 118
batteries 278
becquerel 345
Becquerel, Henri 336
bell, electric 288
beta decay 344
beta particles 336–9
bicycle 63–4
bicycle pump 117
bimetal strip 133
biomass 365
binoculars 177, 193
bouncing ball 100
bourdon gauge 117
Boyle's law 138
braking system 114

bridges 26–7
Brownian motion 8
brushes (in motor) 292
bubble chamber 339

cable rating 271–2
camera 190
    and eye 191
capacitance 323
capacitors 323–4
car breaking system 114
carbon dating 349
cathode rays 311–3
cathode ray oscilloscope 218, 314–17
cells 278–9
Celsius scale 127
centre of gravity 34
centigrade scale 127
chain reaction 350
change of state 6, 9, 11, 158–9
characteristics 260
charge 226
    and current 235, 239
    unit of 239
charging by rubbing 226–7
Charles' law 136
CHP 367
chromatic aberration 185
circuit
    breakers 275
    calculations 253–8
    principles 236–45, 253–8
    symbols 236
circular motion 76
clap–echo method 217
climbing rope 101
cloud chamber 338
coil and magnet experiments 298
collisions 86–90
    elastic 88–9
    inelastic 87–8
colour 179–81
    addition 180, rear cover
    filters 180
    primary 180–1
    television 180
    vision 180
combined heat and power 367
commutator 292
compass 285, 289
components of forces 29
compression 25–7
concave
    lenses 185
    mirrors 171
conduction
    electrical 228, 237
    heat 148–52
conductors
    electrical 228, 237
    heat 148–9, 152
conservation of energy 4, 97–9
conservation of momentum 84–90
convection 146–7
conventional current direction 240
converging lenses 184–7
convex
    lenses 184–7
    mirrors 171
cooling curve 159
coulomb
    unit of charge 239
    definition of 239
couple 33
critical angle 176
critical mass 350

CRO 314–17
crumple zone 83, 101
Curie 336
current
    alternating 240, 270
    and charge 235, 239
    conventional direction 240
    in circuits 235–40, 244–5, 253–9
    induced 297–8
    magnetic effect of 286–8
    magnetic force on 290–1
cycling 63–4

d.c. motor 292
    making a model 293
decibel (dB) scale 222
deflection tube 312
demagnetisation 283, 289
density 19–20, 122
deviation 179
diffraction
    of light 203
    of sound 220
    of water waves 201
diffusion 9
digital systems 331–3
diode
    bridge 322
    p–n junction 321
    thermionic 311
displacement 42
dispersion 179
distance–time graphs 38–9
diverging lens 185
domains 283
domestic wiring 274–5
double insulation 271
double slit experiment 203
dry cell 279
dynamo 299

$E = mc^2$ 352–3
ear 216
earth wire 271
echo sounding 217
echoes 217
eclipses 167
eddy currents 302
efficiency 105
    of a light bulb 266
    of machines 105–7
    of power stations 360, 362
    of transformers 305
elastic limit 24
elastic materials 24
e.l.c.b. 275
electric
    charge 226
    current 235–40, 244–5, 253–8
    fields 231
    motors 292–5
electric circuit
    breakers 275
    calculations 253–8
    principles 236–45, 253–8
    symbols 236
electrical energy 267
    calculations 267–8
    costs 269
    power 264–9
electricity bills 269
electricity generation 360–1
Einstein, Albert 352
electromagnetic
    induction 297–305
    waves 204
    spectrum 207–10

electromagnets 287–8
electromotive force (e.m.f.)
    induced 297–8
    of cell 280
electronic systems 328–32
electronic timer 41
electrons 227
    beams of 311–13
    charge and mass 312–13
    in electrical conduction 240
    in orbits (shells) 340
electroscope 310
    as radiation detector 337
electrostatics 226
electrostatic
    hazards 232
    precipitator 231
    sprayer 231
elements 7, 340–1
e/m for electron 313
energy 1–5
    and life style 368
    and mass 352
    and power 102
    and transport 368
    and work 92
    conservation of 4
    elastic potential 100
    electrical 264, 267–9
    forms of 5
    gravitational potential 94
    in activities 357
    in food 357
    in the home 358–9
    internal 2, 11, 145, 154
    kinetic 94–7
    kinetic and potential 97–9
    nuclear 351–4
    practical sources 359–66
    saving 5
    spreading of 4
    stored 2, 13
    supply in UK 359, 363
    transfer 1, 3, 97–9, 104–6
    unit of 94
equilibrium 31
    conditions for 33
    types of 35
evaporation 161–2
expansion
    of gases 136
    of liquids 134
    of solids 130–2
    of water 134
expansivity, linear 132–3
explosions 84–5
eye 188
    and camera 191

farad 323
Faraday, Michael 297
    coil and magnet experiments 297
    law of electromagnetic induction 298
feedback 329
ferromagnetism 283
Feynman, Richard 6
fields 12–13
    electric 12, 231
    gravitational 13, 71–3
    magnetic 12, 285–7
fission 350
fixed points 127
floating 123–5
fluorescence 210
focal length
    lenses 185

376

focal length
  measurement (convex lens) 186
  mirrors 171
focal plane 185
force 22
  components of 29
  gravitational 71–3
  and acceleration 43, 68–9
  and change of velocity 43
  and motion 57–60
  and momentum 80–1
  and pressure 110
  measurement of 22–3
  on conductors 290–2
  stretching effect of 22–5
  turning effect of 32–3
  unit of 23, 70
forces
  balanced 31, 63–5
  in pairs 66, 84
  vector addition of 28–9
free-fall 50, 79
frequency 197
  measurement with CRO 316
  sound waves 218
  vibrating strings 222–3
friction 59–60
  and heating 60
fuel reserves 363
fuels 363
full-wave rectification 301
fuses 272
fusion, latent heat of 158–9
fusion, nuclear 353

Galileo Galilei 57
Galvani, Luigi 235
galvanometer, moving coil 294
gamma rays 336–9
  emission 344
  in e.m. spectrum 210
gamma camera 349
gas laws 136–42
gases, kinetic theory of 140–1, 342
gears 107
Geiger–Müller tube 338
generator rule (LH rule) 298
generator 301
  making a model 301
generation of mains electricity 306, 360–1
geothermal energy 366
gravitational
  field 13, 71–3
  field strength 71
  force 71–3
gravity, centre of 34
gray (unit of radiation dose) 347
greenhouse effect 212

half-life 345
harmonics 222–3
heat (see also internal energy)
  conduction 148–9
  convection 146–7
  radiation 152, 211–12
heating
  and change of state 11, 159–62
  by friction 60
  elements 266
hertz 197
Hooke's law 23
hot water system 134, 147
house insulation 150–1, 367
hydraulic machines 113–14
hydroelectric energy 361

hydrometers 124

ideal gas 141
images
  in convex lenses 184–7
  in plane mirrors 169–70
  real 187
  virtual 169
impulse 81
induced
  charge 229
  e.m.f. 297–9
  magnetism 282
induction motor 302
inertia 61–2
infra-red radiation 209, 211–12
insulation, of houses 150–1, 367
insulators
  electrical 228, 237
  heat 148–50
interference
  fringes 203
  light 203
  sound 220
  water waves 202
internal energy 2, 11, 145, 154–62
  and temperature 145, 154
internal resistance 280
inverse square law, light 168
inverter 330
ionising radiation 210, 337
  and risk 346–7
  detecting 337–9
  dose 347
ionization 210, 230
  by nuclear radiation 337
  chamber 337
ions 230
isotopes 341

jack
  hydraulic 113–14
  screw 107
jet engine 86
joule 93–5
joulemeter 157
junction diode 321

Kelvin scale 137
kilogram 16
kilowatt 102
kilowatt-hour 269
kinetic energy 94–9
kinetic theory of gases 140–1, 342
  and gas laws 140–1
  and evaporation 161–2
  and latent heat 159

land and sea breezes 147
latent heat
  of fusion 158–61
  of vaporisation 158–61
lateral inversion 170
LDR 251, 259, 328–9
LED 325
left-hand grip rule (solenoid) 287
left-hand rule (generator) 298
length 15
lenses 184–7
Lenz's law 299
levers 32–3, 107
light
  and sound 220
  brightness 168
  interference 203
  nature of 204

  rays 166
  reflection 168–71
  sources 165
  speed of 207
  waves 203
light bulbs 133, 266
light-dependent resistor 251, 259, 328–9
light-emitting diode 325
lightning conductor 230
linear expansivity 132
logic gates 331–2
longitudinal waves 214
long sight 189
loudness 218, 222
loudspeaker 291
low-level radiation 346

machines 104–7
  hydraulic 113–14
Magdeburg hemispheres 115
magnetic fields 285–6
  due to currents 286–8
  Earth's field 289
  force on a coil in 292
  force on currents in 290
magnetic materials 281–3
magnets 281–4
  making 282, 289
  poles of 281
  theory of 283
  uses of 284
mains electricity 270–5
  National Grid system 306–7
  domestic circuits 274–5
magnification
  lenses 188
  of microscope 192
  of telescope 193
magnifying glass 192
mains electricity 270–5
  wiring a plug 273
Maltese Cross tube 312
manometer 117
mass 16, 61–2
  and acceleration 69
  and energy 352
  and weight 70–3
mass number 341
matter
  molecular theory of 7–11
  particulate theory of 6–11
mechanical advantage 106
megawatt 102
meniscus 10
meters
  in circuits 240, 242
  moving coil 294
metre 15
microphone 299
microcomputer 41, 49, 374
microscope
  simple 192
  compound 192
microwaves 208
Millikan's experiment 313
mirage 178
mirrors
  curved 171
  plane 169–70
models of the atom 342–3
  Bohr 343
  nuclear 342–3
  plum-pudding 342
molecular theory of matter 7–11
  evidence for 8–9
  and shaping materials 25

molecules 7
  estimating size 10
moments 32–3
  principle of 32–3
momentum 80–90
  and collisions 86–90
  and explosions 84–5
  and Newton's second law 80–3
  conservation of 84–90
motion
  equations of 50–2
  force and 57–8, 61–6, 68–73
  laws of 61–6, 68–73
motor, d.c. 292
motor rule (RH rule) 290
moving-coil meters 294
musical notes 222–3
mutual induction 303–5

NAND gate 331
National Grid 306–7
neutrino 90, 344
neutron 227, 340
newton 23
Newton's
  first law 61
  second law 62, 70
  second law and momentum 80–1
  third law 66
  thought experiment 78–9
nodes 222–3
noise 218, 222
non-ohmic conductors 260
NOR gate 331
NOT gate 330–1
n-type semiconductor 320
nuclear
  energy 351–4
  fission 350
  fuel cycle 352
  fusion 350
  medicine 349
  radiation 336–9
  reactions 344
  reactors 351, 353
  waste disposal 352
nuclear power 350–4
  arguments for and against 354
nucleus, structure of 340
nuclide 341

ocean thermal energy conversion (OTEC) 364
Oersted, H.C. 286
ohmic conductors 248, 260
Ohm's law 248
oil-drop experiment
  molecular size 10
  Millikan's 313
optical fibre 177–8
optical density 174
OR combination 238
OR gate 331
oscilloscope 218, 314–17

parabolic reflector 171
parallel and series
  circuits 244–5, 253–9
  addition of resistors 255, 257
particulate theory of matter 6–11
pascal 110
periscope 170
photocopying 232
photodetector switch 41, 49–50, 99
photodiode 325

photons 343
pinhole camera 166
pitch 218
plane mirrors 169–70
plastic materials 24
p-n junction diode 321
poles, magnetic 281
   rule for solenoid 287
positive holes 320
potential difference 241
   in circuits 241–2, 253–8
potential divider 259
potential energy
   elastic 100
   electrical 241–2, 245
   gravitational 94
potentiometer 251
power
   and energy 102
   definition 102
   electrical 264–9
   loss in cables 307–8
   mains generation 306–7, 360–2
   ratings 266
   supplies 322, 324
   unit of 102
power stations 360–1
pressure 110
   atmospheric 115–18
   in liquids 111–14
   measuring 117–18
pressure law for gases 139
primary cells 279
primary coil 303
primary colours 180, rear cover
principal focus, lenses 185
prisms
   reflection in 177
   refraction through 179
projectile motion 77–8
projector, slide 191
proton 227, 340
p-type semiconductor 320
pulleys 107

quantum 343

radiation
   electromagnetic 207–212
   ionising 210, 337
   nuclear 336–9
   protection 347
radiation, infra-red 211–12
   absorbers 211
   emitters 211
radioactive decay 344–5
radioactivity 336–345
   uses 348–9
radio waves 208
rainbow 180
ramp 106
ray diagrams
   concave lens 185
   convex lens 184–7
   curved mirrors 171
   plane mirror 169–70
r.c.c.b. 275
reactor, nuclear 351, 353
real and apparent depth 175
real image 187
rectification 322
reflection of light
   by curved mirrors 171
   by plane mirrors 168–70
   by prisms 177
   total internal 176–8
reflection of water waves 199

reflection of sound 219
refraction of light 174–80
   and speed 175
   by lenses 184–7
   by prisms 179–80
   in optical fibres 177–8
refraction of sound 220
refraction of water waves 199–200
refractive index 175
refrigerator 162
relative velocity 54
relativity 352
relay 288, 329
renewable energy sources 364–6
resistance 243, 248–52
   and current 243
   definition and unit 248
   effect of temperature 260
   equation 249
   internal (of cell) 280
   measurement of 252
   of junction diode 321
resistivity 250
resistors 250–1
   addition in parallel 257
   addition in series 255
   colour codes for 250
   in series and parallel circuits 253–9
resonance 223
resultant force 28
reverberation 219
rheostat 251
right-hand rule (motor) 290
ring main 275
ripple tank 198
rocket engine 86
Rutherford's scattering experiment 342

safety helmet 83
satellites 78–9
scalars 28
seat belts 82–3
second 17
secondary cells 279
secondary coil 303
semiconductors 320
series and parallel
   circuits 244–5, 253–9
   addition of resistors 255, 257–8
shadows 167
s.h.c. 155–8
   definition of 155
   measuring 157–8
   using high value for water 156
short circuit 258
short sight 189
sievert (unit of radiation dose equivalent) 347
sky-diving 65
slide projector 191
slip rings 300
smoothing 324
solar
   cells 279, 365
   collector 212, 364
   energy 212, 364–5
   furnace 171, 365
   panel 212, 364
solenoid 287
sonar 217
sound
   and light 220
   and noise 218
   and vacuum 215
   diffraction 220

interference 220
reflection 219
refraction 220
speed of 216
waves 215
spark counter 337
sparking 230
specific heat capacity 155–8
   definition of 155
   measuring 157–8
   using high value for water 156
specific latent heat
   of fusion 160
   of vaporisation 160–1
spectrum
   continuous 179, rear cover
   electromagnetic 207–10
   line 179, rear cover
   of sound 223
   visible 179, rear cover
speed 37
   and velocity 42–3
   average 37
   instantaneous 38
   measurement of 40–1
   of electromagnetic waves 207
   of light 207
   of sound 216–17
   of waves 197
speedometer 44, 302
speed–time graphs (see velocity–time graphs)
split-ring commutator 292
spring, stretching 22–3
stability 35
standing (stationary) waves
   in air columns 223
   on strings 222
state, changes of 6, 9, 11, 158–9
states of matter 6
static electricity 226
stationary (standing) waves
   in air columns 223
   on strings 222
stroboscope 40–1, 48
structures 26–7
switches 237–8
   transistor 327–9

tachograph 47
tape recording, 289, 299
telephone 288
telescopes 193–4
television 180, 208, 317
temperature 127–8
   fixed points 127
tension 26–7
terminal velocity 64–5
theory of matter 6–11
thermionic emission 310–1
thermistors 129, 251, 259, 329
thermocouple 129
thermometers 127, 129
   clinical 130
thermostat 133, 329
Thomson's experiment 313
ticker timer 40, 48
tidal energy 366
time 17
time base 315
torque 33
total internal reflection 176–8
tracer techniques 349
transformers 304–7
transistor 326–7
   amplifier 330
   switch 327–9

transmission of mains electricity 307–8
transverse waves 196
truth tables 238, 330

ultrasound 220
ultraviolet radiation 210
units 15–17
upthrust 122
U-values 151

vacuum 115–16, 118
   flask 212
van de Graaff generator 229
vaporisation
   latent heat of 158–9
vectors 28–9, 42, 53–4
   addition of 28–9, 53–4
VELA 41, 49, 374
velocity 42
   and speed 42–3
   changing 43
velocity–time graphs 45–7
virtual image 169
vision, defects of 189
visual angle 189, 192–3
volt, definition of 241
Volta 235
voltage 241
   in circuits 241–2, 253–8
voltaic pile 235, 278
voltmeters
   use in circuits 242
volume 18

water
   expansion of 134
   specific heat capacity 156
   specific latent heat of fusion 160
   specific latent heat of vaporisation 160–1
water waves 198–202
   diffraction 201
   interference 202
   reflection 199
   refraction 199–200
watt 102
wave equation 197, 219
wave energy 365
waveforms 218, 316
wavefront 198
wavelength 197
   and colour 204
   measurement (light) 204
   sound waves 214–5
wave motion 196
waves
   electromagnetic 207–10
   light 203–4
   longitudinal 214
   sound 215
   stationary (standing) 222–3
   transverse 196
   water 198–202
wavespeed 197
weather 119
weight 70–3
   and mass 70–3
weightlessness 73, 79
wind energy 365
wiring a plug 273
work 92–3
   and energy 92
   unit of 93

X rays 210
X-ray tube 210

Young's slits experiment 203